MS&A

Modeling, Simulation and Applications

Volume 21

MS&A publishes advanced textbooks and research-level monographs that will illustrate the scientific foundations of the modeling and simulation process as well as concrete instances of its role in addressing complex and relevant problems in everyday life. Mathematical modeling aims to describe through mathematics the different aspects of the real world, their dynamics and their interaction. Numerical simulation provides accurate and certified solutions to complex mathematical models by means of scientific computing. Modeling and numerical simulation have become the road-map for mathematics to develop and analyze novel techniques to solve problems in basic sciences (such as physics, chemistry, biology) and engineering, environmental, life and social sciences. The purpose of this series is to host high level contributions describing the interplay among mathematical analysis, numerical analysis and scientific computing, advanced programming techniques, control and optimization, validation, verification and testing. This interplay makes the modeling and numerical simulation process as a whole a unique and effective tool for applied sciences as well as for enhancing technological innovation. The series has already published some successful books (relevant also from a print/online sales perspective), and has planned to publish in the next 2 years about 10 new works focusing on the most significant emerging areas. The Series in indexed in SCOPUS. Volumes of the series are indexed in Web of Science - Thomson Reuters.

Xavier Blanc • Claude Le Bris

Homogenization Theory for Multiscale Problems

An introduction

 Springer

Xavier Blanc
Laboratoire Jacques-Louis Lions
Université Paris Cité
Paris Cédex 13, France

Claude Le Bris
École des Ponts ParisTech & Centre
Inria de Paris
Marne-la-Vallée, France

ISSN 2037-5255 ISSN 2037-5263 (electronic)
MS&A
ISBN 978-3-031-21835-4 ISBN 978-3-031-21833-0 (eBook)
https://doi.org/10.1007/978-3-031-21833-0

Mathematics Subject Classification: 35J15, 35J70, 35B27

Cover illustration: "Schematic representation of a set of defects that become rare at infinity from the origin (model discussed in Chapter 4)"

This Springer imprint is published by the registered company Springer Nature Switzerland AG
The registered company address is: Gewerbestrasse 11, 6330 Cham, Switzerland

Ut per rivulos non statim in mare eligas introire:
quia per faciliora ad difficiliora oportet devenire.
Tommaso D'Aquino (1225–1274)
Epistola de modo studendi.

[You should choose to enter through little streams, and not suddenly into the sea, because it is proper that one come gradually through the easier things to the more difficult ones.
Thomas Aquinas
Letter on a Method of Study, Sixteen Precepts for Acquiring Knowledge]

Foreword

Homogenization theory was born 50 years ago. It is the age of maturity. It is the age of wisdom. The theory has generated many offsprings. It has emerged at the interface between an abstract viewpoint and practical considerations in the specific context of periodic media. It was then increasingly considered for other media: locally periodic media, random media, etc. It was originally considered for linear equations. It now applies to semilinear equations, quasilinear equations such as the p-Laplacian, and even fully nonlinear equations, such as the Hamilton-Jacobi equations. It initially provided a mean to change scale in the bulk of a domain. It then became a manner to study boundary layers and derive wall laws. After addressing only equations, such as the diffusion equation, it was next applied to *systems*, such as the Stokes system. Over the years, it has also addressed spectral problems and not only problems for a given right-hand side, time-dependent problems after steady-state problems, problems posed over perforated domains or domains that have fractal boundaries rather than problems posed on nice and regular domains.

With maturity, a mathematical theory runs a twofold risk.

The first risk is that the theoretical developments become too technical. Most of the simple questions have essentially been solved. Those that remain unsolved are challenging. They require an exceptional creativity and an elaborate technical toolbox. Several recent contributions in the area are extremely technical and their mathematical substance may escape the outsider. If the sophistication may be an incentive for some readers, it might also act as a repellent for some other readers.

The second risk is that the theoretical developments, although originally motivated by applications, progressively get somewhat disconnected from reality. Abstraction sows its own poison.

Our personal motivation for writing this textbook originates from the above observations.

We believe it is time to summarize the essence of the theory and make it accessible to outsiders. Major reference treatises exist. They are mostly authored by the founders of the theory and by some of the landmark contributors. These treatises are a sum. We wish to cite [BLP11, ZKO94, Tar09]. It is out of the question for authors like ourselves to try and compete with these experts and to write such exhaustive treatises. Other monographs also exist, which complement the above treatises by focusing on some given aspects or some given approaches. Examples are [All02, Bra02, PS08, SPSH92]. Some research articles regularly introduce breakthroughs. Some survey articles, such as [ES08], periodically review the state of the art. Some research books aim at presenting new developments: a recent example is [AKM19]. All these contributions bring something to the table. Finally, some didactic and pedagogic presentations such as [CD99] or [BR18] initiate the readers into elaborate techniques. But the latter are not many. It is not forbidden to believe that another book of introduction to the topic could help. It could usefully complement the already existing literature, providing a different perspective or a different take on the various issues. The student, or the outsider to the field, may then cherry pick which pedagogic presentation fits best with their own perception or allows for the most efficient learning curve. The present textbook is precisely meant for this.

It is also ample time to reconcile homogenization theory with its original practical objectives. As we mentioned above, the development of homogenization theory stemmed from practical questions, for instance the wish to better understand various questions arising in material sciences such as phase transitions, mixing properties, etc. The success of the theory has been substantial, both in terms of phenomenological understanding and in terms of theoretical equipment. Homogenization theory has also been useful quantitatively. In its modern variant known as *multiscale methods*, it serves as a theoretical guideline. It provides the necessary theoretical understanding of numerous computational approaches that address various situations. In many actual circumstances, practitioners need a quick and quantitative answer to the questions they have. What if the actual material does not satisfy the idealized assumptions of the theory? One must proceed no matter what. How? What macroscopic properties to expect? The mathematical apparatus provided by classical homogenization theory may then be very frustrating. A practitioner in an industrial context may typically be able to only allocate at most a couple of hours of computational time for the simulation. An approximate quantitative reply is nevertheless expected. At least the order of magnitude of the reply must be correct, or the trend to be anticipated if a change is being operated in the conditions of the experiments or the parameters of the industrial device. It is then unclear how to change the focus, from a theory that primarily aims at accuracy to a methodology that provides coarse but swift replies.

Some of our research works [BLL03, BLL07a, BLL07b, BLL12, BLL15, BLL18, BLL19], performed under the leadership of and in collaboration with Pierre-Louis Lions, have led us to develop a portfolio of modelling strategies, which we like to regroup under the collective name of *"nonperiodic"* problems and approaches. The terminology is meant to emphasize that this allows us to free ourselves from the natural and historical assumption of the theory while not inserting the settings we study into the enormous (yet successful) realm of random modelling.

We aim to revisit these contributions along this textbook. Some of them already date back to two decades ago. We will put them into perspective. We will try and show how they did contribute to solve some of the relevant questions mentioned above. We will present, in passing, homogenization theory in the periodic setting and in the random setting and use both settings as a starting point and a point of reference. We will thereby emphasize, explicitly or between the lines, the differences with the nonperiodic setting we advocate and investigate in detail.

To a certain extent, what we propose here is to revisit a portion of homogenization theory through the prism of a large variety of nonperiodic problems, all arising from practical considerations.

In a more mathematical language (or from a more philosophical standpoint, all this is a matter of taste), we use nonperiodic problems as a battery of testbeds to assess the stability of the results of periodic homogenization with respect to perturbations of the idealized periodicity assumption.

The present textbook is mainly based upon classes we have been teaching at the advanced undergraduate or master level, at University Pierre et Marie Curie, at University Paris Diderot and at the University of Chicago. We only assume from the reader a basic knowledge of functional analysis and of partial differential equations theory. Typically, the level expected is that of the two celebrated and classical textbooks, respectively by Haïm Brezis [Bre11] and Lawrence Craig Evans [Eva10]. Because we have a practical purpose, we also presume that the reader has been *exposed*, and not more than that, to some basic notions of discretization techniques for differential equations. Occasionally, some notions of probability theory may also be useful. In all three domains, analysis, numerical analysis and probability theory, we will in any event make our exposition as self-consistent as possible. If need be, we will recall the necessary notions. Our very idea is to hold the reader's hands and help them through it. We strongly believe that it is important to learn as soon as possible the intuitive ideas that are hidden behind the abstract tools and the technicalities of their proofs. Those ideas are sometimes only accessible to the limited audience of experts who have already understood everything. This is way too late in our considered opinion. We therefore often turn our back to rigorous proofs (which anyway may be read elsewhere) and deliberately choose to present matters in layman's terms that we can all understand.

We are grateful to our close colleagues, and in particular to Yves Achdou, Rutger Biezemans, Rémi Goudey, David Gérard-Varet, Pierre Le Bris, Frédéric Legoll, Tony Lelièvre, Alexei Lozinski, Urbain Vaës, Sylvain Wolf, for their many constructive remarks on the preliminary version and on the French version of this manuscript. We thank the two anonymous referees of the French version for their careful reading and their many suggestions. The following colleagues also provided comments and suggestions for some specific portions of our text: Andrea Braides (Sect. 6.2.2), Adina Ciomaga (Chap. 6), Anthony Patera (Chap. 5) and Daniel Peterseim (Sect. 5.2.4). We wish to thank Rutger Biezemans, Ludovic Chamoin, Gaspard Jankowiak and Alexei Lozinski for graciously providing us with some illustrations and numerical results, and also Alfio Quarteroni and Francesca Bonadei for welcoming the present manuscript at the collection MS&A, Springer.

The url https://www.ljll.math.upmc.fr/~blanc/files/errata-book.pdf contains corrections to errors we noticed after submission of the manuscript, together with some additional remarks. The first author acknowledges support from Inria through a partial "delegation" which allowed him to find the necessary time to write this textbook during the academic years 2019–2020 and 2020–2021. The second author thanks the Department of Mathematics of the University of Chicago for its hospitality during many visits in the past years. The work of the second author is, since 2008, partially supported by the *European Office of Aerospace Research and Development* (currently under Grant FA8655-20-1-7043) and the *Office of Naval Research* (currently under Grant N00014-20-1-2691).

Paris, France
2023

Preliminary Remarks

Except for some incursions that we will allow ourselves in Chap. 2 and in Chap. 6, our main topic of interest throughout this textbook is the equation

$$-\operatorname{div}(a_\varepsilon(x)\,\nabla u^\varepsilon(x)) = f(x). \tag{1}$$

Equation (1) is posed in a bounded domain $\mathcal{D} \subset \mathbb{R}^d$. This domain $\mathcal{D}$ is assumed to have a sufficiently regular boundary $\partial\mathcal{D}$. If need be, we will make precise this regularity later. The ambient dimension is typically $d = 3$, but nothing is set in stone. In particular, we will often consider the one-dimensional setting. We will explain later why this choice of working out homogenization theory in dimension $d = 1$, possibly surprising to experts, is important to us. We will occasionally look also at dimension $d = 2$. The two-dimensional setting is notably convenient to expose computational approaches, because in that dimension geometrical questions such as mesh generations are much easier than in dimension 3. Questions related to functional analysis and partial differential equations theory are, on the other hand, often quite peculiar in dimension 2.

Equation (1) is supplied with a boundary condition that most often in the sequel will be the homogeneous Dirichlet boundary condition

$$u^\varepsilon(x) = 0, \tag{2}$$

on the boundary $\partial\mathcal{D}$ of the domain. It could also be, incidentally, a Neumann boundary condition or another type of boundary condition. We will also make this precise in due course.

The difficulty of the problem owes to the oscillatory character of the coefficient $a_\varepsilon(x)$ in (1). This coefficient encodes, intuitively, the microscopic physical nature of the medium. It oscillates at the fine lengthscale characterized by ε. In sharp contrast, the function $f(x)$ on the right-hand side only varies at scale 1. It models in practice the forces the system under study is subjected to. These forces are usually independent of the structure of the medium, thus of the finest scale ε of the microstructure.

We will most of the time assume, for simplicity of exposition, that the co-efficient a_ε is scalar-valued. We will do so particularly when considering it is matrix-valued only adds unnecessary technicalities. It is however to be borne in mind that, in practice, it is generally matrix-valued. The coefficient $a_\varepsilon(x)$ will also be often considered as the *rescaling* of a fixed coefficient $a(x)$, that is

$$a_\varepsilon(x) \,=\, a\left(\frac{x}{\varepsilon}\right). \tag{3}$$

Homogenization theory is intrinsically meant to address the general case of a coefficient $a_\varepsilon(x)$ that needs not read as in (3). Some of our theoretical developments will however assume this structure, which we will extensively exploit.

We are perfectly aware of the strong limitation imposed by the assumption (3). We will emphasize this as often as necessary. We will likewise emphasize the contexts for which we may free ourselves from this assumption. This will be in particular the case when we introduce computational approaches in Chap. 5 and when we analyze theoretically most of these approaches. Even if the structure (3) may be considered too specific, we do believe that its complete theoretical and numerical study constitutes a major leap forward.

In any event, we will always assume that the coefficient $a_\varepsilon(x)$ enjoys the necessary coerciveness properties so that Eq. (1) may be proven to be well posed using only classical arguments and not sophisticated theories. In the scalar-valued case, this amounts to assuming that, uniformly in ε, the coefficient is bounded away from zero and bounded from above, that is

$$\exists\, 0 < \mu, \quad \exists\, M < +\infty, \quad \forall\, \varepsilon > 0, \quad \mu \le a_\varepsilon(x) \le M, \tag{4}$$

for all $x \in \mathcal{D}$, or at least almost everywhere if the coefficient is not continuous. We will *not* investigate any question regarding possible degeneracy, or blow-up of this coefficient. In those "regular" conditions, an immediate application of the Lax-Milgram Lemma shows that the solution to (1) and (2) uniquely exists, say in the Sobolev space $H_0^1(\mathcal{D})$ when $f \in L^2(\mathcal{D})$ and even for a less regular right-hand side actually. The major *practical* difficulty for the resolution of (1), that is, to approach the exact value of its solution using a numerical discretization whatever the particular method used for this purpose may be, is that the oscillations of the solution $u^\varepsilon(x)$ are expected to replicate the oscillations of the coefficient $a_\varepsilon(x)$. These oscillations, of typical size $\varepsilon \ll 1$, are only captured using a generic computational approach when the discretization parameter h (the meshsize of a finite element method, the stepsize of a finite difference approach) is taken smaller than ε. We are precisely going to deliberately assume that we cannot afford such a tiny parameter, which would lead to prohibitively long computational times or prohibitively large memory requirements that would exceed the performances and capacities of the tools we have access to.

Let us at once say that, if our purpose is indeed to devote all our efforts to study Eq. (1), the task should not be overlooked. Indeed...

> **1st warning to the reader:**
> **Do not underestimate the difficulty of Eq. (1).**
> **Although it is linear and in divergence form,**
>
> - **it presents, for homogenization, considerable mathematical difficulties, some of which are still unsolved,**
> - **it arises, in one form or another, in a myriad of application areas.**

We are going to discover that the linear nature of the equation is, for the questions we investigate, only an *illusion*. The heart of the homogenization process is rather the variation of the solution u_ε in function of the variation of the parameter a_ε. This mechanism outright involves a *nonlinear*, and most often *nonlocal* map. As for our second claim above regarding the practical relevance of the equation, it cannot be overstated. Equation (1) appears as such in heat conduction, in mechanical elasticity, in electrostatics, etc. It also appears, possibly in a slightly different form, in a large portfolio of more complex, possibly time-dependent or nonlinear problems.

The guiding principle of homogenization theory is to embed the specific Eq. (1) for a given small scale ε into a *family* of Eq. (1) for a *sequence* of parameters ε vanishing in the limit. The hope is that, considering the limit "$\varepsilon = 0$" yields a problem, both simpler to understand and simpler to simulate numerically, than the original problem at $\varepsilon \neq 0$ fixed. In the limit problem, the small scale would have disappeared and, correspondingly, all the practical limitation mentioned above as well. The discretization parameter h could be chosen larger than initially expected. This dream will indeed come true in a large variety of situations. We recognize here a frequent characteristics of mathematical problems. Some of them are easier to solve in their asymptotic regimes. For (1), the limit equation, as we shall see, takes the form

$$- \operatorname{div}(a^*(x) \nabla u^*(x)) = f(x), \tag{5}$$

where the whole difficulty hides behind the practical identification of the coefficient a^*, called the homogenized coefficient. In the bulk part of this textbook, the coefficient a^* will be constant. To some extent, homogenization theory "divides and conquers". The difficult resolution of Eq. (1) is traded for a supposedly simpler calculation of the coefficient a^* and the resolution, also expected simpler, of (5). That said, this textbook will show that completing this appealing program for a large variety of coefficients a_ε is not straightforward. Which brings us to a crucial observation. We do not only aim to understand theoretically the mathematical issues. We also look for practical answers.

2nd warning to the reader:
Putting homogenization in action comes at a price.
The approach takes full power

– **either when the purpose is to understand,**
– **or when the purpose is to practically solve (1) repeatedly,**

for instance because Eq. (1) has to be solved successively for many functions $f(x)$ in a row. Put differently, if the only purpose is to solve (1) *once* for some ε small, then it is a much better strategy to employ any computer power available to solve it brute force than to follow the route suggested by homogenization. Placing boundaries is not demeaning a given approach. On the contrary, it is adding value to it.

Given that we are about to focus almost all our attention to Eq. (1) at the expense of all other equations, and given that we are indeed taking the road of homogenization theory which we know is expensive and demanding, what are we going to do? We are going to consider any conceivable variation of the coefficient $a_\varepsilon(x)$ within (1) that would encode a realistic nature of the medium, or the material under study. This is a question of huge theoretical and practical impact.

Contents

Chapter 1
In Dimension "Zero"

Abstract We will really start studying homogenization theory in Chap. 2. For now, we only propose an informal introduction to the problem. How can we expect the solutions of (1) to behave in the limit $\varepsilon \to 0$? What information can we obtain "for free"? What additional price should we pay to answer to more difficult questions? We will see that, specifically, two ingredients show up: weak convergence of sequences of functions, and the (related) notion of mean value of a function. The material we collect in the present chapter will be used throughout this textbook. It will prove useful in (at least) two ways: both in getting an intuition about more elaborate problems and in building mathematical proofs.

Keywords Average · Weak convergence · Quasiperiodic function · Almost periodic function · Stationary function · Particle system

We will really start studying homogenization theory in Chap. 2. For now, we only propose an informal introduction to the problem. How can we expect the solutions of (1) to behave in the limit $\varepsilon \to 0$? What information can we obtain "for free"? What additional price should we pay to answer to more difficult questions? We will see that, specifically, two ingredients show up: weak convergence of sequences of functions, and the (related) notion of mean value of a function. The material we collect in the present chapter will be used throughout this textbook. It will prove useful in (at least) two ways: both in getting an intuition about more elaborate problems and in building mathematical proofs. We will learn the importance of structural assumptions such as periodicity, almost-periodicity, stationary ergodic structure, etc..., and understand how these assumptions modify the way we address a given problem. This chapter is one of the most important in the textbook. It is also the one that is the farthest from the classical presentations of homogenization theory.

1.1   A Simplification to Better Understand

As a first step to understand the structure of Eq. (1), we make a very brutal
simplification: we simply forget about the differential operators in (1), so that the
equation reads as

$$- a_\varepsilon(x)\, u^\varepsilon(x) = f(x), \tag{1.1}$$

in the domain $\mathcal{D}$, that is,

$$u^\varepsilon(x) = -\frac{1}{a_\varepsilon}(x)\, f(x). \tag{1.2}$$

Despite this oversimplified setting, we will see that many properties of Eq. (1.2) are
still valid for (1). This will be clear, first in dimension one in Chap. 2, then in higher
dimension in the following chapters. Of course, things will prove more difficult and
more intricate, but many of the simple observations we make here will remain valid
in a more general setting.

First, as announced in our Preliminary Remarks, it is clear from (1.2) that
the relation between u^ε and a_ε is *nonlinear*, although Eq. (1) is linear (and so
is (1.1)). This nonlinearity is related to the structure of Eq. (1), but neither to the
differential operators present in (1), nor to the dimension considered. Nevertheless,
this nonlinearity will prove to be the main difficulty in our homogenization problem.
We will return to this in the following chapters.

Second, looking at (1.2), one clearly understands that *practically* solving Eq. (1)
promises to be difficult when the parameter ε becomes small.   Indeed, the
oscillations and heterogeneities of the coefficient a_ε are directly transmitted to the
solution u^ε via (1.2). Hence, in order to capture these oscillations, it is necessary
to use a discretization with resolution (at least) finer than ε. Again, this observation
will remain true for Eq. (1). We will already see an example of this in Chap. 2.

Let us now embed Eq. (1.1) into a *family* of equations parametrized by ε. The
bounds (4) imply that the sequence of functions $\frac{1}{a_\varepsilon}(x)$ is bounded in L^∞. Thus, we
can find a subsequence $\varepsilon' \to 0$ and a function, which we denote by $\frac{1}{a}(x)$, such that
the following convergence holds in the weak $L^\infty - \star(\mathcal{D})$ topology:

$$\frac{1}{a_{\varepsilon'}} \xrightarrow{\varepsilon' \to 0} \frac{1}{a}, \tag{1.3}$$

that is,

$$\int_{\mathcal{D}} \frac{1}{a_{\varepsilon'}}(x)\, \varphi(x)\, dx \xrightarrow{\varepsilon' \to 0} \int_{\mathcal{D}} \frac{1}{a}(x)\, \varphi(x)\, dx, \tag{1.4}$$

for any function $\varphi \in L^1(\mathcal{D})$. Note that, *a priori*, the function $\frac{1}{a}(x)$ depends on the subsequence ε'. For simplicity, we assume that $f \in L^2(\mathcal{D})$. We get, using the Cauchy-Schwarz inequality, the following weak convergence in $L^2(\mathcal{D})$:

$$u^{\varepsilon'} \overset{\varepsilon' \to 0}{\longrightarrow} -\frac{1}{a} f. \tag{1.5}$$

Indeed, if $\varphi \in L^2(\mathcal{D})$,

$$\int_{\mathcal{D}} \frac{1}{a_{\varepsilon'}}(x) \, f(x) \, \varphi(x) \, dx \overset{\varepsilon' \to 0}{\longrightarrow} \int_{\mathcal{D}} \frac{1}{a}(x) \, f(x) \, \varphi(x) \, dx,$$

because $f \in L^2(\mathcal{D})$ and $\varphi \in L^2(\mathcal{D})$ imply $f \varphi \in L^1(\mathcal{D})$. We then define $u^*(x) = -\frac{1}{a}(x) \, f(x)$, and we can say (this sounds somewhat pedantic here, but it will be made clear in the sequel) that the weak limit u^* of the sequence $u^{\varepsilon'}$ solution to (1.1) satisfies the limit equation (which we will soon call *homogenized equation*):

$$-\overline{a}(x) \, u^*(x) = f(x). \tag{1.6}$$

Let us sit back and think a little about this.

We have just proved that the limit equation exists, and that it has the same form as the original Eq. (1.1). As we will see below, this is both a general fact (it is actually true for Eq. (1)), and some sort of a miracle (it fails to be true in many other situations, as we will see in an example in Sect. 2.5 of Chap. 2). We insist on the fact that (1.6) is indeed an *equation*, since the coefficient $\overline{a}$ is *independent* of the right-hand side f. Whatever the right-hand side f, the coefficient $\overline{a}$ remains the same, as far as we keep the same subsequence ε' in (1.3). Thus, in some sense, we have obtained the limit behaviour of the solution for small values of ε.

An important point here is that the coefficient $\overline{a}$ of the limit equation has a very special form: it is the *inverse* of the limit (1.3) of the *inverse* $\frac{1}{a_{\varepsilon'}}(x)$ of $a_{\varepsilon'}(x)$. It is *not*, as one could expect at first sight, the limit of the coefficients $a_{\varepsilon'}(x)$. We will come back to this point below.

Note that this result is of importance, despite many shortcomings that we are going to describe. With very few information, namely the bounds (4), we have been able to obtain a limit equation. Moreover, the solution of this equation, which defines the limit behaviour of the system under consideration, depends linearly on the datum f. This is an important piece of information.

However, the main difficulty here is that we know *nothing* about the limit we have obtained:

(i) first, we do not know if the whole sequence u^ε converges. We have only proved convergence for a subsequence. It is therefore hopeless to prove any uniqueness result for the limit;

(ii) second, we do not really know how to compute the coefficient $\bar{a}$, since it is defined by a *compactness method*. The only thing we know (which has allowed us to define the limit coefficient) is that a sequence such as $\dfrac{1}{a_\varepsilon}$, bounded in L^∞, is relatively compact in the weak-$\star$ topology;

(iii) finally, even for a converging subsequence, we are unable to prove any convergence rate of $u^{\varepsilon'}$ to u^*, since we do not even know any rate for the convergence of $\dfrac{1}{a_{\varepsilon'}}$ to $\dfrac{1}{\bar{a}}$ itself.

We cannot hope to obtain further information in full generality. In order to realize this, the reader may build examples of all possible situations. Should we want to know more, we have to use more restrictive assumptions. A way to do this is to *assume a structure* on the coefficient $a_\varepsilon(x)$. The structure that has been mostly used, pedagogically and in terms of the first possible practical applications, is a structure of separation of scales. This may be written as

$$a_\varepsilon(x) = a\left(\frac{x}{\varepsilon}\right) \tag{1.7}$$

for a fixed function a that *does not depend on* ε, and, moreover, that is *periodic*:

$$a_\varepsilon(x) = a\left(\frac{x}{\varepsilon}\right) \qquad \text{where} \quad a \quad \text{is periodic.} \tag{1.8}$$

Let us first study the case (1.8). We will then return, in Sect. 1.4, to an assumption (1.7) where a may not be periodic (but still has a specific structure).

Remark 1.1 Although we "know nothing" about the limit in general, we can still obtain some information indirectly. For instance, the limit satisfies a *comparison principle*. If $a_\varepsilon \leq b_\varepsilon$, in the sense $\displaystyle\int_{\mathcal{D}} \frac{1}{b_\varepsilon(x)} \varphi(x)\, dx \leq \int_{\mathcal{D}} \frac{1}{a_\varepsilon(x)} \varphi(x)\, dx$ for any function $\varphi \geq 0$ in $L^1(\mathcal{D})$, then, for a subsequence such that both $\dfrac{1}{a_{\varepsilon'}}$ and $\dfrac{1}{b_{\varepsilon'}}$ converge (in the weak-$\star$ topology), the convergence (1.4) implies that $\bar{a} \leq \bar{b}$ in the same sense. This comparison principle, also called *monotonicity*, is also valid in higher dimension (in the sense of symmetric matrices). $\qquad\qquad\square$

Remark 1.2 In the above argument, we have simply forgotten the differential operators in (1), in order to simplify the problem and prove that it is equivalent to computing weak limits. Another, slightly less brutal, way to introduce such limits is to replace Eq. (1) by a Schrödinger-type equation:

$$-\Delta u^\varepsilon + V_\varepsilon u^\varepsilon = f, \tag{1.9}$$

still posed in a domain $\mathcal{D}$ of $\mathbb{R}^d$, with, for instance, homogeneous Dirichlet boundary conditions. The potential $V_\varepsilon(x)$ is assumed to be non-negative, bounded, and oscillating (think for instance of $V_\varepsilon(x) = V(x/\varepsilon)$). For the same reasons as above, it converges weakly, up to the extraction of a subsequence. Existence of a solution to (1.9) is easily proved by applying the Lax-Milgram Lemma, or by minimizing the associated energy. Multiplying (1.9) by u^ε and integrating over $\mathcal{D}$, the Green formula gives

$$\int_{\mathcal{D}} |\nabla u^\varepsilon|^2 + \int_{\mathcal{D}} V_\varepsilon \left(u^\varepsilon\right)^2 = \int_{\mathcal{D}} f u^\varepsilon \le \|f\|_{L^2(\mathcal{D})} \|u^\varepsilon\|_{L^2(\mathcal{D})},$$

where we have applied the Cauchy-Schwarz inequality. This estimate and the Poincaré inequality then imply that u^ε is bounded in $H^1_0(\mathcal{D})$. We may thus extract a subsequence that converges weakly in $H^1_0(\mathcal{D})$, and (using the Rellich Theorem) strongly in $L^2(\mathcal{D})$. Hence, the product $V_\varepsilon u^\varepsilon$ is a product of a weak convergence by a strong convergence. We may thus pass to the limit in (1.9) and find

$$- \Delta u^* + \overline{V} u^* = f, \tag{1.10}$$

where $\overline{V}$ is the weak limit of V_ε, and u^* the limit of u^ε. Therefore, finding the limit equation of (1.9) amounts to computing the weak limit of V_ε. Note also that, here again, the limit equation has the same form as the original one.   □

1.2   The Periodic Setting

In this section, we assume that a_ε is periodic, in the sense of (1.8). As it is the case for the one-dimensional setting (exposed below in Chap. 2), this assumption allows for some simplifications, but it may prove misleading. The set of periodic functions is indeed very specific. The periodic setting has been very successful because it was mathematically simpler, practically less costly (at times when computers were not as powerful as today), and well adapted to many industrial applications. In full generality and for "real-life" applications, it however remains an academic simplification. On the other hand, some experts who consider the one-dimensional setting too simple, do use the periodic setting without questioning it.

We are going to repeat the work we have performed on Eq. (1.1), this time with the additional assumption (1.8). It will allow us to pass to the limit in the explicit expression (1.2) of the solution u^ε. This solution now reads as

$$u^\varepsilon(x) = -\frac{1}{a} \left(\frac{x}{\varepsilon}\right) f(x).$$

We recall the following result for periodic functions. For simplicity, we state it in dimension 1, although it is easily extended to higher dimensions.

Proposition 1.1 *Let b be a function in $L^\infty(\mathbb{R})$. Assume that it is periodic of period 1. Then the sequence of functions $b\left(\frac{\cdot}{\varepsilon}\right)$ converges weakly-$\star$ in L^∞ to the* constant function, denoted by $\langle b \rangle$, called the average of b, and equal to

$$\langle b \rangle = \int_0^1 b. \tag{1.11}$$

Proof We want to show that, for any $v \in L^1(\mathbb{R})$, we have

$$\int_{\mathbb{R}} b\left(\frac{x}{\varepsilon}\right) v(x) \, dx \xrightarrow{\varepsilon \to 0} \int_0^1 b \int_{\mathbb{R}} v.$$

We prove this for v being the characteristic function of an interval. The density of piecewise constant functions in $L^1(\mathbb{R})$ then allows to conclude. We need to prove that, for any $\alpha < \beta$,

$$\int_\alpha^\beta b\left(\frac{x}{\varepsilon}\right) dx \xrightarrow{\varepsilon \to 0} (\beta - \alpha) \int_0^1 b.$$

We use the fact that b is periodic, and write:

$$\int_\alpha^\beta b\left(\frac{x}{\varepsilon}\right) dx = \varepsilon \int_{\frac{\alpha}{\varepsilon}}^{\frac{\beta}{\varepsilon}} b(y) \, dy$$

$$= \varepsilon \left(\left[\frac{\beta}{\varepsilon}\right] - \left[\frac{\alpha}{\varepsilon}\right] - 1\right) \langle b \rangle + \varepsilon \int_{\frac{\alpha}{\varepsilon}}^{\left[\frac{\alpha}{\varepsilon}\right]+1} b(y) \, dy + \varepsilon \int_{\left[\frac{\beta}{\varepsilon}\right]}^{\frac{\beta}{\varepsilon}} b(y) \, dy$$

$$= (\beta - \alpha) \langle b \rangle + O(\varepsilon), \tag{1.12}$$

where $[x]$ denotes the integer part of x. This concludes the proof. $\square$

In the limit $\varepsilon \to 0$, from Proposition 1.1 applied to the function $\dfrac{1}{a}$, we infer that the *whole* sequence u^ε converges:

$$u^\varepsilon \longrightarrow u^* = -\left\langle \frac{1}{a} \right\rangle f, \tag{1.13}$$

in $L^2(\mathcal{D})$. This allows us to explicitly determine the coefficient $\bar{a}$ appearing in the limit Eq. (1.6) of (1.1). Hence, we have proved that u^* is solution to

$$-\frac{1}{\left\langle \frac{1}{a} \right\rangle} u^*(x) = f(x). \tag{1.14}$$

In other words, $\bar{a}$ is the *unique* constant coefficient given by

$$\bar{a} = \frac{1}{\left\langle \frac{1}{a} \right\rangle}. \tag{1.15}$$

Note that this coefficient is *never* equal to the natural guess coefficient $\langle a \rangle$, unless a is constant (it is of limited interest to consider a constant coefficient with small scale oscillations (1.8)). We indeed recall the *Jensen inequality*, $\varphi \left(\int_0^1 a(x)\,dx \right) \leq \int_0^1 \varphi(a(x))\,dx$, for any convex function φ. Applying it to $\varphi(t) = \frac{1}{t}$ for $t > 0$, we find that $\bar{a} \leq \langle a \rangle$. Moreover, since φ is *strictly* convex, equality in the Jensen inequality occurs if and only if a is constant. We also note that, in this argument, we do not use the fact that a is periodic. The only important fact is that the coefficient $\bar{a}$ is defined as a mean value. We keep this in mind for further use. Anyhow, in the case (1.1) (and, actually, in the setting of Chap. 2 where the differential operators are restored), our analysis shows that, in order to identify $\bar{a}$, we need to know the *statistics* (this term will be made clear in Sect. 1.6) of $\frac{1}{a}$, and not only that of a... This case is the simplest one. In hardly more difficult situations, or in higher dimensions, guessing the homogenized coefficient will prove very difficult, if even possible.

Assumption (1.8) has allowed us, at least in the periodic case, to get information that was lacking in the general case, as we pointed out in (i) and (ii) above. Let us note that this assumption allows to prove that we have *weak convergence* in $L^p(\mathcal{D})$ (depending on the integrability of the right-hand side f, here we have used $p = 2$), but we do *not* have *strong convergence*. Indeed, the periodic function $\frac{1}{a}\left(\frac{x}{\varepsilon}\right)$ can only converge strongly to its average $\left\langle \frac{1}{a} \right\rangle$ if it is a constant. Moreover, assumption (1.8) brings some information about the convergence rate mentioned in (iii) above. In the periodic case, estimate (1.12) in the proof of Proposition 1.1 implies that

$$\left| \int_0^1 \frac{1}{a}\left(\frac{x}{\varepsilon}\right) dx - \left\langle \frac{1}{a} \right\rangle \right| = O(\varepsilon). \tag{1.16}$$

Formally (at least if $f \equiv 1\ldots$), this indicates a convergence rate in (1.13). But for the weak-$\star$ convergence in L^∞ stated in Proposition 1.1, a proof of such a rate is not possible. To be more precise, if the test function is only assumed to be in L^1, no convergence rate can be proved. Provided we assume further regularity, as for instance C^1, we can rigorously prove a convergence rate of order ε. This exercise is left to the reader.

As a preliminary remark for the next chapters, we wish to emphasize that any periodic function b, say of period 1, satisfies the following property:

(P1) the rescaled functions $b\left(\dfrac{\cdot}{\varepsilon}\right)$ converge at rate $O(\varepsilon)$ to their average, in the sense of (1.16) above.

Moreover, such a periodic function b also satisfies the following four elementary properties:

(P2) the average $\langle|b|\rangle$ defines a norm on the set of periodic functions, which is equivalent to the *uniform L^1* norm $\|b\|_{L^1_{\text{unif}}} = \sup\limits_{x\in\mathbb{R}} \int_{|y-x|\leq 1} |b(y)|\,dy$;

(P3) any periodic function b such that $\langle b\rangle = 0$ is the derivative of a *periodic* function, unique up to the addition of a constant;

(P4) periodic functions satisfy the *Poincaré-Wirtinger inequality*:

$$\int_{[0,1]} |b(x) - \langle b\rangle|^2\,dx \leq C_{PW} \int_{[0,1]} \left|b'(x)\right|^2\,dx,$$

where the constant C_{PW} only depends on the period (here, 1), and not on b;

(P5) any continuous periodic function is bounded, thus in particular it is strictly sublinear at infinity, in the sense that $\dfrac{b(x)}{1+|x|} \overset{|x|\to+\infty}{\longrightarrow} 0$.

All these properties hold true *mutatis mutandis* in higher dimensions.

It is thus natural to hope that the periodic case will allow us to have a good knowledge (in the sense of the properties mentioned in (i)–(ii)–(iii)) of the behaviour of the solution of Eq. (1) for small values of ε. This will indeed be the case, as we will see in the following chapters. For now, we study the possible settings beyond the periodic setting. We start by what *could* be seen as a digression. We will discover in Sects. 1.4–1.6 that it is not.

1.3 Energy of Infinite Systems of Particles

Homogenization theory is by nature linked with the theory of weak convergence. In particular settings such as the periodic case, the question is that of the computation of mean values of functions. The problem happens to be closely related to a question that may seem at first sight rather independent: the average energy of infinite systems of particles. We could explain the link between homogenization of partial differential equations and average energy of systems of particles considering the *discrete* version of the differential operators in (1). In that, we would replace $\dfrac{du}{dx}$ by $\dfrac{u(\cdot + h) - u(\cdot)}{h}$. However, this is not necessary in the present chapter, all the more since, for now, we have eliminated the differential operators!

Our discussion about systems of particles is a useful guideline for all the homogenization settings we are going to study. It will allow us to test geometrical assumptions and to guess what type of asymptotic results we can (or cannot) hope to prove. The analogy we elaborate upon will even be translated into a "real" mathematical tool in the next section.

1.3.1 Choice of a Model for an Infinite Periodic System

Let us consider an infinite system of particles (one could alternatively say *atoms*, or *atomistic sites*), in dimension one for simplicity. All the following can be easily generalized to higher dimensions. Intuitively, this model describes in an oversimplified way a crystalline material. We fix $N \in \mathbb{N}$ and $2N + 1$ particles, located at the points $X_k \in \mathbb{R}$, $k \in \{-N, \ldots, N\}$. These positions are assumed (again for simplicity) to be ordered: $X_{-N} \leq X_{-N+1} \leq \cdots \leq X_N$. The interaction between particles is modelled by a so-called *nearest-neighbour potential*. This means that a particle located at X_k only "sees" those located at X_{k-1} and X_{k+1} (with the convention that if $k = -N$ or $k = N$, it only sees its unique neighbour). The interaction potential is denoted by V and only depends on the distance. If the number N is finite, this model is seen as that of a finite system of particles (some type of "molecule"), and its energy is equal to the sum of all interactions:

$$E_N = \sum_{-N \leq k \leq N-1} V(|X_{k+1} - X_k|). \tag{1.17}$$

We have fixed the particle positions at X_k because we have indeed implicitly assumed that the distances $|X_{k+1} - X_k|$ are exactly equal to the minimum value of the potential V. Put differently, the particles are located so as to minimize the global energy E_N. We now aim at studying the limit $N \to +\infty$. The reader understands that N plays the role of the parameter $\dfrac{1}{\varepsilon}$ in homogenization.

We first make the following point. Assume that the particles are equally spaced, say, $X_{k+1} - X_k = 1$ for all $k \in \{-N, \ldots, N - 1\}$. Assume also that the value $V(1)$ of the potential is negative: $V(1) < 0$. This assumption is physically relevant. Two particles far away from each other are supposed to weakly interact, so $V \to 0$ at infinity. As a consequence the minimum of V is negative. The energy E_N of the system of $2N + 1$ particles reads as $E_N = 2N V(1)$ thus diverges to $-\infty$ as $N \to +\infty$. This is not surprising. The energy is an extensive quantity, that is, it depends linearly on the number of particles considered. One needs to *renormalize* it in order to obtain, in the limit of a large number of particles, a finite value. This phenomenon is similar to the renormalization we apply to the integral of a periodic function over a large domain, in order to define its average. We therefore consider the *energy per particle* $\dfrac{1}{2N+1} E_N$, which, in the present case, trivially satisfies

$$\frac{1}{2N+1} E_N = \frac{2N}{2N+1} V(1) \longrightarrow V(1), \tag{1.18}$$

as $N \to +\infty$. The value $V(1)$ of this limit in a sense defines the energy of the infinite *periodic* system of particles $X_k = k$, $k \in \mathbb{Z}$. Actually, this is the only possible sense in which that energy may be defined, but such considerations are beyond the scope of this textbook. As we will see below, this energy plays a role equivalent to the homogenized coefficient $\bar{a}$ defined by (1.14), or more generally A^* defined by (5). Our aim is now to eliminate the periodic assumption $X_{k+1} - X_k = 1$ that we have used here. Doing so, we will understand how we can shuffle this system of particles (soon we will talk about "defects" or "perturbations") while keeping a well-defined energy per particle $\lim\limits_{N \to +\infty} \dfrac{1}{2N+1} E_N$.

1.3.2 Introduction to (Local) Defects

Let us first break the nice periodic arrangement above by adding a *local defect*. This is easily achieved assuming that a particle is not at its "right place". A simple case is:

$$X_k = \begin{cases} k \text{ for } k \neq 0, \\ \alpha \text{ for } k = 0, \end{cases} \tag{1.19}$$

where, say, $\alpha \in]0, 1[$. The energy of the system then reads as

$$\begin{aligned} E_N &= \sum_{k=-N}^{-2} V(|X_{k+1} - X_k|) + V(|1 + \alpha|) + V(|1 - \alpha|) + \sum_{k=1}^{N-1} V(|X_{k+1} - X_k|) \\ &= (N-1) V(1) + V(1 + \alpha) + V(1 - \alpha) + (N-1) V(1). \end{aligned} \tag{1.20}$$

We find that the energy per particle $\dfrac{1}{2N+1} E_N$ is not modified asymptotically. This should not be a surprise when considering the renormalization performed. As in the periodic case, it converges to $V(1)$ in the limit $N \to +\infty$. Introducing only one defect (or a finite number of defects) does not change the average energy. We make note of this simple fact, we will return to it in the sequel. In order to "see" the defect, we need to use a quantity that captures a finer information than that seen by the average energy. To this end, we consider the *difference* between the periodic system and the perturbed one, that is,

$$\Delta E_N = V(1 + \alpha) + V(1 - \alpha) - 2V(1). \tag{1.21}$$

This quantity obviously has a limit as $N \to +\infty$ in this oversimplified setting: it does not depend on N. We will see in Chap. 2, Sect. 2.2, when discussing the corrector equation in the case of defects, that this way of "zooming in" about the relevant details may be extended to homogenization problems. From the modelling viewpoint, the energy of the infinite perturbed system, in the limit $N \to +\infty$, should be measured more precisely than in average. We use as a *reference* the energy of the underlying periodic system. This can be seen as a way to *normalize* (that is, deal with the fact that the system is infinite) that is different from dividing by the number of particles.

Instead of perturbing the position of the particle indexed by 0, we could of course simply remove that particle, or change the position of any finite number of particles, and find the same result. Only the explicit expression (1.21) would change. We could similarly imagine a displacement of all particles, for instance

$$X_k = k + \exp(-|k|),$$

with suitable properties on V (which we do not want to rigorously define for now), and straightforwardly adapt the above argument. All this is true as far as we manipulate quantities that are formally summable at infinity. The perturbation should *vanish at infinity* sufficiently fast. In any such situation, the energy per particle (1.18) is left unchanged and the energy difference with (1.21) remains finite.

For a perturbation to leave the energy unchanged, vanishing at infinity is however not the only option. A perturbation could simply become "rare" at infinity. Consider as an example the perturbation which leaves all particles at their original positions $X_k = k$, *except* those particles of indices $k = \pm 2^n$ for $n \in \mathbb{N}$. The latter particles are moved to positions $X_k = k + \alpha$, where, say, $\alpha \in]0, 1[$. Given an interval of length N, the number of integers in this interval that are powers of 2 is bounded by $O(\log N) \ll N$. As a consequence, the energy per particle, here again, remains unchanged. Nevertheless, the perturbation does not disappear at infinity. However far we look on the real line, we still find some particles the positions of which have been modified. And actually, the energy difference ΔE_N does not converge as $N \to +\infty$.

One may also think of a "blockwise" perturbation, as in the following example:

$$X_k = \begin{cases} k + 1/2 & \text{for } (2n)^2 < |k| \le (2n+1)^2, \\ k & \text{for } (2n+1)^2 < |k| \le (2n+2)^2. \end{cases} \tag{1.22}$$

In this example, we have a succession of large blocks of particles, each block being either shifted by $1/2$ or not. Since, when computing the energy of a system, only *distances* between neighbouring particles are important, the system defined by (1.22) has exactly the same energy per particles as the original periodic system, although its spatial distribution is fairly different.

Another way to modify a periodic system is to locally change the period. For example, consider $X_k = k$ and $X_{-k} = -k\sqrt{2}$, for $k \in \mathbb{N}$. This is no more a *local* defect. This may be seen as a (crude) modelling of two periodic systems glued together. The two periods on the left and on the right of the origin are incommensurable. In this particular one-dimensional case, the interface between the two systems is reduced to the origin. It may be far more complicated in higher dimensions. The energy per particle is modified: a simple computation shows that it consists of a weighted combination of the energies of the two periodic systems.

The above examples are more and more complicated modifications of the original periodic setting.

They allow to understand the wide diversity of configurations for which we can compute an average energy per particle. Some of these configurations are "small" modifications (which we will soon call *perturbations*) of the simple periodic configuration, while some are much farther from this reference. We could carry on and consider other examples. Instead, we are going to consider general situations in which we change *all* the positions of the particles. We thus conclude this section by studying two settings that aim at being general and including the above cases as particular cases. The first setting is deterministic. The second one is random.

1.3.3 Toward More General Perturbations

We will present in Sect. 1.4.5 below a general deterministic setting for the computation of the average energy of infinite systems of particles. We only give now a simplified version of this setting (adapted to dimension one). Our general setting is based on the following observation: in order to be able to compute an average energy per particle such as (1.18), we only need *most* particles to be well distributed. Indeed, when computing the average energy, one divides by N. If the number of "displaced" particles is of order $o(N)$, their contribution vanishes in the limit $N \to +\infty$. This observation, which was originally made in the article [BLL03], suggests that one should consider systems of particles of positions X_k, for $k \in \mathbb{Z}$, such that the following properties are satisfied:

(H1) there is no arbitrarily large cluster of particles, in the sense that

$$\sup_{x \in \mathbb{R}} \# \{k \in \mathbb{Z};\ |x - X_k| \le 1\} < +\infty; \tag{1.23}$$

(H2) there is no arbitrarily large region containing no particle, in the sense that

$$\exists R > 0 \quad \inf_{x \in \mathbb{R}} \# \{k \in \mathbb{Z};\ |x - X_k| \le R\} > 0; \tag{1.24}$$

(H3) (H3.1) in the limit of large volumes, the number of particles scales as the volume, that is,

$$\exists \lim_{R \to +\infty} \frac{1}{2R} \# \{k \in \mathbb{Z};\ X_k \in [-R, R]\}, \tag{1.25}$$

and (H3.2) the pairs of nearest-neighbour particles are distant from approximate periods, in the following sense: for all $\delta > 0$ sufficiently small, and for all $h > 0$, the following limit exists

$$l(h, \delta) = \lim_{R \to +\infty} \frac{1}{4R\delta} \# \{k \in \mathbb{Z}; \ X_k \in [-R, R] \ ; \ |X_{k+1} - X_k - h| \leq \delta \}. \tag{1.26}$$

For simplicity, the above properties have been stated in dimension one and for nearest-neighbour interaction only. It is clear that, in the periodic case, $X_k = Tk$ for $k \in \mathbb{Z}$, and for some fixed period T, so that the above assumptions are satisfied. A simple computation in particular shows that $l(h, \delta) = \frac{1}{2\delta}$ if $h \in [T - \delta, T + \delta]$, and $l(h, \delta) = 0$ otherwise. In the limit $\delta \to 0$, $l(h, \delta)$ converges to the distribution $l(h)$ equal to the Dirac mass at $h = T$. It is an easy exercise to prove that, in the general setting, $l(h)$ is the measure defined by

$$\forall \varphi \in C_b^0(\mathbb{R}), \quad \int_{\mathbb{R}} \varphi(h) dl(h) = \lim_{R \to +\infty} \frac{1}{2R} \sum_{k \in \mathbb{Z}, |X_k| \leq R} \varphi(X_{k+1} - X_k).$$

It is also possible to check that the examples of Sect. 1.3.2 are particular cases of this general setting, but we will not proceed in this direction. In the above case, one easily computes the average energy per particle using (H3.2) and obtain

$$\lim_{N \to +\infty} \frac{1}{2N+1} E_N = \int_{\mathbb{R}} V(|h|) \, dl(h). \tag{1.27}$$

To close this section, and with a viewpoint that is different from the above modifications of the periodic setting, we briefly mention the random setting. Among the configurations considered so far, at least the periodic ones can be interpreted as particular cases of the random setting, as we will see later in this chapter. The random setting is however more general, in a somewhat different way.

Still using the simplified setting of the present section, we consider an infinite set of particles, indexed by $k \in \mathbb{Z}$, with positions $X_k(\omega) = k + Y_k(\omega)$, where the random variables Y_k are independent and identically distributed. We also assume, in order to keep the ordering of the particles, that $|Y_k| < 1/2$ almost surely, for all k. Thus, the position $X_k(\omega)$ is slightly different from the reference position k, and random. Applying the strong law of large numbers, one easily proves that, in the limit $N \to +\infty$, the energy per particle converges to the expectation $\mathbb{E}(V(1 + Y_1 - Y_0))$. The quantitative modification is obviously substantial compared to the periodic case.

1.4 Periodicity with Defects

We have briefly studied in Sect. 1.3 some aspects of the computation of average energy for infinite systems of particles. Our aim was to explore cases beyond the periodic setting. We now return to the analytic framework of Sects. 1.1 and 1.2. We keep the structural assumption (1.7), but not the periodic assumption (1.8). Let us try and reproduce the variety of perturbations of the periodic system that we have introduced in Sect. 1.3.

1.4.1 Compactly Supported Perturbations

Consider again $\dfrac{1}{a}\left(\dfrac{x}{\varepsilon}\right)$, where a is a "perturbation" of a reference periodic function a_{per}. The notion will be made precise below. Our first example in Sect. 1.3 consisting in shifting one or finitely many particles within the periodic systems corresponds, in the present setting, to modifying a_{per} by adding a compactly supported function. This analogy seems only intuitive. It is actually deeper than that, and will be given a sound mathematical formalism below (Sect. 1.4.4). For now, we assume that $a = a_{per} + \widetilde{a}$ where $\widetilde{a}$ is a bounded, compactly supported function. We also assume that a is bounded away from 0. We have

$$\frac{1}{a_{\varepsilon}}(x) = \frac{1}{a_{per} + \widetilde{a}}\left(\frac{x}{\varepsilon}\right) \xrightarrow{\varepsilon \to 0} \left\langle \frac{1}{a_{per}} \right\rangle. \tag{1.28}$$

Indeed, $c = \dfrac{1}{a_{per} + \widetilde{a}} - \dfrac{1}{a_{per}} = -\dfrac{\widetilde{a}}{a_{per}(a_{per} + \widetilde{a})}$ is a bounded and compactly supported function. It satisfies $\displaystyle\int_{\mathbb{R}} \left| c\left(\frac{x}{\varepsilon}\right) \right| dx = \varepsilon \int_{\mathbb{R}} |c(x)|\, dx = O(\varepsilon)$, from which we infer that $c\left(\dfrac{x}{\varepsilon}\right)$ converges strongly, hence weakly, to 0 in $L^p(\mathcal{D})$ for any $1 \le p < +\infty$. In particular, the convergence rate $O(\varepsilon)$ of $\displaystyle\int_0^1 \frac{1}{a_{per}}\left(\frac{x}{\varepsilon}\right) dx - \left\langle \dfrac{1}{a_{per}} \right\rangle$ to 0 is left unchanged under the addition of $\widetilde{a}$ to a_{per}. Arguing similarly, we have $\left\langle \dfrac{1}{a_{per} + \widetilde{a}} \right\rangle = \left\langle \dfrac{1}{a_{per}} \right\rangle$, where the average of the nonperiodic function $\dfrac{1}{a_{per} + \widetilde{a}}$ is defined by

$$\left\langle \frac{1}{a_{per} + \widetilde{a}} \right\rangle = \lim_{R \to +\infty} \frac{1}{2R} \int_{-R}^{R} \frac{1}{a_{per}(x) + \widetilde{a}(x)}\, dx. \tag{1.29}$$

We also note that, for any function $\varphi \in L^1(\mathcal{D})$, we have $\int_{\mathbb{R}} c\left(\dfrac{x}{\varepsilon}\right) \varphi(x)\, dx =$
$\varepsilon \int_{\mathbb{R}} c(x)\varphi(\varepsilon x)\, dx = O\left(\displaystyle\int_{|x| \le O(\varepsilon)} |\varphi(x)|\, dx\right) = o(1)$. Thus, the sequence $c\left(\dfrac{\cdot}{\varepsilon}\right)$
converges to 0 in $L^\infty(\mathcal{D})$-weak-$\star$. This proves the expected convergence (1.28).

1.4.2   Perturbation in L^p

We have just built the analogue of the perturbation of a finite number of positions
of particles. We can now build slightly more general perturbations that vanish at
infinity, in a sense we are going to make precise. We choose to add to a_{per} a function
$\widetilde{a} \in L^p(\mathbb{R})$, where, for instance, $p = 2$. Somewhat formally, we can say that such
a function vanishes at infinity. In order to make this formal statement rigorous, it
is sufficient to assume that $\widetilde{a}$ is regular. For instance, if $\widetilde{a}$ is uniformly Lipschitz
continuous (in a more elaborate version, $\widetilde{a}$ will be of Hölder or Sobolev regularity),
an assumption that we are going to use in the forthcoming chapters, we do have
that $\widetilde{a}$ vanishes at infinity. We also assume that $\widetilde{a}$ is such that $a = a_{per} + \widetilde{a}$ is
bounded from above, and bounded away from 0. The latter assumption implies that
$\widetilde{a}$ is not too negative, compared to a_{per}.

Let us consider again $c = \dfrac{1}{a_{per} + \widetilde{a}} - \dfrac{1}{a_{per}} = -\dfrac{\widetilde{a}}{a_{per}(a_{per} + \widetilde{a})}$. We know that
$c \in L^2(\mathbb{R})$ (and actually that $c \in L^\infty(\mathbb{R})$ too). Writing

$$\int_{\mathbb{R}} \left| c\left(\frac{x}{\varepsilon}\right) \right|^2 dx = \varepsilon \int_{\mathbb{R}} |c(x)|^2\, dx = O(\varepsilon),$$

we see that the sequence $c\left(\dfrac{x}{\varepsilon}\right)$ converges strongly to 0 in $L^2(\mathcal{D})$ (actually, in
$L^2(\mathbb{R})$), with the rate $O(\sqrt{\varepsilon})$. Thus, it converges weakly in $L^2(\mathcal{D})$. We also have
the same rate for this weak convergence. Applying the Cauchy-Schwarz inequality,
we indeed find that, for any $\varphi \in L^2(\mathcal{D})$,

$$\left| \int_{\mathcal{D}} c\left(\frac{x}{\varepsilon}\right) \varphi(x)\, dx \right| \le \left(\int_{\mathbb{R}} \left| c\left(\frac{x}{\varepsilon}\right) \right|^2 \right)^{1/2} \left(\int_{\mathcal{D}} |\varphi|^2 \right)^{1/2}$$

$$= \sqrt{\varepsilon} \left(\int_{\mathbb{R}} |c|^2 \right)^{1/2} \left(\int_{\mathcal{D}} |\varphi|^2 \right)^{1/2}$$

$$= O(\sqrt{\varepsilon}). \tag{1.30}$$

In particular, this argument shows, taking $\varphi \equiv 1$, that, as above, $\left\langle \dfrac{1}{a_{per} + \widetilde{a}} \right\rangle =$ $\left\langle \dfrac{1}{a_{per}} \right\rangle$. This time, however, the rate of convergence is only $O(\sqrt{\varepsilon})$, whereas it is $O(\varepsilon)$ in the periodic case. We also note that, since $c\left(\dfrac{x}{\varepsilon}\right)$ is uniformly bounded, we still have convergence in $L^\infty(\mathcal{D})$ weak-$\star$.

Of course, a simple adaptation of the above argument allows to prove a similar result when $\widetilde{a} \in L^p(\mathbb{R}) \cap L^\infty(\mathbb{R})$, for $1 \le p < +\infty$ and $p \neq 2$. The main point is, in (1.30), to use the Hölder inequality instead of the Cauchy-Schwarz inequality. The proof is likewise easily adapted to higher dimensions $d \ge 2$. We leave it to the reader to check that for any function $\widetilde{a} \in L^p(\mathbb{R}^d) \cap L^\infty(\mathbb{R}^d)$, the average obtained does not depend on $\widetilde{a}$ (it is equal to that of the periodic case), and that the convergence rate is $O\left(\varepsilon^{\min\left(1, \frac{d}{p}\right)}\right)$.

So far we have gone, step by step, from a_{per} to $a_{per} + \widetilde{a}$, with first $\widetilde{a}$ bounded with compact support, next $\widetilde{a}$ bounded and in L^p, with $p < +\infty$, possibly arbitrarily large. Doing so, we can observe that some properties of the periodic case are preserved, while some other properties are lost. As the vanishing rate of the defect at infinity decreases (which corresponds to taking larger and larger integrability exponents p, and/or decrease the dimension d, as may be seen by computing the integral $\displaystyle\int |f(r)|^p \, r^{d-1} \, dr$ in spherical coordinates), the convergence rate deteriorates. This is to be expected since, as the defect spreads out, it progressively becomes a substantial modification of the periodic structure of the original coefficient, even at infinity. In the next chapters, we will find similar phenomena with qualitatively equivalent properties.

1.4.3 An Example of Non-local Defects

The previous section deals with defects that are localized around the origin. It is possible to modify this assumption in the following way: instead of vanishing at infinity, the defect may become "rare" as we go away from the origin. To be more specific, think of a coefficient reading as $a = a_{per} + \widetilde{a}$, where the perturbation $\widetilde{a}$ takes the following form:

$$\widetilde{a}(x) = \sum_{k \in \mathbb{Z}} a_0 \left(x - \mathrm{sgn}(k) 2^{|k|}\right), \quad a_0 \text{ has compact support.} \tag{1.31}$$

In this setting, as far as one goes at infinity, some defects (that is, non-zero contributions of $\widetilde{a}$) are present. We have already mentioned such a setting (see p. 11).

Here again, we assume that a is bounded from above and bounded away from 0. We apply an argument similar to that of the previous section, setting

$$c = \frac{1}{a_{per} + \widetilde{a}} - \frac{1}{a_{per}} = -\frac{\widetilde{a}}{a_{per}(a_{per} + \widetilde{a})}.$$

This function is bounded, but does not belong to $L^2(\mathbb{R})$, contrary to Sect. 1.4.2. We are interested in the behaviour of $c\left(\frac{x}{\varepsilon}\right)$ in the domain $]0, 1[$, in the limit $\varepsilon \to 0$. In order to study it, we write:

$$\int_0^1 \left| c\left(\frac{x}{\varepsilon}\right) \right| dx \leq C \int_0^1 \left| \widetilde{a}\left(\frac{x}{\varepsilon}\right) \right| dx = C\varepsilon \int_0^{1/\varepsilon} |\widetilde{a}(y)| \, dy,$$

where the constant C depends on the lower bounds a and a_{per} only. Using (1.31), we infer

$$\int_0^1 \left| c\left(\frac{x}{\varepsilon}\right) \right| dx \leq C\varepsilon \sum_{k \in \mathbb{Z}} \int_0^{1/\varepsilon} \left| a_0\left(y - \text{sgn}(k)2^{|k|}\right) \right| dy.$$

In the above sum, only a finite number of terms are different from zero. Indeed, denoting by R_0 the size of the support of a_0, the corresponding integral is zero as far as $2^{|k|} > \frac{1}{\varepsilon} + R_0$, that is,

$$|k| > \frac{\ln\left(\frac{1}{\varepsilon} + R_0\right)}{\ln(2)}.$$

Hence,

$$\int_0^1 \left| c\left(\frac{x}{\varepsilon}\right) \right| dx \leq C\varepsilon \sum_{|k| \leq C_0 |\ln(\varepsilon)|} \int_{\mathbb{R}} |a_0(y)| dy \leq CC_0\varepsilon |\ln(\varepsilon)| \|a_0\|_{L^1(\mathbb{R})},$$

where the constant C_0 depends only on R_0. Thus, the rate of convergence is (at least) $\varepsilon |\ln \varepsilon|$.

We could produce more and more examples, in the spirit of those of Sect. 1.3. For now, we prefer returning to the atomistic system of this Sect. 1.3. We want to formalize its analogy with homogenization theory more rigorously.

1.4.4 Formalizing the Link with Systems of Particles

We start by considering again the periodic system of Sect. 1.3, that is, the system of periodic positions $X_k = k$ for $k \in \mathbb{Z}$. Let φ be a smooth compactly-supported function, and define the series of translates $\sum_{k \in \mathbb{Z}} \varphi(. - X_k)$. This series is well defined (only a finite number of terms in the sum contribute), and is periodic.

We consider the vector space spanned by such series $\sum_{k \in \mathbb{Z}} \varphi(. - X_k)$ as φ varies, and define its closure for the uniform L^2 norm, that is,

$$\|f\|_{L^2_{\text{unif}}(\mathbb{R})}^2 = \sup_{x \in \mathbb{R}} \int_{|y-x| \le 1} |f(y)|^2 \, dy. \tag{1.32}$$

Doing so, we obtain the space

$$L^2_{per}(\mathbb{R}) = \left\{ f \in L^2_{\text{loc}}(\mathbb{R}), \quad f \text{ periodic of period } 1 \right\}. \tag{1.33}$$

This procedure connects an infinite set of points, $X_k, k \in \mathbb{Z}$, and a "usual" function space, $L^2_{per}(\mathbb{R})$. Of course, we restrict ourselves to dimension one for simplicity of exposition, but all this is valid in higher dimensions.

This relation between the set of points $\{X_k\}_{k \in \mathbb{Z}}$ (with, here, $X_k = k$) and the functions of the form

$$\sum_{k \in \mathbb{Z}} \varphi(. - X_k),$$

which are actually periodic, allows to make a link between the computation of the average energy of the periodic system $X_k = k$, $k \in \mathbb{Z}$, and the computation of the average of a periodic function a_{per}, that is, the weak limit of the rescaled function $a_{per} \left(\frac{\cdot}{\varepsilon} \right)$. Indeed, if for instance φ has compact support, then

$$\frac{1}{2R} \int_{[-R,R]} \sum_{k \in \mathbb{Z}} \varphi(x - X_k) \, dx$$

behaves like $\frac{1}{2R} \# \{ k \in \mathbb{Z}; \ X_k \in [-R, R] \}$ (multiplied by $\int \varphi$), that is, like an average energy. The reader has probably noticed that we have implicitly used the fact that a periodic system does satisfy assumption (1.25)!

In this textbook, we want to go beyond the mere existence of an average. We have a specific aim related to homogenization theory of Eq. (1). As we have seen from the crude reasoning we made starting from (1.1), we should expect to manipulate quotients of functions, or inverses of functions, . . . Thus, it is necessary (actually it is sufficient) to know how to deal with *products of functions*. In other words, the notion of *"vector space spanned by..."* is not the right notion to consider. We need to consider the notion of *algebra*. Let us return to the preceding construction (which, actually, is too specific, hence misleading), and consider the *algebra* generated by functions of the form

$$\sum_{k \in \mathbb{Z}} \varphi(. - X_k)$$

as φ varies, being smooth and compactly supported. This algebra is clearly that of smooth periodic functions (of period 1), and we will denote it by $C_{per}^{\infty}(\mathbb{R})$. Next, we can close this algebra for the uniform norm (1.32) and recover $L_{per}^2(\mathbb{R})$ defined by (1.33). In the latter stage of this process, the notion of algebra has been lost. However, we can, in some sense, recover it "collectively" in the following way: we point out, with obvious notation, that the product of two functions in $L_{per}^2(\mathbb{R})$ is in $L_{per}^1(\mathbb{R})$, or that the product of two *bounded* functions of $L_{per}^2(\mathbb{R})$ is in $L_{per}^2(\mathbb{R})$. Thus, the operation "taking the algebra + closing" gives a family of functional spaces that is stable by product. This procedure will prove to be the right one from a functional analysis viewpoint: as we are going to see now, it can be generalized to nonperiodic geometries.

We consider now the first type of perturbation we have studied in Sect. 1.3, defined by (1.19). In this system, only the particle at X_0 is not in place. We have moved it from $X_0 = 0$ to $X_0 = a \neq 0$ but, as we have pointed out in Sect. 1.3, we could as well have bluntly removed it from the system, thus defining

$$X_k = \begin{cases} k & \text{for } k < 0, \\ k+1 & \text{for } k \geq 0, \end{cases} \tag{1.34}$$

(the energy difference (1.21) would then have been $\Delta E_N = V(2) - 2V(1)$). Let us do this for the sake of variety! As an exercise, the reader can study the case (1.19) and check that this changes nothing to our argument. If we follow the procedure we have introduced above for the periodic system, we now consider series of translated functions $\sum_{k \neq 0 \in \mathbb{Z}} \varphi(. - k)$ where φ is smooth and has compact support. First, we need to characterize the algebra spanned by such series. Then, we will close it with respect to the L_{unif}^q norm, say, with $q = 2$. Intuitively, what we expect to find is clear. Indeed, we write $\sum_{k \neq 0 \in \mathbb{Z}} \varphi(. - k) = \sum_{k \in \mathbb{Z}} \varphi(. - k) - \varphi$. The reasoning of the previous paragraph suggests that the first term leads to $L_{per}^2(\mathbb{R})$, while the second, that is, the closure for the $L_{\text{unif}}^2(\mathbb{R})$ norm of smooth, compactly supported functions, is known to be the space

$$L_0^2(\mathbb{R}) = \left\{ f \in L_{\text{loc}}^2(\mathbb{R}), \quad \lim_{|x| \to +\infty} \int_{|y-x| \leq 1} |f(y)|^2 \, dy = 0 \right\} \tag{1.35}$$

of functions which vanish at infinity "with respect to the local L^2 norm". The functional space associated to our atomistic construction (1.19) thus reads as $L^2_{per}(\mathbb{R}) + L^2_0(\mathbb{R})$. The detailed argument we present below will prove this claim.

Only with this simple argument, we however already understand that, upon considering a function of the form $a_{per} + \tilde{a}$ where $\tilde{a} \in L^2(\mathbb{R})$ (a function that will become, in the forthcoming chapters, a coefficient of Eq. (1)), we have "almost" treated the analogue, in terms of functional space, of the system of particles (1.34). Indeed, the space $L^2(\mathbb{R})$ is a natural subspace of $L^2_0(\mathbb{R})$ (but a *strict* subspace: for a function f in $L^2(\mathbb{R})$, the series $\sum f_k$ defined by $f_k = \left(\int_{[k,k+1]} |f(x)|^2 \, dx \right)^{1/2}$, $k \in \mathbb{Z}$, belongs to ℓ^2 whereas, when $f \in L^2_0(\mathbb{R})$, we only have that the sequence $(f_k)_{k \in \mathbb{Z}}$ tends to zero as $|k| \to +\infty$).

We also intuitively understand the following analogy: saying that the average energy of the system of particles (1.34), or (1.19), is equal to the average energy of the periodic system, is similar to saying that $\left\langle \dfrac{1}{a_{per} + \tilde{a}} \right\rangle = \left\langle \dfrac{1}{a_{per}} \right\rangle$ and that this average is the weak limit of the rescaled functions $\dfrac{1}{a_{per} + \tilde{a}} \left(\dfrac{\cdot}{\varepsilon} \right)$. We have thus established the link between the preceding section and the present one. We have an equivalence between the notions of the preceding section and those developed here. This could be adapted to any type of defect.

For the sake of completeness, we now prove that the space obtained from the system of particles (1.34) (or equivalently (1.19)) is indeed, as announced, equal to $L^2_{per}(\mathbb{R}) + L^2_0(\mathbb{R})$.

We fix the set $\{X_k\}_{k \in \mathbb{Z}}$ to be (1.34). The following proof may be easily adapted to the case (1.19). Let $\mathcal{A}$ be the algebra generated by functions of the form $\sum_{k \neq 0 \in \mathbb{Z}} \varphi(. - k)$, for $\varphi \in C_c^\infty(\mathbb{R})$, where $C_c^\infty(\mathbb{R})$ denotes the set of smooth functions having compact support. We denote by $\mathcal{A}^{(2)}$ the closure of $\mathcal{A}$ for the $L^2_{\text{unif}}(\mathbb{R})$ norm, and we aim at proving that

$$\mathcal{A}^{(2)} = L^2_{per}(\mathbb{R}) + L^2_0(\mathbb{R}), \tag{1.36}$$

where L^2_{per} is defined by (1.33) and L^2_0 by (1.35). First, let us prove that $C_c^\infty(\mathbb{R}) \subset \mathcal{A}$. In order to do so, we fix $\varphi \in C_c^\infty(\mathbb{R})$, and define

$$\varphi_1(x) = \varphi(x) - \varphi(x+1), \quad f(x) = \sum_{k \neq 0} \varphi_1(x-k) = \varphi(x+1) - \varphi(x).$$

By definition, $f \in \mathcal{A}$, hence $\varphi_1 \in \mathcal{A}$. By iterating this argument, we infer

$$\forall j \in \mathbb{Z}, \quad \varphi(\cdot + j) - \varphi \in \mathcal{A}.$$

Next, we choose $\psi \in C_c^\infty(\mathbb{R})$ such that $\psi = 1$ in the support of φ. Applying the above argument to ψ, we deduce that $\psi - \psi(\cdot + k) \in \mathcal{A}$, for all $k \in \mathbb{Z}$. Since $\mathcal{A}$ is an algebra, we have

$$\forall j, k \in \mathbb{Z}, \quad (\varphi - \varphi(\cdot + j))(\psi - \psi(\cdot + k)) \in \mathcal{A}.$$

If k, j and $k - j$ are all sufficiently large, we have

$$\forall x \in \mathbb{R}, \quad (\varphi(x) - \varphi(x + j))(\psi(x) - \psi(x + k))$$
$$= \varphi(x)\psi(x) - \underbrace{\varphi(x + j)\psi(x)}_{=0} - \underbrace{\varphi(x)\psi(x + k)}_{=0} + \underbrace{\varphi(x + j)\psi(x + k)}_{=0} = \varphi(x)\psi(x),$$

where we have used that, in all products except $\varphi\psi$, the factors have disjoint supports. Since $\psi = 1$ in the support of φ, we have $\varphi\psi = \varphi$. We have thus proved that $\varphi \in \mathcal{A}$.

We next prove that $C_{per}^\infty \subset \mathcal{A}$, where C_{per}^∞ is the set of smooth $1-$periodic functions defined on $\mathbb{R}$. Let $f \in C_{per}^\infty$. There exists $\varphi \in C_c^\infty(\mathbb{R})$ such that

$$\forall x \in \mathbb{R}, \quad f(x) = \sum_{j \in \mathbb{Z}} \varphi(x - j). \tag{1.37}$$

To see this, it is sufficient to use a partition of unity, that is, a function $\chi \in C_c^\infty(\mathbb{R})$ such that

$$\sum_{k \in \mathbb{Z}} \chi(\cdot - k) = 1,$$

and to define $\varphi = \chi f$. Equality (1.37) also reads as $f(x) = \varphi(x) + \sum_{j \neq 0} \varphi(x - j)$.

We have already proved the first term is in $\mathcal{A}$. So is, by definition, the second term. Thus, $f \in \mathcal{A}$.

We now prove that $\mathcal{A} = C_{per}^\infty + C_c^\infty$. We have just shown that $C_{per}^\infty + C_c^\infty \subset \mathcal{A}$. Moreover, $C_{per}^\infty + C_c^\infty$ is an algebra, and contains all functions of the form $\sum_{j \neq 0} \varphi(\cdot - j)$. Hence, it contains $\mathcal{A}$, by definition of $\mathcal{A}$. We thus have $\mathcal{A} = C_{per}^\infty + C_c^\infty$. Finally, we note that the closure of $C_{per}^\infty + C_c^\infty$ for the L_{unif}^2 norm is $L_{per}^2 + L_0^2$, which eventually implies $\mathcal{A}^{(2)} = L_{per}^2 + L_0^2$.

As we have already said, it is possible to consider other functional settings to build algebras from infinite sets of points $\{X_k\}$, in the same spirit as above for $L_{per}^2(\mathbb{R}) + L_0^2(\mathbb{R})$. We are going to define a general framework, using systems $\{X_k\}$ that are much more "chaotic" than (1.19). To this end, we will need conditions going beyond the simple assumption (H3) (i.e. (1.25)–(1.26)) used so far. This additional condition stems from the need to consider, in the homogenization theory of Eq. (1), products of an arbitrarily large number of factors. If, instead of considering only pair-potential interactions (and *a fortiori* nearest-neighbour interactions), we had used N-body interactions that "mix" N-tuples $V_j(X_{k_1}, \ldots, X_{k_j})$, $j \leq N$, then we would have come across such additional conditions. We now give a brief account of such a construction.

1.4.5   A General Deterministic Framework

The settings we have presented in Sect. 1.3 and the preceding subsections are simplified settings. We are now going to consider an infinite set of points $\{X_k\}_{k \in \mathbb{Z}}$ with stronger assumptions than those we have had so far. More precisely, we again impose (H1)–(H2), but, instead of (H3.1)–(H3.2) (i.e (1.25)–(1.26)), we assume the stronger condition, "at any order":

(H3) for all $n \in \mathbb{N}$, for all $(\delta_0, \delta_1, \ldots \delta_n) \in (\mathbb{R}^{+*})^{n+1}$, the following limit exists

$$
f_n(\delta_0, h_1, \delta_1, h_2, \delta_2, \ldots, h_n, \delta_n) = \lim_{R \to \infty} \frac{1}{2R} \# \Big\{ (k_0, k_1, \ldots, k_n) \in \mathbb{Z}^{n+1},
$$

$$
|X_{k_0}| \leq \delta_0 R, \quad |X_{k_0} - X_{k_1} - h_1| \leq \delta_1, \ldots, |X_{k_0} - X_{k_n} - h_n| \leq \delta_n \Big\},
$$

$$
\tag{1.38}
$$

where the convergence takes place in $L^\infty(\mathbb{R}^n)$.

In the particular cases $n = 0$ and $n = 1$, we recover conditions (H3.1) and (H3.2) of Sect. 1.3. Indeed, taking first $n = 0$ and $\delta_0 = 1$, then (by linearity) any $\delta_0 > 0$, we find (H3.1). Similarly, taking $n = 1$, $\delta_0 = 1$, $h_1 = h$, $\delta_1 = \delta$, $f_1(\delta_0, h_1, \delta_1) = l(h, \delta)$, we find (H3.2). As announced at the end of Sect. 1.4.4, condition (H3) for $n \geq 3$ gives a control on the relative positions of n-tuples of points.

It is actually useful, in view of using it in the context of homogenization, to reformulate condition (H3) above in the following, more *analytical* terms:

(H3) for all $n \in \mathbb{N}$, the following limit exists

$$
\lim_{R \to \infty} \frac{1}{2R} \sum_{X_{k_0} \in [-R, R]} \cdots \sum_{X_{k_n} \in [-R, R]} \delta_{(X_{k_0} - X_{k_1}, \ldots X_{k_0} - X_{k_n})}(h_1, \ldots, h_n)
$$

$$
= l^n(h_1, \ldots, h_n), \tag{1.39}
$$

and is a non-negative, uniformly locally bounded measure.

The equality

$$f_n(\delta_0, h_1, \delta_1, h_2, \delta_2, \ldots, h_n, \delta_n) = |B_{\delta_0}| \, l^n \left[(h_1 + B_{\delta_1}) \times \cdots \times (h_n + B_{\delta_n}) \right],$$

where $B_\delta = [-\delta, \delta]$, allows to connect (1.38) and (1.39).

Following what we have achieved in Sect. 1.4.4, we now build a functional setting that will be useful in our theory of homogenization. Assuming that the set $\{X_k\}_{k \in \mathbb{Z}}$ satisfies conditions (H1)–(H2)–(H3), we define the vector space $\mathcal{A}(\{X_k\})$ spanned by the functions of the form

$$f(x) = \sum_{k_1 \in \mathbb{Z}} \sum_{k_2 \in \mathbb{Z}} \cdots \sum_{k_n \in \mathbb{Z}} \varphi(x - X_{k_1}, x - X_{k_2}, \ldots, x - X_{k_n}), \qquad (1.40)$$

where $\varphi \in \mathcal{D}(\mathbb{R}^n)$. For $s \in \mathbb{N}$ and $p \in [1, +\infty[$, we denote by $\mathcal{A}^{(s,p)}(\{X_k\})$, or simply $\mathcal{A}^{(s,p)}$, the closure of $\mathcal{A}(\{X_k\})$ for the $W_{\text{unif}}^{s,p}$ Sobolev norm. This norm is defined similarly to the uniform L^2 norm in (1.32). In the case $s = 0$, we simply denote $\mathcal{A}^{(0,p)} = \mathcal{A}^{(p)}$. One easily understands that functions of the form (1.40) are prototypes of products of n factors of sums of translated functions $\psi(x - X_k)$. Thus, $\mathcal{A}^{(s,p)}$ is also the closure, for the $W_{\text{unif}}^{s,p}$ norm, of the algebra generated by the functions of the form

$$f(x) = \sum_{k \in \mathbb{Z}} \varphi(x - X_k), \qquad \varphi \in \mathcal{D}(\mathbb{R}).$$

This is consistent with the definition of $\mathcal{A}^{(2)}$ that we have used in Sect. 1.4.4, and thus with formula (1.36). In fact, a proof similar to the one we present on p. 20 and following, allows to generalize (1.36) in $\mathcal{A}^{(s,p)} = W_{per}^{s,p} + W_0^{s,p}$, in the case (1.34), for any integer s and any $p \in [1, +\infty[$. Here, the spaces $W_{per}^{s,p}$ and $W_0^{s,p}$ are equivalent to L_{per}^2 and L_0^2, defined by (1.33) and (1.35), respectively, where the L_{unif}^2 norms are replaced by $W_{unif}^{s,p}$ norms.

Let us point out that, as we have mentioned above, our aim is to build *algebras* of functions, hence the notation $\mathcal{A}$. This may however be misleading. Indeed, except for the case $p = \infty$, which is not explicitly dealt with so far, and requires some adaptation, the spaces $\mathcal{A}^{(s,p)}$ *are not* algebras. They are only *closures* of an algebra. However, as briefly mentioned on p. 19, they do have some nice properties that will be sufficient in the sequel. As an example, these spaces are stable when multiplied by bounded functions, that is, if $(f, g) \in \mathcal{A}^{(\infty)} \times \mathcal{A}^{(p)}$, then $fg \in \mathcal{A}^{(p)}$. Similarly, if $(f, g) \in \mathcal{A}^{(1,\infty)} \times \mathcal{A}^{(1,p)}$, then $fg \in \mathcal{A}^{(1,p)}$, etc. Moreover, the *family* of spaces $\mathcal{A}^{(k,p)}$ is stable by product: if $(f, g) \in \mathcal{A}^{(p)} \times \mathcal{A}^{(p)}$, then $fg \in \mathcal{A}^{(p/2)}$, etc... All these properties are simple consequences of the Hölder inequality.

The aim of the above construction is to obtain functions that have an average.

Let $\{X_k\}_{k \in \mathbb{Z}}$ be a set of points satisfying (H1)-(H2)-(H3). Then any function $f \in \mathcal{A}^{(s,p)}$ indeed has an average in the following sense:

$$\langle f \rangle = \lim_{R \to \infty} \frac{1}{2R} \int_{[-R,R]} f. \qquad (1.41)$$

If f has the particular form (1.40), this average reads as

$$\langle f \rangle = \int_{\mathbb{R}} \int_{\mathbb{R}^{n-1}} \varphi(x, x - h_1, \ldots, x - h_{n-1}) \, dl^{n-1}(h_1, \ldots, h_{n-1}) \, dx. \qquad (1.42)$$

We thus understand the interest of the "shifts" h between particles in Assumption (H3). They may be interpreted as pseudo-periods, that are used in the *explicit* computation of averages of functions.

In order to make a link with homogenization theory, we now insert in (1.1) a coefficient a_ε of the form $a_\varepsilon(x) = a\left(\frac{x}{\varepsilon}\right)$, where $a \in \mathcal{A}^{(p)}$. In addition, we assume that a_ε is bounded from above and away from zero. With our assumptions (H1)-(H2)-(H3), the issue we have in order to make the connection with homogenization theory is the following. We would like a setting where, as in the periodic case that it generalizes, the weak limit of functions coincides with their large volume average. Now, (1.41) defines a scalar number while the weak limit of a rescaled function $f(x/\varepsilon)$ in the algebra needs not be constant. So we modify our Assumption (H3), in which the origin plays too specific a role, as follows:

(H3') for all $n \in \mathbb{N}$, the following limit exists

$$\lim_{\varepsilon \to 0} \mu^n\left(\frac{x}{\varepsilon}, h_1, \ldots, h_n\right) = v^n(h_1, \ldots h_n),$$

where

$$\mu^n(y, h_1, \ldots, h_n)$$

$$= \sum_{i_0 \in \mathbb{Z}} \sum_{i_1 \in \mathbb{Z}} \cdots \sum_{i_n \in \mathbb{Z}} \delta_{(X_{i_0}, X_{i_0} - X_{i_1}, \ldots X_{i_0} - X_{i_n})}(y, h_1, h_2, \ldots h_n).$$

Using the more stringent assumption (H3'), we now have

$$\langle f \rangle = \lim_{\varepsilon \to 0} f\left(\frac{\cdot}{\varepsilon}\right), \qquad (1.43)$$

where *both sides* are constant. This generalizes the periodic case. Here again, if f has the particular form (1.40), this average writes

$$\langle f \rangle = \int_{\mathbb{R}} \int_{\mathbb{R}^{n-1}} \varphi(x, x - h_1, \ldots, x - h_{n-1}) \, dv^{n-1}(h_1, \ldots, h_{n-1}) \, dx. \qquad (1.44)$$

With the assumptions (H1)-(H2)-(H3'), we can now prove that, if $a_\varepsilon(x) = a\left(\dfrac{x}{\varepsilon}\right)$, where $a \in \mathcal{A}^{(p)}$, then the solution u^ε weakly converges (in $L^2(\mathcal{D})$ for instance) to u^*, which is solution to the limit Eq. (1.6), with

$$\bar{a} = \frac{1}{\left\langle \frac{1}{a} \right\rangle}. \qquad (1.45)$$

The average is understood in the sense of (1.43). This is a generalization of the result (1.15). On the other hand, the question of convergence is still open, except in some very special cases. To date, we are still unable to derive a complete homogenization theory in this setting. We need to limit our ambition. Nevertheless, this theoretical apparatus is useful as a test-bed to generate new ideas of settings. In particular, it provides guidelines for certain specific, both relevant and interesting, cases. We will briefly return to these questions in Sect. 4.2.3.

We conclude our brief account on this general setting with some remarks.

First, the above construction, together with the assumptions we have derived, may be adapted to higher dimensions. It is a simple exercise to generalize the assumptions and the formulas we have obtained.

Second, it is clear that the periodic setting is a particular case of the above general framework. Indeed, if we use the set of points $\{X_k\}_{k \in \mathbb{Z}} = T\mathbb{Z}$, for some fixed period $T > 0$, the assumptions (H1)-(H2)-(H3') are satisfied. Moreover, the spaces $\mathcal{A}^{(s,p)}$ consist of periodic functions, and the measures l^n in (1.39), which allow for the computation of averages (1.42), may easily be computed: for instance, $l^1(h)$ is the Poisson measure

$$l^1(h) = \sum_{k \in \mathbb{Z}} \delta(h - kT).$$

Couples of points in $\{X_k\}_{k \in \mathbb{Z}}$ are distant from a multiple of T. Of course, this measure l^1 is non-negative and uniformly locally bounded.

Other examples may be identified as particular instances of the present general setting. This is the case for instance of a locally modified periodic lattice.

Third, a careful examination of the above arguments shows that we have not used condition (H2). The motivation for introducing this assumption was originally the study of atomistic models mentioned in Sect. 1.3. In order to compute average energy of such systems, (H2) was quite natural. Later, we have adapted this setting to homogenization theory. For the sake of consistency, we have kept this condition, thereby giving a common mathematical background to two different (but heuristically similar) mathematical enterprises. Moreover, we think that this condition is natural, if not necessary, in the context of homogenization of elliptic equations. Indeed, even if it is possible to define the functional spaces $\mathcal{A}^{(s,p)}$ without Assumption (H2), the functions in such a space vanish at places where particles are "rare" in the original geometry. Think for instance about $\{X_k\} = \mathbb{N}$ (or any intersection of a periodic lattice with a half-space in higher dimensions). It is clear that, for any $a \in \mathcal{A}^{(2)}$,

$$\|a\|_{L^2([x-1,x+1])} \xrightarrow[x \to -\infty]{} 0.$$

As a consequence, a function of the form (3) (with $a \in \mathcal{A}^{(2)}$) cannot define a uniformly elliptic operator in (1). It is therefore not clear, in such a case, whether (1) is well-posed. Thus, developing homogenization theory for this problem is not immediate, at least with relatively simple analytical tools.

Finally, we emphasize that (H3') implies not only that the rescaled functions $f\left(\frac{\cdot}{\varepsilon}\right)$ converge, but also that their limit is *a constant*. One could imagine a similar setting in which the limit is a measure. In such a case, ν^n would depend on the variable x. Formula (1.44) would still hold, with the measure $d\nu^{n-1}(h_1, \ldots, h_{n-1})dx$ replaced by $d\nu^{n-1}(x, h_1, \ldots, h_{n-1})$. Such a framework goes beyond the scope of the present textbook, and we will not develop it further.

1.5   The Quasi- and Almost Periodic Settings

In line with the preceding section about atomistic systems, we now consider two seemingly harmless modifications of the periodic case: the *quasi-periodic* case on the one hand, and the *almost periodic* case on the other hand. Loosely speaking, the first setting is defined as finite sums of periodic functions with different periods, while the second one is defined as infinite such sums. We will see below the rigorous definitions. Despite the large number of mathematical studies devoted to these settings, it is not clear whether they have any *practical* interest. They are nevertheless particularly useful, in homogenization theory, as examples of perturbations of the periodic case. Studying them shows which properties of the periodic case are preserved, and which other properties are lost in the perturbation process. This in turn allows to understand how exceptional, or not, the periodic setting is. This gives a precious insight on what can be expected in more general (and more demanding) situations.

Throughout the present Sect. 1.5, we consider *complex*-valued functions, in order to define the suitable functional spaces. This is a more natural framework. It is always possible to restrict our considerations to real-valued functions, which gives the corresponding functional spaces of quasi-periodic or almost periodic real-valued functions, respectively.

1.5.1 The Quasi-Periodic Setting

A quasi-periodic function of one variable (we will see in Sect. 4.2.2.1 below the generalization to several variables) is defined as the trace on the diagonal of a continuous periodic function in higher dimensions:

Definition 1.1 Let $f \in C^0(\mathbb{R})$. We say that f is *quasi-periodic* if there exists $m \in \mathbb{N}^*$ and $F : \mathbb{R}^m \to \mathbb{C}$ continuous such that

$$\forall x \in \mathbb{R}, \quad f(x) = F(x, \ldots, x), \tag{1.46}$$

where F is periodic with respect to each of its argument, that is, for all $1 \leq j \leq m$, there exists $T_j > 0$, such that $\forall x \in \mathbb{R}^m$, $F(x_1, \ldots, x_j + T_j, \ldots, x_m) = F(x_1, \ldots, x_m)$. We will denote by $QP(T)$ the set of quasi-periodic functions associated to $T = (T_1, \ldots, T_m)$. The real numbers $T_1, \ldots, T_m$ will be called the quasi-periods of the function f.

Note that in the case $m = 1$, we recover the classical periodic setting. If for instance the periods T_j are all integer multiples of a common period $T_0 > 0$, then f is periodic, of period $\max_j T_j$. On the contrary, if at least two of the periods of F are incommensurable (for instance, if $m = 2$, $T_1 = 1$ and $T_2 = \sqrt{2}$), then the function f is *not* periodic.

Let us mention a common ambiguity about the space of quasi-periodic functions. This space is only closed (for the uniform norm, but our point holds for other more general norms) if we fix *a priori* the quasi-periods $T = (T_1, \ldots, T_m)$. On the other hand, the union of $QP(T)$, for all possible values of T, is *not* closed. Its closure for the uniform norm is in fact the set of almost periodic functions, as we will see below.

The following Lemma allows to only consider the frequencies $\dfrac{1}{T_j}$ linearly independent over $\mathbb{Z}$:

Lemma 1.1 *In Definition 1.1, one can always assume that the frequencies* $\left(\dfrac{1}{T_j}\right)_{1 \leq j \leq m}$ *are linearly independent over $\mathbb{Z}$.*

The proof of this elementary result is left as an exercise to the reader. From now on, we will assume that the frequencies $\dfrac{1}{T_j}$ are always linearly independent over $\mathbb{Z}$.

Note that the set $QP(T)$ is an algebra, in the sense that it is stable by linear combination and products. It is also stable by inversion, in the sense that, if $f \in QP(T)$ satisfies $|f(x)| \geq \alpha > 0$ for all $x \in \mathbb{R}$, then $1/f \in QP(T)$. More generally, if ϕ is a continuous function on the range of $f \in QP(T)$, then $\phi(f) \in QP(T)$.

We also remark the following property, which will prove important in the sequel: the set $QP(T)$ always contains periodic functions. Indeed, for any m-tuple of integers $(k_1, \ldots, k_m)$, the plane wave

$$\mathbb{R} \longrightarrow \mathbb{C}$$

$$x \longmapsto \exp\left(2i\pi\left(\sum_{j=1}^{m} \frac{k_j}{T_j}\right)x\right) \tag{1.47}$$

is periodic of period $\overline{T} = \left|\sum_{j=1}^{m} \dfrac{k_j}{T_j}\right|^{-1}$. Since the frequencies $\left(\dfrac{1}{T_j}\right)_{1 \le j \le m}$ are linearly independent over $\mathbb{Z}$, this period $\overline{T}$ may be *arbitrarily large* (except in the case $m = 1$, which is actually the periodic case). To see this, let us consider the particular case $m = 2$ with $T_1 = 1$ and $T_2 = \sqrt{2}$. *Dirichlet's approximation Theorem* (see for instance [Nat96, Theorem 4.1, page 98]) allows to build a sequence of integers $p \in \mathbb{Z}$ and $q > 0$ that are arbitrarily large and such that $|p - q\sqrt{2}| < q^{-1}$. Thus, taking $k_1 = q$ and $k_2 = -p$, we have

$$\left(\overline{T}\right)^{-1} = \left|q - \frac{p}{\sqrt{2}}\right| = \frac{1}{\sqrt{2}}\left|p - q\sqrt{2}\right| \le \frac{1}{q\sqrt{2}} \xrightarrow[q \to +\infty]{} 0.$$

This implies that $\overline{T}$ can be as large as we want. In the sequel, we will make use of Dirichlet's approximation Theorem several times. In particular, it will be the key point in proving that some of the properties (P1) through (P5), which hold in the periodic case, do not hold any longer in the quasi-periodic or in the almost periodic case.

Before going further, we recall the definition of trigonometric polynomials.

Definition 1.2 We call *trigonometric polynomial* any finite linear combination of plane waves, that is, any function $F : \mathbb{R}^m \longrightarrow \mathbb{C}$ of the form

$$F(x) = \sum_{n=0}^{N} \alpha_n e^{i\lambda_n \cdot x},$$

where $\alpha_n \in \mathbb{C}$ and $\lambda_n \in \mathbb{R}^m$. In the particular case where F is assumed to be periodic of period $T = (T_1, \ldots, T_m)$, we say that F is a T-periodic trigonometric polynomial, and F is of the form

$$F(x) = \sum_{k \in \mathbb{Z}^m} \alpha_k \exp\left(2i\pi \sum_{j=1}^{m} \frac{k_j x_j}{T_j}\right), \tag{1.48}$$

where the sequence $(\alpha_k)_{k \in \mathbb{Z}^m}$ has finite support.

We are interested in the existence of an average for quasi-periodic functions. This is the point of the following result,

Proposition 1.2 *Let f be a quasi-periodic function. Then, it has an average. In addition, for any periodic function F such that f is the trace of F in the sense of Definition 1.1, we have*

$$\langle f \rangle = \frac{1}{|Q(T)|} \int_{Q(T)} F(x_1, \ldots, x_m) dx_1 \ldots dx_m, \qquad (1.49)$$

where $Q(T) =]0, T_1[\times]0, T_2[\times \cdots \times]0, T_m[$.

Proof We first prove the result if F is a trigonometric polynomial, that is, F reads as (1.48), where the sequence $(\alpha_k)_{k \in \mathbb{Z}^m}$ has a finite support:

$$f(x) = \sum_{k \in \mathbb{Z}^m} \alpha_k \exp\left(2i\pi x \sum_{j=1}^m \frac{k_j}{T_j} \right).$$

We define $\lambda_k = \sum_{j=1}^m \frac{k_j}{T_j}$. We know that $\lambda_k = 0 \iff k = 0$. Moreover, for any fixed $R > 0$, we have

$$\frac{1}{2R} \int_{-R}^R f(x) dx = \alpha_0 + \sum_{k \neq 0} \alpha_k \frac{1}{2R} \frac{\sin(2\pi R \lambda_k)}{\pi \lambda_k} \xrightarrow[R \to +\infty]{} \alpha_0 = \frac{1}{|Q(T)|} \int_{Q(T)} F.$$

This proves the result for trigonometric polynomials. Next, we are going to use the *Weierstrass trigonometric Theorem* (see for instance [Zyg02, Theorem 3.6]). This result states that the set of trigonometric polynomials is dense (for the uniform norm) in the set of continuous periodic functions. Assume that f is quasi-periodic. There exists a periodic function F satisfying (1.46), and for any $\delta > 0$, there exists a trigonometric polynomial F_δ such that

$$\forall (x_1, \ldots, x_m) \in \mathbb{R}^m, \quad |F_\delta(x_1, \ldots, x_m) - F(x_1, \ldots, x_m)| \le \delta. \qquad (1.50)$$

We set $f_\delta(x) = F_\delta(x, \ldots, x)$, so that we also have

$$\forall x \in \mathbb{R}, \quad |f_\delta(x) - f(x)| \le \delta. \qquad (1.51)$$

Next, we write:

$$\frac{1}{2R}\int_{-R}^{R}f(x)dx - \frac{1}{|Q(T)|}\int_{Q(T)}F = \frac{1}{2R}\int_{-R}^{R}f(x)dx - \frac{1}{2R}\int_{-R}^{R}f_\delta(x)dx$$

$$+\frac{1}{2R}\int_{-R}^{R}f_\delta(x)dx - \frac{1}{|Q(T)|}\int_{Q(T)}F_\delta$$

$$+\frac{1}{|Q(T)|}\int_{Q(T)}F_\delta - \frac{1}{|Q(T)|}\int_{Q(T)}F.$$

The modulus of the first term is bounded by δ by virtue of (1.51). The last term is dealt with similarly using (1.50). This yields

$$\left|\frac{1}{2R}\int_{-R}^{R}f(x)dx - \frac{1}{|Q(T)|}\int_{Q(T)}F\right|$$

$$\leq 2\delta + \left|\frac{1}{2R}\int_{-R}^{R}f_\delta(x)dx - \frac{1}{|Q(T)|}\int_{Q(T)}F_\delta(x)dx\right|.$$

Keeping $\delta > 0$ fixed, we let R tend to infinity. We thus have

$$\limsup_{R\to+\infty}\left|\frac{1}{2R}\int_{-R}^{R}f(x)dx - \frac{1}{|Q(T)|}\int_{Q(T)}F\right| \leq 2\delta.$$

Since this is valid for any $\delta > 0$, we have proved the desired result. $\square$

Note that we have proved in passing (see (1.50) and (1.51)) that the set of trigonometric polynomials of $QP(T)$ is dense in $QP(T)$ for the uniform norm. Otherwise stated, denoting by $\mathcal{T}$ the set of trigonometric polynomials, $\mathcal{T}\cap QP(T)$ is dense in $QP(T)$. This is a consequence of the Weierstrass trigonometric Theorem. We point out that the closure of the set $\mathcal{T}$ for the uniform norm is the space of almost periodic functions, as we will see below (Theorem 1.2).

We also have the following weak convergence result for rescaled quasi-periodic functions:

Proposition 1.3 *Assume that f is a quasi-periodic function. Then we have*

$$f\left(\frac{x}{\varepsilon}\right) \xrightarrow[\varepsilon\to 0]{\star} \langle f\rangle, \tag{1.52}$$

for the weak $L^\infty(\mathbb{R})$-$\star$ topology.

Proof As for the proof of Proposition 1.2, we use the fact that trigonometric polynomials are dense in the space of continuous periodic functions. We skip the approximation part within the proof (second part of the proof), which is the same as for Proposition 1.2, and we only prove the result when F is a trigonometric polynomial. In such a case, we have

$$F(x_1, \ldots, x_m) = \sum_{k \in \mathbb{Z}^m} \alpha_k \exp \left(2i\pi \sum_{j=1}^m \frac{k_j x_j}{T_j} \right),$$

where the sequence $(\alpha_k)_{k \in \mathbb{Z}^m}$ vanishes except for a finite number of values of k. Thus,

$$f \left(\frac{x}{\varepsilon} \right) = \sum_{k \in \mathbb{Z}^m} \alpha_k \exp \left(2i\pi \frac{x}{\varepsilon} \sum_{j=1}^m \frac{k_j}{T_j} \right).$$

Here again, we denote $\lambda_k = \sum_{j=1}^m \frac{k_j}{T_j}$. As before, $\lambda_k = 0 \iff k = 0$. Moreover, fixing $a < b$, we have

$$\int_a^b f \left(\frac{x}{\varepsilon} \right) dx = (b-a)\alpha_0 + \varepsilon \sum_{k \neq 0} \alpha_k \frac{e^{2i\pi\lambda_k \frac{b}{\varepsilon}} - e^{2i\pi\lambda_k \frac{a}{\varepsilon}}}{2i\pi\lambda_k} \xrightarrow[\varepsilon \to 0]{} (b-a)\alpha_0 = (b-a)\langle f \rangle.$$

Thus, for any piecewise constant compactly supported function φ, we deduce that $\int f \left(\frac{\cdot}{\varepsilon} \right) \varphi \to \langle f \rangle \int \varphi$, as $\varepsilon \to 0$. Using the fact that piecewise constant compactly supported functions are dense in $L^1(\mathbb{R})$, we deduce (1.52). It remains to apply the approximation argument of F by trigonometric polynomials, as in the proof of Proposition 1.2, to conclude the proof. □

Proposition 1.3 is the analogue, for quasi-periodic functions, of Proposition 1.1 for periodic functions. In the quasi-periodic case, property (P1) however does not hold. The convergence rate of (1.52) can be different from $O(\varepsilon)$, despite the equality of averages stated in Proposition 1.2.

Such a result is beyond the scope of the present textbook. The reader is refereed to [BBMM05], where it is proved that, for any $\delta \in]0, 1]$, there exists a quasi-periodic function f such that the convergence rate in (1.52) is *exactly* ε^δ.

We now focus on properties (P2)–(P5) established for the periodic case in Sect. 1.2. We start with property (P2), that is, the equivalence between the L^q_{unif} norm and the norm defined by $\langle | \cdot |^q \rangle^{1/q}$. We are going to prove that this property is no longer true in the quasi-periodic case. More precisely, for any $q \in [1, +\infty[$ and any $T = (T_1, \ldots, T_m)$ such that $m \geq 2$ and the frequencies $\left(\frac{1}{T_1}, \ldots, \frac{1}{T_m} \right)$ are linearly independent over $\mathbb{Z}$ (as mentioned on p. 27), we are going to build, for any $\delta \in]0, 1[$, a function $f_\delta \in QP(T)$ such that

$$\| f_\delta \|_{L^q_{\text{unif}}(\mathbb{R})} \geq \gamma > 0, \quad \langle | f_\delta |^q \rangle^{1/q} \xrightarrow{\delta \to 0} 0, \tag{1.53}$$

which will contradict (P2). We assume that the periods T_j are non-decreasing with respect to j. This is always true up to a permutation of the variables x_j. We thus have $0 < T_1 \le \cdots \le T_m$. We fix $\rho \ge 0$ of class C^∞ with support included in $]-T_1/2, T_1/2[$, and such that $\int_{\mathbb{R}} \rho^q = 1$. We define

$$F_\delta(x_1, \ldots x_m) = \delta^{-1/q} \prod_{j=1}^{m} \rho\left(\frac{x_j}{\delta}\right).$$

We extend F_δ to $\mathbb{R}^m$ by T-periodicity, and set

$$f_\delta(x) = F(x, \ldots, x),$$

which belongs to $QP(T)$. Applying Proposition 1.2, we compute

$$\langle |f_\delta|^q \rangle^{1/q} = \langle |F_\delta|^q \rangle^{1/q}$$

$$= \left(\frac{1}{|Q(T)|} \int_{Q(T)} \delta^{-1} \rho\left(\frac{x_1}{\delta}\right)^q \cdots \rho\left(\frac{x_m}{\delta}\right)^q dx_1 \ldots dx_m \right)^{1/q}$$

$$= \frac{\delta^{\frac{m-1}{q}}}{|Q(T)|^{1/q}}.$$

Unless $m = 1$, which is the periodic case, we have

$$\langle |f_\delta|^q \rangle^{1/q} \xrightarrow{\delta \to 0} 0. \tag{1.54}$$

On the other hand, restricting the support of ρ if necessary, we may assume that it is contained in $[-1, 1]$. Hence, if $\delta \le 1$,

$$\|f_\delta\|_{L^q_{\text{unif}}(\mathbb{R})}^q \ge \int_{-1}^{1} f_\delta(t)^q dt = \int_{-1}^{1} \delta^{-1} \rho\left(\frac{t}{\delta}\right)^{mq} dt$$

$$= \int_{-1/\delta}^{1/\delta} \rho(x)^{mq} dx = \int_{-1}^{1} \rho(x)^{mq} dx.$$

The Hölder inequality applied to ρ^q gives

$$\int_{-1}^{1} \rho(x)^{mq} dx \ge 2^{1-m} \left(\int_{-1}^{1} \rho^q \right)^m = 2^{1-m}.$$

We thus have $\|f_\delta\|_{L^q_{\text{unif}}(\mathbb{R})}^q \ge 2^{1-m}$. This and (1.54) imply (1.53).

Next, we study property (P3). It does not hold either in the quasi-periodic case. For well chosen, but after all quite general quasi-periods T (actually, any $T = (T_1, T_2)$ such that $T_1/T_2 \notin \mathbb{Q}$ is suitable, and more generally any T such that $QP(T)$ is not a space of periodic functions), we now prove that there exists $f \in QP(T)$ such that $\langle f \rangle = 0$, and there does not exist $g \in QP(T)$, differentiable, such that $g' = f$. For this purpose, we use our remark p. 28 regarding the fact that $QP(T)$ contains periodic functions with arbitrarily large periods. Consider an irrational number r. Dirichlet's approximation Theorem implies that there exist two sequences of integers $(p_n)_{n \in \mathbb{N}}$ and $(q_n)_{n \in \mathbb{N}}$ such that $q_n > 0$, $q_n \geq n + 1$, and $|r - p_n/q_n| \leq q_n^{-2}$. We then choose a sequence $(\alpha_n)_{n \in \mathbb{N}}$ such that

$$\sum_{n \in \mathbb{N}} |\alpha_n| < +\infty, \quad \sum_{n \in \mathbb{N}} \alpha_n^2 q_n^2 = +\infty. \tag{1.55}$$

Such a sequence exists: take for instance $\alpha_n = (n + 1)^{-3/2}$, which is the general term of a convergent series and satisfies $\alpha_n^2 q_n^2 \geq (n + 1)^{-3+2} = (n + 1)^{-1}$. We define

$$f(x) = \sum_{n \in \mathbb{N}} \alpha_n e^{2i\pi x(p_n - rq_n)},$$

which is quasi-periodic by definition, and satisfies $\langle f \rangle = 0$. More precisely, $f \in QP(1, r^{-1})$. Assume that there exists $g \in QP(1, r^{-1})$, differentiable, such that $g' = f$. Then $g(x) = G(x, x)$, where G is continuous and periodic. Hence, G reads as

$$G(x, y) = \sum_{(k, j) \in \mathbb{Z}^2} \beta_{kj} e^{2i\pi(kx + rjy)}, \quad \sum_{(j, k) \in \mathbb{Z}^2} |\beta_{kj}|^2 < +\infty, \tag{1.56}$$

where the convergence of the series is understood in the $L^2 \left([0, 1] \times \left[0, \frac{1}{r} \right] \right)$ sense. In addition,

$$\beta_{kj} = \left\langle G(x, y) e^{-2i\pi(kx + rjy)} \right\rangle = \left\langle g(x) e^{-2i\pi(k + rj)x} \right\rangle$$

$$= \lim_{R \to +\infty} \frac{1}{2R} \int_{-R}^{R} g(x) e^{-2i\pi(k + rj)x} dx. \tag{1.57}$$

An integration by parts yields

$$\frac{1}{2R} \int_{-R}^{R} g(x) e^{-2i\pi(k + rj)x} dx = \frac{1}{2R} \left[g(x) \frac{e^{-2i\pi(k + rj)x}}{-2i\pi(k + rj)} \right]_{-R}^{R}$$

$$+ \frac{1}{2R} \int_{-R}^{R} g'(x) \frac{e^{-2i\pi(k + rj)x}}{2i\pi(k + rj)} dx.$$

Since g is bounded, the first term of the right-hand side vanishes as $R \to +\infty$. Since $g' = f$, the second term converges to the average of the function $x \mapsto f(x)\dfrac{e^{-2i\pi(k+rj)x}}{2i\pi\,(k+rj)}$. We insert this into (1.57) and find

$$\beta_{kj} = \left\langle f(x)\frac{e^{-2i\pi(k+rj)x}}{2i\pi\,(k+rj)} \right\rangle = \left\langle F(x,y)\frac{e^{-2i\pi(kx+rjy)}}{2i\pi\,(k+rj)} \right\rangle$$

$$= \begin{cases} \dfrac{\alpha_n}{2i\pi(p_n - rq_n)} & \text{if } k = p_n, \ j = -q_n, \\[2mm] 0 & \text{otherwise.} \end{cases}$$

We use first $(p_n - rq_n)^2 \le q_n^{-2}$ and next (1.55). We obtain

$$\sum_{(k,j)\in\mathbb{Z}^2} |\beta_{kj}|^2 = \sum_{n\in\mathbb{N}} \frac{|\alpha_n|^2}{4\pi^2\,(p_n - rq_n)^2} \ge \frac{1}{4\pi^2}\sum_{n\in\mathbb{N}} |\alpha_n|^2 q_n^2 = +\infty.$$

This contradicts (1.56). Thus, f does not have any primitive function in $QP(T)$.

We turn to property (P4), that is, Poincaré-Wirtinger inequality. This inequality does not hold in the quasi-periodic case. Here again, this is due to the fact that, for any $T = (T_1, \ldots, T_m)$ with $m \ge 2$ and $\mathbb{Z}$-linearly independent frequencies, $QP(T)$ contains functions that are periodic with arbitrarily large periods. Fix an irrational number r, and define $f(x) = \exp(2i\pi(p - qr)x)$, where p, q are integer such that $p \ge 2$ and $|p/q - r| \le q^{-2}$. Such a function is in $QP(T)$ for $T = (1, R^{-1})$ and satisfies $\langle f \rangle = 0$. In addition, $f'(x) = 2i\pi(p - qr)f(x)$, hence

$$\left\langle |f|^2 \right\rangle = 1, \quad \left\langle |f'|^2 \right\rangle = 4\pi^2(p - qr)^2 \le \frac{4\pi^2}{q^2}.$$

Dirichlet's approximation Theorem, which we have already used above, allows us to take q arbitrarily large, which contradicts Poincaré-Wirtinger inequality in $QP(T)$.

However, in the particular case where the quasi-periods satisfy a diophantine condition (in the case $m = 2$, this is equivalent to assuming that the quotient of the quasi-periods is not a *Liouville numbers*), then a somewhat degraded version of the Poincaré-Wirtinger inequality holds true. In order to describe this, we first give the following definition:

Definition 1.3 Let $x \in \mathbb{R} \setminus \mathbb{Q}$. We say that x is a *Liouville number* if, for all $\alpha \in \mathbb{N}$, there exists $j \in \mathbb{Z}$ and an integer $k \ge 2$ such that

$$\left| x - \frac{j}{k} \right| \le \frac{1}{k^\alpha}. \tag{1.58}$$

It is known [BT04, pages 10-11] that all Liouville numbers are transcendental. Indeed, for any algebraic number x, inequality (1.58) is false for $\alpha > 2$ (we recall that a real number is algebraic if it is the root of a polynomial with integer coefficients, and that it is transcendantal if it is not algebraic). Intuitively, Liouville numbers are irrational numbers that can be approximated with arbitrary precision by rational numbers. The set of Liouville numbers is of zero Lebesgue measure in $\mathbb{R}$. The following property, sometimes called *Gårding inequality*, is a weakened version of the Poincaré-Wirtinger inequality:

Lemma 1.2 (Gårding Inequality)   *Let f be a quasi-periodic function, in the sense of Definition 1.1, with $m = 2$. We assume that the periods T_1, T_2 of the function F are such that the quotients $\dfrac{T_1}{T_2}$ and $\dfrac{T_2}{T_1}$ are irrational and are not Liouville numbers. Then there exists $s \geq 1$ such that, if $F \in H^{s+1}(Q)$,*

$$\left\langle (f - \langle f \rangle)^2 \right\rangle = \left\langle (F - \langle F \rangle)^2 \right\rangle \leq C \left\| \partial_{x_1} F + \partial_{x_2} F \right\|_{H^s(Q)}, \tag{1.59}$$

where the constant C only depends on the periods (T_1, T_2) and on s.

One should note that, in the left-hand side, the notation $\langle \cdot \rangle$ stands for the average in the natural functional space, that is, for the first expression, the average of quasi-periodic functions, and for the second one, the average of periodic functions. In addition, the partial derivative $\partial_{x_1} F + \partial_{x_2} F$ of the periodic function F (of two variables) also reads as $\partial_{\frac{x_1 + x_2}{2}} F$ in the coordinate system $\left(\dfrac{x_1 + x_2}{2}, \dfrac{x_1 - x_2}{2} \right)$. Thus, the H^s norm of this derivative implicitly involves derivatives of F in the direction orthogonal to the diagonal $x_1 = x_2$, that is, $\partial_{\frac{x_1 - x_2}{2}} F$. In other words, this H^s norm involves derivatives that do not correspond to derivatives of f.

We also point tout that, in the general case $m > 2$, a Gårding inequality holds, under a Diophantine type condition. This condition is more complex than, but similar to, that of Lemma 1.2 (see [Koz79, Condition (C)]).

Proof We first prove the result when F is smooth. A density argument will then allow us to conclude. Since F is periodic, it may be written as the Fourier series

$$F(x_1, x_2) = \sum_{k \in \mathbb{Z}^2} \widehat{F}_k e^{2i\pi \left(\frac{k_1 x_1}{T_1} + \frac{k_2 x_2}{T_2} \right)},$$

where, since F is smooth, $\widehat{F}_k$ tends to 0 faster than any power of k, as $|k| \to +\infty$. In addition, we have

$$\left\| \partial_{x_1} F + \partial_{x_2} F \right\|_{H^s(Q)}^2 = \sum_{k \in \mathbb{Z}^2} \left(1 + k_1^2 + k_2^2 \right)^s \left(\frac{k_1}{T_1} + \frac{k_2}{T_2} \right)^2 |\widehat{F}_k|^2. \tag{1.60}$$

Since T_1/T_2 is not a Liouville number, we know that there exists $\alpha_{12} \in \mathbb{N}$ and $C > 0$ such that, for all $p \in \mathbb{Z}$ and all $q \in \mathbb{N}^*$,

$$\left| \frac{T_1}{T_2} - \frac{p}{q} \right| \geq \frac{C}{q^{\alpha_{12}}}. \tag{1.61}$$

Indeed, if $q = 1$, this inequality holds true with $C = \text{dist}\left(\frac{T_1}{T_2}, \mathbb{Z}\right) > 0$, and if $q \geq 2$, we simply apply (1.58).

Next, we write, for $k_2 \neq 0$,

$$\left| \frac{k_1}{T_1} + \frac{k_2}{T_2} \right| = \frac{|k_2|}{T_1} \left| \frac{k_1}{k_2} + \frac{T_1}{T_2} \right| \geq \frac{|k_2|}{T_1} \frac{C}{|k_2|^{\alpha_{12}}} = \frac{C}{T_1} \frac{1}{|k_2|^{\alpha_{12}-1}}, \tag{1.62}$$

where we have used (1.61). This inequality is also valid if we exchange the roles of the indices 1 and 2, since T_2/T_1 is not a Liouville number. Hence, for $k_1 \neq 0$,

$$\left| \frac{k_1}{T_1} + \frac{k_2}{T_2} \right| \geq \frac{C'}{T_2} \frac{1}{|k_1|^{\alpha_{21}-1}}, \tag{1.63}$$

where the constant C' is *a priori* different from the constant C in (1.62), but is positive and only depends on T_1 and T_2. Collecting (1.62) and (1.63), we infer that there exists a constant C'' such that

$$\left| \frac{k_1}{T_1} + \frac{k_2}{T_2} \right| \geq \frac{C''}{|k|^{\alpha-1}},$$

where we have chosen $\alpha = \max(\alpha_{12}, \alpha_{21})$. Inserting this inequality into (1.60), we thus have

$$\left\| \partial_{x_1} F + \partial_{x_2} F \right\|_{H^s(Q)}^2 \geq (C'')^2 \sum_{k \in \mathbb{Z}^2 \setminus \{(0,0)\}} (1 + |k|^2)^s \frac{1}{|k|^{2\alpha-2}} |\widehat{F}_k|^2$$

$$\geq (C'')^2 \sum_{k \in \mathbb{Z}^2 \setminus \{(0,0)\}} |\widehat{F}_k|^2 = (C'')^2 \left\langle (F - \langle F \rangle)^2 \right\rangle,$$

for $s = \alpha - 1$. This concludes the proof. $\qquad\qquad\qquad\qquad\qquad\qquad \square$

Finally, we note that property (P5) is clearly true in the present case. Indeed, any quasi-periodic function f is defined as the trace of a function F which is *periodic and continuous*, hence bounded. Thus, f itself is bounded. As a consequence, it is strictly sublinear at infinity.

It is however worth mentioning that the space $QP(T)$ is closed for the uniform norm, but is not closed for the L^q_{unif}, norm, $1 \leq q < +\infty$. This is similar to the fact that, for any domain $\mathcal{D}$, $C^0(\mathcal{D})$ is not closed for the norm $\|\cdot\|_{L^q(\mathcal{D})}$. One may study the closure of $QP(T)$ for the norm $\| \cdot \|_{L^q_{unif}(\mathbb{R})}$. This gives a larger space, which is a subspace of L^q_{unif}, and in which property (P5) holds, in the following sense:

$$\forall f \in L^q_{unif}(\mathbb{R}), \quad \lim_{|x| \to +\infty} \frac{\|f\|_{L^q([x,x+1])}}{1+|x|} = 0.$$

1.5.2   The Almost Periodic Setting

We consider in this section the notion of almost periodicity. Almost periodicity may be seen as a generalization of quasi-periodicity, hence also a generalization of periodicity. We give here only a brief account of this theory, and we refer the reader to [Bes32, Boh18] for more details. Here again, we only consider the one-dimensional case. All the ideas and results can be generalized to higher dimensions. Recall that, as we mentioned at the beginning of Sect. 1.5, we consider complex-valued functions.

Definition 1.4 Let f be a continuous bounded function defined on $\mathbb{R}$, and let $\delta > 0$. The real number τ is a δ-*almost period* of f if

$$\sup_{x \in \mathbb{R}} |f(x+\tau) - f(x)| \le \delta. \tag{1.64}$$

We denote by $T_\delta(f)$ the set of δ-almost periods of f. The function f is said to be *almost periodic* if

$$\forall \delta > 0, \quad \exists L = L(\delta) > 0 \quad \text{such that} \quad \forall a \in \mathbb{R}, \quad T_\delta(f) \cap [a, a+L] \ne \emptyset. \tag{1.65}$$

The set of almost periodic functions on the real line is denoted by $PP(\mathbb{R})$.

We first point out that a periodic function is almost periodic. Indeed, if f is T-periodic, then for any $k \in \mathbb{Z}$, the real number $\tau = kT$ satisfies (1.64), since the left-hand side vanishes. Thus, kT is a δ-almost period of f, for any $\delta > 0$ and any $k \in \mathbb{Z}$. Thus, $T\mathbb{Z} \subset T_\delta(f)$, which implies that $T_\delta(f)$ satisfies (1.65), with, for instance, $L(\delta) = T$.

Likewise, any quasi-periodic function is an almost periodic function, but the converse is not true, as we will see below in (1.66).

Definition 1.4 implies that $PP(\mathbb{R})$ is stable by linear combination. This result is not obvious (see for instance [Bes32, page 5]). However, it is clear from (1.64) that $f \in PP(\mathbb{R})$ implies $1/f \in PP(\mathbb{R})$ (of course if f is bounded away from 0), and that $f \in PP(\mathbb{R})$ implies $f^2 \in PP(\mathbb{R})$. We thus know that $PP(\mathbb{R})$ is stable by product using the polarization identity

$$4fg = (f + g)^2 - (f - g)^2.$$

This proves that $PP(\mathbb{R})$ is an algebra.

Definition 1.4 also allows to prove that if f is almost periodic (hence continuous), then it is uniformly continuous. Indeed, let us fix $\delta > 0$. Then there exists $L > 0$ such that $T_{\delta/3}(f) \cap [a, a + L] \neq \emptyset$, for all $a \in \mathbb{R}$. Since f is continuous, it is uniformly continuous on the compact set $[-L, L]$. It follow that there exists $\eta > 0$ such that for any pair (x_1, x_2) in $[-L, L]$ such that $|x_1 - x_2| \leq \eta$, we have $|f(x_1) - f(x_2)| \leq \delta/3$. Fixing x, y such that $|x - y| \leq \min(\eta, L/2)$, we thus know that there exists $\tau \in [x - L/2, x + L/2] \cap T_{\delta/3}(f)$ such that $\tau > 0$, according to (1.65). Applying the triangle inequality, we find

$$|f(x) - f(y)| \leq |f(x) - f(x - \tau)| + |f(x - \tau) - f(y - \tau)| + |f(y - \tau) - f(y)|.$$

According to the definition of τ, the first and third term of the right-hand side are both bounded by $\delta/3$. Considering the second one, it is clear that $x - \tau$ and $y - \tau$ both lie in the interval $[-L, L]$, and that $|x - \tau - (y - \tau)| = |x - y| \leq \eta$. Thus, this term is also bounded by $\delta/3$, which proves that $|f(x) - f(y)| \leq \delta$. We have therefore shown that f is uniformly continuous.

The following result is a different characterization of almost periodic functions:

Theorem 1.1 (Bochner's Characterization) *A bounded continuous function f is almost periodic if and only if the set of its translates is relatively compact for the uniform convergence topology.*

For the proof of this result, we refer to [Bes32, pages 10–11] or [DS88a, §IV.7]. We also state the following fundamental result:

Theorem 1.2 *The set $PP(\mathbb{R})$ is the closure of the set of trigonometric polynomials for the uniform norm.*

The proof of this result may again be found in [Bes32, pages 29–31], in [Boh18, pages 80–88], or in [DS88a, pages 281–285].

The Weierstrass trigonometric Theorem (see page 29) together with Theorem 1.2 allow to prove that any quasi-periodic function is almost periodic. Indeed, if $f \in QP(T)$ for some T, we can apply the Weierstrass trigonometric Theorem to the function F of which f is the trace (see Definition 1.1). Doing so, we prove that f is the limit, for the uniform convergence, of a sequence of trigonometric polynomials. According to Theorem 1.2, we thus have $f \in PP(\mathbb{R})$.

It is worth mentioning the difference between the space $PP(\mathbb{R})$ and the space $QP(T)$ considered in the preceding Section. The latter is defined for fixed quasi-periods $T = (T_1, \dots, T_m)$, whereas $PP(\mathbb{R})$ contains quasi-periodic functions of any quasi-periods. In this sense, the space $PP(\mathbb{R})$ is "infinitely larger" than $QP(T)$, for a fixed T. Indeed, for any $N \in \mathbb{N}$, for any $T \in \mathbb{R}^N$, $QP(T) \subset PP(\mathbb{R})$. Furthermore, there exist almost periodic functions that are not quasi-periodic, as the following example shows:

$$f(x) = \sum_{n \geq 1} \frac{1}{n^2} e^{\frac{2i\pi x}{n}}. \tag{1.66}$$

Clearly, f is the limit (for the uniform convergence) of a sequence of trigonometric polynomials (the partial sums of the series). We argue by contradiction, and assume that $f \in QP(T)$ for some $T = (T_1, \ldots, T_m)$. Then there exists $F : \mathbb{R}^m \to \mathbb{R}$ such that $f(x) = F(x, \ldots, x)$, where F is continuous and T-periodic. Fixing $\xi \in \mathbb{R}$, the function g defined by $g(x) = f(x)e^{-2i\pi\xi x}$ is quasi-periodic, and its quasi-periods read $\left(T_1, \ldots, T_m, \frac{1}{\xi}\right)$. If the inverses $\left(\frac{1}{T_1}, \ldots, \frac{1}{T_m}, \xi\right)$ of the quasi-periods are linearly independent over $\mathbb{Z}$, then Proposition 1.2 implies that

$$\langle g \rangle = \frac{1}{|Q(T_1, \ldots, T_m, \xi)|} \int_{Q(T_1, \ldots, T_m, \xi)} F(x_1, \ldots, x_m)e^{-2i\pi\xi x_{m+1}} dx_1 \ldots dx_{m+1}$$

$$= 0.$$

But a simple computation, which we leave to the reader, gives

$$\langle g \rangle = \lim_{R \to +\infty} \frac{1}{2R} \int_{-R}^{R} f(x)e^{-2i\pi\xi x} dx = \begin{cases} \frac{1}{n^2} & \text{if } \xi = \frac{1}{n}, \\ 0 & \text{otherwise.} \end{cases}$$

We have thus proved that, for any integer $n \geq 1$, the system $\left(\frac{1}{T_1}, \ldots, \frac{1}{T_m}, \frac{1}{n}\right)$ is necessarily linearly dependent over $\mathbb{Z}$. Since the system $\left(\frac{1}{T_1}, \ldots, \frac{1}{T_m}\right)$ is, on the other hand, linearly independent, we infer that there exists integers $k_j(n)$ such that

$$\sum_{j=1}^{m} \frac{k_j(n)}{T_j} = \frac{1}{n}.$$

Again using that the system $\left(\frac{1}{T_1}, \ldots, \frac{1}{T_m}\right)$ is linearly independent over $\mathbb{Z}$, this implies that $k_j(n) = \dfrac{k_j(1)}{n}$. This is a contradiction, since all $k_j(n)$ are integers. As a consequence, the function f defined by (1.66) does not belong to $QP(T)$, whatever the value of T.

Theorem 1.2 allows to prove that any almost periodic function has an average, as stated in the following proposition.

Proposition 1.4 *Let $f \in PP(\mathbb{R})$. Then f has an average.*

Proof The proof is identical to that of Proposition 1.2. We first note that any trigonometric polynomial has an average. Then, we point out that the average is a linear form which is continuous for the uniform norm. Theorem 1.2 allows to conclude.   □

We also have the equivalent of Proposition 1.3, which reads as follows.

Proposition 1.5 *Let* $f \in PP(\mathbb{R})$. *Then*

$$f\left(\frac{x}{\varepsilon}\right) \xrightarrow[\varepsilon \to 0]{*} \langle f \rangle, \tag{1.67}$$

for the $L^\infty(\mathbb{R})$ *topology.*

The proof is again a simple adaptation of that of Proposition 1.3. We leave it to the reader.

As in the quasi-periodic setting, we could study properties (P1)–(P5) of Sect. 1.2. Let us simply point out that properties (P1)–(P4) do not hold in $PP(\mathbb{R})$, since they already do not hold in the quasi-periodic setting. As for Property (P5), it is clearly true in $PP(\mathbb{R})$, since any function in $PP(\mathbb{R})$ is by definition bounded.

So far, we have dealt with continuous functions only. In order to be able to manipulate weak solutions of partial differential equations, it is useful to consider the generalization of the space $PP(\mathbb{R})$ to functions that are not continuous, but only (locally) integrable. One can achieve this by several methods, which are not equivalent to one another. First, we define the following norms and semi-norms:

Definition 1.5 Let $p \in [1, +\infty[$. We call:

- *Stepanov norm* the L^p_{unif} norm;

- *Weyl semi-norm* the quantity $\|f\|_{W^p} = \lim\limits_{R \to +\infty} \sup\limits_{x \in \mathbb{R}} \left(\frac{1}{R} \int_x^{x+R} |f|^p\right)^{1/p}$, for all $f \in L^p_{loc}(\mathbb{R})$;

- *Besicovitch semi-norm*, the quantity $\|f\|_{B^p} = \lim\limits_{R \to +\infty} \sup \left(\frac{1}{2R} \int_{-R}^R |f|^p\right)^{1/p}$, for all $f \in L^p_{loc}(\mathbb{R})$.

In Definition 1.5, the sup (for the Weyl semi-norm) and the lim sup (for the Besicovitch semi-norm) are understood in $\mathbb{R} \cup \{+\infty\}$. Defining the limit for the Weyl semi-norm is a simple exercise that is left to the reader. The proof can be found in [Bes32, page 72]. The quantities $\|\cdot\|_{W^p(\mathbb{R})}$ and $\|\cdot\|_{B^p(\mathbb{R})}$ are only semi-norms because, for instance, compactly supported functions have semi-norm zero. Among these functions, however, none is almost periodic (see [Boh18, page 63] for the case $p = 2$, but the proof is easily adapted to any value of p, $1 \le p < +\infty$). Otherwise stated, when restricted to $PP(\mathbb{R})$, these semi-norms become norms. In

order to build functional spaces adapted to these norms, we consider the respective
closures

$$S^p(\mathbb{R}) = \left\{ f \in L^p_{loc}(\mathbb{R}), \quad \exists \ (f_n)_{n \in \mathbb{N}} \text{ sequence in } PP(\mathbb{R}), \right.$$

$$\left. \lim_{n \to +\infty} \|f_n - f\|_{L^p_{unif}} = 0 \right\},$$

$$\mathcal{W}^p(\mathbb{R}) = \left\{ f \in L^p_{loc}(\mathbb{R}), \quad \exists \ (f_n)_{n \in \mathbb{N}} \text{ sequence in } PP(\mathbb{R}), \right.$$

$$\left. \lim_{n \to +\infty} \|f_n - f\|_{W^p} = 0 \right\},$$

$$\mathcal{B}^p(\mathbb{R}) = \left\{ f \in L^p_{loc}(\mathbb{R}), \quad \exists \ (f_n)_{n \in \mathbb{N}} \text{ sequence in } PP(\mathbb{R}), \right.$$

$$\left. \lim_{n \to +\infty} \|f_n - f\|_{B^p} = 0 \right\}.$$

As we have just pointed out, the norm L^p_{unif} is indeed a norm on $S^p(\mathbb{R})$, while
$\|\cdot\|_{W^p}$ and $\|\cdot\|_{B^p}$ are semi-norms on $\mathcal{W}^p(\mathbb{R})$ and $\mathcal{B}^p(\mathbb{R})$, respectively. We thus
define the following equivalence relation in each of these spaces:

$$\forall f, g \in \mathcal{W}^p(\mathbb{R}), \quad f \sim g \iff \|f - g\|_{W^p} = 0, \tag{1.68}$$

and

$$\forall f, g \in \mathcal{B}^p(\mathbb{R}), \quad f \sim g \iff \|f - g\|_{B^p} = 0, \tag{1.69}$$

The quotiented spaces built from $\mathcal{W}^p(\mathbb{R})$ and $\mathcal{B}^p(\mathbb{R})$ are closed normed vector
spaces.

Definition 1.6 Let $p \in [1, +\infty[$. We define:

- $S^p(\mathbb{R}) = S^p(\mathbb{R})$ the space of almost periodic functions in the sense of Stepanov,
  equipped with the norm $L^p_{unif}(\mathbb{R})$;
- $W^p(\mathbb{R}) = \mathcal{W}^p(\mathbb{R})/ \sim$ the space of almost periodic functions in the sense of
  Weyl, equipped with the norm $\|\cdot\|_{W^p}$;
- $B^p(\mathbb{R}) = \mathcal{B}^p(\mathbb{R})/ \sim$ the space of almost periodic functions in the sense of
  Besicovitch, equipped with the norm $\|\cdot\|_{B^p}$.

The map $f \mapsto \langle f \rangle$ is a linear form, which is continuous with respect to each
norm and semi-norm in Definition 1.5. A simple density argument allows to prove
that whenever f is in S^p, W^p or B^p, f has an average.

Let us mention that, instead of building the space $S^p(\mathbb{R})$ by completion, it is also possible to define almost-periodicity using the L^p_{unif} norm. We then replace (1.64), in Definition 1.4, by

$$\sup_{x \in \mathbb{R}} \int_x^{x+1} |f(x+\tau) - f(x)|^p dx \leq \delta^p,$$

and leave all the rest of the definition unchanged. We then recover the space $S^p(\mathbb{R})$, as shown in [CLL98, pages 260–261]. A way to prove the equivalence of the two definitions is to check that both spaces are the closure, for the L^p_{unif} norm, of the set of trigonometric polynomials.

In the next section, we are going to study the random setting. We will see that the preceding periodic, quasi-periodic and almost-periodic settings may be seen as particular cases of the random setting. The latter is however more general.

1.6   The Random Setting

We are now going to allow the coefficient a_ε in (1.1) to be random. More precisely, we assume that $a_\varepsilon(x) = a\left(\frac{x}{\varepsilon}, \omega\right)$, where the function $a(x, \omega)$ depends on two variables. The first one is the "usual" space variable x, while the second one $\omega \in \Omega$ models the randomness of the coefficient. We assume that the reader is familiar with the notion of *probability space* and *random variable*, together with their elementary properties. We are going to recall some basic facts. The reader should, if need be, refer to a classical probability textbook such as [Bil95], [Dud02] or [Dur19]. Needless to say, we present the theory in dimension one for simplicity, but it may be easily extended to higher dimensions.

1.6.1   Basic Elements for the Random Setting

We denote by Ω a set (called the sample space and modelling the universe of accessible states) and by $\mathcal{P}(\Omega)$ the collection of subsets of Ω. Let $\mathcal{T} \subset \mathcal{P}(\Omega)$. We say that $\mathcal{T}$ (which models the available information) is a σ-field (one also says σ-algebra) if $\Omega \in \mathcal{T}$, it is closed under countable unions and complement. On the σ-field $\mathcal{T}$, it is possible to define a *probability* measure $\mathbb{P}$, that is, a non-negative measure of total mass one ($\mathbb{P}(\Omega) = 1$). We recall that a (non-negative) measure $\mathbb{P}$ is a map from $\mathcal{T}$ to $\mathbb{R}_+ \cup \{+\infty\}$ such that $\mathbb{P}(\emptyset) = 0$ and, for any disjoint countable family of elements A_i of $\mathcal{T}$, $\mathbb{P}\left(\bigcup_{i=1}^{+\infty} A_i\right) = \sum_{i=1}^{+\infty} \mathbb{P}(A_i)$. We say that a property is satisfied almost surely if the set A of $\omega \in \Omega$ for which this property is not satisfied is of zero measure for $\mathbb{P}$. The triplet $(\Omega, \mathcal{T}, \mathbb{P})$ is called a *probability space*. A (real-valued) *random variable* is a map from Ω to $\mathbb{R}$ which is measurable with respect to the σ-field $\mathcal{T}$, that is, such that for any Borel set $B \subset \mathbb{R}$, the set $\{\omega \in \Omega \, ; \, X(\omega) \in B\}$ (often denoted $\{X \in B\}$) belongs to $\mathcal{T}$. For any $\omega \in \Omega$, $X(\omega)$ is a *realization* of the random variable X. The *expectation* of the random variable X is defined by

$$\mathbb{E}(X) = \int_{\Omega} X(\omega) \, d\mathbb{P}(\omega).$$

The *law* of the random variable X is the measure $\mathbb{P} \circ X^{-1}$ defined by

$$\mathbb{E}(f(X)) = \int_{\mathbb{R}} f(x) \, d(\mathbb{P} \circ X^{-1})(x),$$

for any bounded measurable function f. The random variable X is said to have a *density* $p(x)$ (where p is a non-negative, integrable function of unit mass over $\mathbb{R}$), if, for any bounded measurable function f,

$$\mathbb{E}(f(X)) = \int_{\mathbb{R}} f(x) \, p(x) \, dx.$$

In such a case, the law of X reads as $d(\mathbb{P} \circ X^{-1})(x) = p(x)dx$.

In practice, the expectation of the random variable X can be approximated (this is the so-called *Monte Carlo method*) by taking the mean value of different values of X obtained by a random draw according to the law of X. The mathematical justification of this method is the *strong law of large numbers*: if X_i, $i \in \mathbb{Z}$, is a sequence of independent random variables sharing the same law as X, and if $\mathbb{E}(|X|) < +\infty$, then, almost surely,

$$\mathbb{E}(X) = \lim_{n \to +\infty} \frac{X_1(\omega) + \ldots + X_n(\omega)}{n}. \tag{1.70}$$

The *central limit Theorem* makes precise the convergence rate: under the same conditions, if in addition $\mathbb{E}(X^2) < +\infty$, then the random variable defined by

$$\frac{\sqrt{n}}{\sigma} \left(\frac{X_1(\omega) + \ldots + X_n(\omega)}{n} - \mathbb{E}(X) \right), \tag{1.71}$$

where $\sigma^2 = \mathbb{E}\left((X - \mathbb{E}(X))^2\right) = \mathbb{E}(X^2) - (\mathbb{E}(X))^2$ is the *variance* of the random variable X, *converges in law* to a random variable G. The law of G is the *reduced centered Gaussian law* of density $p(x) = \frac{1}{\sqrt{2\pi}} \exp\left(-\frac{x^2}{2}\right)$. We recall that a sequence of random variables Y_n converges in law to Y if $\mathbb{E}(f(Y_n))$ tends to $\mathbb{E}(f(Y))$ for any continuous bounded function f. This result explains why Gaussian random variables (that is, random variables having a Gaussian law) are so important in probability theory.

We now return to our coefficient $a_\varepsilon(x) = a\left(\dfrac{x}{\varepsilon}, \omega\right)$ of (1.1). We understand that, for a fixed value of x, $a(x, .)$ is a random variable, hence a measurable function of ω. On the other hand, almost surely in ω, the map $x \mapsto a(x, \omega)$ is measurable (in the sense of Lebesgue) with respect to x. Next, we are going to make a *structural assumption* that relates $x \in \mathbb{R}$ and $\omega \in \Omega$.

1.6.2   The Notion of Stationarity

We now assume that the group $(\mathbb{Z}, +)$ *acts on* $(\Omega, \mathcal{T}, \mathbb{P})$. This means that we consider a map τ, called the *action* (or, also, the *shift*), from the product $\mathbb{Z} \times \Omega$ to Ω, which we denote by $\tau(k, \omega) = \tau_k(\omega)$. We assume that this action *preserves the probability measure* $\mathbb{P}$, that is,

$$\forall k \in \mathbb{Z}, \ \forall A \in \mathcal{T}, \quad \tau_k A \in \mathcal{T} \quad \text{and} \quad \mathbb{P}(\tau_k A) = \mathbb{P}(A). \tag{1.72}$$

In addition, we assume that this action is *ergodic*, that is,

$$\forall A \in \mathcal{T}, \quad (\forall k \in \mathbb{Z}, \ \tau_k A = A) \implies (\mathbb{P}(A) = 0 \text{ or } \mathbb{P}(A) = 1). \tag{1.73}$$

This property intuitively states that a subset A of Ω which is invariant under all translations can only be $A = \emptyset$ or $A = \Omega$. These equalities are of course understood *in the sense of the probability measure* $\mathbb{P}$.

We next introduce the notion of *stationary* function. Consider $f : \Omega \times \mathbb{R} \longmapsto \mathbb{R}$ that is locally integrable (for the measure $dx\, d\mathbb{P}$). We say that f is *stationary* if it satisfies the following property

$$\forall k \in \mathbb{Z}, \ f(x + k, \omega) = f(x, \tau_k \omega) \text{ for almost every } x \in \mathbb{R} \text{ and almost surely.} \tag{1.74}$$

Note that a function $f(x, \omega)$ that does not depend on ω and is $\mathbb{Z}$-periodic with respect to x satisfies (1.74). This allows to recover the periodic setting in a simple and natural way. Of course, condition (1.74) is much more general than this, as we will see below. Let us also explain the intuitive meaning of (1.74). We already know that, for any bounded measurable function F, for any $k \in \mathbb{Z}$, and for almost every $x \in \mathbb{R}$, we have

$$\mathbb{E}\left(F(f(x + k, \omega))\right) = \mathbb{E}\left(F(f(x, \tau_k \omega))\right) = \mathbb{E}\left(F(f(x, \omega))\right),$$

where we have used (1.72). Thus, (1.74) implies that the law of the random variable $f(x, \omega)$ is a $\mathbb{Z}$-periodic function of x. It is actually possible to prove that if f is a random field, and if its law is periodic, then there exists a probability space $(\Omega, \mathcal{T}, \mathbb{P})$ and an ergodic group action τ such that f is stationary in this framework, in the sense of (1.74). This theorem is known as Kolmogorov's Theorem (see [Shi95, Theorem 3, p 163]). In other words, stationarity exactly means that $f(x, \omega)$, though it is not necessarily periodic, is "statistically" periodic, that is, periodic *in law*.

The setting we have just defined is called the *stationary ergodic setting*. In order to better understand this framework, we now prove that it includes as a particular case that of *i.i.d. random variables*. Indeed, let us consider a sequence $(X'_k)_{k \in \mathbb{Z}}$ of random variables, defined on a probability space that we denote $(\Omega', \mathcal{T}', \mathbb{P}')$. The reason for using this notation (with a prime) is pedagogical and will become clear below. We assume that all X'_k share the same law, and that they are independent of each other. We are going to realize this sequence as a stationary sequence on a probability space $(\Omega, \mathcal{T}, \mathbb{P})$, for a suitable ergodic group action τ that preserves the measure $\mathbb{P}$.

Let $\Omega = \mathbb{R}^{\mathbb{Z}}$ be the set of real-valued sequences indexed by $\mathbb{Z}$. We define the *product σ-field $\mathcal{T}$*, that is, the smallest σ-field that contains all the elements of the form $A = \prod_{k \in \mathbb{Z}} A_k$ where $A_k = \mathbb{R}$ except for a finite number of indices k. For such a value of k, A_k is a Borel set of $\mathbb{R}$. Next, we define the probability measure $\mathbb{P}$ as the image measure of $\mathbb{P}'$ by X', that is,

$$\mathbb{P}(A) = \prod_{k \in \mathbb{Z}} \mathbb{P}' \left(X'_k \in A_k \right),$$

which also reads as

$$d\,\mathbb{P}(\omega) = d\,\mathbb{P}' \circ \left(X' \right)^{-1} (\omega) = \prod_{k \in \mathbb{Z}} d\mathbb{P}' \left(X'_k \right)^{-1} (\omega_k).$$

Intuitively, for any $\omega' \in \Omega'$, we define the sequence ω by $\omega_k = X'_k(\omega')$, and we set the probability of the event ω to be $\mathbb{P}(\omega) = \mathbb{P}'(\omega')$. This uniquely defines the probability measure $\mathbb{P}$. The fact that $\mathbb{P}$ is indeed a probability measure is a consequence of the fact that the random variables X'_k are independent of each other. We next define the group action of $(\mathbb{Z}, +)$ over Ω. For $j \in \mathbb{Z}$, let $\tau_j \omega$ be the sequence $(\omega_{k+j})_{k \in \mathbb{Z}}$. This action is known as the *Bernoulli shift*. It preserves the probability $\mathbb{P}$, because

$$d\,\mathbb{P}(\tau_1 \omega) = \prod_{k \in \mathbb{Z}} d\,\mathbb{P}' \left(X'_k \right)^{-1} ((\tau_1 \omega)_k) = \prod_{k \in \mathbb{Z}} d\,\mathbb{P}' \left(X'_k \right)^{-1} (\omega_{k+1})$$

$$= \prod_{k \in \mathbb{Z}} d\,\mathbb{P}' \left(X'_{k+1} \right)^{-1} (\omega_{k+1})$$

since the X'_k are identically distributed

$$= \prod_{k \in \mathbb{Z}} d\,\mathbb{P}' \left(X'_k \right)^{-1} (\omega_k) = d\,\mathbb{P}(\omega)$$

It remains to prove that this action is ergodic, which is more delicate. We refer the reader to [Kre85, section 1.4] for a complete proof. We only give here the main ideas. Assume that $A \in \mathcal{T}$ is of the form $A = \{\omega \in \Omega; \omega_k \in A_k \subsetneq \mathbb{R}, 1 \le k \le N\}$. Not all sets $A \in \mathcal{T}$ are of this form. However, the idea is that it is sufficient to consider sets for which only a finite number of components A_k are not equal to $\mathbb{R}$ (here we simply took them to be the components k, $1 \le k \le N$). Assuming that such an A is stable under the action of τ, we infer that $A = \{\omega \in \Omega; \omega_k \in A_{k-1}, 2 \le k \le N+1\}$, and so on. As a consequence, the set $\{1, \ldots, N\}$ is invariant under translation, that is, it is either empty or equal to $\mathbb{Z}$. Put differently, either $A = \Omega$ or $A = \emptyset$. Our argument is of course not completely rigorous, but it gives a reliable idea of the actual proof, which is based on the following disjunction argument: any set A which has the above form (that is, a *cylinder*), satisfies that, for sufficiently large k, the sets A^c and $(\tau_{-k}A)^c$ have disjoint supports. Hence, $\mathbb{P}(A \cap \tau_{-k}A) = \mathbb{P}(A)\,\mathbb{P}(\tau_{-k}A)$. Since τ is measure-invariant, $\mathbb{P}(A)\,\mathbb{P}(\tau_{-k}A) = \mathbb{P}(A)^2$. Now, if A is translation invariant, $\tau_{-k}A = A$ and we get $\mathbb{P}(A) = \mathbb{P}(A)^2$, whence $\mathbb{P}(A) \in \{0, 1\}$. Such sets span all the σ-field, so we conclude by a density argument. This proof actually establishes a property which is stronger than ergodicity, but we do not wish to enter such details.

We next define X_k as the k-th coordinate on Ω, that is, $X_k(\omega)$ is the k-th coordinate of the sequence $\omega \in \Omega$. More intuitively, $X_k(\omega) = \omega_k = X'_k(\omega')$. By construction, the sequence $X_k(\omega)$ is stationary, since $X_k(\tau_1\omega) = \omega_{k+1} = X_{k+1}(\omega)$. As a conclusion, we have just interpreted an i.i.d. sequence as a stationary sequence, by a change of the probability space, with an adapted definition of the group action.

We now briefly introduce a variant of the above setting. This variant is called *continuous* stationarity setting, as opposed to the *discrete stationarity* we have just defined. The modification lies in the group considered for the action on Ω: instead of using the group $(\mathbb{Z}, +)$, we consider an action of $(\mathbb{R}, +)$. The fact that the action preserves the probability measure and that it is ergodic are now expressed in the following way, which is different from (1.72) and (1.73):

$$\forall y \in \mathbb{R}, \ \forall A \in \mathcal{T}, \quad \mathbb{P}(\tau_y A) = \mathbb{P}(A), \tag{1.75}$$

and

$$\forall A \in \mathcal{T}, \quad \left(\forall y \in \mathbb{R}, \ \tau_y A = A\right) \implies \left(\mathbb{P}(A) = 0 \text{ or } \mathbb{P}(A) = 1\right). \tag{1.76}$$

The *continuous* stationarity of a function is then defined as

$$f(x + y, \omega) = f(x, \tau_y\omega) \text{ for almost every } (x, y) \in \mathbb{R} \times \mathbb{R} \text{ and almost surely.} \tag{1.77}$$

instead of (1.74). Arguing as above for the case of discrete stationarity and periodicity in law, it is easy to see that this property expresses that the law of f does not depend on the spatial position $x \in \mathbb{R}$.

Throughout this textbook, we preferably use the notion of discrete stationarity, because this notion is better suited to formalize the perturbations of periodicity that we consider. The unperturbed periodic structure is encoded in the underlying periodic lattice given by the discrete group $(\mathbb{Z}^d, +)$ and the corresponding action τ. The periodic functions are those that do not depend on the variable ω.

Continuous stationarity is more delicate to manipulate in this perspective, as we will see in the next section.

1.6.3 The Periodic, Quasi-Periodic and Almost-Periodic Settings as Particular Cases of the Continuous Random Setting

In order to realize the periodic setting as a particular case of the continuous ergodic stationary setting, we are going to use a specific set Ω, along with an associated probability measure $\mathbb{P}$ and an action τ. The set Ω is defined as the (one-dimensional in the present case) torus $\mathbb{T} = [0, 1]_{per}$ (that is, the interval $[0, 1]$ where the boundaries $\{0\}$ and $\{1\}$ are identified with each other, or equivalently, the unit circle S^1). The probability measure is the Lebesgue measure on $\mathbb{T}$, and the action τ is the addition $\tau_y \omega = \omega + y$, for any $\omega \in \mathbb{T}$, and any $y \in \mathbb{R}$. One easily checks that (1.75) and (1.76) hold. The continuous stationarity property (1.77) is then equivalent to periodicity. Indeed, it implies that if $x' + \omega' = x + \omega \bmod(1)$ then $f(x', \omega') = f(x, \omega' + x' - x) = f(x, \omega)$ for any $x \in \mathbb{R}$, $x' \in \mathbb{R}$, $\omega \in \mathbb{T}$, $\omega' \in \mathbb{T}$. This allows to identify $f(x, \omega)$ to a $\mathbb{Z}$-periodic function of the variable $x + \omega$. (Should we be dealing with continuous functions, we could write this f as $f(0, x + \omega)$, where $x + \omega$ is understood as $x + \omega \bmod(1)$.) Conversely, a $\mathbb{Z}$-periodic function g may be seen as a continuous function by setting $f(x, \omega) = g(x + \omega)$ for all $\omega \in \mathbb{T}$, $x \in \mathbb{R}$.

A similar connection may be established in the quasi-periodic case. We fix the quasi-periods $T = (T_1, \ldots T_m)$, and assume that the associated frequencies $\dfrac{1}{T_i}$ are independent over $\mathbb{Z}$. We define $\Omega = [0, 1]^m$, and we let $\mathbb{P}$ be the Lebesgue measure: $d\mathbb{P}(\omega) = d\omega$. For the action of $\mathbb{R}$ on Ω, we set $\tau_x \omega = \omega + x\frac{1}{T}$, which is to be understood as the componentwise equality $(\tau_x \omega)_i = \omega_i + \frac{x}{T_i}$, for all $1 \leq i \leq m$. It is then clear that this action is ergodic, since the frequencies $\left(\dfrac{1}{T_1}, \ldots, \dfrac{1}{T_m} \right)$ are linearly independent over $\mathbb{Z}$.

It is also possible to identify the almost-periodic setting to a stationary ergodic case, but this is much more involved, and uses the concept of Bohr compactification of the real line $\mathbb{R}$. We refer to [Pan96] or [DS88a, DS88b] for a presentation of this subject. Let us only mention the main two theoretical ingredients used for the connection.

Theorem 1.3 ([DS88b], Theorem XI.2.2)   *There exists a topological compact abelian group* $(\mathbb{G}, +)$ *such that* $(\mathbb{R}, +)$ *is a dense subgroup of* $\mathbb{G}$, *and such that*

$$PP(\mathbb{R}) = \left\{ f_{|\mathbb{R}}, \quad f \in C^0(\mathbb{G}) \right\}.$$

Moreover, the map $f \mapsto f_{|\mathbb{R}}$ *is an algebra isomorphism from* $C^0(\mathbb{G})$ *to* $PP(\mathbb{R})$. *It is also an isometry for the uniform norms. This group* $\mathbb{G}$ *is called the* Bohr compactification *of* $\mathbb{R}$.

Theorem 1.4 ([Fol95], Theorem 2.10 [Haar Theorem])   *Let* $(\mathbb{G}, +)$ *be a locally compact topological group. Then there exists a non-negative locally finite measure* μ *on* $\mathbb{G}$ *which is invariant under the group law of* $\mathbb{G}$. *This measure is called the Haar measure of* $\mathbb{G}$. *This measure is unique up to multiplication by a constant. When* $\mathbb{G}$ *is compact, we normalize it so that* $\mu(\mathbb{G}) = 1$, *and it is unique.*

The above two results allow to define $\Omega = \mathbb{G}$, equipped with $\mathbb{P} = \mu$, the Haar measure, as a probability measure. The action τ is then defined by

$$\forall x \in \mathbb{R}, \quad \forall \omega \in \Omega, \quad \tau_x \omega = \omega + x.$$

It is well-defined according to Theorem 1.3. The Haar Theorem shows that the measure $\mathbb{P}$ is invariant under this action. For any $F \in C^0(\Omega)$, we define $f : \mathbb{R} \times \Omega \to \mathbb{R}$ by

$$\forall x \in \mathbb{R}, \quad \forall \omega \in \Omega, \quad f(x, \omega) = F(\tau_x \omega).$$

Such a function is of course stationary in the sense of (1.77). Moreover, by definition of Ω, the function $\overline{f} : x \mapsto f(x, 0)$ is almost periodic. It remains to prove that the action τ is ergodic, in the sense of (1.77). We will not prove this result, and refer for instance to [AF09, Theorem 2.7] for its proof.

It should be noted that, although $\mathbb{R}$ is dense in $\mathbb{G}$, it has zero-measure (in the sense of the Haar measure of $\mathbb{G}$). Indeed, since $\mathbb{R}$ is the disjoint union of intervals of the form $[k, k+1[$ for $k \in \mathbb{Z}$, and $\mathbb{P}$ is translation invariant, we have

$$\mathbb{P}(\mathbb{R}) = \sum_{k \in \mathbb{Z}} \mathbb{P}([k, k+1[) = \sum_{k \in \mathbb{Z}} \mathbb{P}(\tau_k[0, 1[) = \sum_{k \in \mathbb{Z}} \mathbb{P}([0, 1[),$$

Thus, since $\mathbb{P}$ is a probability measure, $\mathbb{P}([0, 1[) = 0$. Applying the above formula, we infer $\mathbb{P}(\mathbb{R}) = 0$.

We also point out that the average $\langle u \rangle$ of $u \in PP(\mathbb{R})$, which exists according to Proposition 1.4, satisfies

$$\langle u \rangle = \int_{\mathbb{G}} u(z) d\mathbb{P}(z),$$

where, with a slight abuse of notation, we have identified u with its extension as an element of $C^0(\mathbb{G})$, and $\mathbb{P}$ is the Haar measure of $\mathbb{G}$. This is a consequence of the fact that the Haar measure is unique, and of the fact that the linear form $u \longmapsto \langle u \rangle$ satisfies the properties defining the Haar measure.

Similarly, for any $p \in [1, +\infty[$, since $PP(\mathbb{R})$ is an algebra, $|u|^p \in PP(\mathbb{R})$, and

$$\langle |u|^p \rangle = \int_{\mathbb{G}} |u(z)|^p d\mathbb{P}(z).$$

The Hölder inequality allows to recover that the map $u \mapsto \langle u \rangle$ is continuous for the Besicovitch semi-norm B^p (see Definition 1.5). Hence, a density argument allows to prove that the Besicovitch space $B^p(\mathbb{R})$ (see Definition 1.6) is isomorphic to $L^p(\mathbb{G}, d\mathbb{P})$. For the proof of this result, we refer to [AF09, Proposition 2.2].

As a conclusion, we have identified a set Ω (the Bohr compactification $\mathbb{G}$ of $\mathbb{R}$), a probability measure (the Haar measure of $\mathbb{G}$) and an ergodic action τ, such that the space of almost-periodic functions is isomorphic to the set of (continuous) stationary functions. However, this construction is purely abstract, and, to our knowledge, does not allow any simplification in the associated homogenization theory.

1.6.4 Properties of Stationary Functions

We return to the (discrete) stationary ergodic setting introduced in Sect. 1.6.2. We have seen that it includes as a particular case, on the one hand, the case of i.i.d. sequences $\{X_k\}_{k \in \mathbb{Z}}$, and on the other hand, the periodic setting. As a consequence, we expect suitable properties of averages of stationary functions (hence of rescaled such functions). This is indeed the case.

Theorem 1.5 (Ergodic Theorem, Discrete Case, [Shi95, chap V, §3]) *Let* $(\Omega, \mathcal{T}, \mathbb{P})$ *be a probability space equipped with an action τ, which preserves the probability measure $\mathbb{P}$ and is ergodic, in the sense of* (1.72)–(1.73). *Assume that* $g(x, \omega) \in L^\infty(\mathbb{R}, L^1(\Omega))$ *is stationary as in* (1.74). *Then we have*

$$\lim_{N \to +\infty} \frac{1}{2N+1} \sum_{k=-N}^{N} g(x, \tau_k \omega) = \mathbb{E}(g(x, .)), \qquad (1.78)$$

for almost all x, almost surely in ω. In addition, denoting by $Q = [0, 1]$,

$$g\left(\frac{x}{\varepsilon}, \omega\right) \xrightarrow{\varepsilon \to 0} \mathbb{E}\left(\int_Q g(y, .) \, dy\right), \qquad (1.79)$$

in L^∞-weak-$\star$ with respect to x, almost surely in ω.

We note that, assuming that the function $g(x, \omega)$ depends on x only through its integer part $[x]$, and setting $X_k(\omega) = g(k, \omega)$, the convergence (1.78) gives, for the stationary sequence $X_k(\omega)$,

$$\lim_{N \to +\infty} \frac{1}{2N+1} \sum_{k=-N}^{N} X_k(\omega) = \mathbb{E}(X_0).$$

If in particular the sequence $(X_k)_{k \in \mathbb{Z}}$ is i.i.d., we recover the strong law of large numbers (1.70).

The proof of Theorem 1.5 goes beyond the scope of the present textbook, so we will not include it here. We refer the interested reader to [Shi95, pages 409–411], for instance. We only bear in mind that, as it is the case for the particular case of i.i.d. sequences, it is rather natural that a result such as (1.78) holds.

We also note that the convergence (1.79) can be formally deduced from (1.78). Indeed, the L^∞-weak-$\star$ convergence is equivalent to convergence when testing against any characteristic function of an interval. For simplicity, we assume that this interval is $[-1, 1]$ and that $\varepsilon = 1/N$, for some integer $N \in \mathbb{N}$. We thus have

$$\int_{-1}^{1} g\left(\frac{x}{\varepsilon}, \omega\right) dx = \frac{1}{N} \int_{-N}^{N} g(y, \omega) dy = \frac{1}{N} \sum_{k=-N}^{N-1} \int_{k}^{k+1} g(y, \omega) dy$$

$$= \frac{1}{N} \sum_{j=-N}^{N-1} G(0, \tau_k \omega),$$

where $G(x, \omega) = \int_{x}^{x+1} g(y, \omega) dy$. This function G is stationary, hence we may apply (1.78), and find

$$\int_{-1}^{1} g\left(\frac{x}{\varepsilon}, \omega\right) dx \xrightarrow{\varepsilon \to 0} 2\mathbb{E}\left(\int_{Q} g(y, .) \, dy\right),$$

that is, (1.79) tested with the function $\mathbf{1}_{[-1,1]}$. Should the interval be different from $[-1, 1]$ and/or $1/\varepsilon$ not be an integer, a similar proof is possible, but more technical.

A similar result exists in the case of continuous stationary functions:

Theorem 1.6 (Ergodic Theorem, Continuous Case, [Bec81]) *Let $(\Omega, \mathcal{T}, \mathbb{P})$ be a probability space equipped with an action τ, which preserves the measure $\mathbb{P}$ and is ergodic, in the sense of (1.75)–(1.76). Assume that $g(x, \omega) \in L^\infty(\mathbb{R}, L^1(\Omega))$ is stationary, in the continuous sense (1.77). Then we have*

$$\lim_{R \to +\infty} \frac{1}{2R} \int_{-R}^{R} g(x, \tau_y \omega) \, dy = \mathbb{E}\left(g(x, .)\right) = \mathbb{E}(g), \quad \text{almost surely.} \quad (1.80)$$

and

$$g\left(\frac{x}{\varepsilon}, \omega\right) \xrightarrow{\varepsilon \to 0} \mathbb{E}(g), \quad in \ L^{\infty} - weak - \star, \quad almost \ surely. \quad (1.81)$$

Let us mention that, "as usual", we have stated these results in dimension 1, but they are valid in any dimension. In order to adapt them, one only needs to replace the indices $k \in \mathbb{Z}$ by multi-indices $k \in \mathbb{Z}^d$, integrals over $[-R, R]$ by integrals over $[-R, R]^d$, and the interval $Q = [0, 1]$ by the d-dimensional cube $Q = [0, 1]^d$.

We next turn to our original question regarding the convergence of the solution u^{ε} of (1.1), this time in the random setting. Clearly, Theorem 1.5 answers it. Indeed, let us consider the equation

$$-a\left(\frac{x}{\varepsilon}, \omega\right) u^{\varepsilon}(x, \omega) = f(x), \quad (1.82)$$

where the coefficient $a(x, \omega)$ is stationary in the discrete sense (1.74). Here, we note that we have assumed not only that the right-hand side f does not depend on ε, but also that it is deterministic. The solution reads

$$u^{\varepsilon}(x, \omega) = -\frac{1}{a}\left(\frac{x}{\varepsilon}, \omega\right) f(x). \quad (1.83)$$

Recalling that $\dfrac{1}{a}$ is stationary, the convergence (1.79) allows to prove that, for instance if $f \in L^2(\mathcal{D})$,

$$u^{\varepsilon}(x, \omega) \xrightarrow{\varepsilon \to 0} u^*(x) = -\mathbb{E}\left(\int_Q \frac{1}{a}(x, .) \, dx\right) f(x), \quad in \ L^2 - weak,$$

almost surely. It follows that the limit u^* can be interpreted as the solution to the limit Eq. (1.6) with the coefficient

$$\bar{a} = \frac{1}{\mathbb{E}\left(\int_Q \frac{1}{a}(x, .) \, dx\right)}. \quad (1.84)$$

This expression should be compared with (1.15). As we have stated in Sect. 1.2, it is not the *statistics* of a, but that of $\dfrac{1}{a}$, which determines the homogenized limit.

Having proved the convergence, a natural question is again that of the *convergence rate*. This question is the matter of property (P1) in the periodic case. In the i.i.d. case, it is answered by the central limit Theorem. Since the i.i.d. case is a special case of the stationary setting, we expect that, under suitable assumptions, such a convergence rate can be established in general. However, the stationary setting contains very different cases, such as the periodic and almost periodic cases. Thus, we suspect that a large variety of convergence rates may occur, and that the question may be very technical. This is indeed the case. We refer the reader to the notion of *mixing conditions*, developed for instance in [Shi95, Chapter 5], which formalize the variety of phenomena that may happen. We only give here an example that is *genuinely random*. By this we mean that it is not a re-interpretation of a deterministic case.

We consider the stationary function

$$a(x, \omega) = 1 + \sum_{k \in \mathbb{Z}} X_k(\omega) \mathbf{1}_Q(x - k), \tag{1.85}$$

where $\mathbf{1}_Q$ is the characteristic function of the unit interval $Q = [0, 1]$, and X_k is an i.i.d. sequence. We assume that X_k is non-negative and bounded, almost surely. By construction, the coefficient a is bounded away from zero ($a \geq 1$ almost surely) and from above. It is stationary in the discrete sense, since, as we know, an i.i.d. sequence can be interpreted as a stationary sequence. We have

$$a(x + 1, \omega) = 1 + \sum_{k \in \mathbb{Z}} X_k(\omega) \mathbf{1}_Q(x + 1 - k)$$

$$= 1 + \sum_{k \in \mathbb{Z}} X_{k+1}(\omega) \mathbf{1}_Q(x - k) = a(x, \tau_1 \omega).$$

For this function (1.85), we have

$$\frac{1}{a}(x, \omega) = \sum_{k \in \mathbb{Z}} Y_k(\omega) \mathbf{1}_Q(x - k), \quad \text{where} \quad Y_k = (1 + X_k(\omega))^{-1} \tag{1.86}$$

is a bounded i.i.d. sequence. For any fixed $\varphi \in L^1(\mathbb{R})$, we have, applying (1.79) to $g = \frac{1}{a}$,

$$\int_{\mathbb{R}} \varphi(x) \frac{1}{a} \left(\frac{x}{\varepsilon}, \omega \right) dx \xrightarrow[\varepsilon \to 0]{} \int_Q \mathbb{E} \left(\frac{1}{a} \right) \int_{\mathbb{R}} \varphi = \mathbb{E}(Y_k) \int_{\mathbb{R}} \varphi = \mathbb{E}(Y_0) \int_{\mathbb{R}} \varphi, \tag{1.87}$$

and the question of proving a convergence rate for (1.79) amounts, for a fixed φ, to prove a convergence rate for (1.87). To this end, we assume that $\varepsilon = \frac{1}{N}$ where N is an integer. We then have

$$\int_{\mathbb{R}} \varphi(x) \frac{1}{a} \left(\frac{x}{\varepsilon}, \omega \right) dx = \sum_{k \in \mathbb{Z}} Y_k(\omega) \int_{\mathbb{R}} \varphi(x) \mathbf{1}_Q(Nx - k) dx$$

$$= \sum_{k \in \mathbb{Z}} Y_k(\omega) \int_{k/N}^{(k+1)/N} \varphi(x) dx.$$

Assuming (actually this may be done without loss of generality) that the support of φ is included into $[0, 1]$, the question reduces to determining the rate in the following convergence

$$\sum_{k=0}^{N-1} Y_k(\omega) \int_{k/N}^{(k+1)/N} \varphi(x)\,dx \stackrel{N\to+\infty}{\longrightarrow} \mathbb{E}(Y_0) \int_0^1 \varphi(x)\,dx, \qquad (1.88)$$

for a fixed function φ. If $\varphi = \mathbf{1}_Q$, this result is given by the central limit Theorem we recalled above, and is $\dfrac{1}{\sqrt{N}}$. It follows that the convergence in (1.83) for this particular case, that is,

$$\int_0^1 \frac{1}{a}\left(\frac{x}{\varepsilon},\omega\right) dx \stackrel{\varepsilon\to 0}{\longrightarrow} \mathbb{E}\left(\int_Q \frac{1}{a}(x,.)\,dx\right) \qquad (1.89)$$

occurs at rate $\sqrt{\varepsilon}$. This is sufficient to outline the difference with the periodic case, for which we have obtained (see property (P1)) a rate ε in (1.16). Let us see this in more details.

Taking $\varphi \neq \mathbf{1}_Q$ does not modify the convergence rate $\sqrt{\varepsilon}$, provided φ is smooth enough (or integrable enough). We are going to prove it in the case $\varphi \in C^0([0,1])$. Then φ is bounded, hence $\varphi \in L^4([0,1])$. We are going to *briefly* reproduce the proof of the strong law of large numbers and of the central limit Theorem in the special case of L^4 i.i.d. random variables. Changing Y_k into $Y_k - \mathbb{E}(Y_k)$ does not change the fact that the sequence is i.i.d. Thus, we may assume without loss of generality that $\mathbb{E}(Y_k) = 0$. We then consider the quantity $\mathbb{E}\left(\left[\sum_{k=0}^{N-1} h(k,N)\,Y_k(\omega)\right]^4\right)$ where $h(k,N) = \int_{k/N}^{(k+1)/N} \varphi(x)\,dx$. We aim at finding an upper bound for this quantity. Using that $\mathbb{E}(Y_k) = 0$, we infer that the only terms that do not vanish in the expansion of $\mathbb{E}\left(\left[\sum_{k=0}^{N-1} h(k,N)\,Y_k(\omega)\right]^4\right)$ are those involving *even* powers of Y_k. We thus have:

$$\mathbb{E}\left(\left[\sum_{k=0}^{N-1} h(k,N)\,Y_k(\omega)\right]^4\right) = \sum_{k_1}\sum_{k_2} h(k_1,N)^2\, h(k_2,N)^2\, \mathbb{E}\left((Y_{k_1})^2\,(Y_{k_2})^2\right).$$

Using the Hölder inequality, together with the fact that $\varphi \in L^4([0,1])$, we obtain

$$\sum_{k=0}^{N-1} (h(k,N))^4 = \sum_{k=0}^{N-1}\left(\int_{k/N}^{(k+1)/N} \varphi(x)\,dx\right)^4 \leq \frac{1}{N^3} \sum_{k=0}^{N-1} \int_{k/N}^{(k+1)/N} \varphi^4(x)\,dx$$

$$= O\left(\frac{1}{N^3}\right).$$

A similar argument gives $\sum_{k=0}^{N-1}(h(k,N))^2 = O\left(\dfrac{1}{N}\right)$ since, *a fortiori*, $\varphi \in$ $L^2([0,1])$. Using the fact that Y_k is bounded, we get

$$\mathbb{E}\left(\left[\sum_{k=0}^{N-1} h(k,N)\,Y_k(\omega)\right]^4\right) = O\left(\frac{1}{N^2}\right),$$

from which we infer that the expectation value of $\sum_{N\in\mathbb{N}}\left(\left[\sum_{k=0}^{N-1} h(k,N)\,Y_k(\omega)\right]^4\right)$ is finite. Hence, the associated function is almost surely finite. The corresponding series is almost surely convergent. Thus, its general term tends to zero as $N \to +\infty$. This gives a meaning to the convergence (1.88), which generalizes (in some sense) the strong law of large numbers. On the other hand, the convergence rate is obtained by a similar proof to that of the central limit Theorem in simple cases. Using the same notation, we consider, for $u \in \mathbb{R}$, the function $\Phi_N(u) =$ $\mathbb{E}\left(\exp\left[i\,u\,\sqrt{N}\sum_{k=0}^{N-1} h(k,N)\,Y_k(\omega)\right]\right)$. Since the Y_k are independent of each other, we have $\Phi_N(u) = \prod_{k=0}^{N-1}\mathbb{E}\left(\exp\left[i\,u\,\sqrt{N}\,h(k,N)\,Y_k(\omega)\right]\right)$. Using an expansion for small values of u, the kth factor reads as

$$\mathbb{E}\left(\exp\left[i\,u\,\sqrt{N}\,h(k,N)\,Y_k(\omega)\right]\right) = 1 - \frac{N}{2}u^2\,(h(k,N))^2\,\mathbb{E}\left((Y_k)^2\right)$$
$$+ O\left(N^{3/2}\,(h(k,N))^3\right),$$

where the first term of the expansion vanishes since $\mathbb{E}(Y_k) = 0$. We thus have, at least formally, since we take the logarithm of a complex number (the rigorous proof requires a more precise argument)

$$\log \Phi_N(u) = \sum_{k=0}^{N-1}\log\left(1 - \frac{N}{2}u^2\,(h(k,N))^2\,\mathbb{E}\left((Y_0)^2\right) + O\left(N^{3/2}\,(h(k,N))^3\right)\right)$$

$$= -\frac{N}{2}u^2\,\mathbb{E}\left((Y_0)^2\right)\sum_{k=0}^{N-1}(h(k,N))^2 + O\left(N^{3/2}\sum_{k=0}^{N-1}(h(k,N))^3\right)$$

$$= -\frac{N}{2}u^2\,\mathbb{E}\left((Y_0)^2\right)\sum_{k=0}^{N-1}\left(\int_{k/N}^{(k+1)/N}\varphi(x)\,dx\right)^2 + O\left(\frac{1}{\sqrt{N}}\right).$$

Here, we have dealt with the second term of the right-hand side using similar estimates as above and the fact that $\varphi \in C^0([0, 1])$, hence $\varphi \in L^3([0, 1])$. Since φ is continuous, the limit of the right-hand side is equal to $-\dfrac{1}{2} u^2 \, \mathbb{E}\left((Y_0)^2\right) \int_0^1 \varphi^2$.

Indeed, $\displaystyle\int_{k/N}^{(k+1)/N} \varphi(x)\, dx \approx \frac{1}{N} \varphi\left(\frac{k}{N}\right)$ and the Riemann sum $\dfrac{1}{N} \displaystyle\sum_{k=0}^{N-1} \varphi^2\left(\frac{k}{N}\right)$

converges to the integral $\displaystyle\int_0^1 \varphi^2$. Finally, we obtain that $\Phi_N(u)$ converges to

$$\exp\left(-\frac{1}{2} u^2 \, \mathbb{E}\left((Y_0)^2\right) \int_0^1 \varphi^2\right).$$

Put differently,

$$\sqrt{N} \left(\sum_{k=0}^{N-1} Y_k(\omega) \int_{k/N}^{(k+1)/N} \varphi(x)\, dx - \mathbb{E}(Y_0) \int_0^1 \varphi\right)$$

$$\xrightarrow[N \to +\infty]{\mathcal{L}} \mathcal{N}\left(0, \frac{1}{\mathrm{Var}(Y_0) \int_0^1 \varphi^2}\right), \qquad (1.90)$$

where $\xrightarrow{\mathcal{L}}$ denotes the convergence in law, $\mathcal{N}(0, \sigma)$ is the centered normal law of variance σ^2, and $\mathrm{Var}(Y_0) = \mathbb{E}\left[(Y_0 - \mathbb{E}(Y_0))^2\right]$ is the variance of the random variable Y_0. The property (1.90) gives a mathematical meaning to the fact that, intuitively, the convergence (1.88) occurs, in law, up to a Gaussian, with the rate $\dfrac{1}{\sqrt{N}}$, as in the classical case of the central limit Theorem. This implies the convergence rate $\sqrt{\varepsilon}$, as announced.

We have just proved that, in (1.79), the convergence rate is $\sqrt{\varepsilon}$, at least for a stationary function built from an i.i.d. sequence. This is very different from the rate we have obtained in the periodic case, that is, ε, for (1.16). However, the rate $\sqrt{\varepsilon}$ is in some sense the best possible for "genuinely" random cases. It is due to the short *correlation length* in the i.i.d. setting. More generally, the convergence rate in (1.79) is determined by the decay rate of the function

$$R(\tau) = \mathbb{E}\left[\left(g(x, \omega) - \mathbb{E}\left(g(x, \omega)\right)\right)\left(g(x + \tau, \omega) - \mathbb{E}\left(g(x + \tau, \omega)\right)\right)\right],$$

called the *auto covariance function*. In the case (1.85), one easily sees that $R(\tau)$ has compact support since, up to adding 1 to the function f and assuming that $\mathbb{E}(Y_k) = 0$, we have, for the random part g, $\mathbb{E}\left(Y_k(\omega)\, Y_{k+1}(\omega)\right) = 0$, which is the "best" possible decay. When

$$R(\tau) \overset{|\tau| \to +\infty}{\sim} \frac{1}{|\tau|^{\alpha}},$$

where $\alpha > 0$, the convergence rate becomes slower as α decreases. More precisely, the rate is $\sqrt{\varepsilon}$ while $\alpha \geq 1$, then becomes $\varepsilon^{\alpha/2}$ for $\alpha \leq 1$. This is a somewhat paradoxical situation: the periodic setting is a case of an "infinite" correlation length. Nevertheless, the convergence rate suddenly improves, going from $\varepsilon^{\alpha/2}$ for small values or α, to ε for the periodic case where $\alpha = 0$. Although surprising, this result is not contradictory: the periodic case is one case (among many others) for which the correlation length is infinite. It may exist other cases for which the convergence rate is not so good, let alone much worse.

We close this section with some comments that further highlight the difference between stationary functions and periodic functions. We successively study properties (P2)–(P5). Of course, we have already seen in Sect. 1.6.3 that these properties (except (P5)) do not hold in the quasi-periodic or almost periodic case, hence we know that they are in general false for the stationary setting. We however want to study them using the general formalism of the random case.

We start with property (P2), that is, the equivalence between the L^1_{unif} norm and the average norm. This equivalence would imply

$$\exists C > 0, \quad \mathbb{E}\left(\int_0^1 |u|\right) \geq C \sup_{x \in \mathbb{R}} \int_x^{x+1} |u|, \tag{1.91}$$

for any stationary u (in the sense of (1.74)). It is worth mentioning that the right-hand of this inequality, although it looks random, is in fact deterministic. Indeed, if we denote by $Z(\omega) = \sup_{x \in \mathbb{R}} \int_x^{x+1} |u(y, \omega)| dy$, the stationarity of u and a change of variables give, for all $k \in \mathbb{Z}$,

$$Z(\tau_k \omega) = \sup_{x \in \mathbb{R}} \int_x^{x+1} |u(y, \tau_k \omega)| dy = \sup_{x \in \mathbb{R}} \int_x^{x+1} |u(y + k, \omega)| dy$$

$$= \sup_{x \in \mathbb{R}} \int_{x+k}^{x+1+k} |u(y, \omega)| dy = \sup_{x \in \mathbb{R}} \int_x^{x+1} |u(y, \omega)| dy = Z(\omega).$$

Since τ is an ergodic action, we infer that Z is deterministic.

In order to contradict inequality (1.91), we introduce the following function:

$$u(x, \omega) = \sum_{j \in \mathbb{Z}} Y_j(\omega) \varphi(x - j), \tag{1.92}$$

where $(Y_j)_{j \in \mathbb{Z}}$ is an i.i.d sequence, $Y_j \geq 0$ almost surely. The function $\varphi \in L^1$ is non-negative, and has compact support in the interval $]0, 1[$. Thus, $|u| = u$, and inequality (1.91) reads as

$$\mathbb{E}(Y_0) \int_0^1 \varphi \geq C \sup_{x \in \mathbb{R}} \left(\sum_{j \in \mathbb{Z}} Y_j \int_x^{x+1} \varphi(y - j)dy \right)$$

$$\geq C \sup_{x \in \mathbb{Z}} \left(\sum_{j \in \mathbb{Z}} Y_j \int_x^{x+1} \varphi(y - j)dy \right),$$

where we have used that the upper bound over $\mathbb{R}$ is always larger than the upper bound over $\mathbb{Z}$. In the sum, all terms vanish except $j = x$. This is due to the fact that the support of φ is included in $]0, 1[$. Thus, the inequality becomes

$$\mathbb{E}(Y_0) \int_0^1 \varphi \geq C \sup_{j \in \mathbb{Z}} Y_j \int_0^1 \varphi.$$

We choose $\varphi \neq 0$, so that $\int \varphi$ does not vanish. We thus have $\mathbb{E}(Y_0) \geq C \|Y_0\|_{L^\infty(\Omega)}$, for any non-negative bounded random variable Y_0, which cannot be. We have reached a contradiction.

Next, we study property (P3) about primitives of zero-average stationary functions. We consider a stationary sequence $X_k(\omega)$, $k \in \mathbb{Z}$, such that $\mathbb{E}(X_k) = 0$. We define $Y_N(\omega) = \frac{1}{N} \sum_{k=1}^N X_k(\omega)$, for $N \in \mathbb{N}$. It is clear that $Y_N(\tau_1 \omega) =$

$\frac{1}{N} \sum_{k=1}^N X_{k+1}(\omega) = Y_N(\omega) - \frac{X_1(\omega)}{N} + \frac{X_{N+1}(\omega)}{N}$. Hence, unless the sequence X_k is constant, Y_N is not stationary. The best we know is that its expectation vanishes, and that Y_N converges almost surely to 0 as $N \to +\infty$. This is a consequence of the ergodic theorem and of the fact that $\mathbb{E}(X_k) = 0$. Returning to the case of stationary functions, this remark tells us that integrals of the form $F(x, \omega) = \int_0^x f(t, \omega) \, dt$ satisfy

$$\lim_{|x| \to +\infty} \frac{1}{|x|} F(x, \omega) = 0, \quad \text{almost surely,} \quad (1.93)$$

as far as $f(x, \omega)$ is a stationary function which expectation vanishes. To prove this fact, one may apply the ergodic theorem, or equivalently apply the above argument to the stationary sequence $X_k(\omega) = \int_k^{k+1} f(t, \omega) \, dt$ (then, it is necessary to bound integrals of the type $\int_{[x]}^{x} |f|(t, \omega)\,dt$, and we leave such details to the reader). A similar argument is also valid in the continuous stationary setting. As a consequence, property (P3) of Sect. 1.2, which is true in the periodic case, does not hold in general in the stationary setting. The above example proves that *given a stationary function with zero average, its primitives satisfy* (1.93) *only, and are not stationary in general.*

To further study this question, we now provide an even more universal counterexample than that of the i.i.d. case above. For this purpose, we use the *continuous* stationary setting. Let $X_t(\omega)$, $t \in \mathbb{R}$, be a Gaussian stationary process, such that $\mathbb{E}(X_t) = 0$. Such a process is uniquely defined by its covariance K, which reads as

$$\forall (t, s) \in \mathbb{R}^2, \quad K(t, s) = \mathbb{E}(X_t X_s).$$

One easily proves, using that X_t is stationary, that K depends on $|t - s|$ only. We thus note $K(t, s) = K(|t - s|)$, and assume that

$$\int_0^t K(s)\,ds \xrightarrow[t \to +\infty]{} +\infty. \tag{1.94}$$

We define the stationary function $f(t, \omega) = X_t(\omega)$. Its average clearly vanishes. Let g be a primitive of f, that is, a function such that

$$g(t, \omega) - g(s, \omega) = \int_s^t f(u, \omega)\,du,$$

almost everywhere in t, almost surely in ω. Let us assume that g is stationary and in $L^2(\Omega)$. We multiply the above equality by $f(t, \omega)$, compute the expectation of the product, and find

$$\mathbb{E}(g(t, \omega) f(t, \omega)) - \mathbb{E}(g(s, \omega) f(t, \omega)) = \int_s^t K(|t - u|)\,du.$$

Since f and g are stationary (and in L^2), the first term does not depend on t. The Cauchy-Schwarz inequality, together with the fact that X_t is bounded in L^2, uniformly with respect to t, and the fact that g is stationary and is in L^2 imply that the second term is bounded independently of s and t. Thus, the left-hand side is bounded as $t \to +\infty$. This contradicts (1.94). Of course, assuming that $g \in L^q$, $q > 1$, instead of $g \in L^2$ leads to a contradiction too. One only needs to replace the Cauchy-Schwarz inequality by the Hölder inequality in the above argument.

This counter-example is more general than the i.i.d. case, but the reader may note that for some stationary ergodic settings, a Gaussian process has only bounded primitives. This is of course the case for the periodic setting. Indeed, in such a case, the kernel K is periodic, since

$$K(t - s) = \mathbb{E}(X_t X_s) = \int_0^1 f(t + \omega)f(s + \omega)d\omega = \int_0^1 f(t - s + \omega)f(\omega)d\omega,$$

and f is periodic. Since K has zero average (because f has, since $\mathbb{E}(X_t) = 0$) and $K \in L^1_{loc}$ because its law is Gaussian, all its primitives are periodic, hence bounded. Thus (1.94) does not hold.

We turn to the study of property (P4), that is, the Poincaré-Wirtinger inequality, and prove that it does not hold using the random formalism. Should it be true, it would read as

$$\exists C > 0, \quad \mathbb{E}\left(\int_0^1 |\nabla u|^2\right) \geq C\mathbb{E}\left(\int_0^1 |u|^2\right), \tag{1.95}$$

for any stationary (in the sense of (1.74)) function u with zero average. We come back to the example of the form (1.92) above, that is,

$$u(x, \omega) = \sum_{j \in \mathbb{Z}} Y_j(\omega)\varphi(x - j),$$

where the sequence $(Y_j)_{j \in \mathbb{N}}$ are i.i.d. and centered. The function φ is assumed to be differentiable, with zero mean, and compact support. Note that this support is not necessarily included in $[0, 1]$. Inequality (1.95) applied to this function u would give, since $\mathbb{E}(Y_j Y_i) = 0$ whenever $i \neq j$,

$$\sum_{j \in \mathbb{Z}} \mathbb{E}(Y_j^2) \int_0^1 \varphi'(x - j)^2 dx \geq C \sum_{j \in \mathbb{Z}} \mathbb{E}(Y_j^2) \int_0^1 \varphi(x - j)^2 dx.$$

Since $\mathbb{E}(Y_j^2)$ does not depend on j, and assuming that Y_j does not identically vanish, we would find

$$\underbrace{\sum_{j \in \mathbb{Z}} \int_0^1 \varphi'(x - j)^2 dx}_{= \int_\mathbb{R} (\varphi')^2} \geq C \underbrace{\sum_{j \in \mathbb{Z}} \int_0^1 \varphi(x - j)^2 dx}_{= \int_\mathbb{R} \varphi^2}.$$

In other words, the Poincaré-Wirtinger would hold on the real line $\mathbb{R}$, which is known to be false.

We finally study property (P5). We assume that $f \in L^\infty(\mathbb{R}, L^1(\Omega))$ is stationary, and we aim at proving that

$$\frac{f(x, \omega)}{1 + |x|} \xrightarrow{|x| \to +\infty} 0, \quad \text{almost surely.} \tag{1.96}$$

This obviously holds true. Actually, another property, which is more adapted to the sequel, may be stated as follows: assume that f', the derivative of f with respect to x, is stationary, belongs to $L^\infty(\mathbb{R}, L^1(\Omega))$, and has zero average. Then, f satisfies (1.96). This property is a simple consequence of the ergodic theorem (1.79). Indeed, when applied to f', it implies that, for all $R > 0$,

$$\frac{f(R, \omega)}{R} - \frac{f(0, \omega)}{R} = \frac{1}{R} \int_0^R f'(x, \omega) dx$$

$$= \int_0^1 f'(Rx, \omega) dx \xrightarrow{R \to +\infty} \mathbb{E} \int_Q f' = 0,$$

almost surely. This also holds if we replace R by $-R$, and we have proved (1.96).

Chapter 2
Homogenization in Dimension 1

Abstract In Chap. 1, we have removed the differential operators in (1), making our study very specific. Our study of homogenization theory really starts now.

Keywords One-dimensional homogenization · Corrector function · Defect · Transport equation · Advection-diffusion equation · Hamilton-Jacobi equation · Random setting

In Chap. 1, we have removed the differential operators in (1), making our study very specific. Our study of homogenization theory really starts now. We consider (1) in dimension 1, that is,

$$
\begin{cases}
-\dfrac{d}{dx}\left(a_\varepsilon(x)\,\dfrac{d}{dx}u^\varepsilon(x)\right) = f(x) \text{ in } [0,1], \\[2mm]
u^\varepsilon(0) = u^\varepsilon(1) = 0,
\end{cases}
\tag{2.1}
$$

where the coefficient a_ε is assumed to be bounded from above and bounded away from 0, for almost every $x \in [0,1]$, uniformly with respect to ε:

$$
0 < \mu \le a_\varepsilon(x) \le M < +\infty.
\tag{2.2}
$$

Many experts consider the one-dimensional case as too peculiar, if not irrelevant. This is due to the many simplifications occurring in this specific setting. Let us mention some of them.

- In dimensions $d \ge 2$, among vector-valued functions, only those that are curl-free are gradients. In other words, such a function $\mathbf{f}$ should satisfy the *Cauchy relations* $\partial_{x_j}\mathbf{f}_i = \partial_{x_i}\mathbf{f}_j$, $1 \le i, j \le d$. On the contrary, in dimension 1, any function is the derivative of another function. As a consequence, solving a divergence-form equation is very peculiar in this dimension.
- Another miraculous property of the one-dimensional setting is related to the *Green function* $G(x, y)$ (sometimes also called the *elementary solution*) of the

Laplace operator. Such a function G is defined as the solution of $-\Delta_x G(x, y) = \delta_y(x)$ in $\mathbb{R}^d$. As $|x - y|$ tends to infinity, the asymptotics of G are similar to that of the Green function $G_\varepsilon(x, y)$ associated to Eq. (1), that allows to solve (1) via the representation formula $u^\varepsilon(x) = \int_{\mathcal{D}} G_\varepsilon(x, y) f(y) \, dy$. In dimension 1, the Green function G satisfies, up to normalization, $G(x, y) = -\dfrac{|x - y|}{2}$, hence it does not vanish at infinity. It does not either in dimension $d = 2$, since in such a case it behaves like $G(x, y) = -\log|x - y|$. However, in dimensions $d \geq 3$, it decays like $\dfrac{1}{|x - y|^{d-2}}$ as $|x - y| \to +\infty$. This somewhat "*unnatural*" behaviour at infinity is the source of very specific properties of solutions to elliptic equations.

A third point (actually related to the second one above) is the fact that functional analysis in dimension 1 is very peculiar. As an example (among others), let us mention that functions in $H^1(]0, 1[)$ are continuous (they are actually Hölder continuous). Here again, this property is specific to dimension 1. In the present textbook, we study Eq. (1), for which the natural energy space happens to be $H^1(\mathcal{D})$, that is, on the one hand, the space in which the variational formulation of (1) is posed (see (2.10) and Sect. 3.1), and, on the other hand, the space in which the energetic interpretation of (1) is possible (we will return to this in Chap. 6). Thus, the reader understands that, since all solutions we deal with are continuous, the situation is particularly simple in dimension 1. Further, the convergence results we establish are artificially improved (indeed, an H^1 convergence implies, in particular, a convergence in the uniform norm).

Along with these properties, some others that are more subtle will show up in due course. All the above facts may mislead the beginner into hasty (and incorrect) conclusions. Hence the natural mistrust for one-dimensional homogenization.

Nevertheless, the case of dimension 1 is a valuable test bench, owing to the drastic simplifications of the mathematical arguments involved. Since "*He who can do more can do less*", we need at least to understand the one-dimensional case. This will prove a sound basis to address all the possible structures (periodic or not) of the coefficient a in (1).

In the present chapter, we give a complete study of Eq. (1) in dimension 1, that is, (2.1), for various coefficients a. As we will see, the work we did in Chap. 1 will prove useful in this respect. This will allow us to discover unexpected new features of the problem. We will additionally consider the *numerical approximation* of the problem. We will also present some surprising facts, with cases in which things do not go as expected. We will mention what can be generalized to higher dimensions, and what cannot.

Compared to Chap. 1, the main difference is that *we restore the differential operators*. We had removed them until now, and we will thus need for the first time some elements of *functional analysis*.

2.1 Our First One-Dimensional Cases

2.1.1 Solution and Limit of the Elliptic Equation

As announced, we restore the differential operators and consider Eq. (2.1). This equation is to be understood in the sense of distributions, that is,

$$\forall \varphi \in C_c^\infty(]0, 1[), \quad \int_0^1 a_\varepsilon \frac{du^\varepsilon}{dx} \frac{d\varphi}{dx} dx = \int_0^1 f\varphi \, dx.$$

It is clear from this formulation that it is not necessary that a_ε be differentiable. Moreover, a simple application of the Lax-Milgram Lemma in the Hilbert space $H_0^1(]0, 1[)$ allows to prove that this variational formulation has a unique solution u^ε in $H_0^1(]0, 1[)$, for any $f \in L^2(]0, 1[)$ (note that $f \in H^{-1}(]0, 1[)$ would be sufficient to proceed). Thus, u^ε is continuous, since, in dimension one, $H^1(]0, 1[) \subset C^0(]0, 1[)$ with continuous injection. Hence, the boundary conditions have a classical meaning. One easily computes u^ε by integrating the equation twice. Defining $F(x) = \int_0^x f(y) \, dy$, it reads

$$u^\varepsilon(x) = -\int_0^x (a_\varepsilon(y))^{-1} (c_\varepsilon + F(y)) \, dy, \tag{2.3}$$

where the constant c_ε is

$$c_\varepsilon = -\frac{\int_0^1 (a_\varepsilon(y))^{-1} F(y) \, dy}{\int_0^1 (a_\varepsilon(y))^{-1} \, dy}. \tag{2.4}$$

As in Chap. 1, the bounds on a_ε allow us to conclude that, at least for an extraction ε', we have the weak L^∞-$\star$ convergence (1.3). We infer that $c_{\varepsilon'}$ converges in $\mathbb{R}$ to

$$c_* = -\frac{\int_0^1 (\overline{a}(y))^{-1} F(y) \, dy}{\int_0^1 (\overline{a}(y))^{-1} \, dy}. \tag{2.5}$$

Thus, $c_{\varepsilon'} + F$ strongly converges to $c_* + F$, and $u^{\varepsilon'}$ converges for all $x \in [0, 1]$ to

$$u^*(x) = -\int_0^x (\bar{a}(y))^{-1} (c_* + F(y)) \, dy. \tag{2.6}$$

This limit is called the *homogenized solution*. It is actually the solution to

$$\begin{cases} -\dfrac{d}{dx}\left(\bar{a}(x)\dfrac{d}{dx}u^*(x)\right) = f(x) \text{ in } [0, 1], \\[2mm] u^*(0) = u^*(1) = 0. \end{cases} \tag{2.7}$$

The latter equation is called the *homogenized equation*. We recall that $(\bar{a}(x))^{-1}$ is the weak limit of a convergent subsequence of $(a_\varepsilon(y))^{-1}$. In full generality, we are unable to identify the coefficient $\bar{a}(x)$. *A fortiori*, we cannot prove that the whole sequence u^ε converges to the solution u^* of Eq. (2.7). We have at least been able to identify a limit equation, up to extraction of a subsequence. We also note that this limit equation has the same form as the original one (2.1).

As in Chap. 1, in order to prove more, we will need to assume some additional properties about the coefficient a_ε. This is what we are going to do in Sect. 2.1.3. For now, let us investigate three different points.

The first point concerns the boundary conditions in (2.1). We may temporarily replace the *homogeneous Dirichlet boundary conditions* by *homogeneous Neumann boundary conditions*, thereby changing (2.1) into

$$\begin{cases} -\dfrac{d}{dx}\left(a_\varepsilon(x)\dfrac{d}{dx}u^\varepsilon(x)\right) = f(x) \text{ in } [0, 1] \\[2mm] (u^\varepsilon)'(0) = (u^\varepsilon)'(1) = 0. \end{cases} \tag{2.8}$$

In order to have existence of a solution, we need to impose the condition $\displaystyle\int_0^1 f = 0$ (a simple way to see this is to integrate the first line of (2.8) on the interval $[0, 1]$ and use the boundary conditions). The solution is then defined up to an additive constant. In order to make it unique, we impose, for instance, $u^\varepsilon(0) = 0$. Solving explicitly (2.8) may be done as easily as for (2.1), and we obtain for $u^\varepsilon(x)$ the expression (2.3), with, this time, $c_\varepsilon = 0$. Actually, the computation is simpler than in the Dirichlet case, and is similar to those of Chap. 1 without the differential operators. We obtain:

$$u^\varepsilon(x) = -\int_0^x (a_\varepsilon(y))^{-1} F(y) \, dy. \tag{2.9}$$

One easily checks that this is a solution to (2.8) under the condition $F(1) = \int_0^1 f = 0$. Thus, the solution to (2.8) has similar properties as that of (2.1). In particular, up to extraction of a subsequence, the limit equation is (2.7), except for the boundary conditions, which have changed from Dirichlet to Neumann $(u^*)'(0) = (u^*)'(1) = 0$. In other words, the homogenized operator (that appears in the left-hand side of (2.7)) does not depend on the boundary conditions. We will return to this point in Chap. 3, Lemma 3.1 with more generality.

The second point is related to *inverse problems*. We have identified the limit Eq. (2.7) of (2.1), and have observed that the coefficient $\bar{a}$ is the weak limit of (a subsequence of) the inverse of the original coefficient a_ε. This is good news. However, the fact that the coefficient is defined as a weak limit is in some sense "bad" news. Indeed, although $(a_\varepsilon)^{-1}$ may be far from its weak limit $(\bar{a})^{-1}$ (think about the case of a periodic oscillating function—as we have seen in Chap. 1—and its weak limit, which is its average; the function may be far away from its average, as Fig. 2.1 shows), the solution u^ε is close to the homogenized solution u^*. As a consequence, in an equation like (2.1) (thus, more generally, like (1)) it is possible to have two coefficients that are far from each other, but that give solutions close to each other. In other words, the inverse problem of finding the coefficient from the solutions of (1) is unstable, so far as we allow arbitrarily small oscillations! Such an inverse problem is ubiquitous in the engineering sciences. This instability is clearly explained in the article [Lio78], to which the interested reader is referred.

The third point we want to make concerns numerical methods and is the matter of the next section.

Fig. 2.1 Example of a piecewise constant function $g\left(\frac{x}{\varepsilon}\right)$ that converges weakly to its average $\langle g \rangle$, but satisfies $|g_\varepsilon(x) - \langle g \rangle| = 1$, for all x

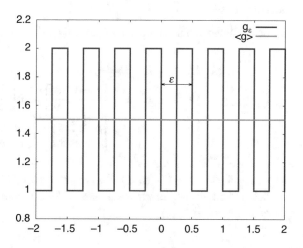

2.1.2 What About Numerics?

We give in this section a quick introduction to elementary notions about numerical methods. These notions will be complemented in Chap. 5. First, we recall that the solution u^ε to Eq. (2.1) is the unique solution to the following *variational formulation: Find* $u^\varepsilon \in H_0^1(]0, 1[)$ *such that, for any function* $v \in H_0^1(]0, 1[)$,

$$\int_0^1 a_\varepsilon(x)\, (u^\varepsilon)'(x)\, v'(x)\, dx = \int_0^1 f(x)\, v(x)\, dx. \qquad (2.10)$$

The fact that (2.10) is equivalent to (2.1) is classical and we will not prove it here. It holds for $f \in L^2(]0, 1[)$. It is even true in the case $f \in H^{-1}(]0, 1[)$. In the latter case, the right-hand side of (2.10) is to be understood as the duality bracket $\langle f, v \rangle_{H^{-1}, H_0^1}$. For the details, we refer, for instance, to [Bre11, Section 8.4, Example 2, page 223] in dimension 1, and to [Bre11, Section 9.5, Example 3, page 294] in higher dimensions. We will return to this in Sect. 3.1. This variational formulation will prove useful in computing a numerical approximation of the solution u^ε, using a so-called *finite element method*. Should the reader be unfamiliar with such methods, he may consult [Qua18, Chapter 4]. The method we are going to describe here is called the $\mathbb{P}^1$ *finite element method*. In such a method, the unknown function is approximated by a piecewise affine function (a polynomial of degree 1 on each element). To this end, we first introduce the mesh of the domain. Here, this domain is [0, 1], and by mesh we mean a partition of [0, 1] into N sub-intervals. For the sake of simplicity, we choose these intervals of equal length $H = \dfrac{1}{N}$. These intervals thus read as $\left[\dfrac{i}{N}, \dfrac{i+1}{N}\right]$ for $i = 0, \ldots, N-1$. We define the space V_H as the finite-dimensional vector space spanned by the continuous functions χ_i:

$$\chi_i(x) = \begin{cases} 1 & \text{if } x = \dfrac{i}{N}, \\ \text{affine} & \text{if } x \in \left[\dfrac{i-1}{N}, \dfrac{i}{N}\right] \text{ and } \left[\dfrac{i}{N}, \dfrac{i+1}{N}\right], \\ 0 & \text{if } x \in \left[0, \dfrac{i-1}{N}\right] \cup \left[\dfrac{i+1}{N}, 1\right], \end{cases} \qquad (2.11)$$

where $i = 1, \ldots, N-1$. Clearly, V_H is a subspace of $H_0^1(]0, 1[)$. On this subspace, we define the so-called *discrete variational formulation: Find* $u_H^\varepsilon \in V_H$ *such that, for all function* $v_H \in V_H$,

$$\int_0^1 a_\varepsilon(x)\, (u_H^\varepsilon)'(x)\, (v_H)'(x)\, dx = \int_0^1 f(x)\, v_H(x)\, dx. \qquad (2.12)$$

The vector space V_H, indexed by $H > 0$, is usually called an *internal approximation* of $V = H_0^1(]0, 1[)$. The method consisting in replacing $V = H_0^1(]0, 1[)$ by V_H in the variational formulation (2.10) is called the *Galerkin approximation*. We will come back to these notions in Chap. 5. Any $v_H \in V_H$ reads $v_H(x) = \sum_{j=1}^{N-1} (v_H)_j \chi_j(x)$ for some scalar coefficients $(v_H)_j$. The numerical solution is sought under this form, that is, $u_H^\varepsilon(x) = \sum_{j=1}^{N-1} (u_H^\varepsilon)_j \chi_j(x)$. With this notation, (2.12) becomes

$$\sum_{j=1}^{N-1} \left[\int_0^1 a_\varepsilon(x) (\chi_i)'(x) (\chi_j)'(x) \, dx \right] (u_H^\varepsilon)_j = \int_0^1 f(x) \chi_i(x) \, dx, \qquad (2.13)$$

for all $i = 1, \dots, N-1$. This can be written as the following linear system

$$[A_\varepsilon] \left[u_H^\varepsilon \right] = [f_H], \qquad (2.14)$$

where the matrix $[A_\varepsilon]$ and the column vectors $\left[u_H^\varepsilon \right]$ et $[f_H]$ are defined, respectively, by

$$[A_\varepsilon]_{ij} = \int_0^1 a_\varepsilon(x) (\chi_i)'(x) (\chi_j)'(x) \, dx,$$

$$\left[u_H^\varepsilon \right]_j = (u_H^\varepsilon)_j, \qquad [f_H]_j = \int_0^1 f(x) \chi_j(x) \, dx, \qquad (2.15)$$

for $1 \le i, j \le N-1$. We are now in position to start our argument.

We have seen in Chap. 1 that, forgetting about the differential operators, the solution of (2.1) becomes $u^\varepsilon(x) = (a_\varepsilon(x))^{-1} f(x)$. Hence, at least formally, we expect that the solution has oscillations of the same type as $a_\varepsilon(x)$. Actually, this is corroborated by the explicit formula (2.3). Assume that the coefficient $a_\varepsilon(x)$ oscillates at scale ε (think about the prototypical case $a_{\mathrm{per}}\left(\dfrac{x}{\varepsilon}\right)$), and that this scale is small. We wish to capture these oscillations with linear combinations of piecewise affine functions of the form (2.11). This naturally leads to imposing that H is smaller than ε, as, for instance, $H = \dfrac{\varepsilon}{10}$. The linear system we hence need to solve, that is, (2.14), is of size $O\left(\dfrac{1}{\varepsilon}\right)$, which is large. If $\varepsilon = 10^{-3}$, solving such a system is still doable. But this is in dimension 1. Intuitively, in dimension d, the system is of size $O\left(\dfrac{1}{\varepsilon^d}\right)$. This is a prohibitively large cost for solving (2.14).

So far, so good. Let us imagine that we choose to ignore the above considerations about the size of H, and use a value of H larger (or even much larger) than ε. The reason could be that we cannot afford such a computational cost imposed by small values of H, or simply that we want to see for ourselves what happens if H is small, but remains much larger than ε. Mathematically, this may be formalized by fixing H, and letting $\varepsilon \to 0$. The interest of the notion of limit is precisely to understand a situation in which a parameter is "much smaller" than another parameter. Up to extracting a subsequence (we do not make it explicit for the sake of clarity), the terms i, j of the matrix A_ε satisfy the following convergence:

$$[A_\varepsilon]_{ij} = \int_0^1 a_\varepsilon(x)\,(\chi_i)'(x)\,(\chi_j)'(x)\,dx$$

$$\xrightarrow{\varepsilon \to 0} \int_0^1 \bar{b}(x)\,(\chi_i)'(x)\,(\chi_j)'(x)\,dx =: \left[\bar{B}\right]_{ij}, \qquad (2.16)$$

where the function $\bar{b}$ is defined by

$$a_\varepsilon(x) \longrightarrow \bar{b}(x) \quad \text{in} \quad L^\infty - \text{weak} - \star. \qquad (2.17)$$

Hence, the column vector $\left[u_H^\varepsilon\right]$ solution to (2.14) converges to the vector $[\overline{u_H}]$, solution to

$$\left[\bar{B}\right][\overline{u_H}] = [f_H]. \qquad (2.18)$$

Using (2.16), we infer that (2.18) is the linear system associated with the discrete variational formulation: *Find $u_H^\varepsilon \in V_H$ such that, for any function $v_H \in V_H$,*

$$\int_0^1 \bar{b}(x)\,(\overline{u_H})'(x)\,(v_H)'(x)\,dx = \int_0^1 f(x)\,v_H(x)\,dx, \qquad (2.19)$$

which is the approximation, using the *same* $\mathbb{P}_1$ finite element method, of

$$\begin{cases} -\dfrac{d}{dx}\left(\bar{b}(x)\,\dfrac{d}{dx}\bar{u}(x)\right) = f(x) \text{ in } [0,1] \\[4mm] u^*(0) = u^*(1) = 0. \end{cases} \qquad (2.20)$$

We understand that, by choosing H much larger than ε (actually, this corresponds to forgetting about the oscillatory nature of the problem), we have designed a numerical method that cannot be expected to produce a valid approximation of the exact solution. Indeed, for small values of ε, our discrete solution u_H^ε is close to the solution $\bar{u}$ to (2.20), whereas the exact solution u^ε is close to the homogenized solution u^* to (2.7). These two solutions $\bar{u}$ and u^* are unrelated, because the weak limit $\bar{b}$ defined by (2.17) is not equal to the weak limit $\bar{a}$ defined by (1.3). Recall that the weak limit of the inverse of a sequence is not the inverse of its weak limit, except in trivial cases, as we have seen in Chap. 1. As a conclusion, let us say that the approximation we have built is *incorrect*. It is not even *approximately* correct, it is *exactly incorrect*! Since $\overline{a} \neq \overline{b}$, the error is $O(1)$ as H tends to 0. The considerations we have just developed are reminiscent of those of [BO00].

At this stage, the reader having basic notions of numerical analysis of the methods of discretization of partial differential equations, and therefore already familiar with finite element methods, can be legitimately confused. Finite element methods are known to be very efficient when applied to elliptic equations. So what happens here? In order to explain this quantitatively, let us return to a classical result in the numerical analysis of finite element methods. This result is called an *a priori* error estimate, and is based on the celebrated *Céa's Lemma*..

Subtracting the variational formulations (2.10) and (2.12), we have, for any function $v_H \in V_H$,

$$\int_0^1 a_\varepsilon(x)\,(u^\varepsilon - u_H^\varepsilon)'(x)\,(v_H)'(x)\,dx = 0,$$

that is,

$$\int_0^1 a_\varepsilon(x)\,\left|(u^\varepsilon - u_H^\varepsilon)'(x)\right|^2 dx = \int_0^1 a_\varepsilon(x)\,(u^\varepsilon - u_H^\varepsilon)'(x)\,(u^\varepsilon - v_H)'(x)\,dx.$$

For the left-hand side, we use the fact that a_ε is bounded away from 0. For the right-hand side, we use that it is bounded from above. Applying the Cauchy-Schwarz inequality, we find

$$\left\|(u^\varepsilon - u_H^\varepsilon)'\right\|_{L^2(]0,1[)} \leq C \left\|(u^\varepsilon - v_H)'\right\|_{L^2(]0,1[)}.$$

This is valid for any $v_H \in V_H$, and the constant C does not depend on ε nor on H. It only depends on the bounds on a_ε. Therefore,

$$\left\|(u^\varepsilon - u_H^\varepsilon)'\right\|_{L^2(]0,1[)} \leq C \inf_{v_H \in V_H} \left\|(u^\varepsilon - v_H)'\right\|_{L^2(]0,1[)}. \tag{2.21}$$

This inequality is the famous Céa's Lemma. Intuitively, it states the following: the *numerical error* (difference between the exact solution u^ε and the approximated solution u_H^ε) is bounded by the *approximation error* (difference between the exact solution u^ε and its best approximation in the discretization space V_H). So far, this space V_H can be any subspace of $H_0^1(]0, 1[)$. Thus, we have no information on the approximation error. We are now going to use the specific form of V_H. This will allow us to derive an estimation of the approximation error. Here, V_H is a space of piecewise affine functions. In such a setting, we have

$$\inf_{v_H \in V_H} \|v - v_H\|_{H^1(]0,1[)} \leq C\,H\,\|v''\|_{L^2(]0,1[)}, \tag{2.22}$$

for any function $v \in H^2(]0, 1[)$, where the constant C does not depend on v nor on H. Assuming that the coefficient a_ε is differentiable, we know that u^ε is in H^2. Hence applying (2.22) to $v = u^\varepsilon$, we get

$$\left\| (u^\varepsilon - u_H^\varepsilon)' \right\|_{L^2(]0,1[)} \leq C H \left\| (u^\varepsilon)'' \right\|_{L^2(]0,1[)}. \tag{2.23}$$

Now we understand! The explicit expression (2.3) of u^ε indeed shows that the second derivative $(u^\varepsilon)''$ involves the first derivative of a_ε ... And the derivative of an oscillating function is very large. For instance, if $a_\varepsilon(x) = a\left(\dfrac{x}{\varepsilon}\right)$, the derivative reads $(a_\varepsilon)'(x) = \dfrac{1}{\varepsilon} a'\left(\dfrac{x}{\varepsilon}\right)$. Then estimate (2.23) becomes

$$\left\| (u^\varepsilon - u_H^\varepsilon)' \right\|_{L^2(]0,1[)} \leq C \frac{H}{\varepsilon}, \tag{2.24}$$

where the constant C may be different from above, but is still independent of H and ε. This proves that the method does converge as $H \to 0$, which is well-known. But the estimate is very poor, so far as the quotient $\dfrac{H}{\varepsilon}$ is not small! Everything becomes clear.

As we have just shown, when designing numerical methods for equations of the type (1), we will have two choices: either using a very fine mesh, which may prove (very) costly, or design an adapted approximation method, since a "generic" method such as a standard finite element method is not efficient in such a case. We will come back to this subject in Chap. 5.

Let us illustrate estimate (2.24) by a numerical example. We define, on the interval $]0, 1[$, the problem (2.1) where the coefficient a_ε is defined by

$$a_\varepsilon(x) = a_{\text{per}}\left(\frac{x}{\varepsilon}\right), \qquad a_{\text{per}}(x) = \begin{cases} 1 & \text{if } x \in \left]0, \frac{1}{2}\right[, \\ 10 & \text{if } x \in \left]\frac{1}{2}, 1\right[, \end{cases}$$

and a_{per} is periodic of period 1. Thus, $a^* = \frac{20}{11} \approx 1,818$ and the coefficient $\overline{b}$ in (2.20) is equal to $\overline{b} = \frac{11}{2} = 5,5$. As we see on Fig. 2.2, when H is of moderate size (that is, not too small), the solution converges to the solution to (2.20), which is far from the homogenized solution. On the contrary, when $H \ll \varepsilon$, the solution does converge to the homogenized solution. Still for this example, we have plotted, in Fig. 2.3, for $\varepsilon = 0.1$ and $H = 0.01$, on the one hand, the numerical solution and the homogenized solution, and on the other hand, their derivatives. Even if u^* is close to u^ε, $(u^*)'$ is not a good approximation of $(u^\varepsilon)'$. We will study this point in Sect. 2.2 below (see (2.30)). We will show that, if we add a well suited "correction" to $(u^*)'$ (see (2.34) and (2.37)), then we get an accurate approximation of $(u^\varepsilon)'$.

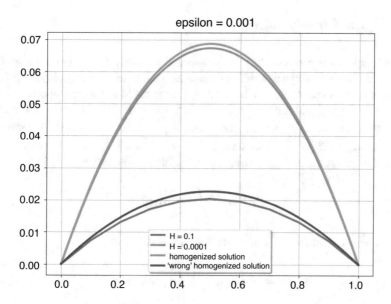

Fig. 2.2 Examples of numerical simulations for $\varepsilon = 10^{-3}$, with different values of the mesh size H. The green curve is the homogenized solution u^*. The red one is the solution $\bar{u}$ to the homogenized equation, where a^* is replaced by $\langle a \rangle$. As far as $H \gg \varepsilon$, even if H is small, the solution u_H^ε (blue curve, corresponding to $H = 10^{-1}$) is close to $\bar{u}$. On the contrary, if $H \ll \varepsilon$, the solution u_H^ε (orange curve, $H = 10^{-4}$) converges to u^*

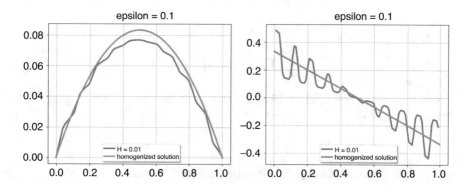

Fig. 2.3 Example of a numerical simulation with $\varepsilon = 0.1$ and $H = 0.01$: on the left-hand side the numerical solution and the homogenized solution. On the right-hand side their derivatives. Even if the solution u^ε and the homogenized solution u^* are close to each other, their derivatives are not

2.1.3   The Periodic Case

As in Chap. 1, we now are going to assume that the coefficient $a_\varepsilon(x)$ has a specific *structure*. This will allow us to learn some more about the weak limit $\bar{a}$, hence about the limit of the solution u^ε. The structure we are going to use is the fact that a_ε is the rescaling of a periodic function a_{per}: it has the form (1.7), with $a = a_{per}$. In this case, we know from Chap. 1 that the whole sequence $(a_\varepsilon(x))^{-1}$ weakly converges to the average $\langle (a_{per})^{-1} \rangle$. This limit is thus equal to $(\bar{a})^{-1}$, which is constant and explicit in this context. As a consequence, the limit of the sequence c_ε is

$$c_* = - \int_0^1 F(y)\, dy, \tag{2.25}$$

and the whole sequence u^ε converges, for all $x \in [0, 1]$, to

$$u^*(x) = - \left\langle (a_{per})^{-1} \right\rangle \int_0^x (c_* + F(y))\, dy. \tag{2.26}$$

This function is solution to the homogenized equation

$$\begin{cases} -a^* \dfrac{d^2}{dx^2} u^*(x) = f(x) \text{ in } [0, 1] \\[2mm] u^*(0) = u^*(1) = 0 \end{cases} \tag{2.27}$$

where

$$\left(a^* \right)^{-1} = \left\langle (a_{per})^{-1} \right\rangle, \tag{2.28}$$

which is a special case of (2.7). Hence, in this particular case, the limit equation is known and explicit.

2.2   The Quality of Approximation: The Corrector

Let us return to the general case. For the sake of clarity, we replace the sequence ε' by ε. We want to know in which sense we can prove that the solution u^ε (satisfying (2.3)) converges to the homogenized solution (2.6). It is actually easier to deal with the first derivatives of these functions. The difference $u^\varepsilon - u^*$ is called the *residual* (some authors call it the *corrector*, but this term is misleading since it could be mistaken with the corrector w defined below by (2.59)–(2.60); this is why we prefer not to use the latter terminology). The derivative of the residual reads as

$$(u^\varepsilon)'(x) - (u^*)'(x)$$
$$= -\left((a_\varepsilon)^{-1}(x) - (\overline{a})^{-1}(x)\right)(c_* + F(x)) + (a_\varepsilon)^{-1}(x)(c_* - c_\varepsilon). \qquad (2.29)$$

Given the bounds on a_ε and the convergence of c_ε to c_*, it is clear that the second term of the right-hand side tends to 0 in L^∞. On the contrary, the first term vanishes in L^∞-weak-$\star$ only. Actually, there is no reason for this sequence to converge strongly in any Lebesgue space, since the convergence of $(a_\varepsilon)^{-1}$ to $(\overline{a})^{-1}$ is not strong. Integrating this equality, we deduce that u^ε converges strongly to u^* in (at least) $L^2(\mathcal{D})$, but its derivative converges only weakly.

We now aim at understanding how we can *improve* this convergence. In particular, we would like to obtain a *strong* convergence of u^ε in $H^1(\mathcal{D})$. In order to do so, we need to remove the first term of the right-hand side of (2.29). As we have just seen, this term is indeed responsible for the lack of strong convergence. We proceed in the periodic case, but the argument may be adapted to more general settings (and actually it *will* be adapted in Sect. 3.3 of the next Chapter).

In the periodic setting, the difference (2.29) writes (this time the whole sequence converges, so we do not need to use an extracted sequence ε')

$$(u^\varepsilon)'(x) - (u^*)'(x) = -\left(a_{per}\left(\frac{x}{\varepsilon}\right)^{-1} - \langle(a_{per})^{-1}\rangle\right)\left(-\int_0^1 F + F(x)\right)$$
$$+ \left(a_{per}\left(\frac{x}{\varepsilon}\right)\right)^{-1}\left(-\int_0^1 F - c_\varepsilon\right). \qquad (2.30)$$

We now introduce the solution w_{per} of the following equation (this solution is defined up to an additive constant that we may fix by assuming that the integral of $w_{per}(0) = 0$):

$$\begin{cases} -\dfrac{d}{dy}\left(a_{per}(y)\left(1 + \dfrac{d}{dy}w_{per}(y)\right)\right) = 0, & \text{in } [0, 1], \\ w_{per} \text{ periodic} & \text{of period } 1. \end{cases} \qquad (2.31)$$

It is possible to use a general argument to prove that such a solution exists and is unique. Actually, we will do this in higher dimensions in Chap. 3, Sect. 3.4. In dimension 1, the solution may be computed explicitly. A simple computation indeed gives

$$w_{per}(y) = -y + \frac{1}{\langle(a_{per})^{-1}\rangle}\int_0^y (a_{per})^{-1}. \qquad (2.32)$$

Its derivative reads as

$$w'_{\text{per}}(y) = -1 + \frac{1}{\left\langle (a_{\text{per}})^{-1} \right\rangle} \left(a_{\text{per}}(y) \right)^{-1}. \tag{2.33}$$

The uniqueness is straightforward in this one-dimensional context: taking two solutions $w_{\text{per},1}$ and $w_{\text{per},2}$, we write the equation satisfied by their difference $v_{\text{per}} = w_{\text{per},1} - w_{\text{per},2}$, that is, $-\dfrac{d}{dy}\left(a_{\text{per}}(y) \dfrac{d}{dy} v_{\text{per}}(y) \right) = 0$. Solving it explicitly, we infer that v_{per} is constant. Recalling that $v_{\text{per}}(0) = 0$, this constant must be 0. However, this uniqueness is actually not needed for our present considerations. It will prove useful in Chap. 5, when we study numerical methods for computing w_{per}. An error estimate for a given numerical method is indeed often an extension of the uniqueness result for the exact solution of the corresponding problem.

If, instead of using $(u^*)'$ as an approximation of $(u^\varepsilon)'$, we replace it by $(u^*)'(x)\left(1 + w'_{\text{per}}\left(\frac{x}{\varepsilon}\right)\right)$, then the difference (2.30) becomes

$$(u^\varepsilon)'(x) - (u^*)'(x)\left(1 + w'_{\text{per}}\left(\frac{x}{\varepsilon}\right)\right) = \left(a_{\text{per}}\left(\frac{x}{\varepsilon}\right)\right)^{-1}\left(-\int_0^1 F - c_\varepsilon\right), \tag{2.34}$$

which now converges *strongly* to 0.

We have *modulated* the function $(u^*)'$ with oscillations of the function $w'_{\text{per}}\left(\frac{x}{\varepsilon}\right)$. These oscillations are closely related to those of the coefficients a_ε. Doing so, we have eliminated the weakly converging term in (2.30). This is the reason why Eq. (2.31) is called the *corrector equation*, and its solution w_{per} the *corrector*. We have defined and used it here in a periodic one-dimensional case, but we will soon extend this to general settings.

We also note the following point: integrating (2.34) suggests to consider $u^{\varepsilon,1}(x) = u^*(x) + \varepsilon (u^*)'(x) w_{\text{per}}\left(\frac{x}{\varepsilon}\right)$, which is an approximation of u^ε under the form of an expansion in powers of ε.

Another question that may be asked regards the accuracy of this expansion. To which order in ε does $u^{\varepsilon,1}$ approximate u^ε in H^1? Computing

$$(u^{\varepsilon,1})'(x) = (u^*)'(x)\left(1 + w'_{\text{per}}\left(\frac{x}{\varepsilon}\right)\right) + \varepsilon (u^*)''(x) w_{\text{per}}\left(\frac{x}{\varepsilon}\right),$$

(we assume that $u^* \in H^2$, which is true, as we have seen above, as soon as f is sufficiently regular) and subtracting it to (2.34), we obtain

$$(u^\varepsilon)'(x) - (u^{\varepsilon,1})'(x)$$

$$= \left(a_{\text{per}}\left(\frac{x}{\varepsilon}\right)\right)^{-1}\left(-c_\varepsilon - \int_0^1 F\right) - \varepsilon (u^*)''(x) w_{\text{per}}\left(\frac{x}{\varepsilon}\right). \tag{2.35}$$

In the present periodic setting, (2.4) implies

$$- c_\varepsilon - \int_0^1 F = \frac{\int_0^1 (a_\varepsilon(y))^{-1} F(y)\, dy}{\int_0^1 (a_\varepsilon(y))^{-1}\, dy} - \int_0^1 F.$$

Assuming, for instance, that F is of class C^1, as we have seen in Sect. 1.2, we obtain

$$- c_\varepsilon - \int_0^1 F = O(\varepsilon), \tag{2.36}$$

On the other hand, since the function w_{per} reads as (2.32), we note that w_{per} is bounded on $\mathbb{R}$. The calculation is the same as the one that leads to (1.12) in the proof of Proposition 1.1 in Chap. 1. Inserting this bound and (2.36) into (2.35), we find

$$\left\| u^\varepsilon - u^{\varepsilon,1} \right\|_{H^1(]0,1[)} = O(\varepsilon), \tag{2.37}$$

in the present periodic setting.

A few remarks are in order.

First, we would like to give the intuition about the origin of Eq. (2.31). Its rigorous derivation will be given in Chap. 3, Sect. 3.1.1, in a general multidimensional setting. We have observed that $(u^\varepsilon)'$ converges only weakly to $(u^*)'$ in $L^2(\mathcal{D})$. This is due to the oscillations of $a_\varepsilon(x) = a\left(\dfrac{x}{\varepsilon}\right)$, as we see in expression (2.4). In order to better capture these oscillations, we need to *modulate* $(u^*)'(x)$ with functions having oscillations of the same nature. Formally, we replace $(u^*)'(x)$ by $(u^*)'(x) (\overline{w})'\left(\dfrac{x}{\varepsilon}\right)$, where the mean value of the function $(\overline{w})'$ is assumed to be one, in order to recover $(u^*)'(x)$ in average (that is, in the limit $\varepsilon \to 0$). We apply the operator of Eq. (2.1) to this product, and assume that the variables x and $y = \dfrac{x}{\varepsilon}$ are independent (this is true because they live at different scales in the limit $\varepsilon \to 0$), we obtain $-\dfrac{d}{dy}\left(a_{\text{per}}(y)\left(1 + \dfrac{d}{dy} w_{\text{per}}(y)\right)\right)$, where we have set $(\overline{w})' = 1 + (w_{\text{per}})'$. This gives (2.31). This equation is thus, in some sense, a *zoom* of Eq. (2.1) at the small scale ε. A rigorous version of this hand-waving argument will be given in Chap. 3.

Actually, the above remark also allows to understand why the additive constant up to which w_{per} is defined is irrelevant. Indeed, such a constant, once inserted into $u^{\varepsilon,1}$, does not modify the above convergence, up to order ε.

We also point out that the fact that $w_{\mathrm{per}} \in L^\infty(\mathbb{R})$ is not essential to prove that the difference (2.35) vanishes in the limit $\varepsilon \to 0$. It is actually sufficient that w_{per} be *strictly sublinear at infinity*, that is,

$$\lim_{|x| \to +\infty} \frac{w_{\mathrm{per}}(x)}{1 + |x|} = 0. \tag{2.38}$$

Indeed, (2.38) implies that $\varepsilon\, w\left(\dfrac{x}{\varepsilon}\right)$ vanishes, hence so does (2.35). On the other hand, the fact that w_{per} is bounded allows to *quantify* the convergence rate, and prove (2.37).

The convergence rate (2.37) is actually unexpected, and is due to two different things: the periodic setting, and the fact that the dimension is 1. In such a case, no boundary effects are present. Indeed, the approximation $u^{\varepsilon,1} = u^*(x) + \varepsilon\,(u^*)'(x)\,w_{\mathrm{per}}\left(\dfrac{x}{\varepsilon}\right)$ does not satisfy the homogeneous Dirichlet boundary conditions, contrary to u^ε. In higher dimensions, this implies a poor approximation near the boundary, which slows down the convergence. We will return to this point in the next chapters.

Of course, the rate ε is optimal. To realize this, one may choose specific (non-trivial) functions f and a, and check that the convergence rate is *exactly* equal to ε.

2.3 One-Dimensional Defects

We now ask the following question. We have performed in the preceding section a "correction" in the very specific, periodic case. Does this correction extend to the general setting of Sect. 2.1.1? To this end, we would like to solve, for $a_\varepsilon(x) = a(x/\varepsilon)$,

$$\begin{cases} -\dfrac{d}{dy}\left(a(y)\left(1 + \dfrac{d}{dy}\,w(y)\right)\right) = 0 \text{ in } \mathbb{R}, \\[4mm] \lim_{|y| \to +\infty} \dfrac{w(y)}{1 + |y|} = 0. \end{cases} \tag{2.39}$$

Here, we need to solve the equation on the whole real line (at least because we no more have any periodic cell for a), and we thus add the condition of strict sublinearity at infinity (2.38).

In the present one-dimensional setting, finding the solutions to the first line of (2.39) is easy. These solutions are *exactly* the functions w the derivatives of which read as $w'(y) = -1 + c\,a^{-1}(y)$, where $c \in \mathbb{R}$ is an arbitrary constant. Thus, the solutions read as $w(y) = d - y + c \displaystyle\int_0^y a^{-1}(t)\,dt$ where the constant d is irrelevant. So far, so good. Now, condition (2.38) amounts to prove that there exists a particular constant c for which the solution w is strictly sublinear at infinity. This may not be possible given an arbitrary function a satisfying the bounds (2.2). As an example, one may build a smooth function a such that $1 \leq a \leq 2$ and

$$a(t) = \begin{cases} 1 & \text{if } t < -1, \\ 2 & \text{if } t > 1. \end{cases}$$

In such a case, the strict sublinearity of w at $+\infty$ implies $\dfrac{c}{2} - 1 = 0$, and the strict sublinearity at $-\infty$ implies $c - 1 = 0$.

In Chap. 3 (Proposition 3.4), we will explain how, without the existence of a strictly sublinear corrector, it is still possible to improve weak convergence into strong convergence. In general, we are not able, however, to prove a convergence rate such as (2.37) without using a corrector. For now, we wish to study cases that are much more general than the periodic setting, and which allow to prove existence of a solution to (2.39). In order to define these nonperiodic cases, we use the nonperiodic settings introduced in Chap. 1. Owing to the fact that we work in dimension 1, the existence of a solution to (2.39) reduces to proving that there exists a suitable constant c in

$$w'(y) = -1 + c\, a^{-1}(y),$$

so that the strict sublinearity at infinity holds true. This is related to the specific properties of the considered setting: existence of an average for a^{-1}, and strict sublinearity for functions the derivatives of which have zero mean value. We will take advantage of the material we have accumulated in Chap. 1. In higher dimensions, solving the general corrector equation corresponding to (2.39) will prove, however, to be a real challenge, as we will see in the next chapters.

The periodic case we have studied in Sect. 2.1.3, although academic, is a suitable setting in which the general results of Sect. 2.1.1 can be made precise. We however know from Chap. 1 that a large variety of settings may be considered. These fall into two categories:

- *local* defects to periodicity: such defects change only slightly the (periodic) setting at infinity. See Sect. 1.4 of Chap. 1, in which defects may have compact support (Sect. 1.4.1) or be in L^p (Sect. 1.4.2). In both cases, they do not change the average of the underlying periodic function.
- *Global* defects to periodicity: such defects, although they only slightly change the function everywhere, may modify its average. The first such example is studied in Sect. 1.4.5. We also studied the quasi-periodic case (Sect. 1.5.1), the almost periodic case (Sect. 1.5.2) and the random case (Sect. 1.6).

One may also define an intermediate type of defect, namely defects that are spread out, but that do not change the average of the background. We saw an example of such defects in Sect. 1.4.3.

We will now study local defects. We will only briefly consider the intermediate category at the end of the section. We leave aside the second type of defects. The quasi-periodic case is a straightforward extension of the periodic setting. It is indeed an easy exercise to prove that (2.39) has a solution. The almost periodic case is only slightly different and hardly brings anything new. In addition, as we have seen in Chap. 1, it is a particular case of the random setting (important differences will be however mentioned in Chap. 4, Sect. 4.3.2), a case that will be studied in details in Sect. 2.4 below. Finally, the general assumptions exposed in Sect. 1.4.5 may be treated similarly. Indeed, these assumptions imply that functions have an average, and that, at least *in dimension 1*, a function which derivative has zero mean value is strictly sublinear at infinity. Hence, it is sufficient to fix the constant $c = \left(\langle a^{-1}\rangle\right)^{-1}$ in $w'(y) = -1 + c\, a^{-1}(y)$ in order to prove that (2.39) has a solution, and then proceed "as in the periodic case". The convergence rate (analogue of (2.37)) is however not fixed in full generality. This is to be expected, given the wide variety of cases covered by such assumptions). This convergence rate may be made precise in specific situations.

For local defects, we choose as an example

$$a = a_{\mathrm{per}} + \widetilde{a}, \quad \text{with } \widetilde{a} \in L^p(\mathbb{R}), \text{ for some } 1 \le p < +\infty, \tag{2.40}$$

where a_{per} is periodic. We assume that a_{per} and a both satisfy the condition (2.2). Of course, the homogenized equation is that of the periodic case, since the defect does not change the mean value that defines the homogenized coefficient. One easily checks, indeed, that: $\left\langle (a_{\mathrm{per}} + \widetilde{a})^{-1}\right\rangle = \left\langle (a_{\mathrm{per}})^{-1}\right\rangle$, where the average of the left-hand side is to be understood as the large-volume limit, as defined in Chap. 1, (1.29).

Assuming (2.40), the difference (2.29) takes a form similar to (2.30), and writes

$$(u^\varepsilon)'(x) - (u^*)'(x) = -\left((a_{\mathrm{per}} + \widetilde{a}) \left(\frac{x}{\varepsilon}\right)^{-1} - \left\langle (a_{\mathrm{per}} + \widetilde{a})^{-1}\right\rangle \right) \left(-\int_0^1 F + F(x) \right)$$

$$+ \left((a_{\mathrm{per}} + \widetilde{a}) \left(\frac{x}{\varepsilon}\right) \right)^{-1} \left(-\int_0^1 F - c_\varepsilon \right),$$

where the constant c_ε is given by (2.4). The above difference is only weakly convergent. Instead of using the periodic corrector w_{per} defined by (2.32), we now consider the corrector "with defect" defined by

$$w'(x) = -1 + a^* \left(a_{\mathrm{per}} + \widetilde{a}\right)^{-1}(x),$$

$$\text{whence} \quad w(x) = -x + a^* \int_0^x \left(a_{\mathrm{per}} + \widetilde{a}\right)^{-1}(y)\, dy. \tag{2.41}$$

A simple computation shows that it is a solution to

$$
\begin{cases}
-\dfrac{d}{dy}\left(a(y)\left(1+\dfrac{d}{dy}w(y)\right)\right) = 0 \quad \text{in } \mathbb{R}, \\[2mm]
w = w_{\text{per}} + \widetilde{w} \qquad\qquad\qquad \text{with } (\widetilde{w})' \in L^p(\mathbb{R}).
\end{cases}
\tag{2.42}
$$

It is obvious that if $(\widetilde{w})' \in L^p(\mathbb{R})$, then the average $\langle(\widetilde{w})'\rangle$ exists and vanishes. This may be shown by the following application of the Hölder inequality:

$$
\left|\frac{1}{2R}\int_{-R}^{R}(\widetilde{w})'\right| \le \frac{1}{2R}(2R)^{1/p'}\left(\int_{-R}^{R}|(\widetilde{w})'|^p\right)^{1/p}
$$

$$
\le (2R)^{-1/p}\left\|(\widetilde{w})'\right\|_{L^p(\mathbb{R})} \xrightarrow{R\to+\infty} 0,
$$

where p' is the conjugate exponent to p. As a consequence, $\widetilde{w}$ is strictly sublinear at infinity. Hence so is w. Should w be not regular enough to be defined pointwise, this property would then take the weaker sense

$$
\lim_{|x|\to+\infty}\frac{1}{1+|x|}\int_{x-1}^{x+1}|w(t)|\,dt = 0.
\tag{2.43}
$$

Here, $w' \in L^\infty$, hence w is uniformly continuous. Thus, this property is equivalent to $w(x)/|x| \to 0$ as $|x| \to +\infty$.

We note that, according to the equality $a = a_{\text{per}} + \widetilde{a}$ and the form $w = w_{\text{per}} + \widetilde{w}$, the Eqs. (2.31) and (2.42) imply that $\widetilde{w}$ is solution to

$$
-\frac{d}{dy}\left((a_{\text{per}}+\widetilde{a})(y)\frac{d}{dy}\widetilde{w}(y)\right) = \frac{d}{dy}\left(\widetilde{a}(y)\left(1+\frac{d}{dy}w_{\text{per}}(y)\right)\right).
\tag{2.44}
$$

This form of Eq. (2.42) will prove useful when we consider multidimensional cases (see for instance (4.6)). Using (2.41) and (2.31), we obtain the following explicit expression

$$
\widetilde{w}'(x) = a^*\left((a_{\text{per}}+\widetilde{a})^{-1}(x) - (a_{\text{per}})^{-1}(x)\right)
$$

$$
= -a^*\widetilde{a}(x)\,(a_{\text{per}})^{-1}(x)\,(a_{\text{per}}+\widetilde{a})^{-1}(x).
\tag{2.45}
$$

This validates our choice of looking for $(\widetilde{w})'$ in $L^p(\mathbb{R})$: it is clear from this relation that $(\widetilde{w})'$ and $\widetilde{a}$ belong to the same functional space. Integrating the above expression, we find an explicit formula for $\widetilde{w}$.

Defining next

$$u^{\varepsilon,1}(x) = u^*(x) + \varepsilon (u^*)'(x) \, w \left(\frac{x}{\varepsilon} \right),$$ (2.46)

we find the analogue of (2.34), that is,

$$\left(u^\varepsilon \right)'(x) - \left(u^{\varepsilon,1} \right)'(x) = \left(a \left(\frac{x}{\varepsilon} \right) \right)^{-1} \left(-\int_0^1 F - c_\varepsilon \right).$$ (2.47)

This difference is strongly convergent. More importantly, if, in (2.46), we use the periodic corrector w_{per} instead of w, then (2.47) becomes

$$(u^\varepsilon)'(x) - (u^*)'(x) \left(1 + w_{\mathrm{per}}' \left(\frac{x}{\varepsilon} \right) \right) = \left(a \left(\frac{x}{\varepsilon} \right) \right)^{-1} \left(-\int_0^1 F - c_\varepsilon \right)$$
$$+ \left(\left(a_{\mathrm{per}} \right) \left(\frac{x}{\varepsilon} \right)^{-1} - \left(a_{\mathrm{per}} + \widetilde{a} \right) \left(\frac{x}{\varepsilon} \right)^{-1} \right) \left(-\int_0^1 F + F(x) \right).$$ (2.48)

This difference also converges strongly to 0, for instance in L^p, $p < +\infty$. But the main contrast between (2.47) and (2.48) is that the first expression vanishes in the L^∞ norm, while the second does not (except in the case $\widetilde{a} \equiv 0$, that is, the periodic case). The L^∞ norm captures all the details, at any scale, and the suitable choice of corrector (that is, w instead of w_{per}) is reflected in this norm.

Another way to highlight the difference between (2.47) and (2.48) is to write these relations at scale $\varepsilon \, x$ rather than at scale x. This intuitively means zooming in at the microscopic scale. The equality (2.48) then becomes

$$(u^\varepsilon)'(\varepsilon \, x) - (u^*)'(\varepsilon \, x) (1 + w_{\mathrm{per}}'(x)) = (a(x))^{-1} \left(-\int_0^1 F - c_\varepsilon \right)$$
$$+ \left(\left(a_{\mathrm{per}}(x) \right)^{-1} - \left((a_{\mathrm{per}} + \widetilde{a})(x) \right)^{-1} \right) \left(-\int_0^1 F + F(\varepsilon \, x) \right).$$ (2.49)

Because of the second term of the right-hand side, this difference only converges to zero if $\widetilde{a} \equiv 0$ (that is, in the periodic case). To illustrate this fact, we for instance consider the case $a_{\mathrm{per}} \equiv 1$ and $F(x) = x$. Then the second term of (2.49) reads as

$$\left(1 - \frac{1}{1 + \widetilde{a}(x)} \right) \left(-\frac{1}{2} + \varepsilon x \right).$$

It converges to $-\dfrac{\widetilde{a}(x)}{2(1 + \widetilde{a}(x))}$ as $\varepsilon \to 0$. On the other hand, (2.48) writes

$$(u^\varepsilon)'(\varepsilon \, x) - (u^{\varepsilon,1})'(\varepsilon \, x) = (a(x))^{-1} \left(-\int_0^1 F - c_\varepsilon \right),$$ (2.50)

and vanishes as $\varepsilon \to 0$. This shows that, only if we use the suitable corrector w, taking into account the defect, the convergence to zero holds at the microcopic scale too.

Put differently, the corrector $w = w_{\text{per}} + \widetilde{w}$ is adapted to the coefficient $a_{\text{per}} + \widetilde{a}$. It has allowed us to restore strong convergence, which the periodic corrector also achieves! More importantly, w has allowed us to restore the convergence obtained in the periodic case, at different scales and for various topologies, which is *not* completed using w_{per}. Similar phenomena will be studied in Chap. 4.

Concerning the convergence rate in a given topology, we can easily determine it as a function of the integrability $L^p(\mathbb{R})$ of the defect $\widetilde{a}$. Indeed, studying the convergence rate of terms of the type (2.47) or (2.48) reduces to studying the convergence rate of integrals of the form

$$\int_0^1 \frac{1}{a\left(\frac{x}{\varepsilon}\right)} g(x) dx = \int_0^1 \frac{g(x)}{a_{\text{per}}\left(\frac{x}{\varepsilon}\right) + \widetilde{a}\left(\frac{x}{\varepsilon}\right)} dx$$

$$= \int_0^1 \frac{g(x)}{a_{\text{per}}\left(\frac{x}{\varepsilon}\right)} dx - \int_0^1 g(x) \frac{\widetilde{a}\left(\frac{x}{\varepsilon}\right)}{a_{\text{per}}\left(\frac{x}{\varepsilon}\right) a\left(\frac{x}{\varepsilon}\right)} dx.$$

The first term of the right-hand side corresponds to the periodic case. The second term is dealt with using the Hölder inequality, together with the facts that $\widetilde{a} \in L^p(\mathbb{R})$ and that a is bounded away from 0:

$$\left| \int_0^1 g(x) \frac{\widetilde{a}\left(\frac{x}{\varepsilon}\right)}{a_{\text{per}}\left(\frac{x}{\varepsilon}\right) a\left(\frac{x}{\varepsilon}\right)} dx \right|$$

$$\leq \|g\|_{L^{p'}(]0,1[)} \left\| \frac{1}{a_{\text{per}}} \right\|_{L^\infty(\mathbb{R})} \left\| \frac{1}{a} \right\|_{L^\infty(\mathbb{R})} \left\| \widetilde{a}\left(\frac{\cdot}{\varepsilon}\right) \right\|_{L^p(]0,1[)}.$$

Changing variables, we have $\left\| \widetilde{a}\left(\frac{\cdot}{\varepsilon}\right) \right\|_{L^p(]0,1[)} \propto \varepsilon^{1/p} \|\widetilde{a}\|_{L^p(\mathbb{R}^+)}$, which yields the rate of convergence.

To conclude this section, we briefly address, as announced above, the case of defects that are no more localized, but are "rare" at infinity. This may be formalized *for example* by (1.31), that is,

$$\widetilde{a}(x) = \sum_{k \in \mathbb{Z}} a_0 \left(x - \text{sgn}(k) 2^{|k|} \right), \quad a_0 \text{ has compact support.} \tag{2.51}$$

In such a case, and assuming the usual coercivity conditions, it is possible to solve the corrector equation as we have done previously. We find $w = w_{\text{per}} + \widetilde{w}$, where w_{per} is the periodic corrector, and we again obtain the expression (2.45), namely:

$$\widetilde{w}'(x) = -a^* \frac{\widetilde{a}(x)}{a_{\text{per}}(x) \left(a_{\text{per}}(x) + \widetilde{a}(x) \right)}.$$

We see from this expression that, although it does not vanish, $\widetilde{w}'$ becomes "rare" at infinity, as $\widetilde{a}$ does. On the other hand, it does not share the simple form (2.51) of the original coefficient $\widetilde{a}$. It is however possible to build the algebra spanned by functions of the form (2.51) and close it for some suitable norm (we do not make it precise here, and refer to [Gou22] for the details). It is then simple to prove that $\widetilde{w}'$ belongs to the functional space thus constructed. The above computations of the case of a localized defect carry over to this case, showing that the corrector may be used to obtain convergence properties of $u^\varepsilon - u^{\varepsilon,1}$ similar to those of the case of localized defects [Gou22]. Here again, using the periodic corrector w_{per} instead of $w_{per} + \widetilde{w}$ would not allow such convergence properties in all the topologies considered.

As a conclusion, the corrector w solution to the suitable equation accounting for the presence of the defects allows to recover the approximation qualities of the periodic case. This is an important feature of this corrector, which will prove crucial when studying defects in the multidimensional case in Chap. 4.

2.4   The 1D Random Case

In this section, we again consider the one-dimensional setting, but the coefficient is now random. The equation reads

$$- \frac{d}{dx} \left(a_\varepsilon(x, \omega) \frac{d}{dx} u^\varepsilon(x, \omega) \right) = f(x), \tag{2.52}$$

on the interval $]0, 1[$. Equation (2.52) is supplied with the usual boundary conditions $u^\varepsilon(0, \omega) = u^\varepsilon(1, \omega) = 0$. Of course, these conditions are to be understood *almost surely*.

Consistently with our considerations in Sect. 1.6 of Chap. 1, we assume that the coefficient $a_\varepsilon(x, \omega)$ writes

$$a_\varepsilon(x, \omega) = a\left(\frac{x}{\varepsilon}, \omega\right) \tag{2.53}$$

where $a(x, \omega)$ is assumed to be stationary (in the discrete sense), in an ergodic setting previously defined. We assume that the coefficient a is bounded from above and away from 0, almost surely. This ensures that Eq. (2.52) is well-posed. The right-hand side f of (2.52) is, say, in $L^2(]0, 1[)$, knowing that addressing a lower regularity is possible. We insist on the fact that f is neither oscillating, nor random: it does not depend on ε nor on ω.

The question we ask here is that of the limit of the solution u^ε as $\varepsilon \to 0$. This is identified as *stochastic homogenization* (or *random homogenization*).

Remark 2.1 Let us mention that the terms *stochastic homogenization* (or *random homogenization*) may be misleading. In some other scientific contexts, it may refer to the homogenization of the deterministic Eq. (2.1) *interpreted* in stochastic terms. Such a diffusion equation, and more generally its multidimensional version (1), can indeed be studied using the theory of stochastic processes. The question of homogenization may be addressed in this language. This is not what we wish to do here. We will however consider this alternative notion of *stochastic homogenization* in Sect. 6.3. □

Here again, the fact that we work in dimension 1 allows to compute explicitly the solution u^ε of (2.52). Again denoting by $F(x) = \int_0^x f(y) \, dy$, we have

$$u^\varepsilon(x, \omega) = - \int_0^x (a_\varepsilon(y, \omega))^{-1} (c_\varepsilon(\omega) + F(y)) \, dy, \qquad (2.54)$$

where the constant $c_\varepsilon(\omega)$ reads as

$$c_\varepsilon(\omega) = - \frac{\int_0^1 (a_\varepsilon(y, \omega))^{-1} F(y) \, dy}{\int_0^1 (a_\varepsilon(y, \omega))^{-1} \, dy}. \qquad (2.55)$$

Using the form (2.53) of the coefficient, we infer the following homogenized limits

$$u^*(x) = - (a^*)^{-1} \int_0^x (c_* + F(y)) \, dy, \qquad (2.56)$$

$$(a^*)^{-1} = \mathbb{E} \int_0^1 a^{-1}(y, \omega) \, dy, \qquad (2.57)$$

$$c_* = - \int_0^1 F(y) \, dy. \qquad (2.58)$$

Of course, the function u^* defined by (2.56) is the solution to the homogenized Eq. (2.27), where the homogenized coefficient a^* is defined by (2.57). This is the equivalent of (2.26) in the periodic case.

Similarly to the periodic case studied in Sect. 2.2, we next determine the corrector. Here, it is defined as the solution $w(x, \omega)$ to

$$- \frac{d}{dx} \left(a(x, \omega) \left(1 + \frac{d}{dx} w(x, \omega) \right) \right) = 0 \quad \text{in} \quad \mathbb{R}, \qquad (2.59)$$

with the additional condition

$$w'(x, \omega) \text{ is a stationary function with zero average: } \mathbb{E} \int_0^1 w'(x, \omega) \, dx = 0. \qquad (2.60)$$

The latter condition is similar to the zero average condition in the periodic case. As we have already seen in Sect. 1.6.4 of Chap. 1, property (2.60) does not imply that w itself is a stationary function. It only implies that it is (almost surely) strictly sublinear at infinity, in the sense of (1.93), or in a weaker sense if f is not sufficiently regular with respect to x. This property is sufficient, as was discussed in the previous section (see (2.38)).

The formal derivation of Eq. (2.59) is similar to that for the other settings discussed so far. Roughly speaking, (2.59) is a *zoom* of the original equation, aiming at capturing the oscillations at the small scale. As in Sect. 2.3, this equation is posed on the *whole* real line, contrary to the periodic case, in which knowing the microscopic environment in the periodic cell is sufficient to know it everywhere. As a consequence, substantial difficulties arise, both on the theoretical and on the numerical sides. We will see this in Sects. 4.3 and 5.1.5. In the simple one-dimensional case we consider here, it is however possible to bypass these difficulties.

As we have pointed out in Sect. 2.2 for the periodic setting, we could give an abstract proof of the existence and uniqueness (up to an additive deterministic constant) of the solution to (2.59)–(2.60). This proof happens to be much more involved in the present stochastic setting. This is mainly due to the fact that, as we have just remarked, (2.59)–(2.60) is posed on the whole real line. We will give such a proof in the mutlidimensional case in Sect. 4.3.1 of Chap. 4. Here, we circumvent the difficulty of solving (2.59)–(2.60) by only exhibiting a solution. The reader can easily check that

$$w'(x, \omega) = -1 + a^* a^{-1}(x, \omega),$$

$$\text{whence} \quad w(x, \omega) = w(0, \omega) - x + a^* \int_0^x a^{-1}(y, \omega) \, dy, \qquad (2.61)$$

is a solution. Note that the constraint $\mathbb{E} \int_0^1 w'(x, \omega) \, dx = 0$ holds because $(a^*)^{-1} = \mathbb{E} \int_0^1 a^{-1}(x, \omega) \, dx$. Using the second relation of (2.61), this implies that w is strictly sublinear at infinity, almost surely. In order to show uniqueness, one defines two solutions w_1 and w_2, and remark that $v = w_1 - w_2$ satisfies $-\dfrac{d}{dy}\left(a(y, \omega) \dfrac{d}{dy} v(y, \omega))\right) = 0$. This implies that $a(y, \omega) v'(y, \omega)$ is independent of y. We denote it by $c(\omega)$. This random variable is stationary and independent of y (recall that a, w'_1 and w'_2 are all stationary). Thus, ergodicity allows to infer that c is deterministic. This is indeed an elementary property of stationary random variables [Shi95]. It may be proved by the following argument: for any $\lambda \in \mathbb{R}$, we define $A_\lambda = \{\omega \in \Omega, \quad c(\omega) < \lambda\}$, and $\phi(\lambda) = \mathbb{P}(A_\lambda)$. Since $c \in L^1(\Omega)$, the set A_λ and the function ϕ are well-defined. Moreover, $0 \le \phi \le 1$, ϕ is non-decreasing, $\phi(\lambda) \longrightarrow 0$ as $\lambda \to -\infty$ and $\phi(\lambda) \longrightarrow 1$ as $\lambda \to +\infty$. In addition, A_λ is invariant under the action τ_k for all $k \in \mathbb{Z}$, hence, using ergodicity,

ϕ can only take the values 0 and 1. As a consequence, there exists a real number λ^* such that $\phi(\lambda) = 0$ if $\lambda < \lambda^*$ and $\phi(\lambda) = 1$ if $\lambda > \lambda^*$. Given the definition of A_λ, this implies that $c = \lambda^*$ almost surely. We have thus proved uniqueness up to an additive constant, as announced. As we have already pointed out (see Sect. 2.2, p. 74), such a uniqueness is however not needed in the present theoretical study.

As in the periodic setting studied in Sect. 2.2, we can now compute the approximation $u^{\varepsilon,1}(x, \omega) = u^*(x) + \varepsilon\, (u^*)'(x)\, w\left(\frac{x}{\varepsilon}, \omega\right)$ and prove that, almost surely, such a function satisfies $u^\varepsilon - u^{\varepsilon,1} \to 0$ in $H^1(\mathcal{D})$. Here, contrary to the periodic setting, we need to understand the behaviour of this difference with respect to ω. In order to do so, we could study $u^\varepsilon - u^{\varepsilon,1}$ itself in H^1 norm. It turns out that the difference (or *residual*) $u^\varepsilon - u^*$, which reads as

$$\left(u^\varepsilon - u^*\right)(x, \omega) = -\left(c_\varepsilon(\omega) - c_*\right) \int_0^x (a_\varepsilon(y, \omega))^{-1}\, dy$$

$$- \int_0^x \left((a_\varepsilon(y, \omega))^{-1} - \left(a^*\right)^{-1}\right)(c_* + F(y))\, dy, \quad (2.62)$$

already involves the key terms, that is, terms of the type

$$\int_0^1 a_\varepsilon^{-1}(y, \omega)\, g(y)\, dy,$$

for functions g such that $g = 1$ or $g = F$. Understanding both the differences $u^\varepsilon - u^{\varepsilon,1}$ and (2.62) therefore amounts to understanding and studying the convergence rate of

$$\left(\int_0^1 a_\varepsilon^{-1}(y, \omega)\, g(y)\, dy\right) \overset{\varepsilon \to 0}{\longrightarrow} \left(\mathbb{E} \int_0^1 a^{-1}(y, \omega)\, dy\right) \int_0^1 g(y)\, dy, \quad (2.63)$$

As we have already noticed in Sect. 1.6.4 of Chap. 1, in such a convergence, various rates may occur, due to the large diversity of stationary settings. The particular cases studied in Sect. 1.6.4, that is, (2.53) with a stationary function $a(x, \omega)$ reading as (1.85), give a convergence rate $O(\sqrt{\varepsilon})$ for (2.63), still assuming that g is bounded and continuous. Thus, the same convergence rate holds for the residual defined in (2.62). The rigorous analysis of this one-dimensional situation has been performed only recently (see [BP99, BGMP08]). As one may expect, the multidimensional case is considerably more difficult. It is mainly an open mathematical question. We do not wish to further develop this point here.

Anyway, we note the difference in the convergence rate (even in the simple random case of i.i.d. variables) between the random and the periodic setting. This is an additional point that we will need to keep in mind when dealing with random perturbations of periodic geometries.

2.5 Some "Bad" Cases

We end this chapter by giving a selection of settings in which homogenization does not give the result one may naively expect. Although one-dimensional, theses cases are indeed not as simple as one may think.

In Sects. 2.5.1 and 2.5.2, we show that, contrary to what can be expected from the studies we have developed so far in this chapter, *the homogenized equation may not be of the same form as the original equation*, or may even not exist. Some other cases will be mentioned, showing that such unconventional situations are more common than one could think.

Next, we concentrate on periodic settings with defects. We have seen in Chap. 1 and in Sect. 2.3 that a defect (at least for a "nice" defect, that is, a localized defect that vanishes at infinity) does not change the average of functions involved in the value of the homogenized coefficient. If it has a non-trivial effect, as we have noticed above, this influence is only at the microscopic scale. We have studied this in detail in Sect. 2.3. In sharp contrast, we will present in Sect. 2.5.3 a particular setting in which *even a defect as nice as possible (we will choose a compactly supported defect) may change the homogenized macroscopic limit.*

2.5.1 The Homogenized Equation May Take a Different Form

2.5.1.1 A Simple Example

Our first example is a time-dependent problem and is motivated by absorption questions in electromagnetism. It is alternatively possible to build similar examples in a stationary setting stemming from porous media or particle physics. The evolution problem we consider is the following

$$\begin{cases} \partial_t u^\varepsilon(t, x) + a\left(\dfrac{x}{\varepsilon}\right) u^\varepsilon(t, x) = 0 \quad \text{for} \quad t > 0, \\[2mm] u^\varepsilon(t = 0, x) \qquad\qquad = v(x), \end{cases} \tag{2.64}$$

where $v \in L^2(]0, 1[)$ is fixed, and a is periodic of period 1. In addition, as we have done so far, we assume that $0 < c_1 \le a(x) \le c_2$ for two constants c_i independent of $x \in \mathbb{R}$. Extrapolating on the homogenization theory we have developed so far, we expect a limit equation of the form

$$\begin{cases} \partial_t u^*(t, x) + a^* u^*(t, x) = 0 \quad \text{for} \quad t > 0, \\[2mm] u^*(t = 0, x) \qquad = v(x). \end{cases} \tag{2.65}$$

Now, the solution to (2.64) can be computed explicitly and writes

$$u^\varepsilon(t, x) = v(x) \exp\left(-t\, a\left(\frac{x}{\varepsilon}\right)\right). \tag{2.66}$$

It is then clear that, in the limit $\varepsilon \to 0$,

$$u^*(t, x) = \lim_{\varepsilon \to 0} u^\varepsilon(t, x) = v(x) \langle e^{-ta} \rangle, \tag{2.67}$$

in the sense of the weak $L^2([0, 1)]$ topology, for all $t > 0$, since the function $x \to \exp\left(-t\, a\left(\frac{x}{\varepsilon}\right)\right)$ is, for each time $t > 0$, a periodic function of x. On the other hand, the solution to (2.65) reads as $u^*(t, x) = v(x) \exp(-ta^*)$. For this function to be equal to (2.67), it is necessary that there exist real numbers λ and a^* such that

$$\langle e^{-ta} \rangle = \lambda\, e^{-ta^*} \tag{2.68}$$

for all $t > 0$. This is possible only if a is constant, as we will see shortly. Thus, it is not in general possible that the limit $u^*(t, x) = v(x) \langle e^{-ta} \rangle$ satisfies an equation of the same type as (2.64). In the limit $\varepsilon \to 0$, the *equation takes a different form*.

The proof that (2.68) for all $t > 0$ implies that a is constant is a simple consequence of the following argument. Assuming (2.68), we infer that the function $\log \langle e^{-ta} \rangle$ is affine with respect to t, hence

$$0 = \frac{d^2}{dt^2}\left(\log \langle e^{-ta} \rangle\right) = \frac{d^2}{dt^2}\left(\log\left(\int_0^1 e^{-ta(x)}\, dx\right)\right)$$

This is equivalent to

$$\left(\int_0^1 a^2(x)\, e^{-ta(x)}\, dx\right)\left(\int_0^1 e^{-ta(x)}\, dx\right) - \left(\int_0^1 a(x)\, e^{-ta(x)}\, dx\right)^2 = 0.$$

This is the equality case in the Cauchy-Schwarz inequality for the functions $e^{-ta(x)/2}$ and $a(x)\, e^{-ta(x)/2}$, and is equivalent to the fact that a is constant. Our claim is proven.

For the sake of completeness, we are going to determine the equation actually satisfied by the limit solution $u^*(t, x) = v(x) \langle e^{-ta} \rangle$. We are going to prove that, for a specific function a, the limit u^* satisfies a so-called *evolution equation with delay* of the type

$$\partial_t u^*(t, x) + \langle a \rangle\, u^*(t, x) = \int_0^t K(t - s) u^*(s, x)\, ds, \tag{2.69}$$

with the same initial data as (2.64). The function $K \neq 0$ may be expressed as a function of a (see (2.71)–(2.72) below). In order to carry out the calculation with simple tools (actually, the qualitative result leading from (2.64) to (2.69) is more general), we choose for a the piecewise constant function

$$a(x) = \begin{cases} a_1 \text{ if } x \in [0, \alpha_1], \\ a_2 \text{ if } x \in]\alpha_1, 1], \end{cases} \tag{2.70}$$

where a_1, a_2 are two positive constants, $\alpha_1 \in]0, 1[$. We define $\alpha_2 = 1 - \alpha_1$.

For this particular value of a, Eq. (2.66) allows to compute the limit function $u^*(t, x) = v(x) \left(\alpha_1 e^{-ta_1} + \alpha_2 e^{-ta_2} \right)$. We thus have

$$\partial_t u^* + \langle a \rangle u^*(t, x) = v(x) \left(-a_1 \alpha_1 e^{-ta_1} - a_2 \alpha_2 e^{-ta_2} \right)$$

$$+ v(x) (\alpha_1 a_1 + \alpha_2 a_2) \left(\alpha_1 e^{-ta_1} + \alpha_2 e^{-ta_2} \right)$$

$$= v(x) \alpha_1 \alpha_2 (a_1 - a_2) \left(-e^{-ta_1} + e^{-ta_2} \right).$$

Setting

$$K(t - s) = \lambda e^{-\mu(t-s)}, \tag{2.71}$$

with

$$\lambda = \langle a^2 \rangle - \langle a \rangle^2 \quad \mu = \frac{\langle a^2 \rangle - \langle a \rangle^2}{\langle a \rangle - \frac{1}{\langle \frac{1}{a} \rangle}}, \tag{2.72}$$

it is possible to check (the calculation is tedious but not difficult) that u^* satisfies Eq. (2.69). Such a result may be obtained by a general method avoiding the explicit computation of the solution. To this end, we multiply (2.64) by e^{-pt}, for $p > 0$, and integrate the result from $t = 0$ à $t = +\infty$. Doing so, we see the *Laplace transform*

$$Lu^\varepsilon(p, x) = \int_0^\infty e^{-pt} u^\varepsilon(t, x) \, dt, \quad \text{for } x \in [0, 1], \quad p > 0, \tag{2.73}$$

of u^ε arising. We then study the limit as $\varepsilon \to 0$ of this Laplace transform Lu^ε, which, given (2.64), reads as $Lu^\varepsilon(p, x) = \dfrac{1}{p + a\left(\frac{x}{\varepsilon}\right)} v(x)$. The method is described in detail (in French) for this particular case in [Le 05, p 83-ff]. This computation is more general (recall that the Laplace transform is a natural tool to deal with evolution problems and change them into stationary problems), and applies to various cases in which an equation of the form (2.69) is the homogenized limit of an equation similar to (2.64). We refer to [Gol91] for many more details.

2.5.1.2 Related Examples

A similar phenomenon is observed on the transport equation

$$
\begin{cases}
\partial_t u^\varepsilon(t, x) + a\left(\dfrac{x_2}{\varepsilon}\right) \partial_{x_1} u^\varepsilon(t, x) = 0 & \text{for} \quad t > 0 \\[2mm]
u^\varepsilon(t = 0, x) \qquad\qquad = v(x),
\end{cases}
$$

where $x = (x_1, x_2) \in [0, 1]^2$ is a two-dimensional space variable, but the coefficient a depends only on the second coordinate x_2, and the spatial derivative of the solution u^ε is taken only in the direction x_1. Actually, the solution to this equation is explicit and reads $u^\varepsilon(t, x) = v\left(x_1 - t\,a\left(\dfrac{x_2}{\varepsilon}\right), x_2\right)$. It is possible to prove (see [EH09, p 210] and [Tar89] for more details), using the Laplace transform again, but with more elaborate methods, that the homogenized equation is of the form

$$
\partial_t u^\varepsilon(t, x) + \langle a \rangle\, \partial_{x_1} u^\varepsilon(t, x) = \int_0^t K(t, u(s, .))\, ds,
$$

where $K(t, u)$ is a linear function of u, that remains to be identified.

Such unexpected terms appearing in the homogenized equation may occur in different contexts. For instance, the phenomenon may arise in stationary problems in periodically perforated domains. A Poisson equation $-\Delta u = f$ can become, in the limit, the equation $-\Delta u + u = f$. We refer the reader to [CM82] for more information. Under some particular conditions, the example we are going to consider in Sect. 2.5.2 below will also lead to a similar situation.

Rather than extensively exploring all the possible cases, we wish to make a remark about Eq. (2.64).

2.5.1.3 An Unstable Phenomenon

As we have seen, the fact that additional terms may appear in the homogenization process and change the nature of the equation is not unusual. However, these phenomena happen to be *unstable*. Upon slightly modifying the equation in a suitable way, we may indeed recover a more conventional homogenization process, without any additional term. To realize this, we consider the *parabolic* equation

$$
\partial_t u^\varepsilon(t, x) + a\left(\dfrac{x}{\varepsilon}\right) u^\varepsilon(t, x) - \partial_x^2 u^\varepsilon(t, x) = 0, \tag{2.74}
$$

which may be seen as a slight modification of (2.64). We assume that the initial condition is $v(x)$, as in (2.64). Since we have introduced a second-order differential operator, we need to add boundary conditions. We assume that they read as $u^\varepsilon(t, 0) = u^\varepsilon(t, 1) = 0$ for all $t > 0$. We claim that, in the limit $\varepsilon \to 0$, (and in contrast to (2.64)) Eq. (2.74) converges to an equation of the same form, that is,

$$\partial_t u^\varepsilon(t, x) + \langle a \rangle \, u^\varepsilon(t, x) - \partial_x^2 u^\varepsilon(t, x) = 0. \tag{2.75}$$

In order to prove our claim, we can no more rely on an explicit expression of the solution u^ε. We are thus going to develop arguments that we will need again in the sequel of this textbook, as for instance in Sect. 3.1. These arguments are more elaborate than those we have used so far, but they are useful to know. We assume (this can be proved using standard tools of analysis of partial differential equations) that the solution u^ε exists and is sufficiently regular, so that the following computations make sense. We multiply (2.74) by u^ε and integrate over $[0, 1]$ with respect to x. Integrating by parts, we find

$$\frac{1}{2} \frac{d}{dt} \int_0^1 |u^\varepsilon(t, x)|^2 \, dx + \int_0^1 a\left(\frac{x}{\varepsilon}\right) |u^\varepsilon(t, x)|^2 \, dx$$

$$+ \int_0^1 \left|\partial_x u^\varepsilon(t, x)\right|^2 \, dx = 0, \tag{2.76}$$

from which we first deduce that

$$\frac{1}{2} \frac{d}{dt} \int_0^1 |u^\varepsilon(t, x)|^2 \, dx + \int_0^1 a\left(\frac{x}{\varepsilon}\right) |u^\varepsilon(t, x)|^2 \, dx \leq 0.$$

This inequality, together with the fact that a is bounded, allows to apply the Gronwall lemma. This leads to

$$\int_0^1 |u^\varepsilon(t, x)|^2 \, dx \leq e^{Ct} \int_0^1 |v(x)|^2 \, dx, \tag{2.77}$$

for all $t > 0$, with a constant C independent of t. On the other hand, integrating (2.76) from $t = 0$ to $t = T$, we find

$$\frac{1}{2} \int_0^1 |u^\varepsilon(T, x)|^2 \, dx + \int_0^T \int_0^1 a\left(\frac{x}{\varepsilon}\right) |u^\varepsilon(t, x)|^2 \, dx \, dt$$

$$+ \int_0^T \int_0^1 \left|\partial_x u^\varepsilon(t, x)\right|^2 \, dx \, dt = \frac{1}{2} \int_0^1 |v(x)|^2 \, dx.$$

Thus, we obtain

$$\int_0^T \int_0^1 \left|\partial_x u^\varepsilon(t, x)\right|^2 \, dx \, dt \leq C_T, \tag{2.78}$$

for some constant C_T that depends only on T, v and a, but not on ε. Estimates (2.77) and (2.78) do not depend on ε. Using the Poincaré inequality, they imply that the sequence u^ε is bounded in the functional space $L^\infty \left([0, T], L^2(]0, 1[) \right) \cap L^2 \left([0, T], H_0^1(]0, 1[) \right)$. Up to extracting a subsequence, we can thus assume that this sequence converges weakly to some u^*, in $L^p \left([0, T], L^2(]0, 1[) \right) \cap L^2 \left([0, T], H_0^1(]0, 1[) \right)$ for all $p < +\infty$. Moreover, since Eq. (2.74) may be written as $\partial_t u^\varepsilon(t, x) = -a \left(\dfrac{x}{\varepsilon} \right) u^\varepsilon(t, x) + \partial_x^2 u^\varepsilon(t, x)$, this bound allows to prove that the sequence $\partial_t u^\varepsilon(t, x)$ is bounded in the functional space $L^2 \left([0, T], H^{-1}(]0, 1[) \right)$. Recall now that, by definition, $H^{-1}(]0, 1[)$ is the dual space of $H_0^1(]0, 1[)$. It follows that, applying a classical result (which may be seen as an extension of the Rellich theorem for time-dependent functions, see [Lio69, Théorème 5.1, p 58]— in French—, or [Sim87, Corollary 4]), up to extracting a subsequence again, the sequence u^ε converges *strongly* to u^* in $L^2 \left([0, T], L^2(]0, 1[) \right)$. This result may be understood formally by pointing out that the time and space derivatives are both bounded, in some weak space in time and in space, respectively. This allows to derive compactness in *both* variables. Returning to Eq. (2.74), we may thus pass to the limit in each term, at least in the sense of distributions (and actually in a stronger sense that it is not useful to make precise here). The linear terms $\partial_t u^\varepsilon(t, x)$ and $\partial_x^2 u^\varepsilon(t, x)$ pass to the limit due to the weak convergences we have just proved. As for the "nonlinear" term $a \left(\dfrac{x}{\varepsilon} \right) u^\varepsilon(t, x)$, we have that $a \left(\dfrac{x}{\varepsilon} \right)$ weakly converges in $L^2 \left([0, T], L^2(]0, 1[) \right)$ to $\langle a \rangle$, while $u^\varepsilon(t, x)$ strongly converges to u^* in $L^2 \left([0, T], L^2(]0, 1[) \right)$. We thus find the limit Eq. (2.75), as we have claimed. A careful inspection of the proof reveals that the presence of the second-order term $\partial_x^2 u^\varepsilon(t, x)$ implies (through estimate (2.76)) the strong convergence in the L^2 norm, and thus allows to control the nonlinear term that, in Eq. (2.64), was the delicate term at the origin of the delay term.

2.5.2 The Homogenized Equation May Not Exist, and/or be of a Different Nature

Following up on Sect. 2.5.1, we give here an example in which the homogenized equation has a different form than the original one and may even not exist in some cases. The interest of such a case is that it is *static*, contrary to Sect. 2.5.1. This shows that the phenomena we investigate throughout this Sect. 2.5 are not restricted to time-dependent problems.

We consider here the so-called *advection-diffusion equation*

$$- \frac{d^2}{dx^2} u^\varepsilon(x) + \frac{1}{\varepsilon} b\left(\frac{x}{\varepsilon}\right) \frac{d}{dx} u^\varepsilon(x) = f(x), \qquad (2.79)$$

in dimension 1. Note, however, that the issues we are going to study on this equation are not linked with the dimension. We will proceed with a complete study, in higher dimensions, in Sect. 6.1. Typically, such an equation models a concentration transported at velocity b in a viscous fluid (hence the first term, the "laplacian"), with a source term f. The velocity field b (in higher dimensions it is vector-valued, although here it is a scalar valued function) is assumed to oscillate at scale ε. For reasons that will become clear below, we assume that it is locally integrable. We also assume, and this is essential, that b is *periodic, with zero average* $\langle b \rangle = 0$, and that it is *amplified* by the coefficient $\dfrac{1}{\varepsilon}$ (this might be questionable from a practical viewpoint, but let us consider that our study is purely theoretical). We also assume that the right-hand side f is integrable, that is, $f \in L^1(]0, 1[)$. Boundary conditions, as for instance $u^\varepsilon(0) = u^\varepsilon(1) = 0$, supplement Eq. (2.79). We are interested in the behaviour of the solution u^ε in the limit $\varepsilon \to 0$. In dimensions $d \geq 2$, it is not straightforward to prove existence and uniqueness of this solution u^ε. In the present one-dimensional setting, one easily solves explicitly Eq. (2.79), first computing $(u^\varepsilon)'(x)$, then integrating. Doing so, one obtains

$$(u^\varepsilon)'(x) = \lambda_\varepsilon \, \exp\left(B\left(\frac{x}{\varepsilon}\right)\right)$$

$$- \left(\int_0^x f(y) \exp\left(-B\left(\frac{y}{\varepsilon}\right)\right) dy\right) \exp\left(B\left(\frac{x}{\varepsilon}\right)\right), \qquad (2.80)$$

where we have set $B(x) = \displaystyle\int_0^x b(y)\, dy$, the primitive of b which vanishes at 0, and

$$\lambda_\varepsilon = \left(\int_0^1 \exp\left(B\left(\frac{y}{\varepsilon}\right)\right) dy\right)^{-1}$$

$$\times \int_0^1 \left(\int_0^x f(y) \exp\left(-B\left(\frac{y}{\varepsilon}\right)\right) dy\right) \exp\left(B\left(\frac{x}{\varepsilon}\right)\right) dx. \qquad (2.81)$$

Next, we use that the function b is periodic, bounded, with zero average. As a consequence, its primitive B is periodic too, hence the function $\exp\left(B\left(\frac{\cdot}{\varepsilon}\right)\right)$ converges weakly to $\langle e^B \rangle$. Similarly, the function $\exp\left(-B\left(\frac{\cdot}{\varepsilon}\right)\right)$ converges weakly to $\langle e^{-B} \rangle$. These weak convergences, together with the dominated convergence theorem, allow to prove that

$$\lambda_\varepsilon \xrightarrow{\varepsilon \to 0} \langle e^B \rangle^{-1} \int_0^1 F(x) \langle e^{-B} \rangle dx \langle e^B \rangle = \langle e^{-B} \rangle \int_0^1 F(x) dx,$$

where $F(x) = \int_0^x f(t)\,dt$ is the primitive of f that vanishes at the origin, and

$$(u^\varepsilon)'(x) \xrightarrow{\varepsilon \to 0} \langle e^B \rangle \langle e^{-B} \rangle \int_0^1 F - \langle e^B \rangle \langle e^{-B} \rangle F(x) =: (u^*)'(x). \qquad (2.82)$$

This shows that the limit u^* of u^ε is solution to

$$-a^* \frac{d^2}{dx^2} u^*(x) = f(x), \quad \text{with} \quad a^* = \left(\langle e^B \rangle \langle e^{-B} \rangle \right)^{-1}, \qquad (2.83)$$

with boundary conditions $u^*(0) = u^*(1) = 0$. The homogenized limit of the advection-diffusion Eq. (2.79), with a transport field b and a diffusion coefficient equal to 1 is therefore a pure diffusion equation, with no transport term, but with a diffusion coefficient different from 1. This is clearly unexpected.

The above explicit expressions actually allow to explain our specific choice of a velocity field of the form $\frac{1}{\varepsilon} b \left(\frac{x}{\varepsilon} \right)$, with $\langle b \rangle = 0$. Consider first the example $f \equiv 1$ and $b \equiv 1$ (of course, it contradicts our assumption $\langle b \rangle = 0$). Using (2.80), we obtain

$$(u^\varepsilon)'(x) = (\lambda_\varepsilon - \varepsilon) \exp \left(\frac{x}{\varepsilon} \right) + \varepsilon \quad \text{with} \quad \lambda_\varepsilon = \varepsilon - \left(\exp \left(\frac{1}{\varepsilon} \right) - 1 \right)^{-1}.$$
$$(2.84)$$

This proves that $(u^\varepsilon)'(x)$ vanishes (and since u^ε satisfies homogeneous Dirichlet boundary conditions, so does u^ε), hence this limit cannot be a solution to an equation of the form (2.79) with a right-hand side $f = 1$. Here, there is no homogenized equation, in the sense of an equation of the form $Lu = f$, where L is a linear operator and f the right-hand side of the original equation. Should such an equation exists, it would be $u^* = 0$, which is somewhat disappointing. Assuming that b has zero average, which is in some sense the opposite of the property $b = 1$, is certainly important. Similarly, and somewhat symmetrically, if we now assume that b has zero average but eliminate the factor $\frac{1}{\varepsilon}$, the above computation is only slightly modified, in the sense that B is replaced by εB in (2.80), hence the limit

$$(u^*)'(x) = \int_0^1 F - F(x).$$

Thus, instead of (2.83), the homogenized equation now reads as

$$-\frac{d^2}{dx^2} u^*(x) = f(x),$$

which has a form different from the original equation. Furthermore, the velocity field has completely disappeared from the equation.

The watchful reader has certainly understood that the choice of combining the assumptions of a zero average for b and an amplification factor $\dfrac{1}{\varepsilon}$ allows both differential terms in (2.79) to be of the same order. Thinking of the case $u^\varepsilon(x) = u\left(\dfrac{x}{\varepsilon}\right)$, one realizes that the term $\dfrac{d^2}{dx^2}u^\varepsilon(x)$ behaves like $\dfrac{1}{\varepsilon^2}$ and is thus probably not affected by a first-order term $\dfrac{d}{dx}u^\varepsilon(x)$, which behaves like $\dfrac{1}{\varepsilon}$ ... *except* if the latter is amplified by a factor $\dfrac{1}{\varepsilon}$. However, in order for this perturbation to be not too severe, the velocity field needs to be small in average. This is the intuition that guides the choice of the equation and the calculations above.

We will return to the advection-diffusion Eq. (2.79) in higher dimensions and in a more general setting in Chap. 6.

2.5.3   A Small Defect in a Specific Nonlinear Equation

Up to now, the (local) defects we have studied do not modify the homogenized equation. They only modify the microscopic behaviour. We are now going to see an example of a local perturbation to some periodic background that has important consequences on the limit equation, *even* in dimension 1. The equation is of course different from (1), but remains rather natural and simple, although nonlinear (the reader should not infer from this that any nonlinear equation behaves differently from (1)).

We consider the one-dimensional equation

$$u^\varepsilon(x) + \left|(u^\varepsilon)'(x)\right| = \tilde{V}\left(\frac{x}{\varepsilon}\right), \qquad (2.85)$$

set on the whole real line $\mathbb{R}$. This is a particular case of the so-called *first-order Hamilton-Jacobi equation*

$$u + H(x, u, \nabla u) = 0, \qquad (2.86)$$

for some *Hamiltonian* $H(x, u, p)$, here equal to $H(x, u, p) = |p| - \tilde{V}(x/\varepsilon)$. Equation (2.85) should be thought as the perturbation by the *potential* $\tilde{V}$ of the following equation

$$u^\varepsilon(x) + \left|(u^\varepsilon)'(x)\right| = V_{\text{per}}\left(\frac{x}{\varepsilon}\right),$$

where the *periodic potential* V_{per} is taken to be zero, that is,

$$u(x) + |u'(x)| = 0. \tag{2.87}$$

The only bounded C^1 solution of (2.87) is $u = 0$. This may be proved as follows. Consider a solution u, and denote by $]\alpha, \beta[$ a maximal interval in which $u' > 0$. The bounds α and β may be equal to $-\infty$ and $+\infty$, respectively. On the interval $]\alpha, \beta[$, we have $u(x) = -Ae^{-x}$ for some constant $A > 0$. Necessarily, $\alpha > -\infty$, otherwise u cannot be bounded. Using that u' is continuous and that the interval $]\alpha, \beta[$ is maximal, we infer that $u'(\alpha) = 0$. Hence, $Ae^{-\alpha} = 0$. This contradicts the fact that $A > 0$. Thus, the maximal interval $]\alpha, \beta[$ does not exist. This means that $u'(x) \leq 0$ for all $x \in \mathbb{R}$. A similar proof shows that $u'(x) \geq 0$. Hence, $u' = 0$, which, in view of (2.87), implies that $u = 0$. It is the perturbation of this unique C^1 solution that we are going to study.

We choose a potential $\widetilde{V}$ of class C^∞ (a lower regularity would be sufficient, but it is not our point here), such that

$$\left(\widetilde{V}(x) \leq 0, \quad \forall x \in \mathbb{R}\right), \quad \left(\widetilde{V}(0) = \inf_{x \in \mathbb{R}} \widetilde{V}(x) < 0\right), \quad \mathrm{supp}\left(\widetilde{V}\right) = [-1, 1].$$
$$\tag{2.88}$$

We also assume that

$$\left(\widetilde{V}\right)'(x) < 0, \quad \forall x \in]-1, 0[\quad \text{and} \quad \left(\widetilde{V}\right)'(x) > 0, \quad \forall x \in]0, 1[. \tag{2.89}$$

Under such conditions, we are able to compute a solution to (2.85) explicitly. We consider separately the regions $(u^\varepsilon)' \leq 0$ and $(u^\varepsilon)' \geq 0$, and note that

$$u^\varepsilon(x) = \begin{cases} e^x \left(\widetilde{V}(0) + \displaystyle\int_x^0 e^{-t} \widetilde{V}\left(\frac{t}{\varepsilon}\right) dt\right) & \text{for } x < 0, \\[4mm] e^{-x}\left(\widetilde{V}(0) + \displaystyle\int_0^x e^t \widetilde{V}\left(\frac{t}{\varepsilon}\right) dt\right) & \text{for } x > 0, \end{cases} \tag{2.90}$$

is a solution. Indeed, a simple computation allows to check that, for instance if $x > 0$, the function

$$g(x) = e^x b(x) - b(0) - \int_0^x e^t b(t)\, dt$$

is such that $g(0) = 0$ and $g'(x) = e^x b'(x)$. Hence, if b reaches its global minimum at $x = 0$, is non-increasing for $x < 0$ and non-decreasing for $x > 0$, then the function g satisfies $g(0) = 0$, is non-increasing for $x < 0$ and non-decreasing for $x > 0$. This clearly implies that $g \geq 0$ on $\mathbb{R}$. Applying this to $b = \widetilde{V}\left(\frac{\cdot}{\varepsilon}\right)$ we deduce that

$$e^x \, \widetilde{V}\left(\frac{x}{\varepsilon}\right) - \widetilde{V}(0) - \int_0^x e^t \, \widetilde{V}\left(\frac{t}{\varepsilon}\right) \, dt \geq 0.$$

This means that, u^ε being defined by (2.90), $(u^\varepsilon)'(x)$ is non-negative for $x > 0$, hence $(u^\varepsilon)'(x) = |(u^\varepsilon)'(x)|$. In addition, since (2.90) satisfies $u^\varepsilon(x) + (u^\varepsilon)'(x) = \widetilde{V}\left(\frac{x}{\varepsilon}\right)$, we obtain Eq. (2.85) for $x > 0$. A similar argument holds for $x < 0$.

We have built a bounded C^1 solution to (2.85). We now prove that this solution is unique. First, note that a bounded C^1 solution vanishes at infinity. Indeed, outside the support of $\widetilde{V}\left(\frac{\cdot}{\varepsilon}\right)$, such a solution satisfies (2.87). We first study this solution on the right of the support of $\widetilde{V}\left(\frac{\cdot}{\varepsilon}\right)$. If $u'_\varepsilon \leq 0$ on this interval, then Eq. (2.87) reads as $u^\varepsilon - u'_\varepsilon = 0$, hence $u^\varepsilon(x) = Ae^x$. Unless $A = 0$, u^ε is unbounded as $x \to +\infty$, which is not possible. Thus, either $u^\varepsilon \equiv 0$ on the right of the support of $\widetilde{V}\left(\frac{\cdot}{\varepsilon}\right)$, or there exists an interval on which $u'_\varepsilon > 0$. In the latter case, consider a maximal interval $]\alpha, \beta[$ where $(u^\varepsilon)' > 0$. Then, for all $x \in]\alpha, \beta[$, $u^\varepsilon(x) = -Ae^{-x}$ for some $A > 0$. If this interval is bounded, that is, if $\beta < +\infty$, the fact that it is maximal implies that $(u^\varepsilon)'(\beta) = u^\varepsilon(\beta) = 0$, hence $A = 0$. This contradicts the definition of $]\alpha, \beta[$. Thus, $\beta = +\infty$, and $u^\varepsilon(x) \to 0$ as $x \to +\infty$. A similar argument proves that u^ε vanishes at $-\infty$. Next, we consider two solutions u^ε and v^ε, both bounded, of class C^1, and vanishing at $\pm\infty$. Let $w = u^\varepsilon - v^\varepsilon$. It satisfies $w + |(u^\varepsilon)'| - |(v^\varepsilon)'| = 0$ on $\mathbb{R}$. Applying the triangle inequality, we obtain

$$|w(x)| \leq |w'(x)|, \quad \forall x \in \mathbb{R}. \tag{2.91}$$

Since w is bounded, of class C^1 and vanishes at infinity, assuming that $w \neq 0$, it has a local extremum at some point that we denote by x_0: $w(x_0) \neq 0$ and $w'(x_0) = 0$, which is in contradiction with (2.91). Thus, $w = 0$ and we have proved uniqueness.

Using the explicit expression (2.90) of the solution to (2.85), we are now in position to compute its limit as $\varepsilon \to 0$. Since the function $\widetilde{V}$ has compact support, the sequence of functions $\widetilde{V}\left(\frac{\cdot}{\varepsilon}\right)$ converges strongly to zero in L^p for any $p < \infty$. Hence, in any such space, the (strong) limit u^* of u^ε, is

$$u^*(x) = \begin{cases} e^x \, \widetilde{V}(0) & \text{for } x < 0, \\[2mm] e^{-x} \, \widetilde{V}(0) & \text{for } x > 0. \end{cases} = e^{-|x|} \, \widetilde{V}(0). \tag{2.92}$$

This limit is not a solution to the unperturbed homogenized equation, that is, (2.87). Actually, u^* is solution to

$$\begin{cases} u^*(x) + \left|(u^*)'(x)\right| = 0 & \text{for } x \neq 0, \\ u^*(0) = \widetilde{V}(0), \\ u^*(x) \to 0 & \text{as } |x| \to +\infty. \end{cases} \qquad (2.93)$$

The compactly supported defect $\widetilde{V}$ has therefore considerably modified the homogenized equation!

Note that, in all the above arguments, the fact that, in (2.88), the infimum of $\widetilde{V}$ is reached at the origin, that $\widetilde{V}$ is smooth, non-increasing on $\mathbb{R}_-$ and non-decreasing on $\mathbb{R}_+$, that the unperturbed potential is $V_{\text{per}} \equiv 0$, and most of all, that the dimension is equal to 1, are irrelevant. We have made use of these assumptions in order to simplify our proofs. More generally, a non-trivial, non-positive, compactly supported defect has a macroscopic effect on the homogenized limit of the Hamilton-Jacobi Eq. (2.86) with periodic potential. The notion of solution needed to address this general case is that of *viscosity solution* of the Hamilton-Jacobi Eq. (2.86). Such a solution is unique, and the bounded C^1 solution defined above is a viscosity solution. As a consequence, it is the unique viscosity solution, hence the unique bounded C^1 solution. The above explicit computations (2.90) have allowed us to compute this unique viscosity solution. For more details about viscosity solutions, we refer to [Bar94] (in French) or to the first chapter of [BCE97].

The homogenization result we have established on Eq. (2.85) that converges to an equation of the type (2.93) which is not any longer a differential equation throughout the entire real line, but is only "piecewise" such an equation, complemented with a specific point condition at the origin $x = 0$, is actually a particular instance of a much more general result for Hamilton-Jacobi type equations in the presence of defects. An oscillatory equation such as (2.86), namely for instance

$$u_\varepsilon + H\left(\frac{x}{\varepsilon}, \nabla u_\varepsilon\right) = 0,$$

converges in the homogenization limit to an equation of the following form, which clearly generalizes (2.93),

$$\begin{cases} u^*(x) + H_+^*\left(u^*(x)\right) = 0 & \text{for } x > 0, \\ u^*(x) + H_-^*\left(u^*(x)\right) = 0 & \text{for } x < 0, \\ u^*(0) + F^*\left(\nabla u^*(0^-), \nabla u^*(0^+)\right) = 0, \\ u^*(x) \longrightarrow 0 & \text{as } |x| \longrightarrow +\infty. \end{cases}$$

where H_+^* and H_-^* respectively denote the homogenized Hamiltonians on each of the two half-lines considered, and F^* is a specific function, depending upon the Hamiltonian H, reminiscent of the defect localized at the origin and defining a *junction* condition. Such problems typically appear in problems for discrete networks, for problems involving oscillatory interfaces between media, etc... We refer to [AT15, AT19] for examples of such recent works, which are far too sophisticated to be considered in our introductory text.

Chapter 3
Dimension ≥ 2: The "Simple" Cases: Abstract or Periodic Settings

Abstract The main tools we have used in the first chapter were weak convergence and average of functions. In the second chapter, differential operators came into play. We are now going to see that geometry is important (of course in dimensions higher than 2). We will also need to introduce some abstract arguments, hence in particular functional analysis. In these higher dimensions, we indeed cannot integrate analytically partial differential equations.

Keywords Periodic homogenization · General result · Concentration-compactness · Div-curl lemma · Two-scale expansion · Corrector function · Convergence rate

The main tools we have used in the first chapter were weak convergence and average of functions. In the second chapter, differential operators came into play. We are now going to see that geometry is important (of course in dimensions higher than 2). We will also need to introduce some abstract arguments, hence in particular functional analysis. In these higher dimensions, we indeed cannot integrate analytically partial differential equations.

Keeping in mind the information collected in dimensions 0 and 1, we now address Eq. (1) in dimensions $d \geq 2$. For the convenience of the reader, we recall that (1), supplied with homogeneous Dirichlet boundary conditions, reads as

$$\begin{cases} -\operatorname{div}(a_\varepsilon(x)\,\nabla u^\varepsilon(x)) = f(x) \text{ in } \mathcal{D} \subset \mathbb{R}^d, \\ u^\varepsilon = 0 \qquad\qquad\qquad\qquad \text{on } \partial\mathcal{D}. \end{cases} \tag{1}$$

The domain $\mathcal{D}$ is a subset of $\mathbb{R}^d$. Hence, for $d \geq 2$, u^ε has several arguments $x_1, \ldots, x_d$. We however point out that the unknown function u^ε is *scalar*-valued. Equation (1) is an *equation*, not a *system* of equations. For further details on the case of *systems*, the reader is referred to Remark 3.13 below. On the other hand, the coefficient a_ε is now *matrix-valued*. For consistency, we keep the *lowercase* notation a_ε, but the reader should not be mistaken by this. From now on, the

X. Blanc, C. Le Bris, *Homogenization Theory for Multiscale Problems*, MS&A 21,
https://doi.org/10.1007/978-3-031-21833-0_3

bounds (4) are to be understood in the sense of matrices, that is,

$$\exists\, 0 < \mu, \ M < +\infty, \ \forall\, \varepsilon > 0, \ x \in \mathcal{D}, \ \xi \in \mathbb{R}^d,$$

$$\mu\, |\xi|^2 \ \leq\ (a_\varepsilon(x)\xi)\,.\,\xi \ \leq\ M\, |\xi|^2, \qquad (3.1)$$

where $\eta\,.\,\xi$ denotes the euclidean scalar product between the vectors $\xi, \eta \in \mathbb{R}^d$. When the function $a_\varepsilon(x)$ is only defined almost everywhere in $\mathcal{D}$, inequality (3.1) is likewise to be understood almost everywhere. We note that, as far as the matrix a_ε is symmetric, the upper bound in (3.1) implies that $\|a_\varepsilon\|_{L^\infty(\mathcal{D})}$ is bounded independently of ε. This implication does not hold if a_ε is not symmetric. Indeed, for any skew-symmetric matrix a, for all $\xi \in \mathbb{R}^d$, $(a\xi)\,.\,\xi = 0$. Thus, (3.1) only implies a bound on the *symmetric part* of a. In what follows, we mostly consider that a_ε is symmetric, or even scalar-valued. Whenever this is not the case, we will need to assume that $\|a_\varepsilon\|_{L^\infty}$ is bounded independently of ε, in addition to (3.1).

We start this chapter with a celebrated abstract result regarding the limit of u^ε as $\varepsilon \to 0$. This result is general, but gives few information about the limit. Without any additional assumption on the structure of a_ε, we can hardly know more. We next proceed step by step. We first prove in Sect. 3.2 that, as announced, geometry plays an important role. Then, in Sect. 3.3, we generalize the notion of *corrector*, which has been introduced in Chap. 2. We give in Sect. 3.4 a collection of possible proofs to make precise the convergence of u^ε. Our arguments are developed in the periodic setting, but a large part of them carry over to much more general cases. This allows us to collect some methods and information that will prove useful in Chap. 4, where we study the same problems in the presence of defects to periodicity. As briefly mentioned above, we conclude this chapter with Remark 3.13 about systems of elliptic equations.

3.1 The Abstract Setting and Its Proof

Let us write the *variational formulation* of (1) (this notion was already used in dimension 1, in Chap. 2, Eq. (2.10)). Given a right-hand side $f \in H^{-1}(\mathcal{D})$, we consider the following problem: *Find $u^\varepsilon \in H_0^1(\mathcal{D})$ such that, for any $v \in H_0^1(\mathcal{D})$,*

$$\int_{\mathcal{D}} a_\varepsilon(x)\, \nabla u^\varepsilon(x)\,.\, \nabla v(x)\, dx \ =\ \langle f\,,\, v \rangle_{H^{-1}(\mathcal{D}), H_0^1(\mathcal{D})}, \qquad (3.2)$$

where $\langle\,.\,\rangle_{H^{-1}(\mathcal{D}), H_0^1(\mathcal{D})}$ denotes the duality bracket in the corresponding spaces. Of course, in the case $f \in L^2(\Omega)$, this duality bracket is equal to the $L^2(\mathcal{D})$ scalar product $\int_{\mathcal{D}} f(x)\, v(x)\, dx$. It is a classical result that, under suitable assumptions, this variational formulation is equivalent to the form (1), sometimes called the *strong form*. If $f \in L^2(\mathcal{D})$, and if $a_\varepsilon \in W^{1,1}(\mathcal{D})$, then equation $-\mathrm{div}(a_\varepsilon(x)\, \nabla u^\varepsilon(x)) = f(x)$, is to be understood almost everywhere in $x \in \mathcal{D}$. The boundary condition is encoded in the variational formulation (3.2) *via* $u^\varepsilon \in H_0^1(\mathcal{D})$. When the boundary of the domain is regular, and if u^ε is sufficiently regular (which is the case if f is regular), then the boundary condition may be written in the usual sense $u^\varepsilon(x) = 0$ for (almost) all $x \in \partial\mathcal{D}$. Should f be only in $H^{-1}(\mathcal{D})$, the equation is then understood in the weak sense (hence in the distribution sense), and the boundary condition then only reads as $u^\varepsilon \in H_0^1(\mathcal{D})$.

Classical analysis tools, such as the Lax-Milgram Lemma, allow to prove that, under condition (3.1), there exists a unique solution to the variational formulation (we refer to Appendix A and the literature cited there), hence Eq. (1) has a unique solution in the sense made precise above. In the case of a symmetric matrix a_ε, the proof may alternatively be performed using optimization techniques (see Sect. 6.2.1). The latter approach is in particular valid in the case of a scalar-valued coefficient a_ε.

Choosing $v = u^\varepsilon$ in (3.2) (in a regular setting, this is equivalent to multiplying (1) by u^ε, integrating over $\mathcal{D}$ and using the Green formula), we find

$$\int_{\mathcal{D}} a_\varepsilon(x) \nabla u^\varepsilon(x) . \nabla u^\varepsilon(x) \, dx = \langle f , u^\varepsilon \rangle_{H^{-1}(\mathcal{D}), H_0^1(\mathcal{D})}. \tag{3.3}$$

We next use (3.1) to bound the left-hand side from below, and duality to bound the right-hand side from above. We thus have

$$\int_{\mathcal{D}} |\nabla u^\varepsilon(x)|^2 \, dx \leq \frac{1}{\mu} \|f\|_{H^{-1}(\mathcal{D})} \|u^\varepsilon\|_{H_0^1(\mathcal{D})},$$

where (this is crucial) the constant $\dfrac{1}{\mu}$ does not depend on ε. Applying the Poincaré inequality, we deduce

$$\|u^\varepsilon\|_{H_0^1(\mathcal{D})} \leq C \|f\|_{H^{-1}(\mathcal{D})}, \tag{3.4}$$

where C depends only on μ and on the domain $\mathcal{D}$. In particular, it does not depend on ε. Thus,

$$u^\varepsilon \text{ is bounded in } H_0^1(\mathcal{D}), \quad \text{uniformly with respect to } \varepsilon. \tag{3.5}$$

Extracting a subsequence ε', we infer that $u^{\varepsilon'}$ converges weakly in $H_0^1(\mathcal{D})$. Applying the Rellich Theorem (here again we extract a subsequence, but we keep the same notation ε'), it converges strongly in $L^2(\mathcal{D})$. On the other hand, since the sequence of coefficients a_ε is bounded in $L^\infty(\mathcal{D})$, we can choose the extraction ε' such that $a_{\varepsilon'}$ converges weakly-$\star$ in $L^\infty(\mathcal{D})$. However, we know nothing about the limit of the sequence of products $a_{\varepsilon'} \nabla u^{\varepsilon'}$ that are present in (1). We only know that this sequence is bounded, hence converges weakly in $L^2(\mathcal{D})$, up to extracting a subsequence... It is well-known that, in a product of weak converging sequences, the limit of the product is in general not the product of the limits.

In order to obtain more information, we need to use Eq. (1) and its specific properties. This is the aim of the result we expose in the following section.

3.1.1 An Abstract Result

We now state the main theoretical result in the general theory of homogenization of Eq. (1). We next give some comments. The proof is the matter of Sect. 3.1.2.

Proposition 3.1 *Let $\mathcal{D}$ be a bounded open subset of $\mathbb{R}^d$. Assume that $a_\varepsilon \in L^\infty(\mathbb{R}^d)$ satisfies* (3.1), *uniformly with respect to ε. In the case of a non-symmetric matrix a_ε, we also assume that $\|a_\varepsilon\|_{L^\infty(\mathcal{D})}$ is bounded independently of ε. Then, there exists a matrix a^* sharing the properties of a_ε, and there exists a subsequence $a_{\varepsilon'}$ of a_ε, such that, for any function $f \in H^{-1}(\mathcal{D})$, if $u^{\varepsilon'}$ denotes the solution to* (1) *with $a_{\varepsilon'}$, the following convergences hold*

$$u^{\varepsilon'} \xrightarrow{\varepsilon' \to 0} u^*, \quad a_{\varepsilon'} \nabla u^{\varepsilon'} \xrightarrow{\varepsilon' \to 0} a^* \nabla u^*,$$

$$a_{\varepsilon'} \nabla u^{\varepsilon'} . \nabla u^{\varepsilon'} \xrightarrow{\varepsilon' \to 0} a^* \nabla u^* . \nabla u^*, \qquad (3.6)$$

weakly in $H_0^1(\mathcal{D})$, weakly in $L^2(\mathcal{D})$, and in the distribution sense in $\mathcal{D}$, respectively. Moreover,

$$\int_{\mathcal{D}} a_{\varepsilon'} \nabla u^{\varepsilon'} . \nabla u^{\varepsilon'} \, dx \xrightarrow{\varepsilon' \to 0} \int_{\mathcal{D}} a^* \nabla u^* . \nabla u^* \, dx, \qquad (3.7)$$

and u^* is the solution in $H_0^1(\mathcal{D})$ of

$$- \operatorname{div} \left(a^* \nabla u^* \right) = f, \qquad (3.8)$$

Remark 3.1 Let us make precise what we mean, in Proposition 3.1, by "a matrix a^* sharing the properties of a_ε". If the matrices a_ε are symmetric, we assume (3.1), for some constants μ and M that do not depend on ε. The homogenized matrix a^* is then symmetric, and satisfies (3.1) with *the same constants μ and M*. In contrast, when the matrices are not symmetric, we assume the lower bound of (3.1), while the upper bound is replaced by $\|a_\varepsilon\|_{L^\infty} \leq M$. Then, the homogenized matrix satisfies the lower bound of (3.1) with the same constant μ, but it satisfies *only* $\|a^*\|_{L^\infty} \leq \frac{M^2}{\mu}$. We are not going to enter the details of these proofs here, and refer the interested reader to [MT97] or [Tar09, p 81 ff]. $\qquad\square$

It is important to understand the theoretical scope of Proposition 3.1:

- The matrix a^* and the subsequence ε' do not depend on the right-hand side f of the equation. In the context of modelling of materials, this means that there exists an equivalent material (we say a *homogenized* material), and this material is the same, no matter the stress applied to the material. We have already observed this property in the one-dimensional cases of Chap. 2, where the homogenized coefficient depends only on the averages (in various possible meanings) of $\dfrac{1}{a}$, and not on the right-hand side.

- The fact that the limit exists up to extracting a subsequence of u^ε only owes to the existence of a priori bounds on the problem. These bounds have been established at the beginning of the preceding section (note however that the extraction a priori depends on the right-hand side f, hence the importance of the previous comment). The point is that the limit is solution to a linear partial differential equation (the homogenized equation) with right-hand side f. In particular, this limit depends linearly on f. This is an important piece of information.
- As we have pointed out at the end of Chap. 2, the homogenized equation may not be, in general, of the same form as the original equation. The above result states that, in the particular case of a diffusion equation (1), and under the conditions of Proposition 3.1, the limit equation does read as the original one.

On the other hand, the major weakness of this theoretical result is that, in spite of the fact that we know that the homogenized matrix a^* exists, we have an explicit expression neither for a^* nor for the solution u^*. As we will see below in Sect. 3.3, the abstract result can nevertheless be supplemented by another one, that will make slightly more precise the matrix a^*, although still not providing for an explicit expression. For now, we will only be able to compute a^* (and u^*) in some particular cases. When this is possible, we will likewise establish uniqueness of the limit and convergence of the whole sequence of solutions.

We now remark that, in Proposition 3.1, the boundary condition, encoded in the space H_0^1 of the solutions considered, is actually irrelevant. This is stated in the following result, which we will use repeatedly in the sequel, and which we will prove in the next section.

Lemma 3.1 *Assume that the conditions of Proposition 3.1 are satisfied. Let $\widetilde{D} \subset D$ be an open set, $f \in H^{-1}(\widetilde{D})$ and let $v^{\varepsilon'}$ be a sequence in $H^1(\widetilde{D})$ such that*

$$- \operatorname{div}\left(a_{\varepsilon'} \nabla v^{\varepsilon'}\right) = f \quad in \ \widetilde{D}, \tag{3.9}$$

and

$$v^{\varepsilon'} \xrightarrow{\varepsilon' \to 0} v \quad in \ H^1(\widetilde{D}). \tag{3.10}$$

Then,

$$a_{\varepsilon'} \nabla v^{\varepsilon'} \xrightarrow{\varepsilon' \to 0} a^* \nabla v \quad in \ L^2(\widetilde{D}), \quad hence \ -\operatorname{div}\left(a^* \nabla v\right) = f \quad in \ \widetilde{D}. \tag{3.11}$$

Remark 3.2 A careful inspection of the proof (performed in Sect. 3.1.2 below) shows that the results of Proposition 3.1 and Lemma 3.1 carry over to the case where the right-hand side f depends on ε, provided that we have: $f_\varepsilon \longrightarrow f$ as $\varepsilon \to 0$, strongly in $H^{-1}(\tilde{\mathcal{D}})$. The result of Lemma 3.1 is actually stated with this generality in [All02, Proposition 1.2.19], for instance.   □

3.1.2   Proof of the Abstract Result Using the Compactness Method

In this section, we successively prove Proposition 3.1 and Lemma 3.1.

Proof of Proposition 3.1   The proof falls in several, distinct steps.

Step 1: Bounds and Extraction

We first recall that $H^{-1}(\mathcal{D})$ is a separable Hilbert space. Hence, there exists a countable dense subset $\mathcal{F} \subset H^{-1}(\mathcal{D})$. For each $f \in \mathcal{F}$, we consider the function $u^\varepsilon \in H_0^1(\mathcal{D})$ solution to (1). We have already proved the bound (3.5). Since a_ε is bounded, $a_\varepsilon \nabla u^\varepsilon$ is also bounded in $L^2(\mathcal{D})$. It is thus possible to extract a subsequence ε' such that $u^{\varepsilon'}$ converges weakly in $H_0^1(\mathcal{D})$ and $a_{\varepsilon'} \nabla u^{\varepsilon'}$ converges weakly in $L^2(\mathcal{D})$. Applying the Rellich Theorem, we can further assume that $u^{\varepsilon'}$ converges strongly in $L^2(\mathcal{D})$. Each extraction depends on the right-hand side f. Since $\mathcal{F}$ is countable, a diagonal extraction process allows to build a subsequence $\varepsilon' \to 0$ such that

$$\forall f \in \mathcal{F}, \quad a_{\varepsilon'} \nabla u^{\varepsilon'} \longrightarrow r^* \text{ in } L^2(\mathcal{D}),$$
$$u^{\varepsilon'} \longrightarrow u^* \quad \text{in } H^1(\mathcal{D}), \qquad (3.12)$$
$$u^{\varepsilon'} \longrightarrow u^* \quad \text{in } L^2(\mathcal{D}).$$

Given (3.4), the map $f \mapsto u^\varepsilon$ is linear and continuous from $H^{-1}(\mathcal{D})$ to $H_0^1(\mathcal{D})$, uniformly with respect to ε. Thus, since $\mathcal{F}$ is dense in $H^{-1}(\mathcal{D})$, we have convergence for all $f \in H^{-1}(\mathcal{D})$. Since u^ε is a linear function of f, each of the limits r^* and u^* is a linear function of f. In addition, the continuity properties of u^ε and ∇u^ε with respect to f being uniform with respect to ε, they are satisfied by the corresponding limits, that is,

$$u^* = S(f), \quad r^* = R(f),$$

where $S : H^{-1}(\mathcal{D}) \to H_0^1(\mathcal{D})$ and $R : H^{-1}(\mathcal{D}) \to (L^2(\mathcal{D}))^d$ are linear and continuous. Passing to the limit in the distribution sense in (1), we obtain $-\operatorname{div}(r^*) = f$, that is,

$$\forall f \in H^{-1}(\mathcal{D}), \quad -\operatorname{div}(R(f)) = f. \qquad (3.13)$$

The remainder of the proof consists in proving that $r^* = a^* \nabla u^*$, which reads as $R(f) = a^* \nabla[S(f)]$, that is, $RS^{-1}(u^*) = a^* \nabla u^*$, provided S is invertible. In the following step, we prove that this is the case.

Step 2: The Operator S Is Invertible

In order to prove this fact, we use that, by definition of S, for all $f \in H^{-1}(\mathcal{D})$,

$$\langle f, S(f) \rangle_{H^{-1}, H_0^1} = \lim_{\varepsilon' \to 0} \langle f, u^{\varepsilon'} \rangle = \lim_{\varepsilon' \to 0} \int_{\mathcal{D}} a_{\varepsilon'} \nabla u^{\varepsilon'} . \nabla u^{\varepsilon'}$$

$$\geq \liminf_{\varepsilon' \to 0} \mu \int_{\mathcal{D}} \left| \nabla u^{\varepsilon'} \right|^2, \qquad (3.14)$$

where the rightmost inequality stems from (3.1). In addition, we also have, still using (3.1),

$$\|f\|_{H^{-1}(\mathcal{D})} = \left\| \operatorname{div}\left(a_{\varepsilon'} \nabla u^{\varepsilon'} \right) \right\|_{H^{-1}(\mathcal{D})}$$

$$= \sup_{\substack{\varphi \in H_0^1(\mathcal{D}), \\ \|\varphi\|_{H_0^1(\mathcal{D})} = 1}} \int_{\mathcal{D}} a_{\varepsilon'} \nabla u^{\varepsilon'} . \nabla \varphi \leq M \left\| \nabla u^{\varepsilon'} \right\|_{L^2(\mathcal{D})},$$

where the constant M should be replaced by $\sup_{\varepsilon > 0} \|a_\varepsilon\|_{L^\infty(\mathcal{D})}$ when a_ε is not symmetric. Inserting this inequality into (3.14), we find

$$\forall f \in H^{-1}(\mathcal{D}), \quad \langle f, S(f) \rangle_{H^{-1}, H_0^1} \geq \frac{\mu}{M^2} \|f\|_{H^{-1}(\mathcal{D})}^2 .$$

This and the fact that S is continuous allow to apply the Lax-Milgram Lemma in the Hilbert space $H^{-1}(\mathcal{D})$, thereby proving that S is invertible, with a continuous inverse. To see this, we write equation $S(f) = u$ as follows: $\forall g \in H^{-1}(\mathcal{D})$, $\langle g, S(f) \rangle_{H^{-1}, H_0^1} = \langle g, u \rangle_{H^{-1}, H_0^1}$. The estimates we have just proved allow to show that the left-hand side is bilinear, continuous and coercive in $H^{-1}(\mathcal{D})$, while the right-hand side is clearly linear and continuous in $H^{-1}(\mathcal{D})$.

Step 3: Pointwise Lower Bound on $RS^{-1}(v) . \nabla v$

In this step, we fix $v \in H_0^1(\mathcal{D})$ and define $g = -\operatorname{div}(RS^{-1}(v))$. According to Step 2, RS^{-1} is linear and continuous from $H_0^1(\mathcal{D})$ to $L^2(\mathcal{D})$, hence $g \in H^{-1}(\mathcal{D})$. Applying (3.13) to $f = S^{-1}(v)$, we have $-\operatorname{div}(RS^{-1}(v)) = S^{-1}(v)$. Hence,

$$g = S^{-1}(v).$$

Next, we define v^ε as the solution to

$$-\operatorname{div}\left(a_\varepsilon \nabla v^\varepsilon \right) = g, \quad v^\varepsilon \in H_0^1(\mathcal{D}).$$

Step 1 implies that

$$v^{\varepsilon'} \longrightarrow S(g) = v \text{ in } H^1(\mathcal{D}), \quad a_{\varepsilon'} \nabla v^{\varepsilon'} \longrightarrow R(g) = R S^{-1}(v) \text{ in } L^2(\mathcal{D}),$$

the first limit being strong in $L^2(\mathcal{D})$. For any $\varphi \in C^\infty(\overline{\mathcal{D}})$ such that $\varphi \geq 0$, we use $\varphi\, v^{\varepsilon'}$ as a test-function in the variational formulation defining $v^{\varepsilon'}$. We thus obtain

$$\left\langle g, \varphi\, v^{\varepsilon'} \right\rangle_{H^{-1}, H_0^1} = \int_{\mathcal{D}} a_{\varepsilon'} \nabla v^{\varepsilon'} . \nabla \left(\varphi\, v^{\varepsilon'} \right)$$

$$= \int_{\mathcal{D}} \varphi\, a_{\varepsilon'} \nabla v^{\varepsilon'} . \nabla v^{\varepsilon'} + \int_{\mathcal{D}} v^{\varepsilon'} a_{\varepsilon'} \nabla v^{\varepsilon'} . \nabla \varphi.$$

Since $v^{\varepsilon'}$ converges weakly in H^1, we can pass to the limit in the left-hand side. The second term of the right-hand side is a product of a weak L^2 convergence (of $a_{\varepsilon'} \nabla v^{\varepsilon'}$) by a strong L^2 convergence (of $v^{\varepsilon'} \nabla \varphi$). Therefore,

$$\lim_{\varepsilon' \to 0} \int_{\mathcal{D}} \varphi\, a_{\varepsilon'} \nabla v^{\varepsilon'} . \nabla v^{\varepsilon'} = \langle g, \varphi\, v \rangle_{H^{-1}, H_0^1} - \int_{\mathcal{D}} v\, R S^{-1}(v) . \nabla \varphi.$$

Using $g = -\operatorname{div}(R S^{-1}(v))$, this gives

$$\lim_{\varepsilon' \to 0} \int_{\mathcal{D}} a_{\varepsilon'} \nabla v^{\varepsilon'} . \nabla v^{\varepsilon'} \varphi = \int_{\mathcal{D}} R S^{-1}(v) \nabla(\varphi v) - \int_{\mathcal{D}} v\, R S^{-1}(v) . \nabla \varphi$$

$$= \int_{\mathcal{D}} \varphi\, R S^{-1}(v) . \nabla v. \qquad (3.15)$$

The coefficient a_ε being coercive, we infer

$$\lim_{\varepsilon' \to 0} \int_{\mathcal{D}} a_{\varepsilon'} \nabla v^{\varepsilon'} . \nabla v^{\varepsilon'} \varphi \geq \mu \liminf_{\varepsilon' \to 0} \int_{\mathcal{D}} |\nabla v^{\varepsilon'}|^2 \varphi \geq \mu \int_{\mathcal{D}} |\nabla v|^2 \varphi,$$

where we have used the weak convergence of $\nabla v^{\varepsilon'}$ to ∇v in $L^2(\mathcal{D})$ and the convexity of the map $v \mapsto \int |\nabla v|^2 \varphi$. We thus have

$$\int_{\mathcal{D}} \varphi\, R S^{-1}(v) . \nabla v \geq \mu \int_{\mathcal{D}} |\nabla v|^2 \varphi,$$

for all non-negative $\varphi \in C^\infty(\overline{\mathcal{D}})$. Hence, the function $\psi = R S^{-1}(v) . \nabla v - \mu |\nabla v|^2$, which belongs to $L^1(\mathcal{D})$, satisfies

$$\forall \varphi \in C^\infty(\overline{\mathcal{D}}), \quad \varphi \geq 0, \quad \int_{\mathcal{D}} \varphi\, \psi \geq 0.$$

It is a simple exercise (it suffices to take for φ an approximation of the characteristic function of the set $\{\psi < 0\}$) to prove that this implies that $\psi \geq 0$, that is,

$$RS^{-1}(v).\nabla v \geq \mu|\nabla v|^2, \quad \text{almost everywhere in } \mathcal{D}. \tag{3.16}$$

Step 4: Pointwise Upper Bound on $RS^{-1}(v).\nabla v$

We return to (3.15), and now use the fact that a_ε is bounded by M in the left-hand side. Still assuming that $\varphi \in C^\infty(\overline{\mathcal{D}})$, $\varphi \geq 0$, this gives

$$\lim_{\varepsilon' \to 0} \int_{\mathcal{D}} a_{\varepsilon'} \nabla v^{\varepsilon'}.\nabla v^{\varepsilon'} \varphi \geq \frac{1}{M} \liminf_{\varepsilon' \to 0} \int_{\mathcal{D}} |a_{\varepsilon'} \nabla v^{\varepsilon'}|^2 \varphi$$

$$\geq \frac{1}{M} \int_{\mathcal{D}} |RS^{-1}(v)|^2 \varphi, \tag{3.17}$$

where, as in Step 3, we have used the weak convergence of $a_{\varepsilon'} \nabla u^{\varepsilon'}$ to $RS^{-1}(v)$ and the convexity of $v \mapsto \int |\nabla v|^2 \varphi$. Note that the above estimate uses that $\forall \xi \in \mathbb{R}^d$, $(a_\varepsilon \xi).\xi \geq \frac{1}{M}(a_\varepsilon \xi).(a_\varepsilon \xi)$, which is implied by (3.1) if a_ε is symmetric. If this is not the case, we need to replace M by $\sup_{\varepsilon>0} \|a_\varepsilon\|_{L^\infty}^2 \mu^{-1}$. Although not difficult, this bound is not obvious. We refer to [Tar09, p 81 ff] or [MT97] for the details.

The convergence (3.15) and the inequality (3.17) together imply

$$\int_{\mathcal{D}} \varphi \, RS^{-1}(v).\nabla v \geq \frac{1}{M} \int_{\mathcal{D}} |RS^{-1}(v)|^2 \varphi.$$

Here again, φ can be any non-negative function of class C^∞. This shows

$$RS^{-1}(v).\nabla v \geq \frac{1}{M}|RS^{-1}(v)|^2, \quad \text{almost everywhere in } \mathcal{D}.$$

Applying the Cauchy-Schwarz inequality in $\mathbb{R}^d$, we infer

$$|RS^{-1}(v)| \leq M|\nabla v|, \quad \text{almost everywhere in } \mathcal{D}. \tag{3.18}$$

Step 5: Existence of a^*

Applying (3.18) to $v - w$, we get that, for any open set $O \subset \mathcal{D}$,

$$\forall v, w \in H_0^1(\mathcal{D}), \quad \nabla v = \nabla w \text{ a.e. in } O \quad \Rightarrow \quad RS^{-1}(v) = RS^{-1}(w) \text{ a.e. in } O.$$

$$\tag{3.19}$$

We fix an open set $O \subset\subset \mathcal{D}$ (this notation means that O is an open set the closure of which satisfies $\overline{O} \subset \mathcal{D}$). Let χ be a smooth cut-off function having compact support in $\mathcal{D}$, and such that $\chi_{|O} = 1$. For all $\xi \in \mathbb{R}^d$, we define $v(x) = (\xi \cdot x)\chi(x)$, which satisfies $\nabla v = \xi$ in O. The map $\xi \mapsto v$ is linear, hence so is $\xi \mapsto RS^{-1}v$. We deduce that there exists a function $a^* : O \mapsto \mathbb{R}^{d \times d}$ such that

$$\forall x \in O \quad RS^{-1}(v)(x) = a^*(x)\xi.$$

This function a^* is measurable, and does not depend on the choice of χ, in view of (3.19). In addition, the same arguments allow to prove that, if $O' \subset O$, then the matrix $a^*(x)$ defined using O' coincides with the matrix defined using O on the set O'. Thus, the map a^* is well-defined. We are now going to prove that

$$\forall v \in H_0^1(\mathcal{D}), \quad RS^{-1}v = a^*\nabla v, \quad \text{almost everywhere.} \tag{3.20}$$

We have already proved this property under the following particular form: if $\nabla v = \xi$ in O, then $RS^{-1}v = a^*\xi$ almost everywhere in O. Hence, if v is piecewise affine, this allows to prove (3.20) for this specific v. Next, we use the fact that the set of piecewise affine functions is dense in $H_0^1(\mathcal{D})$, and that both sides of (3.20) are continuous for the $H_0^1(\mathcal{D})$ norm, thereby proving (3.20).

Step 6: Conclusion Equality (3.20), together with the definitions of the operators R and S, allow to prove that $r^* = RS^{-1}(u^*) = a^*\nabla u^*$. This and the convergences (3.12) allow to pass to the limit in the weak formulation of $-\operatorname{div}\left(a_{\varepsilon'}\nabla u^{\varepsilon'}\right) = f$. We obtain

$$\forall \varphi \in H_0^1(\mathcal{D}), \quad \int_\mathcal{D} a^*\nabla u^*.\nabla\varphi = \lim_{\varepsilon' \to 0} \int_\mathcal{D} a_{\varepsilon'}\nabla u^{\varepsilon'}.\nabla\varphi = \langle f, \varphi\rangle_{H^{-1}, H_0^1}.$$

We have thus proved (3.8). It remains to prove the third convergence of (3.6), together with (3.7). We fix $\varphi \in C^\infty(\overline{\mathcal{D}})$. Using $u^{\varepsilon'}\varphi$ as a test function in the variational formulation satisfied by $u^{\varepsilon'}$, we have

$$\int_\mathcal{D} \varphi\, a_{\varepsilon'}\nabla u^{\varepsilon'}.\nabla u^{\varepsilon'} = \int_\mathcal{D} a_{\varepsilon'}\nabla u^{\varepsilon'}.\nabla\left(u^{\varepsilon'}\varphi\right) - \int_\mathcal{D} u^{\varepsilon'}\, a_{\varepsilon'}\nabla u^{\varepsilon'}.\nabla\varphi$$

$$= \left\langle f, u^{\varepsilon'}\varphi\right\rangle - \int_\mathcal{D} u^{\varepsilon'}\, a_{\varepsilon'}\nabla u^{\varepsilon'}.\nabla\varphi.$$

The second line of (3.12) allows to pass to the limit in the first term of the right-hand side, since f and φ are fixed. In order to pass to the limit in the second term, we use that $u^{\varepsilon'}$ converges strongly to u^* in L^2, and the first line of (3.12). Hence, using (3.8),

$$\lim_{\varepsilon' \to 0} \int_\mathcal{D} \varphi\, a_{\varepsilon'}\nabla u^{\varepsilon'}.\nabla u^{\varepsilon'} = \left\langle f, u^*\varphi\right\rangle - \int_\mathcal{D} u^*\, a^*\nabla u^*.\nabla\varphi$$

$$= \int_\mathcal{D} a^*\nabla u^*.\nabla(u^*\varphi) - \int_\mathcal{D} u^*\, a^*\nabla u^*.\nabla\varphi = \int_\mathcal{D} \varphi\, a^*\nabla u^*.\nabla u^*.$$

This limit holds for any smooth function φ, hence we have proved the third convergence of (3.6). Finally, using $\varphi = 1$, we obtain (3.7). $\square$

We now give the

Proof of Lemma 3.1 In order to prove this result, we temporarily anticipate the result of Lemma 3.2, which we will only prove below. The reader can easily check that Lemma 3.1 is not used in the proof of Lemma 3.2.

The sequence $a_{\varepsilon'} \nabla v^{\varepsilon'}$ is bounded in $L^2(\widetilde{\mathcal{D}})$, hence it converges weakly to some $\eta \in \left(L^2(\widetilde{\mathcal{D}})\right)^d$, up to extraction of a subsequence. Since we are going to prove that the only possible limit is $a^* \nabla v^*$, we do not make this extraction explicit. We have

$$a_{\varepsilon'} \nabla v^{\varepsilon'} \xrightarrow[\varepsilon' \to 0]{} \eta \text{ in } L^2(\widetilde{\mathcal{D}}). \tag{3.21}$$

In order to prove that $\eta = a^* \nabla v^*$, we fix $\lambda \in \mathbb{R}^d$, $\varphi \in C^\infty(\widetilde{\mathcal{D}})$ with compact support, and define $u^{\varepsilon'}$ as the solution to

$$\begin{cases} - \operatorname{div}\left(a_{\varepsilon'} \nabla u^{\varepsilon'}\right) = - \operatorname{div}\left[a^* \nabla \left(\lambda . x \, \varphi(x)\right)\right], \\ u^{\varepsilon'} \in H_0^1(\widetilde{\mathcal{D}}). \end{cases} \tag{3.22}$$

Applying Proposition 3.1, we infer the following convergences:

$$u^{\varepsilon'} \longrightarrow u^* \text{ in } L^2(\widetilde{\mathcal{D}}), \quad u^{\varepsilon'} \longrightarrow u^* \text{ in } H^1(\widetilde{\mathcal{D}}),$$

$$a_{\varepsilon'} \nabla u^{\varepsilon'} \longrightarrow a^* \nabla u^* \text{ in } L^2(\widetilde{\mathcal{D}}), \tag{3.23}$$

where the homogenized limit u^* is the unique solution to equation $- \operatorname{div}(a^* \nabla u^*) = - \operatorname{div}\left[a^* \nabla \left(\lambda . x \, \varphi(x)\right)\right]$ in $H_0^1(\widetilde{\mathcal{D}})$, hence

$$u^*(x) = \lambda . x \, \varphi(x).$$

Since the matrix $a_{\varepsilon'}$ is coercive, we have

$$\left(a_{\varepsilon'} \nabla v^{\varepsilon'} - a_{\varepsilon'} \nabla u^{\varepsilon'}\right) . \left(\nabla v^{\varepsilon'} - \nabla u^{\varepsilon'}\right) \geq 0,$$

almost everywhere in $\widetilde{\mathcal{D}}$. We apply Lemma 3.2, which allows to pass to the limit in the above product, since $\nabla v^{\varepsilon'}$ converges weakly to ∇v^*, $a_{\varepsilon'} \nabla u^{\varepsilon'}$ converges weakly to $a^* \nabla u^*$, its divergence being bounded (actually, it is constant) in $H^{-1}(\widetilde{\mathcal{D}})$. Thus,

$$\left(\eta - a^* \nabla u^*\right) . \left(\nabla v - \nabla u^*\right) \geq 0, \tag{3.24}$$

almost everywhere in $\widetilde{\mathcal{D}}$. We now fix $x_0 \in \widetilde{\mathcal{D}}$ where (3.24) is satisfied, and choose φ such that $\varphi = 1$ in a neighbourhood of x_0, and $\lambda = \nabla v(x_0) + t\mu$, for a fixed $\mu \in \mathbb{R}^d$. Inequality (3.24) becomes

$$\left(\eta(x_0) - a^*(x_0)\nabla v(x_0) - a^*(x_0)t\mu\right) \cdot (-t\mu) \geq 0.$$

This is valid for any $t \in \mathbb{R}$, so, letting $t \to 0$, we obtain

$$\left(\eta(x_0) - a^*(x_0)\nabla v(x_0)\right) \cdot \mu = 0.$$

This equality is valid for any $\mu \in \mathbb{R}^d$. Thus, we have $\eta(x_0) = a^*(x_0)\nabla v(x_0)$. Since the latter equality holds for almost all $x_0 \in \widetilde{\mathcal{D}}$, this concludes the proof.   □

3.2   Interlude: When Geometry Comes into Play

The aim of this section is to prove that, in dimensions higher than or equal to 2 and for the problems we consider, geometry plays an important role in the limit $\varepsilon \to 0$. We compare three situations, all in dimension 2, modelling some *composite* materials, that is, materials composed of several (here, two) distinct phases at the microscopic scale. Although the proportion of each material is the same in all the situations we consider, we will see that the geometry according to which the phases are combined with one another may change the homogenized material, that is, the matrix a^*.

3.2.1   A Laminated Material

We start with a two-dimensional case that looks like a one-dimensional situation. Such a material is called *laminated*.
   We consider the following partial differential equation:

$$- \operatorname{div} \left(a_{per} \left(\frac{x_1}{\varepsilon}\right) \nabla u^\varepsilon(x_1, x_2)\right) = f. \tag{3.25}$$

This equation is posed on the domain $Q =]0, 1[^2$, with homogeneous Dirichlet boundary conditions.
   In (3.25), the function a_{per} is assumed to be periodic of period 1 and scalar-valued. It is also assumed to satisfy (4). The point is that it only depends on the first coordinate x_1 of $x = (x_1, x_2) \in \mathbb{R}^2$. This therefore models a two-dimensional

material the properties of which only depend on x_1. A possible example is

$$
a_{per}(x_1) = \begin{cases} \alpha \text{ if } 0 \le x_1 \le 1/2, \\[2mm] \beta \text{ if } 1/2 < x_1 \le 1, \end{cases} \tag{3.26}
$$

where the constants α and β are positive. In such a case, (3.25) is understood as a composite material composed of two constituents, with respective coefficients α and β, each layer of constituent having width $\varepsilon/2$ in the direction x_1, see Fig. 3.1 (left).

The homogenized problem we obtain from (3.25) by letting ε tend to 0 is described in the following proposition:

Proposition 3.2 *As ε vanishes, the solution u^ε to (3.25) converges (strongly in L^2 and weakly in H^1) to the solution u^* of*

$$
-\operatorname{div}\left(\begin{pmatrix} \dfrac{1}{\langle \frac{1}{a_{per}} \rangle} & 0 \\[4mm] 0 & \langle a_{per} \rangle \end{pmatrix} \nabla u^*\right) = f, \tag{3.27}
$$

with $u^* = 0$ on the boundary ∂Q, that is,

$$
-\operatorname{div}\left(\dfrac{1}{\langle \frac{1}{a_{per}} \rangle} \partial_{x_1} u^*(x_1, x_2)\, e_1 + \langle a_{per} \rangle \partial_{x_2} u^*(x_1, x_2)\, e_2\right) = f,
$$

where (e_1, e_2) denotes the canonical basis of $\mathbb{R}^2$.

Let us note that, although the coefficient a_ε is scalar-valued, the homogenized coefficient a^* is a (diagonal) non-trivial matrix. In dimensions $d \ge 2$, this is a generic situation. It is even possible, starting with a scalar-valued (periodic or not) coefficient, to obtain a homogenized matrix that is not diagonal, if only because the property of being diagonal is not stable under a change of basis (see for instance [Tar09, page 73]).

We now explain (3.27) intuitively. In the direction x_1, the material is exactly identical to the one studied in the one-dimensional case in Chap. 2. It is thus natural that the quantity $\dfrac{1}{\langle \frac{1}{a_{per}} \rangle}$ shows up as the homogenized coefficient. In the direction x_2, the material is not heterogeneous at the scale ε. It is thus also natural that the response in this direction be the average (in the usual sense, that is, $\langle a_{per} \rangle$) of the microscopic materials. An analogy with an electrical system is even more convincing: the reader probably knows that electrical *resistances* add up when connected in series, while it is electrical *conductances* (the inverses of resistances) that add up when connected in parallel. It is a similar phenomenon which we observe here.

We now give the:

Proof of Proposition 3.2 Since a_{per} satisfies the bounds (4), a simple argument allows to prove that the sequences $\partial_{x_i} u^\varepsilon$ are bounded in $L^2(Q)$. In other words, u^ε is bounded in $H_0^1(Q)$. Up to extracting a subsequence (we will no longer make this extraction explicit, since in the end, the limit obtained is unique –we have already used such an argument in the preceding chapter, and in the proof of Lemma 3.1–), we may assume that u^ε converges to u^*, weakly in $H_0^1(Q)$. Applying the Rellich Theorem, we may also assume that this convergence is strong in $L^2(Q)$. Let us define, for $i = 1, 2$,

$$\sigma_i^\varepsilon = a_{per}\left(\frac{x_1}{\varepsilon}\right)\partial_{x_i} u^\varepsilon(x_1, x_2).$$

We clearly have

$$\frac{1}{a_{per}\left(\frac{x_1}{\varepsilon}\right)}\sigma_1^\varepsilon = \partial_{x_1} u^\varepsilon \xrightarrow{\varepsilon \to 0} \partial_{x_1} u^*. \tag{3.28}$$

On the other hand, using the bounds on $\partial_{x_1} u^\varepsilon$ and on a_{per}, we also know that σ_1^ε is bounded in $L^2(Q)$. Next, we write (3.27) as follows

$$-\partial_{x_1}\sigma_1^\varepsilon = f + \partial_{x_2}\sigma_2^\varepsilon.$$

From this we deduce that σ_1^ε is bounded $L_{x_1}^2(H_{x_2}^{-1}) = \left(L_{x_1}^2(H_{0,x_2}^1)\right)'$, uniformly with respect to ε. We next use the Aubin-Lions lemma (Lemma A.1), which implies that, σ_1^ε being bounded in $L^2(Q)$ and $\partial_{x_1}\sigma_1^\varepsilon$ being bounded dans $L_{x_1}^2(H_{x_2}^{-1})$, up to extraction, σ_1^ε converges strongly in $L_{x_1}^2(H_{x_2}^{-1})$ to some σ_1. We infer from this the weak convergence

$$\frac{1}{a_{per}\left(\frac{x_1}{\varepsilon}\right)}\sigma_1^\varepsilon \xrightarrow{\varepsilon \to 0} \left\langle\frac{1}{a_{per}}\right\rangle\sigma_1. \tag{3.29}$$

In the left-hand side, $\frac{1}{a_{per}}\left(\frac{\cdot}{\varepsilon}\right)$ indeed converges weakly to its average, while σ_1^ε converges strongly to σ_1. Collecting (3.28) and (3.29) we thus have

$$\sigma_1 = \frac{1}{\left\langle\frac{1}{a_{per}}\right\rangle}\partial_{x_1} u^*.$$

Using that a_{per} does not depend on x_2, we write (this is the key point of the proof)

$$\sigma_2^\varepsilon = a_{per}\left(\frac{x_1}{\varepsilon}\right)\partial_{x_2} u^\varepsilon = \partial_{x_2}\left(a_{per}\left(\frac{x_1}{\varepsilon}\right)u^\varepsilon\right).$$

The sequence u^ε converges strongly in L^2, while the sequence $a_{per}\left(\dfrac{x_1}{\varepsilon}\right)$ converges weakly-$\star$ in L^∞. Thus,

$$a_{per}\left(\frac{x_1}{\varepsilon}\right)u^\varepsilon \xrightarrow{\varepsilon \to 0} \langle a_{per}\rangle u^*,$$

in L^2, whence the weak convergence

$$\sigma_2^\varepsilon \xrightarrow{\varepsilon \to 0} \sigma_2 = \langle a_{per}\rangle \partial_{x_2} u^*,$$

in H^{-1}, which concludes the proof. $\qquad\square$

3.2.2 Checkerboard Materials

Let us now successively consider two *checkerboard* materials.

3.2.2.1 The Periodic Checkerboard

First, we choose a function $a_{per}(x_1, x_2)$ that is Q-periodic, piecewise constant, alternately taking the values $\alpha > 0$ and $\beta > 0$, according to Fig. 3.1 (center), that is,

$$a_{per}(x_1, x_2) = \begin{cases} \alpha & \text{if } (x_1, x_2) \in \left]0, \frac{1}{2}\right[^2 \cup \left]\frac{1}{2}, 1\right[^2, \\ \beta & \text{otherwise.} \end{cases} \qquad (3.30)$$

Fig. 3.1 The three different materials considered in Sect. 3.2: a laminated material (left), a checkerboard material (center), and a random material (right). All materials share the same proportions of the phases (coloured in red and blue, respectively). However, the homogenized coefficient a^* may be different

We define the matrix $A_\varepsilon(x) = A_{per}\left(\dfrac{x_1}{\varepsilon}, \dfrac{x_2}{\varepsilon}\right) = a_{per}\left(\dfrac{x_1}{\varepsilon}, \dfrac{x_2}{\varepsilon}\right)$ Id, and the solution u^ε in $H_0^1(\mathcal{D})$ of

$$- \operatorname{div}\left(A_\varepsilon \nabla u^\varepsilon\right) = f, \tag{3.31}$$

which also writes

$$- \operatorname{div}\left(\begin{pmatrix} a_{per}\left(\dfrac{x_1}{\varepsilon}, \dfrac{x_2}{\varepsilon}\right) & 0 \\ 0 & a_{per}\left(\dfrac{x_1}{\varepsilon}, \dfrac{x_2}{\varepsilon}\right) \end{pmatrix} \nabla u^\varepsilon(x_1, x_2)\right) = f,$$

that is,

$$- \operatorname{div}\left(a_{per}\left(\dfrac{x_1}{\varepsilon}, \dfrac{x_2}{\varepsilon}\right)\left(\partial_{x_1} u^\varepsilon(x_1, x_2)e_1 + \partial_{x_2} u^\varepsilon(x_1, x_2)e_2\right)\right) = f.$$

We have

Proposition 3.3 *The solution* $u^\varepsilon \in H_0^1(\mathcal{D})$ *to (3.31), where* $A_{per} = a_{per}$ Id *and* a_{per} *is defined by (3.30), converges to* $u^* \in H_0^1(\mathcal{D})$ *solution to*

$$- \operatorname{div}\left(a^* \nabla u^*\right) = f, \tag{3.32}$$

where the matrix a^* *reads as*

$$a^* = \sqrt{\alpha\beta} \text{ Id} . \tag{3.33}$$

Obviously, this result suffices to *prove* our claim above: in dimensions higher than or equal to 2, geometry is important. Computing averages of the coefficients, under the form $\langle a_{per}\rangle$ or $\langle 1/a_{per}\rangle$, is not sufficient to determine the matrix a^*. Indeed, these averages depend only on the proportion of the phases. The matrix a^* defined by (3.33) is different from the matrix appearing in (3.27), although the constituents of the composite material, that is, the coefficients α and β, are present in the same proportion (1/2 for each) in coefficients (3.26) and (3.30)... For the sake of completeness, we give below the proof of Proposition 3.3. The reader may skip it at first, since it is not fundamental in the sequel. It uses, in a clever way, several specific properties of the problem: the fact that we work in dimension 2, the fact that we have exactly 2 phases, and the fact that the square checkerboard has certain particular symmetries.

Proof of Proposition 3.3 Obviously, the matrix A_ε defined above satisfies the assumptions of Proposition 3.1. Hence, a homogenized matrix a^* exists. We will now prove that a^* may be made explicit, thereby proving its uniqueness, and the convergence of the whole sequence (and not only of an extraction).

Let $\lambda \in \mathbb{R}^2$ and let u_{per} be a function in $H^1(Q)$ satisfying periodic boundary conditions on the boundary of Q and $-\mathrm{div}\,(A_{per}\nabla u_{per}) = \mathrm{div}\,(A_{per}\lambda)$. The existence of such a function u_{per} may be proved using the Lax-Milgram Lemma, as we will do below for the corrector (see (3.67) and the subsequent paragraph). We define $\mathbf{v}_{per} = \nabla u_{per} + \lambda$, which is periodic too. We note that $\langle \mathbf{v}_{per} \rangle = \lambda$ because $\langle \nabla u_{per} \rangle = 0$, since u_{per} is periodic.

We first prove that

$$a^* \langle \mathbf{v}_{per} \rangle = \langle A_{per}\,\mathbf{v}_{per} \rangle. \tag{3.34}$$

In order to do so, we consider the sequence $w^\varepsilon(x) = (\lambda, x) + \varepsilon\, u_{per}(\frac{x}{\varepsilon})$. Applying Proposition 1.1, we deduce that $\nabla w^\varepsilon(x) = \mathbf{v}_{per}(\frac{x}{\varepsilon})$ converges to $\langle \mathbf{v}_{per} \rangle = \lambda$, weakly in $L^2(Q)$. Likewise, $u_{per}(\frac{x}{\varepsilon})$ converges to $\langle u_{per} \rangle$ weakly in $L^2(Q)$. Thus, w^ε converges to $w_0(x) = (\lambda, x)$, weakly in $H^1(Q)$. Using the fact $-\mathrm{div}\,(A_\varepsilon \nabla u^\varepsilon) = -\mathrm{div}\,(A_\varepsilon \mathbf{v}_{per}(\frac{x}{\varepsilon})) = 0$, we may apply Lemma 3.1, which gives that $A_\varepsilon \nabla w^\varepsilon$ converges weakly to $a^* \nabla w_0 = a^* \lambda = a^* \langle \mathbf{v}_{per} \rangle$. On the other hand, applying Proposition 1.1 to the periodic function $A_{per}\mathbf{v}_{per}$, we have that $A_\varepsilon \nabla w^\varepsilon = (A_{per}\,\mathbf{v}_{per})(\frac{x}{\varepsilon})$ weakly converges to $\langle A_{per}\mathbf{v}_{per} \rangle$. This proves (3.34).

We now return to the matrix A_{per} that we have defined above, and which models the checkerboard structure. Let σ be the rotation of angle $\pi/2$. For all $x \in Q$, we have

$$A_{per}(x) A_{per}(\sigma(x)) = \alpha\beta\,\mathrm{Id}\,.$$

This allows us to write

$$A_{per}(\sigma(x))\mathbf{v}_{per}(\sigma(x)) = \alpha\beta\,(A_{per}(x))^{-1}\,\mathbf{v}_{per}(\sigma(x)), \tag{3.35}$$

whence

$$a^*\langle \mathbf{v}_{per} \rangle = \langle A_{per}\mathbf{v}_{per} \rangle, \quad \text{given (3.34)},$$
$$= \langle (A_{per}\mathbf{v}_{per}) \circ \sigma \rangle, \quad \text{since } \sigma \text{ does not change the average},$$
$$= \alpha\beta \left\langle A_{per}^{-1}\mathbf{v}_{per} \circ \sigma \right\rangle, \quad \text{in view of (3.35)}.$$

Moreover,

$$\mathrm{div}\,(A_{per}(x)(A_{per}(\sigma(x))\mathbf{v}_{per}(\sigma(x))))$$
$$= \alpha\beta\,\mathrm{div}\,(\mathbf{v}_{per}(\sigma(x))) = \alpha\beta\,\mathrm{div}\,(\nabla u_{per}(\sigma(x))) = 0, \tag{3.36}$$

and

$$\mathrm{curl}\,(A_{per}(\sigma(x))\mathbf{v}_{per}(\sigma(x))) = \mathrm{div}\,(A_{per}(x)\mathbf{v}_{per}(x)) = 0. \tag{3.37}$$

Using the fact that we work in *dimension 2*, Eq. (3.37) implies that the function $\mathbf{w}_{per}(x) = A_{per}(\sigma(x))\mathbf{v}_{per}(\sigma(x))$ can be written as $\mathbf{w}_{per}(x) = \nabla h_{per}(x) + \langle \mathbf{w}_{per} \rangle$ where h_{per} is periodic and scalar-valued. On the other hand, (3.36) reads as div $(A_{per}\mathbf{w}_{per}) = 0$. Thus, (3.34) applies, replacing $\mathbf{v}_{per}$ by $\mathbf{w}_{per}$. This gives:

$$a^*\langle A_{per}(\sigma(x))\mathbf{v}_{per}(\sigma(x)) \rangle = \langle A_{per}(x)(A_{per}(\sigma(x))\mathbf{v}_{per}(\sigma(x))) \rangle$$

$$= \alpha\beta\langle \mathbf{v}_{per}(\sigma(x)) \rangle.$$

In addition, $a^*\langle A_{per}(\sigma(x))\mathbf{v}_{per}(\sigma(x)) \rangle = a^*\langle A_{per}\mathbf{v}_{per} \rangle = a^*a^*\langle \mathbf{v}_{per} \rangle$, where we have used (3.34) again. Hence

$$a^*a^*\langle \mathbf{v}_{per} \rangle = \alpha\beta\langle \mathbf{v}_{per}(\sigma(x)) \rangle,$$

$$= \alpha\beta\langle \mathbf{v}_{per} \rangle, \quad \text{since } \sigma \text{ does not change the average.}$$

We have thus proved that $(a^*)^2 \lambda = \alpha\beta\lambda$, for all $\lambda \in \mathbb{R}^2$. This implies $(a^*)^2 = \alpha\beta$ Id. Consequently, the polynomial $P(X) = X^2 - \alpha\beta$ satisfies $P(a^*) = 0$. Since it has two simple roots, a^* is diagonalizable. Recalling that the eigenvalues of a^* are positive, the only possible eigenvalue is $\sqrt{\alpha\beta}$. Thus, $a^* = \sqrt{\alpha\beta}$ Id.   □

3.2.2.2   The Random Checkerboard

We now modify the preceding setting in the following way: instead of considering a *periodic* checkerboard, we now define a *random* checkerboard. More precisely, the coefficient is randomly chosen, with probability 1/2, between values α and β, independently in each cell. Of course, we rescale this structure by a factor ε (see Fig. 3.1, on the right). This gives a random coefficient $a_\varepsilon(x, \omega)$ that we insert in an equation similar to (3.31), namely

$$- \text{div} \left(a_\varepsilon(x, \omega)\nabla u^\varepsilon(x, \omega) \right) = f(x). \tag{3.38}$$

We know that the coefficient a_ε is stationary ergodic, because of a classical construction using the Bernoulli shift that has been introduced in Chap. 1.

Arguing as for the periodic checkerboard, it is possible to prove that, here again, the homogenized matrix reads as (3.33), that is, $a^* = \sqrt{\alpha\beta}$ Id. The argument again uses the fact that the dimension is equal to $d = 2$, and that we have exactly two phases in the material. It also uses, now, that the two phases are distributed independently in each cell, with probability 1/2. The main ingredient is, as above, the rotation σ of angle $\pi/2$, and the fact that some quantities are invariant under this rotation. For the proof, we refer to [ZKO94, p 235 ff] (see also [AKM19, exercise 2.10]). Of course, a preliminary stage needed for the actual proof of equality (3.33) is the extension of Proposition 3.1 to the stochastic setting. We thus need to anticipate on Proposition 4.4 below, which states that the matrix a^* indeed exists and is deterministic.

As a conclusion, we find that both the periodic and the random checkerboard give the same homogenized matrix, although they are different. In a sense, these materials are only identical "on average", that is, the proportion of the different phases are equal. This homogenized matrix is different from the homogenized matrix obtained in the case of a laminated material, with the same proportions of materials.

Put differently, our point is that, for dimensions $d \geq 2$, anything may happen!

Remark 3.3 Equality (3.34) teaches us that, in order to compute the matrix a^*, we need two computations: one with $\langle \mathbf{v}_{per} \rangle = (1, 0)$, and a second one with $\langle \mathbf{v}_{per} \rangle = (0, 1)$. The second part of the proof shows that, however, both computations are in fact unnecessary: they may be bypassed using the (very) specific geometry of the problem. In the more elaborate cases that we will see below in Sect. 3.3, such a bypass will not be possible. We will see that determining a^* indeed requires d computations in dimension d. ☐

Remark 3.4 As we have just seen, phases of identical proportions within a given material may give rise to different macroscopic behaviours. This raises an interesting question: given a number of phases, and given a fixed proportion of each one, what are the possible values of the homogenized matrix, depending upon the geometrical repartition of these phases? This question has been the subject of many research works. The interested reader may consult [LC84b, LC84a, LC87, FM87, Mil90, Gra93, FM94, GMS00] (this list of references is not meant to be exhaustive). ☐

3.3 Correction in the General Setting

Recall that, for simplicity, we concentrate on the case where the matrix a_ε is symmetric. The results we prove here are however valid in the case of non-symmetric matrices, up to some modifications of the arguments that we skip for brevity.

We now have to address two important issues:

- In general, the homogenized matrix a^*, the existence of which is stated in Proposition 3.1, is not given by an explicit formula. Actually, the only cases known in which a^* may be analytically computed are more or less the three cases we have considered so far, namely the one-dimensional setting and the two specific cases of the laminated and the checkerboard materials.
- Even if we could compute a^*, we would use it to define the solution of the homogenized Eq. (3.8). But this solution is only an approximation of u^ε in the *weak H^1* topology, as we have seen in (3.6).

We begin to answer these two issues in the present section. In Sect. 3.3.1, we present a formal approach which, in favorable cases, solves both of them. We obtain an expression for a^*, and we define *correctors* (as we have done in dimension 1 in Sect. 2.2 of Chap. 2) that allow us to build a *strong H^1* approximation of u^ε. We will see that, in "good" cases, this formal computation may be made rigorous. Anyhow, it gives an intuition of what we can expect in full generality and how to solve the above two issues in more general cases, as we will next see in Sect. 3.3.2.

3.3.1   Formal Intuition of the Corrector: Two-Scale Expansion

We start with an approximation of u^ε in the *strong $H^1(\mathcal{D})$* topology. Such an approximation will be more precise than the weak approximation (3.6) given by the homogenized solution. In order to build such an approximation, we examine in more details the behaviour of u^ε as $\varepsilon \to 0$. This is achieved by *postulating* that u^ε is of a specific form (in numerical analysis and in physics, this assumption is sometimes referred to as an *Ansatz*). We write u^ε as an expansion in powers of ε, called a *two-scale expansion*:

$$u^\varepsilon(x) = u_0\left(x, \frac{x}{\varepsilon}\right) + \varepsilon\, u_1\left(x, \frac{x}{\varepsilon}\right) + \varepsilon^2\, u_2\left(x, \frac{x}{\varepsilon}\right) + \dots, \tag{3.39}$$

where each function u_k is assumed to depend on two variables: a macroscopic variable x, and a microscopic variable $\dfrac{x}{\varepsilon}$ (hence the terminology *"two-scale"*).

We are now going to assume *scale separation*, that is, we consider x and x/ε as independent variables. This is not strictly speaking true, but we expect that, in the limit $\varepsilon \to 0$, such an assumption becomes valid. This is indeed the case in our setting if the functions under consideration are periodic with respect to the variable x/ε. Then, the variable x/ε takes all the values of the periodic cell $]0, 1[^d$ in the limit $\varepsilon \to 0$. In the other cases where a form of translation invariance holds, the scale separation turns out to be also valid. However, in most situations, such an assumption may be only (at best) confirmed a posteriori, upon rigorously proving that the limit given by the formal expansion (3.39) is indeed the limit of u^ε. In particular, there exists settings where this strategy fails, but in our introductory exposition, we do not consider these situations.

As a first step, we suppose that the coefficient a_ε is a *periodic* rescaled function: $a_\varepsilon(x) = a_{per}\left(\dfrac{x}{\varepsilon}\right)$. Other cases will be considered later on in the present textbook. Correspondingly, we assume that each function u_k in the expansion (3.39) is periodic with respect to its second variable $y = \dfrac{x}{\varepsilon}$, that is,

$$y \longmapsto u_k(x, y) \text{ is } Q - \text{periodic, with } Q =]0, 1[^d. \tag{3.40}$$

In a way, the function is modulated, at each macroscopic point x, by a periodic function varying at the small scale ε. One may think of a product $f(x)g\left(\dfrac{x}{\varepsilon}\right)$, though the form assumed here is more general. As a matter of fact, this particular product form is exact in the one dimensional case, as we see in formula (2.3), namely $(u^\varepsilon)'(x) = \dfrac{c_\varepsilon + F(x)}{a\left(\frac{x}{\varepsilon}\right)}$. Let us now insert (3.39) into (1) and derive the necessary conditions satisfied by the functions u_k. This computation is not difficult, although somewhat tedious. Applying the chain rule, the gradient of a function of the form $v\left(x, \dfrac{x}{\varepsilon}\right)$ reads as:

$$\nabla\left(v\left(x, \frac{x}{\varepsilon}\right)\right) = (\nabla_x v)(x, y) + \frac{1}{\varepsilon}(\nabla_y v)(x, y), \quad \text{où } y = \frac{x}{\varepsilon}, \tag{3.41}$$

where ∇_x and ∇_y are the partial derivatives with respect to x and y, respectively. Each is a d-uple of partial derivatives of the type $(\partial_{x_1}, \dots, \partial_{x_d})$. This derivation rule precisely encodes the fact that we consider the variables x and x/ε as independent variables.

We thus have, using a instead of a_{per} in order to lighten the notation:

$$-\operatorname{div}\left(a\left(\frac{x}{\varepsilon}\right)\nabla u^\varepsilon\right) = -\frac{1}{\varepsilon^2}\operatorname{div}_y(a(y)\,\nabla_y u_0(x, y))$$

$$-\frac{1}{\varepsilon}\Big[\operatorname{div}_x(a(y)\,\nabla_y u_0(x, y)) + \operatorname{div}_y(a(y)\,\nabla_x u_0(x, y))$$

$$+\operatorname{div}_y(a(y)\,\nabla_y u_1(x, y))\Big]$$

$$-\Big[\operatorname{div}_x(a(y)\,\nabla_x u_0(x, y)) + \operatorname{div}_y(a(y)\,\nabla_x u_1(x, y))$$

$$+\operatorname{div}_x(a(y)\,\nabla_y u_1(x, y)) + \operatorname{div}_y(a(y)\,\nabla_y u_2(x, y))\Big]$$

$$+O(\varepsilon), \tag{3.42}$$

where the remainder term $O(\varepsilon)$ depends on the higher-order derivatives of u_0, u_1, u_2, etc. As already mentioned, we argue as if the two variables x and y were independent. First, (1) implies that the coefficient of $\dfrac{1}{\varepsilon^2}$ vanishes, that is,

$$\operatorname{div}_y(a(y)\,\nabla_y u_0(x, y)) = 0. \tag{3.43}$$

We multiply this equation by u_0 and integrate over the periodic cell Q. This gives

$$\mu\int_Q |\nabla_y u_0(x, y)|^2 \leq \int_Q (a(y)\nabla_y u_0(x, y)) \cdot \nabla_y u_0(x, y)\,dy$$

owing to the coercivity (3.1) of a

$$= - \int_Q \text{div}_y(a(y) \nabla_y u_0(x, y)) u_0(x, y) dy$$

$$+ \int_{\partial Q} (a(y) \nabla_y u_0(x, y)) . \mathbf{n} \, u_0(x, y),$$

where the first term of the right-hand side vanishes because of (3.43), while the second term vanishes owing to the periodicity of $u_0(x, y)$ with respect to y. In the above formula, $\mathbf{n}$ denotes the outer unit normal to ∂Q. We thus have

$$\nabla_y u_0(x, y) = 0, \tag{3.44}$$

in Q. Put differently, the function u_0 only depends on the macroscopic variable x:

$$u_0 = u_0(x). \tag{3.45}$$

Next, we return to (3.42). We insert (3.44) into the term of order $\dfrac{1}{\varepsilon}$, and find

$$- \text{div}_y(a(y) (\nabla_x u_0(x) + \nabla_y u_1(x, y))) = 0.$$

The equation satisfied by u_1 is thus

$$\begin{cases} -\text{div}_y(a(y) (\nabla_x u_0(x) + \nabla_y u_1(x, y))) = 0, & \text{in } Q, \\[2mm] u_1 \text{ periodic with respect to } y. \end{cases} \tag{3.46}$$

Temporarily considering u_0 as given (we will compute it below...), we observe that the solution of this equation may be computed explicitly. Indeed, since (3.46) is linear, the solution is given by

$$u_1(x, y) = \sum_{i=1}^{d} \partial_{x_i} u_0(x) \, w_i(y), \tag{3.47}$$

where the functions w_i are solutions to the so-called *subcell problems*, or *corrector problems*

$$\begin{cases} -\text{div}_y(a(y) (e_i + \nabla_y w_i(y))) = 0, & \text{in } Q, \\[2mm] w_i \text{ periodic.} \end{cases} \tag{3.48}$$

where e_i, $i = 1, \ldots d$ denotes the i-th vector of the canonical basis of $\mathbb{R}^d$. We recognize (3.48) as the generalization to higher dimensions of the periodic one-dimensional corrector equation (2.31). This equation has been introduced (without much justification then) in Sect. 2.2 of Chap. 2. We will prove below (see (3.67) and the paragraph that follows) that w_i solution to (3.48) exists and is unique, up to the addition of a constant.

Remark 3.5 In Eq. (3.46), the unknown function $u_1(x, y)$ is only present through its gradient with respect to y. As a consequence, this equation only determines u_1 up to the addition of a function $v(x)$ independent of y. Similarly, the functions w_i are determined by (3.48) up to the addition of a constant. We may nevertheless assume that this function $v(x)$ vanishes, hence the definition (3.47). This modifies neither the homogenized matrix a^* ($v(x)$ disappears in (3.49) below) nor the value of u_0.   □

Equation (3.46) allows to define u_1. As we will see below, it also allows to define the homogenized matrix a^*. In (3.46), y is considered as the spatial variable, and x as a parameter. Recalling that $y = \dfrac{x}{\varepsilon}$, we understand that, in principle, it should be posed in the domain $\dfrac{1}{\varepsilon} \mathcal{D}$. Here, we consider that it is posed in Q, and that x is a parameter that is independent of y. Our strategy for the above string of derivations has therefore relied upon the following three important simplifications:

(i) we consider that x and y as independent, whereas they are related by $y = \dfrac{x}{\varepsilon}$;

(ii) we replace $\dfrac{1}{\varepsilon} \mathcal{D}$ by $\mathbb{R}^d$, so that (3.46) is posed in the whole space, and

(iii) we use the periodicity (3.40), restricting this equation to the domain Q.

Actually, none of these three simplifications is obvious. They are validated by a rigorous proof of the fact that the expansion we use is correct, as we will see below. It is however important to bear in mind that Eq. (3.46) is posed in a very large domain. This is not unexpected: at the microscopic scale, the macroscopic domain is large! We will see, in settings more complicated than the periodic setting (as for instance the periodic setting with defects), how relevant this remark is, and what mathematical difficulties it raises. The point is, periodicity allows to reduce the corrector problem to a *bounded* domain, that is, the periodic cell Q.

We are now going to determine u_0. Of course, as the reader has already guessed, it is equal to u^*. We return to the expansion (3.42) and consider its zeroth order term. This term must be equal to f, according to (1). We thus have

$$-\mathrm{div}_y(a(y)\,(\nabla_x u_1(x, y) + \nabla_y u_2(x, y))) =$$
$$\mathrm{div}_x(a(y)\,(\nabla_y u_1(x, y) + \nabla_x u_0(x))) + f, \qquad (3.49)$$

with periodic boundary conditions on Q for the function $u_2(x, .)$.

A necessary condition for the function u_2 to exist and be periodic is that the integral of the left-hand side over the periodic cell Q vanishes. Indeed, if $\mathbf{g}$ is a periodic vector-valued function, we necessarily have

$$\int_Q \mathrm{div}\,\mathbf{g}(y)\,dy = \int_{\partial Q} \mathbf{g}(y) \cdot \mathbf{n} = 0.$$

Thus, the integral of the right-hand side of (3.49) over Q also vanishes, which writes

$$- \text{div}_x \left(\int_Q a(y) \left(\nabla_y u_1(x, y) + \nabla_x u_0(x) \right) dy \right) = f(x), \tag{3.50}$$

since the derivatives with respect to x and the integration with respect to y commute. This necessary condition (3.50) is actually sufficient for u_2 to exist. Fixing x, (3.49) is an equation in the variable y of the form $-\text{div}_y(a(y) \nabla z(y)) = h(y)$, where h is periodic. A periodic solution z to this equation exists whenever the integral of h over Q vanishes. This is equivalent to condition (3.50), since the integral of the term $\text{div}_y(a(y) \nabla_x u_1(x, y))$ over Q does vanish, owing to periodicity. Should the reader be unfamiliar with such arguments, they will find the corresponding details at the beginning of Sect. 3.4, page 132.

We insert the expression (3.47) for u_1 into (3.50) and find:

$$- \text{div}_x \left(\int_Q a(y) \sum_{j=1}^d \partial_{x_j} u_0(x)(\nabla_y w_j(y) + e_j) dy \right) = f(x). \tag{3.51}$$

We note that

$$\int_Q a(y) \sum_{j=1}^d \partial_{x_j} u_0(x)(\nabla_y w_j(y) + e_j) dy$$

$$= \int_Q a(y) \sum_{j=1}^d (\nabla u_0(x))_j (\nabla_y w_j(y) + e_j) dy$$

$$= \sum_{j=1}^d (\nabla u_0(x))_j \left(\int_Q a(y) (\nabla_y w_j(y) + e_j) dy \right)$$

$$= \sum_{i=1}^d \left(\sum_{j=1}^d \sum_{k=1}^d \int_Q a_{ik}(y)(\nabla_y w_j(y) + e_j)_k \, dy \, (\nabla u_0(x))_j \right) e_i$$

$$= \sum_{i=1}^d \left(\sum_{j=1}^d \mathfrak{a}_{ij} (\nabla u_0(x))_j \right) e_i$$

$$=: \mathfrak{a} \nabla u_0(x)$$

where the coefficients of the matrix $\mathfrak{a}$ are given by

$$\mathfrak{a}_{ij} = \sum_{k=1}^{d} \int_{Q} a_{ik}(y)(\nabla_y w_j(y) + e_j)_k \, dy$$

$$= \int_{Q} (a(y)(\nabla_y w_j(y) + e_j)) \cdot e_i \, dy, \tag{3.52}$$

for $i, j = 1 \ldots N$. Equation (3.51) may be recast into the *homogenized problem*

$$\begin{cases} -\mathrm{div}\,(\mathfrak{a}\,\nabla u_0) = f, & \text{in } \mathcal{D}, \\ \qquad\qquad u_0 = 0, & \text{on } \partial \mathcal{D}. \end{cases} \tag{3.53}$$

The expansion (3.39) formally implies that $u^\varepsilon \longrightarrow u_0$ as $\varepsilon \to 0$. We thus realize that u_0 must be equal to the homogenized solution u^*, that $\mathfrak{a}$ is the homogenized matrix a^*, and that (3.53) is the homogenized equation (3.8). Hence, we have *formally* obtained the following two crucial pieces of information: first, owing to (3.52), which reads as

$$a_{ij}^* = \int_{Q} (a(y)(\nabla_y w_j(y) + e_j)) \cdot e_i \, dy, \tag{3.54}$$

the homogenized matrix may be expressed as a function of the integral over Q of the correctors w_i, $1 \le i \le d$, defined in (3.48). Second, given the explicit expression (3.47) of u^1, the difference

$$u^\varepsilon(x) - u^*(x) - \varepsilon \sum_{i=1}^{d} \partial_{x_i} u^*(x)\, w_i\left(\frac{x}{\varepsilon}\right) \tag{3.55}$$

converges strongly to 0 in $H^1(\mathcal{D})$ (formally, it is equal to $\varepsilon^2 u_2\left(x, \dfrac{x}{\varepsilon}\right) + \ldots$.)

One should bear in mind that, for now, these informations are only *formal*, except in the one-dimensional setting. Indeed, in such a case, the expressions we have obtained for u^* and a^* reduce to those of Chap. 2, which were proved rigorously. To see this, we insert the expression (2.32) of the corrector into (3.54), and recover the formula $a^* = \left(\langle a_{per}^{-1} \rangle\right)^{-1}$ for the homogenized coefficient. Moreover, the difference $u^\varepsilon - u^{\varepsilon,1}$, where $u^{\varepsilon,1}(x) = u^*(x) + \varepsilon\,(u^*)'(x)\,w\left(\dfrac{x}{\varepsilon}\right)$, converges to 0 in H^1, as it was proved in (2.37). In some other cases (including, in the present chapter, the periodic case), we will give a rigorous meaning to the above two properties.

We conclude this section by some considerations on numerical approaches. In practice, how does the above expansion allow to obtain an approximation of the solution u^ε, for small values of ε? The computational procedure goes as follows:

 (i) determine w_i by solving (3.48) in the periodic cell;
 (ii) compute the coefficients of the matrix a^* using (3.54);
 (iii) solve the homogenized problem (3.53) in order to compute u_0, and
 (iv) if a first-order term is needed, calculate u_1 using (3.47).

If the approximation computed happens to be unsatisfactory, it is possible to proceed and compute the second-order term of the expansion upon solving (3.49), and so on and so forth.

Note that steps (i) and (ii) are preliminary computations that allow, as in the simple cases of the previous sections, to compute the homogenized matrix a^*. These computations actually consist in solving d boundary-value problems, whereas, in the one-dimensional case, the task reduces to the computation of the average of some periodic function. Such computations, together with the assembling of the matrix a^*, are, in full generality, not for free! Their cost should be carefully considered before applying this strategy.

Let us emphasize that the computational workload is particularly light here: periodicity implies that the matrix a^* does not depend on the macroscopic variable x. In a more elaborate case, with a matrix of the form $a\left(x, \dfrac{x}{\varepsilon}\right)$, one would need to solve problems of the type (3.48) *at each macroscopic point* x (see Fig. 3.2). This is of course much more costly, even if such computations may be performed in parallel, once and for all values of the right-hand side f.

The benefit of the above described approach is obvious (although, as we have just said, it is not completely free): we *do not need* to discretize the macroscopic domain at a scale smaller than ε. We "only" need pre-computations at a small scale (that is, the resolution of (3.48)), the result of which is used in the numerical resolution of (3.53) on a coarse mesh. Such a mesh would not allow for the same accuracy in the standard discretization of the original equation. Actually, as shown in Sect. 2.1.2, it would even provide with an incorrect result!

To summarize, solving d problems (3.48) posed on the periodic cell allows to obtain a good approximation of the solution u^ε to (1), for *any right-hand side* f, by only solving (3.53) at the macroscopic scale. Put differently, instead of one computation on a fine mesh, a good approximation of u^ε is obtained by $d + 1$ computation on a standard mesh. Further, the method is all the more efficient if we want to solve (1) for many right-hand sides f. We will return in more details to the approach we have briefly described above in Chap. 5, which is dedicated to numerical questions (see in particular Sect. 5.1).

Remark 3.6 The *two-scale expansion* method we have just described is based on (3.39) and is only a *formal* method. It should not be mistaken for the rigorous method of *two-scale convergence* introduced in [Ngu89] and formalized in [All92]. Such a method is a mathematical proof of homogenization. We will mention it briefly in Sect. 3.4.4.   □

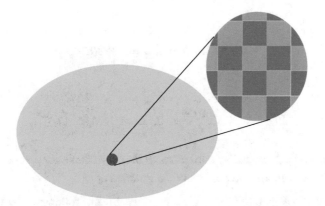

Fig. 3.2 For each macroscopic cell, we solve a subcell problem in order to compute the coefficients of the homogenized matrix

3.3.2   The Correction Theorem

The computations of Sect. 3.3.1 were only formal. In particular, all the functions manipulated were assumed sufficiently regular so as to compute all the derivatives we needed. We have shown that, provided the expansion (3.39) of u^ε was correct and the functions u_k were sufficiently regular, the first terms of the expansion were equal to the functions u_0 and u_1 that we have computed. In the one-dimensional setting, we have rigorously proved that the expansion is indeed correct, in Sect. 2.2 of Chap. 2. In full generality, this mathematical proof now remains to be done. We will perform it in the periodic case in Sect. 3.4, and in some other settings in the subsequent chapters.

For the time being, the formal expansion we have performed gives us an intuition both on the structure of the homogenized matrix using (3.54), and on the tentative equation satisfied by the corrector, namely (3.48). This intuition motivates the following two results (Proposition 3.4 and Proposition 3.5) that we present now, and prove at the end of the present section.

The first of these two results is an extension of Proposition 3.1, where we only have weak convergence of $u^{\varepsilon'}$ to u^* in H^1. In particular, we do not have strong convergence of the derivatives in L^1 for instance. We do not have their convergence almost everywhere either. In order to improve this, we need to add corrections terms to u^*.

Proposition 3.4 (Existence of Correctors) *Under the assumptions of Proposition 3.1, there exist d sequences of functions $w_i^{\varepsilon'}$ in $H^1(\mathcal{D})$ that satisfy*

$$w_i^{\varepsilon'} \xrightarrow{\varepsilon' \to 0} 0, \ \text{weakly in } H^1(\mathcal{D}), \tag{3.56}$$

and

$$- div\, a_{\varepsilon'} \left(e_i + \nabla w_i^{\varepsilon'} \right) = - div(a^* e_i), \tag{3.57}$$

such that

$$\nabla u^{\varepsilon'} - \left(Id + \nabla w^{\varepsilon'} \right) \nabla u^* \xrightarrow{\varepsilon' \to 0} 0, \;\; strongly\; in\; (L^1 (\mathcal{D}))^d. \tag{3.58}$$

These functions $w_i^{\varepsilon'}$ *are called* correctors, *in the sense that their presence in* (3.58) *allows to obtain strong convergence of the gradient of* $u^{\varepsilon'}$.

Remark 3.7 Note that (3.57) may be replaced by the following, less stringent, condition:

$$- div\, a_{\varepsilon'} \left(e_i + \nabla w_i^{\varepsilon'} \right) \xrightarrow{\varepsilon' \to 0} - div(a^* e_i), \;\; strongly\; in\; H^{-1} (\mathcal{D}). \tag{3.59}$$

This convergence, together with (3.56), is indeed sufficient to prove (3.58).   □

Let us point out that, quite often (this will almost always be the case in the present textbook), the homogenized matrix a^* is a constant. Hence, the right-hand side of (3.57) vanishes. This is in particular true in the periodic case, in the periodic case with defect, in the almost periodic case, and in the stationary ergodic case.

In (3.58), the notation $\left(Id + \nabla w^{\varepsilon'} \right)$ stands for the matrix which coefficient (i, j) reads as $\delta_{ij} + \partial_{x_i} w_j^{\varepsilon'}$. Put differently, $\left(Id + \nabla w^{\varepsilon'} \right) \nabla u^*$ denotes the vector the component i of which writes

$$\left[\left(Id + \nabla w^{\varepsilon'} \right) \nabla u^* \right]_i = \sum_{j=1}^{d} \left(\delta_{ij} + \partial_{x_i} w_j^{\varepsilon'} \right) \partial_{x_j} u^*.$$

Of course, the left-hand side of (3.57) has the same structure as the formal periodic corrector equation (3.48). In order to see this, we define $w_i^{\varepsilon}(x) = \varepsilon\, w_i \left(\dfrac{x}{\varepsilon} \right)$ and replace the convergence to zero by an equality (recall that in such a case a^* is constant). This is actually how one proves this result, as we will see below. On the other hand, the convergence (3.58) is a generalization of the formal estimate of the convergence of the difference (3.55).

In the same spirit as Proposition 3.4, the following result gives a general expression for the homogenized matrix a^*. In contrast, Proposition 3.1 only states that a^* exists, without providing any explicit expression for that matrix.

Proposition 3.5 *Still assuming the conditions of Proposition 3.1, the homogenized matrix a^* reads as*

$$a^* = \underset{\varepsilon' \to 0}{\text{weak lim}} \left[a_{\varepsilon'} \left(\text{Id} + \nabla w^{\varepsilon'} \right) \right] \quad in \quad (L^2 (\mathcal{D}))^{d \times d}, \tag{3.60}$$

where $w^{\varepsilon'}$ is defined in Proposition 3.4.

It is obvious that the left-hand side of the convergence (3.59) has the same structure as the "formal" equation of the periodic corrector (3.48). To realize this, it suffices to write $w_i^\varepsilon (x) = \varepsilon \, w_i \left(\dfrac{x}{\varepsilon} \right)$ and replace the convergence to zero (recall that a^* is a constant) by an equality. This is actually, loosely speaking, how one proves this result, as we will see below. The quality of approximation given by the convergence (3.58) is a generalization of the formal estimate of convergence of the difference (3.55).

In the same vein, it is clear that, in the periodic setting and with the same choice of functions w_i^ε, (3.60) reduces to (3.54). We have indeed seen in Chap. 1 that re-scaled periodic functions converge to their average.

The above two results give a mathematical grounding for, and an extension to more general cases of, the formal two-scale expansion we have performed in the simple periodic case. Loosely speaking, they indicate that the computation of a^* and of the correctors (or at least of functions having asymptotically the same properties as correctors if the latter existed) *may be within reach*. Hence, at least in principle, this gives, for each function f, an accurate approximation of u^ε via the solution of the homogenized problem. This is exactly what we have seen using the two-scale expansion.

One should however not be overly optimistic. The expression (3.60) is not really *explicit*. Computing a limit is indeed not so simple in general. Furthermore, in the present case, the right-hand side contains a product of functions, and one of these functions, namely w^ε, is not known explicitly. In the periodic case, the two-scale expansion allows to make this formula *really explicit*, and independent of the extraction. We have seen this formally above, and we will prove it in the next section. This allows to simultaneously obtain the correctors and the homogenized matrix. This is however a miracle due to the periodic structure. In the general case, things are not so easy.

Remark 3.8 The weak limit (3.60) does not depend on the functions w_i^ε as far as they satisfy the conditions of Proposition 3.4. This may be proved easily in dimension $d = 1$. In higher dimensions, we will admit it.   □

We now give the

Proof of Proposition 3.4 As we have done for the proof of Lemma 3.1, we anticipate and use Lemma 3.2 proved in Sect. 3.3.2 below. Here again, the reader may easily check that there is no circular argument: neither the present proof nor its result will be used in the proof of Lemma 3.2.

We consider a domain $\widetilde{\mathcal{D}}$ that contains the closure $\overline{\mathcal{D}}$ of the domain $\mathcal{D}$. There exists a function ϕ of class C^∞, with compact support in $\widetilde{\mathcal{D}}$, and such that $\phi = 1$ in $\mathcal{D}$. We extend $a_{\varepsilon'}$ to $\widetilde{\mathcal{D}} \setminus \mathcal{D}$, for instance by taking it equal to the identity matrix Id in this set. We denote this extension by $\widetilde{a}_{\varepsilon'}$. We apply Proposition 3.1 in the domain $\widetilde{\mathcal{D}}$. Up to extraction of a subsequence (we still denote this extraction by ε'), we obtain a homogenized matrix, that we denote $\widetilde{a}^*$, in $\widetilde{\mathcal{D}}$. In view of Lemma 3.1, this matrix agrees with the homogenized matrix a^* in the domain $\mathcal{D}$. We now solve the problem

$$\begin{cases} -\operatorname{div}\left(\widetilde{a}_{\varepsilon'}\nabla\chi_i^{\varepsilon'}\right) = -\operatorname{div}\left(\widetilde{a}^*\nabla\psi_i\right) & \text{in } \widetilde{\mathcal{D}}, \\ \chi_i^{\varepsilon'} = 0 & \text{on } \partial\widetilde{\mathcal{D}}, \end{cases} \tag{3.61}$$

where ψ_i is defined by

$$\forall x \in \widetilde{\mathcal{D}}, \quad \psi_i(x) = x_i\phi(x). \tag{3.62}$$

The right-hand side of the first line of (3.61) belongs to $H^{-1}(\widetilde{\mathcal{D}})$, hence Proposition 3.1 applies. We thus have

$$\chi_i^{\varepsilon'} \longrightarrow \chi_i^*, \quad \nabla\chi_i^{\varepsilon'} \longrightarrow \nabla\chi_i^*, \quad \widetilde{a}_{\varepsilon'}\nabla\chi_i^{\varepsilon'} \longrightarrow \widetilde{a}^*\nabla\chi_i^*, \tag{3.63}$$

in $L^2(\widetilde{\mathcal{D}})$, hence in $L^2(\mathcal{D})$. Moreover, the solution of the homogenized problem being unique, $\chi_i^* = \psi_i$. In particular, $\chi_i^*(x) = x_i$ in $\mathcal{D}$. We then define $w_i^{\varepsilon'} = \chi_i^{\varepsilon'} - x_i$, which clearly satisfies both the convergences (3.56) and (3.57).

We next prove (3.58). For this purpose, we temporarily assume that u^* is of class C^∞ and has compact support in $\mathcal{D}$. We thus have

$$\int_{\mathcal{D}}\left[a_{\varepsilon'}\left(\nabla u^{\varepsilon'} - \left(\operatorname{Id}+\nabla w^{\varepsilon'}\right)\nabla u^*\right)\right] \cdot \left[\nabla u^{\varepsilon'} - \left(\operatorname{Id}+\nabla w^{\varepsilon'}\right)\nabla u^*\right]$$

$$= \int_{\mathcal{D}}\left(a_{\varepsilon'}\nabla u^{\varepsilon'} \cdot \nabla u^{\varepsilon'}\right) - 2\int_{\mathcal{D}}\left(a_{\varepsilon'}\nabla u^{\varepsilon'}\right) \cdot \left(\left(\operatorname{Id}+\nabla w^{\varepsilon'}\right)\nabla u^*\right)$$

$$+ \int_{\mathcal{D}}\left[a_{\varepsilon'}\left(\operatorname{Id}+\nabla w^{\varepsilon'}\right)\nabla u^*\right] \cdot \left[\left(\operatorname{Id}+\nabla w^{\varepsilon'}\right)\nabla u^*\right]. \tag{3.64}$$

According to (3.7) of Proposition 3.1, we may pass to the limit in the first term of the right-hand side. For the second term, we apply Lemma 3.2. To this end, we consider ∇u^* as a test function, and we clearly have that $\operatorname{div}\left(a_{\varepsilon'}\nabla u^{\varepsilon'}\right)$ is bounded in $H^{-1}(\mathcal{D})$ and that $\left(\operatorname{Id}+\nabla w^{\varepsilon'}\right)_j = e_j+\nabla w_j^{\varepsilon'}$ is curl free. Finally, in order to pass to the limit in the last term of (3.64), we apply (3.7) to the sequence $\chi_j^{\varepsilon'} = x_j + w_j^{\varepsilon'}$ defined above. Recall indeed that it satisfies (3.61). Thus, in the case where u^* is of class C^∞ and has compact support, we have proved that

$$\int_{\mathcal{D}} \left[a_{\varepsilon'} \left(\nabla u^{\varepsilon'} - \left(\mathrm{Id} + \nabla w^{\varepsilon'} \right) \nabla u^* \right) \right] \cdot \left[\nabla u^{\varepsilon'} - \left(\mathrm{Id} + \nabla w^{\varepsilon'} \right) \nabla u^* \right]$$

$$\xrightarrow[\varepsilon' \to 0]{} \int_{\mathcal{D}} a^* \nabla u^* \cdot \nabla u^* - 2 \int_{\mathcal{D}} a^* \nabla u^* \cdot \nabla u^* + \int_{\mathcal{D}} a^* \nabla u^* \cdot \nabla u^* = 0.$$

The fact that $a_{\varepsilon'}$ is coercive then allows to prove the convergence (3.58).

We next address the general case where u^* is not smooth with compact support. We argue by density. We fix $\delta > 0$, which will eventually vanish, and consider $\varphi \in (C^\infty(\mathcal{D}))^d$, compactly supported, such that

$$\| \nabla u^* - \varphi \|_{L^2(\mathcal{D})} \le \delta.$$

Then, recalling that $\mathrm{Id} + \nabla w^{\varepsilon'} \in L^2(\mathcal{D})$, and using the Cauchy-Schwarz inequality we have

$$\left\| \left(\mathrm{Id} + \nabla w^{\varepsilon'} \right) \nabla u^* - \left(\mathrm{Id} + \nabla w^{\varepsilon'} \right) \varphi \right\|_{L^1(\mathcal{D})} \le \left\| \mathrm{Id} + \nabla w^{\varepsilon'} \right\|_{L^2(\mathcal{D})} \delta \le C\delta,$$
(3.65)

where the constant C neither depends on ε' nor on δ. Applying the above argument to φ instead of ∇u^*, we obtain that

$$\int_{\mathcal{D}} \left[a_{\varepsilon'} \left(\nabla u^{\varepsilon'} - \left(\mathrm{Id} + \nabla w^{\varepsilon'} \right) \varphi \right) \right] \cdot \left[\nabla u^{\varepsilon'} - \left(\mathrm{Id} + \nabla w^{\varepsilon'} \right) \varphi \right]$$

$$\xrightarrow[\varepsilon' \to 0]{} \int_{\mathcal{D}} a^* \nabla u^* \cdot \nabla u^* - \int_{\mathcal{D}} a^* \nabla u^* \cdot \varphi - \int_{\mathcal{D}} a^* \varphi \cdot \nabla u^* + \int_{\mathcal{D}} a^* \varphi \cdot \varphi$$

$$= \int_{\mathcal{D}} a^* (\nabla u^* - \varphi) \cdot (\nabla u^* - \varphi).$$

We use the fact that a_ε is coercive and that a^* is bounded, and find

$$\mu \limsup_{\varepsilon' \to 0} \left\| \nabla u^{\varepsilon'} - \left(\mathrm{Id} + \nabla w^{\varepsilon'} \right) \varphi \right\|_{L^2(\mathcal{D})}^2 \le M \| \nabla u^* - \varphi \|_{L^2(\mathcal{D})}^2 \le C\delta^2,$$

where the constant C does not depend on δ. Applying the Cauchy-Schwarz inequality again, we obtain

$$\limsup_{\varepsilon' \to 0} \left\| \nabla u^{\varepsilon'} - \left(\mathrm{Id} + \nabla w^{\varepsilon'} \right) \varphi \right\|_{L^1(\mathcal{D})}$$

$$\le |\mathcal{D}|^{1/2} \limsup_{\varepsilon' \to 0} \left\| \nabla u^{\varepsilon'} - \left(\mathrm{Id} + \nabla w^{\varepsilon'} \right) \varphi \right\|_{L^2(\mathcal{D})} \le C\delta.$$

This estimate, together with (3.65) and the triangle inequality, imply

$$\limsup_{\varepsilon' \to 0} \left\| \nabla u^{\varepsilon'} - \left(\mathrm{Id} + \nabla w^{\varepsilon'} \right) \nabla u^* \right\|_{L^1(\mathcal{D})} \leq C\delta,$$

where the constant C is independent of δ. Since this holds for any $\delta > 0$, we have proved (3.58).

□

Remark 3.9 If we assume more regularity on the correctors w^ε, then the topology in the convergence (3.58) can be improved. Indeed, if for instance $\nabla w^\varepsilon \in L^\infty(\mathcal{D})$, then the bound (3.65) may be enhanced into $\left\| \left(\mathrm{Id} + \nabla w^{\varepsilon'} \right) \nabla u^* - \left(\mathrm{Id} + \nabla w^{\varepsilon'} \right) \varphi \right\|_{L^2(\mathcal{D})} \leq C\delta$. This allows to prove that the convergence (3.58) holds in $L^2(\mathcal{D})$.

□

Proof of Proposition 3.5 Using the function $\chi_i^{\varepsilon'}$ defined above, that is, $\chi_i^{\varepsilon'}(x) = w_i^{\varepsilon'}(x) + x_i$, the convergence (3.60) may be written as follows:

$$a_{\varepsilon'} \nabla \chi_i^{\varepsilon'} \longrightarrow a_i^* \quad \text{in} \quad \left(L^2(\mathcal{D}) \right)^d, \tag{3.66}$$

where $a_i^* = a^* e_i$ denotes the i-th column of the matrix a^*. We then recall that, as we have seen above, χ_i^* is equal to ψ_i, defined by (3.62). Hence, applying (3.63), we deduce

$$\nabla \chi_i^{\varepsilon'} \longrightarrow \nabla \psi_i, \quad \tilde{a}^{\varepsilon'} \nabla \chi_i^{\varepsilon'} \longrightarrow \tilde{a}^* \nabla \psi_i.$$

These convergences hold in $L^2(\tilde{\mathcal{D}})$, hence in $L^2(\mathcal{D})$. In addition, in $\mathcal{D}$, $\tilde{a}_{\varepsilon'} = a_{\varepsilon'}$, $\tilde{a}^* = a^*$ and $\nabla \psi_i = e_i$. This implies (3.66).

□

3.4   Some Possible Proofs in an Explicit Case: The Periodic Setting

Given the two-scale expansion we have introduced in Sect. 3.3.1, we know that

  (i) *if* an expansion of the form (3.39) is valid for the function u^ε;
 (ii) *if* we can differentiate the functions u_k appearing in (3.39) as much as we need;
(iii) *if* we are allowed to make integrations by parts, averages, etc;

then we *necessarily* have that

 (a) the leading order term of the expansion (coefficient of the term ε^0) is u^*, solution to the homogenized equation with coefficient a^*;
 (b) the matrix a^* may be expressed using some functions w_i called correctors, that we can identify as solutions to problems of the type (3.48);

(c) the next term of the expansion (coefficient of the term ε^1) may be written as a function of ∇u^* and of the correctors w_i, and

(d) we may carry on this argument up to any order, computing the higher-order terms of the expansion.

Our aim is now to prove that, in suitable settings, the expansion we have obtained formally is indeed correct, at least up to the second order in ε. We already know that this is true in the one-dimensional (periodic, or even beyond periodic) case. We are now going to address the periodic setting in general dimensions, first assuming as much regularity as necessary to simplify the arguments. Then we will indicate the modifications needed to extend our work to more general settings. We will in passing collect results and material that will be useful to address the settings of Chap. 4, which contain defects and/or involve randomness.

Let us point out that, by proving that the expansion is *valid*, we mean proving that the remainder vanishes as $\varepsilon \to 0$. This may however not happen with the expected rate (put bluntly, the remainder of the truncation at order ε^p may not converge at order ε^{p+1}), and in all topologies. We are going to make this fact precise in the sequel.

In this respect, we only mention the following point, which may sound quite surprising at first sight. Something has gone unnoticed in the analysis of Sect. 3.3.1: the boundary condition $u^\varepsilon = 0$ is satisfied by $u_0 = u^*$, but not by u_1 (see formula (3.47)). In general, $w_j \left(\dfrac{\cdot}{\varepsilon} \right)$ does not vanish on the boundary of the domain $\mathcal{D}$.

It follows that, on the boundary, $u^\varepsilon - u_0(\cdot) - \varepsilon u_1 \left(\cdot, \dfrac{\cdot}{\varepsilon} \right)$ is of order ε, except in some miraculous cases. As a consequence, it cannot be of order ε^2 throughout the domain. Recall indeed that, for instance, because of the properties or the *Trace* map, the $H^1(\mathcal{D})$ norm controls the $L^2(\partial\mathcal{D})$ norm, hence if the convergence on the boundary is slow, then so is the convergence of the gradient on the *whole* domain. Due to boundary conditions, an additional term appears in the expansion (3.39). We need to take this term into account in order to recover the expected rate of convergence. At higher orders, we may expect similar difficulties. We also mention that, independently of the difficulty arising at the boundary, when we consider the gradient of $u^\varepsilon - u_0(\cdot) - \varepsilon u_1 \left(\cdot, \dfrac{\cdot}{\varepsilon} \right)$, the term that is formally of order 1 in fact contains a contribution of order 0, and the term that is formally of order 2 contains a contribution of order 1. Thus in the remainder, it is a priori not possible to expect a term of order ε^2. We rather expect terms of order ε only.

Before proving that the expansion is valid, we need to prove that the functions it involves indeed exist. In the one-dimensional case, at least in the linear case we consider here, many differential equations are explicitly solvable. In dimensions $d \geq 2$, the situation is totally different. Formally, w_i is indeed solution to (3.48), but we do not even know whether such a solution exists, let alone whether it is sufficiently regular for our purpose. The same remark holds for u_2.

As a preliminary step, we now prove that, in the periodic case, a corrector exists, namely that (3.48) has a solution, and that some of the regularity assumed for the coefficient a carries over to the corrector. Let us write (3.48) as follows:

$$- \operatorname{div} \left(a_{per}(y) \left(p + \nabla w_{p,per}(y) \right) \right) = 0 \qquad (3.67)$$

where $p \in \mathbb{R}^d$ is fixed.

Clearly, Eq. (3.67) is of the general (periodic) form

$$- \operatorname{div}(a_{per}(y) \nabla z_{per}(y)) = h_{per}(y), \qquad (3.68)$$

where one should set $h_{per}(y) = \operatorname{div}(a_{per}(y) p)$ in order to recover (3.67). We note the following point, which will prove crucial in the sequel: the right-hand side h_{per} satisfies, at least formally, the condition

$$\int_Q h_{per}(y) \, dy = 0. \qquad (3.69)$$

Indeed, an integration by parts (or the Green formula) on the periodic cell allows to write $\int_Q \operatorname{div}(a_{per}(y) p) \, dy = \int_{\partial Q} (a_{per}(y) p) \cdot \mathbf{n} = 0$. We choose to assume that $h_{per} \in H_{per}^{-1}(Q)$, which is clearly the case for $h_{per}(y) = \operatorname{div}_y(a_{per}(y) p)$, since we actually have $h_{per} \in W^{-1,\infty}(Q)$, given the assumptions satisfied by a_{per}. We then make precise the formal condition (3.69) as follows:

$$\langle h_{per}, \ 1 \rangle_{H_{per}^{-1}(Q), H_{per}^1(Q)} = 0, \qquad (3.70)$$

in the sense of the duality between $H_{per}^{-1}(Q)$ and $H_{per}^1(Q)$. We next write the variational formulation for (3.68): *Find $z_{per} \in H_{per,0}^1$ such that, for all $v_{per} \in H_{per,0}^1$, we have*

$$\int_Q a_{per}(y) \nabla z_{per}(y) \cdot \nabla v_{per}(y) \, dy = \langle h_{per}, v_{per} \rangle_{H_{per}^{-1}(Q), H_{per}^1(Q)}, \qquad (3.71)$$

where the functional space $H_{per,0}^1$ is defined by

$$H_{per,0}^1 = \left\{ v_{per} \in H^1(Q); \ v_{per} \text{ periodic}, \ \int_Q v_{per}(y) \, dy = 0 \right\} \qquad (3.72)$$

Since only the gradient of z_{per}, and not z_{per} itself, appears in (3.68), it is only defined up to the addition of a constant. We choose to fix this constant by imposing that the integral of z_{per} over Q vanishes. Hence the definition of the functional space $H_{per,0}^1$. To obtain (3.71), we formally multiply (3.68) by v_{per} and integrate by parts over Q:

$$-\int_Q \operatorname{div}_y(a_{per}(y) \nabla z_{per}(y)) \, v_{per}(y) \, dy$$

$$= \int_Q a_{per}(y) \nabla z_{per}(y) . \nabla v_{per}(y) \, dy$$

$$- \int_{\partial Q} a_{per}(y) \nabla z_{per}(y) . \mathbf{n} \, v_{per}(y) \, dy, \qquad (3.73)$$

where the boundary term vanishes due to periodicity. It is an easy exercise to prove existence and uniqueness of the solution to the variational formulation (3.71). For instance, in order to apply the Lax-Milgram Lemma, the key argument is to prove that the bilinear form $\int_Q a_{per}(y) \nabla z_{per}(y) . \nabla v_{per}(y) \, dy$ is coercive over the space (3.72). This is a consequence of the Poincaré-Wirtinger inequality, that is, the multi-dimensional version of property (P4) of Chap. 1, and of the coercivity (3.1) of a_{per}. All other necessary assumptions in order to apply the Lax-Milgram Lemma are immediate and left to the reader. Next, we want to prove that the unique solution z_{per} of (3.71) is in fact solution to (3.68). To this end, we are going to use condition (3.70). A priori, equality (3.71) holds only if $v_{per} \in H^1_{per}$ satisfies condition (3.70). In fact, it is possible to remove this restriction. Changing v_{per} into $v_{per} - \int_Q v_{per}$ indeed changes neither the left-hand side of (3.71), where only ∇v_{per} appears, nor the right-hand side, since, formally,

$$\int_Q h_{per}(y) \, v_{per}(y) \, dy = \int_Q h_{per}(y) \left(v_{per}(y) - \int_Q v_{per}(y') \, dy' \right) dy,$$

according to (3.69). This may be rigorously formulated as

$$\langle h_{per}, \, v_{per} \rangle_{H^{-1}_{per}(Q), H^1_{per}(Q)} = \left\langle h_{per}, \, v_{per} - \int_Q v_{per} \right\rangle_{H^{-1}_{per}(Q), H^1_{per}(Q)}$$

because of (3.70). From now on, the classical argument (recalled in Sect. A.2) that allows to recover the original Eq. (3.68) from the variational formulation (3.71) may be adapted without any difficulty. Choosing a test function v that is infinitely differentiable and has compact support in Q, we first obtain that Eq. (3.68) holds *in the sense of distributions* in Q. The periodic boundary conditions are included in the definition of the functional space (3.72), so they are automatically satisfied by the solution z_{per}. We thus have a unique (up to the addition of a constant) solution to (3.68), hence a unique (up to a constant) solution to (3.67).

If we assume more regularity on the coefficient a_{per} and on h_{per}, we can deduce that z_{per} (hence $w_{p,per}$) is likewise more regular. For instance, if a_{per} is of class

$C^{0,\sigma}$, for some $\sigma \in]0, 1[$, then applying Theorem A.12, we have that $\nabla w_{p,per} \in C^{0,\sigma}(Q)$. Recalling that $\nabla w_{p,per}$ is periodic, this clearly implies that $\nabla w_{p,per} \in L^\infty(\mathbb{R}^d)$.

We wish to make another point, this time related to $w_{p,per}$ itself rather then its gradient. Even without assuming any regularity on a_{per} (apart from being bounded and coercive), the Nash-Moser theory allows to prove that $w_{p,per} \in C^{0,\nu}$, for some $\nu \in]0, 1[$ depending only on the coercivity constant of a_{per} and on its L^∞ norm. This is sufficient to apply Theorem A.11 to the function $x \mapsto w_{p,per}(x) + p.x$. Since $w_{p,per}$ is periodic, it actually belongs to $C^{0,\nu}_{unif}(\mathbb{R}^d)$. Proving this in the case of systems (as opposed to the scalar equation considered here) requires the coefficient a_{per} to be Hölder continuous, as it is assumed for instance in [AL87].

The latter point implies, since $w_{p,per} \in C^{0,\nu}_{unif}(\mathbb{R}^d)$, that $w_{p,per}$ is bounded, hence strictly sublinear at infinity, in the sense defined in Sect. 2.3 of Chap. 2. This is condition (2.38) (that is, the second line of (2.39)). Here again, this remark is based on Nash-Moser theory, which holds only for scalar equations, not for elliptic systems in general. In order to obtain such a bound in the case of systems, one needs to assume further regularity on the coefficients of the system.

We point out here, and this will hold true throughout the rest of Chaps. 3 and 4, that, in many of the results and proofs exposed here, the Hölder regularity of the coefficient may be replaced by a weaker assumption. The works [LN03] and [LV00] indeed established that Schauder type elliptic regularity results such as, for instance, Theorem A.12, may be in some sense extended to the case of discontinuous piecewise regular coefficients. We thank an anonymous referee for this remark.

3.4.1   Proof in the (Very) Regular Case

The arguments that follow are borrowed from [BLP11, Chapter 1, Section 2] and from [ES08], to which we refer for the details. We assume that we have built the solutions w_i of (3.48), the solution u^* of (3.8) for some homogenized matrix a^* given by (3.54), and u_2 satisfying (3.49). We also assume that u_1 is given by (3.47), that all these functions are sufficiently regular so that all the computations we are going to do make sense in a classical way, and that the functions $w_i(y)$ and $u_2(x, y)$ are bounded uniformly with respect to $y \in \mathbb{R}^d$. Note however that the latter bounds may be replaced by the fact that w_i and u_2 are strictly sublinear at infinity in y, as we saw in (2.38) of Chap. 2. We will return to this point below.

We *know* since the beginning of this section that, if the coefficient in (1) writes $a_\varepsilon = a_{per}\left(\dfrac{\cdot}{\varepsilon}\right)$ with a_{per} *periodic* and *regular*, and if the right-hand side f is also regular, then the existences and regularities we have assumed above all hold true. Actually, these functions could be *given*, and we would be able to carry out the following strategy. In particular, *periodicity* is not essential in this step, which could thus be adapted to other settings. Only *regularity* is crucial.

We define the approximation function

$$u^{\varepsilon,2}(x) = u^*(x) + \varepsilon \sum_{i=1}^{d} \partial_{x_i} u^*(x)\, w_i\left(\frac{x}{\varepsilon}\right) + \varepsilon^2 u_2\left(x, \frac{x}{\varepsilon}\right), \tag{3.74}$$

that is the second order truncation of the two-scale expansion defined above. Such a function is the generalization to higher dimensions and to second order of the approximation (2.46) of Chap. 2, or of (3.55). We now compute the difference

$$z^{\varepsilon} = u^{\varepsilon} - u^{\varepsilon,2}, \tag{3.75}$$

and use the computation (3.42) to obtain

$$-\operatorname{div}\left(a_{per}\left(\frac{x}{\varepsilon}\right)\nabla z^{\varepsilon}\right) = O(\varepsilon) \text{ in the domain } \mathcal{D}, \tag{3.76}$$

where we recall that the remainder $O(\varepsilon)$ depends on the functions u_k, on their derivatives, and on their second-order derivatives. It is indeed of order ε because we have assumed that all these derivatives are bounded. The expression (3.76) means that there exists a constant C such that, for all $x \in \mathcal{D}$, $\left|\operatorname{div}\left(a_{per}\left(\frac{x}{\varepsilon}\right)\nabla z^{\varepsilon}\right)\right| \leq C\varepsilon$.

In addition, we know that, on the boundary $\partial\mathcal{D}$ of the domain, both u^{ε} and u^* vanish. We thus have, again using the bounds we have assumed on the functions u_k,

$$z^{\varepsilon} = O(\varepsilon) \text{ on } \partial\mathcal{D}. \tag{3.77}$$

An argument based on the *maximum principle* (see Theorem A.8) allows to deduce from (3.76) and (3.77) that

$$z^{\varepsilon} = O(\varepsilon) \text{ throughout the domain } \mathcal{D}. \tag{3.78}$$

If we were dealing with the operator $-\Delta$, that is, if a_{per} were equal to the identity matrix, a (very simple) possible proof of (3.75) would be the following. Up to translating the domain $\mathcal{D}$, we may assume that it contains the origin, and that there exists $R > 0$ such that $\mathcal{D} \subset B(R, 0)$, the ball of center 0 and of radius $R > 0$. We then define

$$U^{\varepsilon}(x) = K\varepsilon\left(R^2 + 1 - |x|^2\right), \quad K > 0.$$

Such a function by construction satisfies

$$-\Delta U^{\varepsilon} \geq 2dK\varepsilon \quad \text{in} \quad \mathcal{D}, \tag{3.79}$$

and $U^\varepsilon \geq K\varepsilon$ on $\partial\mathcal{D}$. Actually, (3.79) is an equality, but only the above inequality matters. Choosing K sufficiently large (independently of ε), we have

$$- \Delta \left(z^\varepsilon - U^\varepsilon \right) \leq 0 \quad \text{in} \quad \mathcal{D}, \qquad z^\varepsilon - U^\varepsilon \leq 0 \quad \text{on} \quad \partial\mathcal{D}.$$

The maximum principle (Theorem A.7) then implies that $z^\varepsilon \leq U^\varepsilon$. A similar argument shows that $z^\varepsilon \geq -U^\varepsilon$. This allows us to conclude that $|z^\varepsilon| \leq U^\varepsilon \leq K\left(R^2 + 1\right)\varepsilon$, that is, (3.78). The point is that the constant K in the right-hand side depends only on $\mathcal{D}$ and on the coercivity constant of a_ε, and not on ε. This method, which we have applied here to the simple case of the Laplace operator, is in fact far more general and is called the *super-solution method*. Owing to inequality (3.79), the function U^ε is called a super-solution of the equation satisfied by z^ε. Such a terminology is clear *a posteriori* since the comparison principle implies that this function lies "above" the solution. For similar reasons, the function $-U_\varepsilon$ is called a *sub-solution*. In the present case, where the elliptic operator is not simply the Laplacian, it is not possible to use an argument as simple as (3.79). However, Theorem A.8, the proof of which is based on a similar comparison technique with well-chosen super-solution and sub-solution (see [GT01, Theorem 8.16, p 191]), still allows to prove (3.78).

With (3.78), we thus have a proof of the fact that the two-scale expansion we introduced in Sect. 3.3.1 is rigorously correct. This proof is valid in a general (*though* regular) setting. We are now going to give a different proof, which is valid in a less regular case, but which relies on *periodicity*. This proof is not so easily adapted to other situations. It is however possible to extend it to *some* nonperiodic cases, as we will see in the rest of this textbook.

3.4.2 Identification of the Homogenized Limit and Convergence via the Div-Curl Lemma

The *energy method*, also called the *oscillating test functions method,* was introduced by François Murat and Luc Tartar in the articles [Mur78, Tar79]. This method is based on the *compensated compactness principle*, or the *compensated compactness method*. It is different from the *concentration-compactness method*, developed by Pierre-Louis Lions in [Lio84, Lio85]. As a matter of fact, we are going to use the latter in the sequel... The two methods are frequently mistaken for each other. We will see both of them in detail below, making clear the terminology and the different nature of the two methods. The main tool of the compensated compactness method is the *div-curl Lemma*.

For pedagogical purposes and for clarity, we apply this method in the periodic setting, although it is by no means restricted to such a setting. It was indeed originally designed to apply to far more general cases, such as the setting we have studied in Sect. 3.3.2. The periodic setting is however sufficient to understand the essential elements of the method. We moreover assume that a_{per} is scalar-valued, although our arguments may easily be adapted to matrix-valued coefficients. Such an adaptation requires at some particular places (notably when we define the corrector w_p) to consider the transpose a_{per}^T of the matrix a_{per}.

As we mentioned above, we have already used Lemma 3.2, first in Sect. 3.1, next in Sect. 3.3, in a general (not necessarily periodic) setting. We are going to state it and prove it in such a general setting, although here, we shall apply it only to the periodic case.

Owing to the bound we have proved at the beginning of Sect. 3.1, and to Proposition 3.1, we know that, up to extracting a subsequence which we do not make explicit henceforth, we may assume that u^ε converges weakly in $H_0^1(\mathcal{D})$ to some function u^*, and that, likewise, $a_{per}(x/\varepsilon)\nabla u^\varepsilon$ converges weakly in $L^2(\mathcal{D})$, to some vector-valued function r^*. Using that Eq. (1) is linear, we can pass to the (weak) limit, and find that r^* satisfies

$$-\operatorname{div} r^* = f. \tag{3.80}$$

On the other hand, for any $p \in \mathbb{R}^d$, the periodic corrector w_p satisfies

$$\nabla w_p\left(\frac{x}{\varepsilon}\right) \longrightarrow \langle \nabla w_p \rangle = 0, \tag{3.81}$$

weakly in L^2, hence

$$p + \nabla w_p\left(\frac{x}{\varepsilon}\right) \longrightarrow p. \tag{3.82}$$

In addition, again using periodicity and the bounds proved in Sect. 3.1, we know that

$$a_{per}\left(\frac{x}{\varepsilon}\right)\left(p + \nabla w_p\left(\frac{x}{\varepsilon}\right)\right)$$

is bounded in L^2, hence converges weakly to $\langle a_{per}(p + \nabla w_p)\rangle$. This average is obviously a *linear* function of $p \in \mathbb{R}^d$, since ∇w_p is itself linear with respect to p (recall that the corrector is unique, up to the addition of a constant). We can thus define the constant matrix $(a)^T$ such that

$$a_{per}\left(\frac{x}{\varepsilon}\right)\left(p + \nabla w_p\left(\frac{x}{\varepsilon}\right)\right) \longrightarrow \langle a_{per}(p + \nabla w_p)\rangle = (a)^T p \tag{3.83}$$

Of course, this matrix a is equal to the matrix (3.52) introduced in the two-scale expansion. We now consider the scalar quantity

$$\left[a_{per}\left(\frac{x}{\varepsilon}\right)\left(p + \nabla w_p\left(\frac{x}{\varepsilon}\right)\right)\right] \cdot \nabla u^\varepsilon = \left[a_{per}\left(\frac{x}{\varepsilon}\right)\nabla u^\varepsilon\right] \cdot \left(p + \nabla w_p\left(\frac{x}{\varepsilon}\right)\right). \tag{3.84}$$

We write this equation as

$$\mathbf{f}^\varepsilon . \mathbf{g}^\varepsilon = \mathbf{r}^\varepsilon . \mathbf{s}^\varepsilon, \qquad (3.85)$$

where we have introduced

$$
\begin{cases}
\mathbf{f}^\varepsilon = a_{per}\left(\dfrac{x}{\varepsilon}\right)\left(p + \nabla w_p\left(\dfrac{x}{\varepsilon}\right)\right), \\[2mm]
\mathbf{g}^\varepsilon = \nabla u^\varepsilon, \\[2mm]
\mathbf{r}^\varepsilon = a_{per}\left(\dfrac{x}{\varepsilon}\right)\nabla u^\varepsilon, \\[2mm]
\mathbf{s}^\varepsilon = p + \nabla w_p\left(\dfrac{x}{\varepsilon}\right).
\end{cases}
\qquad (3.86)
$$

We note that the sequences $\mathbf{f}^\varepsilon$, $\mathbf{g}^\varepsilon$, $\mathbf{r}^\varepsilon$, $\mathbf{s}^\varepsilon$ are all bounded in $L^2(\mathcal{D})$, uniformly with respect to ε. Moreover, $\mathbf{g}^\varepsilon$ and $\mathbf{s}^\varepsilon$ are gradients, while, owing to the corrector equation, $\operatorname{div}\mathbf{f}^\varepsilon = 0$, and because of the original Eq. (1), $\operatorname{div}\mathbf{r}^\varepsilon = -f$. We are thus in position to apply the celebrated following Lemma, which we state in a somewhat simplified form.

Lemma 3.2 (Div-Curl Lemma) *Let $\mathbf{f}^\varepsilon$ and $\mathbf{g}^\varepsilon$ be two vector-valued functions in $\mathcal{D}$ that converge weakly in $\left(L^2(\mathcal{D})\right)^d$, to $\mathbf{f}$ and $\mathbf{g}$, respectively. Assume that the sequence $\operatorname{div}\mathbf{f}^\varepsilon$ is bounded in $L^2(\mathcal{D})$ and $\mathbf{g}^\varepsilon = \nabla w^\varepsilon$. Then the scalar products $\mathbf{f}^\varepsilon . \mathbf{g}^\varepsilon$ converge in the sense of distributions to the scalar product $\mathbf{f} . \mathbf{g}$.*

The interest of such a result is clear. As the reader knows, in general, without any additional assumption, the product of two weakly converging sequences does not converge to the product of the limits. If *all* the derivatives of $\mathbf{f}^\varepsilon$ or $\mathbf{g}^\varepsilon$ were bounded, say, in $L^2(\mathcal{D})$, then the Rellich Theorem would imply strong convergence in $L^2(\mathcal{D})$, allowing to pass to the limit in the scalar product, in the sense of distributions. We would first prove this up to extraction, and then use the uniqueness of the limit to prove that the whole sequence converges. Here, in the statement of Lemma 3.2, only *some* combinations of partial derivatives are bounded, and this nevertheless allows to pass to the limit. This is why this result is so important. It may be in spirit compared to elliptic regularity results: if the Laplacian (that is, a *particular combination* of some of the second-order partial derivatives) is regular, then, provided some suitable conditions are satisfied, *all* the second-order derivatives are regular. Let us point out that this lemma has been generalized in many directions. First, in the original lemma proved by François Murat and Luc Tartar, the assumption $\mathbf{g}^\varepsilon = \nabla w^\varepsilon$ is replaced by the fact that $\operatorname{curl}\mathbf{g}^\varepsilon$ is bounded in $\left(L^2\left(\mathbb{R}^d\right)\right)^d$. Second, it is possible to prove it in the suitable couple of spaces $\left(L^q, L^{q'}\right)$, $1 < q < +\infty$, $\dfrac{1}{q} + \dfrac{1}{q'} = 1$, instead of L^2. We give the proof of

Lemma 3.2 at the end of the present section. This proof will confirm the above considerations, and will explain the terminology "compensation". For now, let us simply *use* this lemma.

We apply Lemma 3.2, first to the couple of sequences ($\mathbf{f}^\varepsilon$, $\mathbf{g}^\varepsilon$), then to the couple ($\mathbf{r}^\varepsilon$, $\mathbf{s}^\varepsilon$), all of them being defined by (3.86).

Passing to the limit in (3.84), and taking into account the weak limits of $\mathbf{f}^\varepsilon$ (see (3.83)), of $\mathbf{g}^\varepsilon$ (since $\nabla u^\varepsilon \rightharpoonup \nabla u^*$ owing to Proposition 3.1), of $\mathbf{r}^\varepsilon$ (owing to (1)) and of $\mathbf{s}^\varepsilon$ (see (3.82)), we obtain

$$\left[(\mathfrak{a})^T p \right] . \nabla u^* = r^* . p,$$

that is,

$$\left[\mathfrak{a} \nabla u^* \right] . p = r^* . p.$$

Since this holds for any $p \in \mathbb{R}^d$, we conclude that $r^* = \mathfrak{a} \nabla u^*$. Passing to the limit in (1) and deducing (3.80), this implies that the matrix $\mathfrak{a}$ defined by (3.83), equal to the "tentative" matrix (3.52), is indeed the homogenized matrix a^*. As a by-product, we also obtain that u^* is solution to $-\text{div}\,(a^* \nabla u^*) = f$.

As a conclusion, we point out that the above proof is based on the same elements as those of the proof of Lemma 3.1. As a matter of fact, and as mentioned at that time, the latter uses Lemma 3.2.

Two important remarks are in order.

First, the above argument proves that u^ε converges to the solution u^* to the homogenized equation, and allows to identify the associated homogenized matrix a^*. However, it gives no information about the rate of convergence of u^ε to u^*. We will need to work more in order to determine this rate of convergence, as we have done in the regular case in Sect. 3.4.1. This is the purpose of Sect. 3.4.3.

Second, and as we have already pointed out above, although the proof has been performed in the periodic setting for simplicity, it actually applies to much more general situations. In particular, in many cases (such as those we have studied in Chap. 1), the property (3.81) would hold true for a suitable strictly sub-linear corrector. Likewise, the weak convergence (3.83) is sufficient to characterize $\mathfrak{a}$ (without resorting to periodicity), and may thus be readily generalized to other settings. The reader understands that we are not far from having a *proof* of formula (3.60) of Proposition 3.5. This being understood, all the arguments we have performed amount to bounds and weak convergences. This clearly allows for generalizations. Even the existence of a corrector could in fact be replaced by the existence of a *sequence* of correctors in the sense of Proposition 3.4. Throughout these generalizations, one would a priori need to consider a subsequence ε'. Should the limit fail to be given by an independent formula, it would not be unique, and one would not be able to prove that the whole sequence converges. We however do not wish to develop this aspect, which is well documented in the usual bibliographical references.

As announced, we conclude this section with the

Proof of Lemma 3.2 We first note that w^ε is only defined up to the addition of a constant, so we may assume without loss of generality that $\int_{\mathcal{D}} w^\varepsilon = 0$. The Poincaré-Wirtinger inequality implies that the sequence w^ε is bounded in $H^1(\mathcal{D})$. Up to the extraction of a subsequence, it converges weakly in $H^1(\mathcal{D})$ and strongly in $L^2(\mathcal{D})$, to some $w \in H^1(\mathcal{D})$. Weak convergence in H^1 implies convergence in the sense of distribution. Since differential operators are continuous for this topology, we have $\mathbf{g} = \nabla w$. Next, we fix $\varphi \in C^\infty(\mathcal{D})$, with compact support. Using an integration by parts, we have

$$\int_{\mathcal{D}} \mathbf{f}^\varepsilon.\mathbf{g}^\varepsilon \, \varphi = \int_{\mathcal{D}} \mathbf{f}^\varepsilon.\nabla w^\varepsilon \, \varphi$$

$$= -\int_{\mathcal{D}} w^\varepsilon \operatorname{div}(\mathbf{f}^\varepsilon \varphi)$$

$$= -\int_{\mathcal{D}} w^\varepsilon \operatorname{div}(\mathbf{f}^\varepsilon) \, \varphi - \int_{\mathcal{D}} w^\varepsilon \mathbf{f}^\varepsilon . \nabla\varphi. \qquad (3.87)$$

Since $\operatorname{div}(\mathbf{f}^\varepsilon)$ is bounded in $L^2(\mathcal{D})$, it converges (again up to an extraction) weakly in $L^2(\mathcal{D})$. Here again, this implies convergence in the sense of distributions, and that the limit of $\operatorname{div}(\mathbf{f}^\varepsilon)$ is $\operatorname{div}(\mathbf{f})$. Now, w^ε converges strongly in $L^2(\mathcal{D})$, so we can pass to the limit in the first term of the right-hand side of (3.87). Similarly, since $\mathbf{f}^\varepsilon$ converges weakly and w^ε converges strongly, we can pass to the limit in the second term of the right-hand side of (3.87). Thus,

$$\lim_{\varepsilon \to 0} \int_{\mathcal{D}} \mathbf{f}^\varepsilon.\mathbf{g}^\varepsilon \, \varphi = -\int_{\mathcal{D}} w \operatorname{div}(\mathbf{f}) \, \varphi - \int_{\mathcal{D}} w \, \mathbf{f}.\nabla\varphi = \int_{\mathcal{D}} \varphi \, \mathbf{f}.\nabla w = \int_{\mathcal{D}} \varphi \, \mathbf{f}.\mathbf{g},$$

which proves the expected convergence in the sense of distributions. The limit being unique, the whole sequence converges.   □

3.4.3 Convergence and Rate of Convergence

As announced above, we now consider a periodic (and "slightly" regular, in a sense made precise below) setting, though most of the arguments we are going to develop may be adapted to situations beyond the periodic case. In particular we will see that our arguments apply to perturbations (in the sense seen in the preceding chapters) of the periodic setting. We are going to prove that the first two terms of the two-scale expansion are correct. We will also be able to establish a rate of convergence.

We assume for simplicity that the coefficient a_ε is scalar-valued, which in particular implies that a_{per} itself is scalar-valued. Adapting the following to the case of a (possibly non-symmetric) matrix-valued coefficient is not difficult. It is not so easy, however, to extend it to *systems* (as opposed to equations) that is, if u^ε is vector-valued. We again refer to Remark 3.13 for such issues.

3.4.3.1 Some Preliminary Computations

Let $w_{p,per}$ be the periodic corrector, that is, the solution to (3.67). Consider the homogenized matrix defined by

$$[a^*]_{ij} = \operatorname*{weak-lim}_{\varepsilon \to 0} a_{per}(./\varepsilon) \, (\delta_{ij} + \partial_j w_{i,per}(./\varepsilon)) \tag{3.88}$$

$$= \int_Q a_{per}(y) \, (\delta_{ij} + \partial_j w_{i,per}(y)) \, dy \tag{3.89}$$

for $1 \le i, j \le d$. We deliberately keep at hand both expressions (3.88) and (3.89). The first one is a direct consequence of (3.60), the second one is (3.54). The reason is that (3.88) is valid in much more general contexts than the periodic case, but is sufficient for most arguments. On the other hand, (3.89) is more explicit, but valid only in the periodic case. It will prove however adaptable to some more general settings.

We now define

$$u_{per}^{\varepsilon,1}(x) = u^*(x) + \varepsilon \sum_{i=1}^{d} \partial_i u^*(x) \, w_{i,per}(x/\varepsilon), \tag{3.90}$$

as is suggested in (3.55) by the two-scale expansion.

In order to simplify the notation, we drop the subscript $\cdot_{per}$ in all the functions we consider. This is also consistent with the fact that most of the arguments below are not restricted to the present periodic case. We will restore it temporarily when periodicity plays a crucial role. In the same spirit, we note a^* rather than $a^*(x)$ although a priori the homogenized matrix may depend on the spatial variable x.

We have

$$\partial_j u^{\varepsilon,1} = \sum_{i=1}^{d} (\delta_{ij} + \partial_j w_i(./\varepsilon)) \, \partial_i u^* + \varepsilon \sum_{i=1}^{d} w_i(./\varepsilon) \, \partial_{ij} u^*, \tag{3.91}$$

for $1 \le j \le d$. With a more concise notation, this reads as

$$\nabla u^{\varepsilon,1} = (\operatorname{Id} + \nabla w(./\varepsilon)) \nabla u^* + \varepsilon \, w(./\varepsilon) \, \nabla \nabla u^*. \tag{3.92}$$

We are now going to study the difference $u^\varepsilon - u^{\varepsilon,1}$, expecting that it is "small" in the limit $\varepsilon \to 0$. The main idea is described in [ZKO94, p26-27]: although we do not explicitly know this difference (no more than we explicitly know u^ε), we expect that, applying the operator $-\operatorname{div}(a_\varepsilon \nabla .)$ to this difference, we will find quantities that we know better (such as f, or ...). We thus write

$$-\operatorname{div}\left(a_\varepsilon \nabla(u^\varepsilon - u^{\varepsilon,1})\right) = -\operatorname{div}\left(a_\varepsilon \nabla u^\varepsilon - a_\varepsilon (\operatorname{Id} + \nabla w(./\varepsilon))\, \nabla u^*(x)\right)$$

$$+\varepsilon\, \operatorname{div}\left(a_\varepsilon\, w(./\varepsilon)\, \nabla\nabla u^*\right)$$

$$= -\operatorname{div}\left((a^* - a_\varepsilon (\operatorname{Id} + \nabla w(./\varepsilon))\, \nabla u^*(x))\right)$$

$$+\varepsilon\, \operatorname{div}\left(a(./\varepsilon)\, w(./\varepsilon)\, \nabla\nabla u^*\right), \quad (3.93)$$

where we have used $-\operatorname{div}(a_\varepsilon \nabla u^\varepsilon) = -\operatorname{div}(a^* \nabla u^*) = f$. Equality (3.93) is written in the same condensed form as (3.92). Its expanded form reads as

$$-\sum_{i=1}^{d} \partial_i \left(a_\varepsilon \partial_i \left(u^\varepsilon - u^{\varepsilon,1}\right)\right) = -\sum_{i=1}^{d}\sum_{j=1}^{d} \partial_i \left[\left(a_{ij}^* - a_\varepsilon \left(\delta_{ij} + \partial_i w_j(./\varepsilon)\right)\right) \partial_j u^*\right]$$

$$+ \varepsilon \sum_{i=1}^{d}\sum_{j=1}^{d} \partial_i \left[a_\varepsilon w_j(./\varepsilon)\partial_{ji} u^*\right].$$

We introduce the matrix-valued function

$$M = a^* - a\, (\operatorname{Id} + \nabla w), \qquad (3.94)$$

which, componentwise, writes $[M(x)]_{ij} = [a^*]_{ij} - a(x)\left(\delta_{ij} + \partial_i w_j(x)\right)$. The argument we are going to develop is valid in any dimension. However, for clarity, we henceforth assume that we work in dimension $d = 3$, where things are easier to understand. We refer to Remark 3.10 below for the general case.

For each $1 \le j \le 3$, the j-th column of the matrix M, that we denote by M_j, is divergence-free, that is,

$$\operatorname{div} M_j(x) = \partial_i [M(x)]_{ij} = \partial_i \left([a^*]_{ij} - a(x)\left(\delta_{ij} + \partial_i w_j(x)\right)\right)$$

$$= 0,$$

in view of the corrector equation $-\operatorname{div}(a(e_j + \nabla w_j)) = \operatorname{div}(a^* e_j)$. This is a generalization of (3.67), the latter being valid only if a^* is constant. As a consequence, for all $1 \le j \le 3$, we claim that there exists a vector-valued function B_j such that $M_j = \operatorname{curl} B_j$. We more concisely write this as $M = \operatorname{curl} B$.

The above property is well-known and valid in general situations. It is often called the *de Rham Lemma*, or the *Hodge (or Helmholtz-Hodge) decomposition*. It is valid whenever M_j is a distribution. For a proof, we refer for instance to [Lio69, pp 67–69] (in French) or [Tem79, Chapter I, Proposition 1.1]. In the following, we will need information about the structure of the function B_j (for instance, in the periodic case, the fact that B_j is periodic), its regularity, and its behaviour at infinity.

The constructive proof of the existence of B_j that we present here will give us, under suitable conditions, all the desired information. We also note that, eventually, this proof will be adapted to cover more general cases than the periodic setting.

In order to simplify the notation, during our proof of the existence of a suitable function B, we denote by $\mathbf{m}$ (which is to be thought as M_j for some $j \in \{1, 2, 3\}$) a map from $\mathbb{R}^3$ into itself. We assume that this function $\mathbf{m}$ is periodic. In this notation, we want to build a map $\mathbf{b}$, from $\mathbb{R}^3$ into itself, such that $\mathbf{m} = \operatorname{curl} \mathbf{b}$. Let us further assume that the average of $\mathbf{m}$ vanishes. This is indeed the case in our situation since $\mathbf{m} = M_j$, because of the definition (3.89) of the homogenized matrix a^*. Because of this property, it is possible to search for $\mathbf{b}$ as a periodic function. Fixing $j \in \{1, 2, 3\}$, we solve the periodic problem $-\Delta \mathbf{a}_j = \mathbf{m}_j$, as we have solved the periodic corrector problem at the beginning of Sect. 3.4. The solution is a periodic function $\mathbf{a}_j$, unique up to the addition of a constant (this constant may be set to zero, but it is not important here). This existence holds true because $\mathbf{m}_j$ satisfies the necessary and sufficient condition $\int_Q \mathbf{m}_j = 0$ for solvability of the periodic Poisson problem. Given all the vector-valued functions $\mathbf{a}_j$, for $j \in \{1, 2, 3\}$, we denote by $\mathbf{a}$ the matrix-valued function the columns of which are $\mathbf{a}_j$. Of course, $\mathbf{a}$ is different from the coefficient a in Eq. (1). We note that

$$ -\Delta \operatorname{div} \mathbf{a} = -\Delta \left(\partial_j \, \mathbf{a}_j \right) = \partial_j \, \mathbf{m}_j = \operatorname{div} \mathbf{m} = 0. $$

Thus, since $\operatorname{div} \mathbf{a}$ is periodic, it follows that $\operatorname{div} \mathbf{a}$ is a constant. Integrating over Q, we have, using periodicity, $\int_Q \operatorname{div} \mathbf{a} = \int_{\partial Q} \mathbf{a} . \mathbf{n} = 0$. Hence, $\operatorname{div} \mathbf{a} = 0$. Consequently, we have $\mathbf{m} = -\Delta \mathbf{a} = \operatorname{curl} \operatorname{curl} \mathbf{a}$. Defining $\mathbf{b} = \operatorname{curl} \mathbf{a}$, which is periodic by construction, regular when $\mathbf{m}$ is regular, and *bounded*, we obtain the desired function. Note that $\mathbf{m}$ is indeed regular in our specific setting since $\mathbf{m} = M_j$ and M is regular because both a and w are. This shows that $\mathbf{b}$ has the same regularity as w. The regularity of $\mathbf{b}$ thus obtained will prove crucial in the sequel. More precisely, in the present periodic case, if $\mathbf{a}$ is of class $C^{0,\alpha}$, Theorem A.12 proves that $w \in C^{1,\alpha}$, hence by definition of $\mathbf{m}$, that $\mathbf{m} \in C^{0,\alpha}$. Again applying Theorem A.12, we deduce that $\mathbf{a} \in C^{2,\alpha}$, thus that $\mathbf{b} \in C^{1,\alpha}$.

Having constructed a suitable function B, we are in position to (formally) study the convergence of $u^\varepsilon - u^{\varepsilon,1}$. Since $M = \operatorname{curl} B$ implies $M(./\varepsilon) = \varepsilon \operatorname{curl}(B(./\varepsilon))$, we may write the first term of the right-hand side of (3.93) under the form

$$ -\operatorname{div}\left(\left(a^* - a_\varepsilon \, (\operatorname{Id} + \nabla w(./\varepsilon)) \, \nabla u^*(x)\right)\right) = -\operatorname{div}\left(M(./\varepsilon) \, \nabla u^*(x)\right) $$
$$ = -\operatorname{div}\left(\varepsilon \operatorname{curl}(B(x/\varepsilon)) \, \nabla u^*\right) $$
$$ = -\varepsilon \operatorname{div}\left[\operatorname{curl}\left(B(x/\varepsilon) \, \nabla u^*\right)\right. $$
$$ \left. -B(x/\varepsilon) \times \nabla \, \nabla u^*\right] $$
$$ = \varepsilon \operatorname{div}\left(B(x/\varepsilon) \times \nabla \, \nabla u^*\right). $$

Thus, (3.93) reads as

$$- \operatorname{div} \left(a_\varepsilon \nabla (u^\varepsilon - u^{\varepsilon,1}) \right) = \varepsilon \operatorname{div} \left(B(./\varepsilon) \times \nabla \nabla u^* \right)$$

$$+ \varepsilon \operatorname{div} \left(a(./\varepsilon) \, w(./\varepsilon) \, \nabla \nabla u^* \right). \qquad (3.95)$$

This shows, at least formally, that $u^\varepsilon - u^{\varepsilon,1}$ is of order $O(\varepsilon)$. We are now going to prove this fact in an adequate functional space.

Remark 3.10 The above argument has been carried out in dimension $d = 3$. We explain in this remark how to extend it in any arbitrary ambient dimension. If $d = 1$, divergence-free functions are constant functions, so the question is trivial. In dimensions $d \geq 2$, we claim that, for any periodic vector-valued function $\mathbf{m}$ such that $\langle \mathbf{m} \rangle = 0$ and $\operatorname{div}(\mathbf{m}) = 0$, there exists a periodic matrix-valued function $B = \left(B_{ij} \right)_{1 \leq i, j \leq d}$ such that $B_{ij} = -B_{ji}$, and $\operatorname{div}(B) = \mathbf{m}$, in the sense that

$$\forall j, \quad \sum_{i=1}^{d} \partial_i B_{ij} = \mathbf{m}_j.$$

A simple way to build such a function B is to solve the equation $-\Delta B_{ij} = \partial_i \mathbf{m}_j - \partial_j \mathbf{m}_i$. This equation has a solution because the right-hand side is a periodic function with zero average. We refer, for instance, to [ZKO94, pages 6-7] for more details. We then apply this result to each column $\mathbf{m} = M_j$ of the matrix M defined by (3.94). All our three-dimensional arguments may then be easily adapted. To end this remark, we note that if $d = 3$, we recover using the general argument that $\mathbf{m}$ is the curl of some vector-valued function. To see this, we write the skew-symmetric matrix B as

$$B = \begin{pmatrix} 0 & -b_3 & b_2 \\ b_3 & 0 & -b_1 \\ -b_2 & b_1 & 0 \end{pmatrix}, \quad \text{whence} \quad \operatorname{div}(B) = \operatorname{curl} \begin{pmatrix} b_1 \\ b_2 \\ b_3 \end{pmatrix}.$$

□

3.4.3.2   A Series of Progressively Less Formal Arguments

We start with a completely formal argument. Eliminating the divergence operators in Eq. (3.95) gives:

$$- a_\varepsilon \nabla (u^\varepsilon - u^{\varepsilon,1}) = \varepsilon \, B(./\varepsilon) \times \nabla \nabla u^* + \varepsilon \, a(./\varepsilon) \, w(./\varepsilon) \, \nabla \nabla u^*. \qquad (3.96)$$

We observe on (3.96) that $\nabla (u^\varepsilon - u^{\varepsilon,1})$ converges to zero as $O(\varepsilon)$. This formal manipulation, which is of course not correct mathematically, may be usefully compared to the one-dimensional setting of Chap. 2. In that chapter, we have established (2.35), that is,

$$- a_{per}\left(\frac{x}{\varepsilon}\right)\left(u^{\varepsilon} - u^{\varepsilon,1}\right)'(x) = \left(\int_0^1 F + c_{\varepsilon}\right)$$

$$+ \varepsilon\, a_{per}\left(\frac{x}{\varepsilon}\right)(u^*)''(x)\, w_{per}\left(\frac{x}{\varepsilon}\right), \qquad (3.97)$$

and we have seen that $\int_0^1 F + c_{\varepsilon} = O(\varepsilon)$. The formal similarity between (3.96) and (3.97) is obvious. The second term of the right-hand side of the two expressions is identical. As for the first one, it differs because in dimension 1, divergence free functions are constant. Thus, the term containing B of (3.96) cannot be expected to be present in (3.97), while, on the other hand, the constant in the first term of (3.97) is the integration constant. Up to these "minor" details, Eqs. (3.96) and (3.97) are identical.

We are unfortunately not allowed to simply omit the divergence operators. We thus now give a proof that is "a little more correct".

We denote by $R_{\varepsilon} = u^{\varepsilon} - u^{\varepsilon,1}$ and by

$$F_{\varepsilon} = B(./\varepsilon) \times \nabla\nabla u^* + a(./\varepsilon)\, w(./\varepsilon)\,\nabla\nabla u^*, \qquad (3.98)$$

so that (3.95) now reads as

$$- \operatorname{div} a_{\varepsilon} \nabla R_{\varepsilon} = \varepsilon\, \operatorname{div} F_{\varepsilon}. \qquad (3.99)$$

We point out that, if, for instance,

$$f \in L^2(\mathcal{D}), \qquad (3.100)$$

then elliptic regularity (see Theorem A.13) implies that $u^* \in H^2(\mathcal{D})$, hence $\nabla\nabla u^* \in L^2(\mathcal{D})$. On the other hand, a (hence w too, due to Theorem A.11) is bounded, and so is B (recall that B has the same regularity as w). We emphasize at this point that the periodic structure is crucial for our arguments. In the following, we will use periodicity again at several stages. Without such an assumption, it is not possible a priori to prove that w and B are bounded. This implies that $F_{\varepsilon} \in L^2(\mathcal{D})$, and more precisely that

$$\|F_{\varepsilon}\|_{L^2(\mathcal{D})} \le \|B\|_{L^{\infty}(\mathbb{R}^d)}\|u^*\|_{H^2(\mathcal{D})} + \|a\|_{L^{\infty}(\mathbb{R}^d)}\|w\|_{L^{\infty}(\mathbb{R}^d)}\|u^*\|_{H^2(\mathcal{D})}$$

$$\le C\|f\|_{L^2(\Omega)},$$

where the constant C only depends on the coercivity constant of a, on $\|a\|_{L^{\infty}}$, on a^* and on $\mathcal{D}$. In particular, it does not depend on ε nor on f. Hence,

$$F_{\varepsilon} \text{ is bounded in } L^2(\mathcal{D}), \text{ uniformly with respect to } \varepsilon. \qquad (3.101)$$

The right-hand side of (3.99) is thus bounded in $H^{-1}(\mathcal{D})$, and Eq. (3.99) is to be understood in the weak sense

$$\forall \varphi \in H_0^1(\mathcal{D}), \quad \int_{\mathcal{D}} a_\varepsilon \nabla R_\varepsilon . \nabla \varphi = \varepsilon \int_{\mathcal{D}} F_\varepsilon . \nabla \varphi \tag{3.102}$$

We temporarily assume that the function R_ε vanishes on the boundary of $\mathcal{D}$. Strictly speaking, it does not, since it is equal there to $R_\varepsilon = \varepsilon \sum_{i=1}^d \partial_{x_i} u^*(x) w_i(x/\varepsilon)$ (recall that both u^ε and u^* vanish on the boundary). This does not matter for the time being. Then, since R_ε *vanishes* on the boundary, we may use it as a test function in (3.102) and find (we recall that we argue with a scalar-valued coefficient a_ε for simplicity, but the generalization to matrix-valued coefficients is not difficult)

$$\int_{\mathcal{D}} a_\varepsilon \, |\nabla R_\varepsilon|^2 = \varepsilon \int_{\mathcal{D}} F_\varepsilon . \nabla R_\varepsilon. \tag{3.103}$$

We next use the Cauchy-Schwarz inequality to bound the right-hand side from above, and the coercivity (3.1) to bound the left-hand side from below. We obtain, for a constant C that does not depend on ε,

$$\int_{\mathcal{D}} |\nabla R_\varepsilon|^2 \leq C \varepsilon^2 \int_{\mathcal{D}} |F_\varepsilon|^2 \tag{3.104}$$

(recall that, by definition, F_ε is vector-valued). Next, we use (3.101), which gives $\int_{\mathcal{D}} |\nabla R_\varepsilon|^2 = O(\varepsilon^2)$. Since we have temporarily assumed that R_ε vanishes on the boundary, the Poincaré inequality then allows to conclude that

$$\|R_\varepsilon\|_{H^1(\mathcal{D})} = O(\varepsilon). \tag{3.105}$$

This is the strong convergence we have claimed on $u^\varepsilon - u^{\varepsilon,1}$, and which was proved in the one-dimensional case in (2.37). It seemingly confirms that the formal two-scale expansion of Sect. 3.3.1 is correct, *except* that we have two serious issues:

(i) we have rigorously proved (3.105) under the assumption that R_ε vanishes on the boundary $\partial \mathcal{D}$, which is not the case, hence

(ii) this result, although it will eventually turn out to be correct as far as far as the $H^1(\mathcal{D})$ convergence is concerned, in general yields an *incorrect* rate of convergence in ε: the order $O(\varepsilon)$ holds only *away from* the boundary $\partial \mathcal{D}$. If one takes the boundary into account, the rate is typically only of order $O(\sqrt{\varepsilon})$.

Let us still temporarily postpone the rigorous arguments that will (at last) prove the correct results. We are now going to slightly digress and study a special setting in which, *somewhat serendipitously*, the result (3.105) holds true.

We choose the domain $\mathcal{D}$ to be the open unit cube $\mathcal{D} =]0, 1[^d$. We assume that $\varepsilon = \dfrac{1}{N}$, where N is an integer, and that the right-hand side f is periodic of period 1 in each direction, with the additional condition

$$\int_{\mathcal{D}} f = 0.$$

As it is the case throughout the present section, we suppose that the coefficient a is periodic. On the other hand, in the original problem (1), we replace the Dirichlet homogeneous condition $u^\varepsilon = 0$ by *periodic boundary conditions*. In other words, we look for a solution u^ε that is periodic of the cube $\mathcal{D} =]0, 1[^d$ and solution in the *whole* space $\mathbb{R}^d$. Such a solution is only defined up to the addition of a constant (as the periodic corrector is, see (3.67) and the subsequent paragraph). To make it unique, we assume for instance that $\int_{\mathcal{D}} u^\varepsilon = 0$. Such a solution, if the coefficient a is continuous, is of class H^2 (see Theorem A.13). The associated homogenized solution is, according to Lemma 3.1, the periodic solution u^* of (3.8). Such a solution is again unique due to the condition $\int_{\mathcal{D}} u^* = 0$. It is thus a solution posed in the *whole* space $\mathbb{R}^d$. Since the coefficient a^* is constant, u^* is obviously in H^2_{unif}. The derivatives of both functions u^ε and u^* are of course periodic of the cube $\mathcal{D} =]0, 1[^d$. Moreover, all the functions a, w, and B are periodic of the cube $]0, 1[^d$, hence, since ε is the inverse of an integer, the rescaled functions $a(./\varepsilon)$, $w(./\varepsilon)$, $B(./\varepsilon)$ are all periodic of the cube $\mathcal{D}$. And since $\nabla \nabla u^*$ is periodic too, so is F_ε. The integration by parts leading from (3.99) to (3.103) is thus entirely valid since the boundary terms cancel by periodicity. The conclusion

$$\int_{\mathcal{D}} |\nabla R_\varepsilon|^2 = O(\varepsilon^2)$$

follows, and is rigorous. Applying the Poincaré-Wirtinger inequality and using the fact that $\int_{\mathcal{D}} R_\varepsilon = O(\varepsilon)$, since the integral of the term $u^\varepsilon - u^*$ vanishes by assumption, we rigorously deduce (3.105). The reader will however acknowledge that our setting is peculiar.

We are now going to give two different proofs of the actual, general, rate of convergence, in the case of the homogeneous Dirichlet conditions, for any domain $\mathcal{D}$ and any sequence $\varepsilon \to 0$. The proofs we are about to present estimate the H^1 norm of the residual R_ε. They give a rigorous meaning (in two different ways) to the formal computation we have performed above. We will then mention some elements of proof for estimating R_ε in $W^{1,q}$ norms for $q \neq 2$, or in Hölder norms.

3.4.3.3 Interior H^1 Convergence

Our first proof uses simpler tools than the second one, but proves a different result (compare (3.115) below to (3.131) in the next section). We split $R_\varepsilon = u^\varepsilon - u^{\varepsilon,1}$ given by (3.99) into $R_\varepsilon = R_{1,\varepsilon} + R_{2,\varepsilon}$, where

$$
\begin{cases}
-\operatorname{div} a_\varepsilon \, \nabla R_{1,\varepsilon} = \varepsilon \, \operatorname{div} F_\varepsilon & \text{in } \mathcal{D}, \\[2mm]
\qquad\quad R_{1,\varepsilon} \quad = \quad 0 & \text{on } \partial\mathcal{D},
\end{cases}
\tag{3.106}
$$

$$
\begin{cases}
-\operatorname{div} a_\varepsilon \, \nabla R_{2,\varepsilon} = 0 & \text{in } \mathcal{D}, \\[2mm]
\qquad\quad R_{2,\varepsilon} \quad = \varepsilon \, T_\varepsilon & \text{on } \partial\mathcal{D}.
\end{cases}
\tag{3.107}
$$

The function T_ε is defined by

$$
T_\varepsilon = \sum_{i=1}^{d} \partial_{x_i} u^*(x) \, w_i(x/\varepsilon).
\tag{3.108}
$$

If, in addition to the assumptions we have imposed so far (that is, (3.1)), we assume that

$$
f \in L^\infty(\mathcal{D}),
\tag{3.109}
$$

and that the boundary $\partial\mathcal{D}$ of the domain $\mathcal{D}$ is of class C^2, then, elliptic regularity theory (we apply Theorem A.14) implies that $\nabla\nabla u^* \in L^q(\mathcal{D})$ for all $q \in [1, +\infty[$. Choosing q sufficiently large, we infer that $\nabla u^* \in C^0(\overline{\mathcal{D}})$. Hence, since, as we have already seen, $w_i \in L^\infty(\mathbb{R}^d)$, we have

$$
\|T_\varepsilon\|_{L^\infty(\partial\mathcal{D})} \leq C,
\tag{3.110}
$$

for some constant C independent of ε.

Since, by definition, the function $R_{1,\varepsilon}$ vanishes on the boundary, the proof of Sect. 3.4.3.2 rigorously applies, showing that

$$
\|R_{1,\varepsilon}\|_{H^1(\mathcal{D})} = O(\varepsilon).
\tag{3.111}
$$

As expected, and as our formal proofs indicate, as far as functions vanish on the boundary, convergence in H^1, at rate $O(\varepsilon)$, holds true.

We now concentrate on the presumably difficult part $R_{2,\varepsilon}$. As $R_{2,\varepsilon}$ is the solution of an elliptic *equation* with zero right-hand side, we may apply the maximum principle (Theorem A.7), and deduce from (3.110) that

$$\left| R_{2,\varepsilon}(x) \right| \leq \varepsilon \, \|T_\varepsilon\|_{L^\infty(\partial D)} \leq C \varepsilon, \tag{3.112}$$

for all $x \in D$, where the constant C only depends on the uniform bound on $\|T_\varepsilon\|_{L^\infty(\partial D)}$, and is independent of ε since the maximum principle uses only the fact that the operator $-\operatorname{div} a_\varepsilon \nabla$ is elliptic.

Next, we use the *Caccioppoli inequality* (see Theorem A.5), which writes

$$\int_{B(x,\rho)} \left| \nabla R_{2,\varepsilon} \right|^2 (y)\, dy \leq \frac{C}{\rho^2} \int_{B(x,2\rho)} \left| R_{2,\varepsilon} \right|^2 (y)\, dy, \tag{3.113}$$

where C is a constant that depends only on the coercivity constant of a and on $\|a\|_{L^\infty}$, and where $x \in D$ and $\rho > 0$ are such that $B(x, 2\rho) \subset D$. We fix an open set D_0 strictly contained in D, in the sense that the closure $\overline{D_0}$ of D_0 is included in D. Choosing ρ sufficiently small, and covering D_0 by balls of radius ρ, we deduce

$$\int_{D_0} \left| \nabla R_{2,\varepsilon} \right|^2 \leq \frac{C}{\rho^2} \int_D \left| R_{2,\varepsilon} \right|^2 \leq C \varepsilon^2, \tag{3.114}$$

where the second inequality is due to (3.112). Collecting (3.112), (3.111), and (3.114), we thus have the convergence

$$\|R_\varepsilon\|_{H^1(D_0)} = O(\varepsilon). \tag{3.115}$$

We emphasize that this order of convergence is true only in an *interior* domain D_0. The analysis we give below in (3.131) will show, as already mentioned above, that the convergence rate is slower near the boundary.

The above proof gives a convergence rate in $H^1(D_0)$ but may also be used to prove the same convergence rate in $L^2(D)$ (that is, in a weaker norm, but up to the boundary). Indeed, estimate (3.112) implies

$$\left\| R_{2,\varepsilon} \right\|_{L^2(D)} \leq C \varepsilon,$$

for some constant C independent of ε. On the other hand, estimate (3.111) gives a similar estimate for $R_{1,\varepsilon}$, so we have

$$\|R_\varepsilon\|_{L^2(D)} = O(\varepsilon).$$

We also note that, using Sobolev imbeddings, we obtain this convergence rate in $L^q(D)$, for all $q < 2^* = \dfrac{2d}{d-2}$.

3.4.3.4 Convergence in H^1 up to the Boundary

Our aim is now to estimate $\|R_\varepsilon\|_{H^1(\mathcal{D})}$ in the *whole* domain $\mathcal{D}$ and not only $\|R_\varepsilon\|_{H^1(\mathcal{D}_0)}$ for an *interior* domain $\mathcal{D}_0$ as in (3.115).

We start with a relatively simple argument that allows to understand that we can indeed deal with the boundary, and that the H^1 convergence rate is (at worst) $\sqrt{\varepsilon}$. This argument is borrowed from [ZKO94, pages 27–28], and is valid under the assumption that $u^* \in C^2(\overline{\mathcal{D}})$, $w_j \in W^{1,\infty}(\mathbb{R}^d)$, and that the open domain $\mathcal{D}$ is of class C^2. The latter property has already been used to prove (3.110). It implies that there exists a smooth cut-off function τ^ε, with compact support in $\mathcal{D}$, such that

$$0 \leq \tau^\varepsilon \leq 1, \quad \varepsilon \left|\nabla \tau^\varepsilon\right| \leq C, \quad \tau^\varepsilon(x) = 1 \text{ if } d(x, \partial\mathcal{D}) > \varepsilon, \tag{3.116}$$

where the constant C only depends on $\mathcal{D}$. The setting we have just defined is more regular than what we have assumed so far, that is, $u^* \in W^{2,q}$ for all $q < +\infty$ and $w_j \in L^\infty(\mathbb{R}^d)$. Note that $\nabla w_j \in L^\infty(\mathbb{R}^d)$ is actually, as we will see below, a consequence of the fact that the coefficient a is Hölder continuous. We use the cut-off function τ^ε of (3.116) to apply a "boundary correction" to the residual, defining

$$Q_\varepsilon = u^\varepsilon - u^* - \varepsilon \tau^\varepsilon \sum_{i=1}^d w_i \left(\frac{\cdot}{\varepsilon}\right) \partial_i u^*. \tag{3.117}$$

Thus, for $R_\varepsilon = u^\varepsilon - u^{\varepsilon,1}$ where $u^{\varepsilon,1}$ is, we recall, defined by (3.90), we have

$$R_\varepsilon - Q_\varepsilon = \varepsilon \left(1 - \tau^\varepsilon\right) \sum_{i=1}^d w_i \left(\frac{\cdot}{\varepsilon}\right) \partial_i u^*.$$

Hence,

$$\|R_\varepsilon - Q_\varepsilon\|_{L^2(\mathcal{D})} \leq \varepsilon \|\nabla u^*\|_{C^0(\mathcal{D})} \sum_{i=1}^d \|w_i\|_{L^\infty(\mathbb{R}^d)} = O(\varepsilon), \tag{3.118}$$

and, for any index $1 \leq j \leq d$,

$$\partial_j \left(R_\varepsilon - Q_\varepsilon\right) = -\varepsilon \partial_j \tau^\varepsilon \sum_{i=1}^d w_i \left(\frac{\cdot}{\varepsilon}\right) \partial_i u^* + \left(1 - \tau^\varepsilon\right) \sum_{i=1}^d \partial_j w_i \left(\frac{\cdot}{\varepsilon}\right) \partial_i u^*$$

$$+ \varepsilon \left(1 - \tau^\varepsilon\right) \sum_{i=1}^d w_i \left(\frac{\cdot}{\varepsilon}\right) \partial_{ij} u^*. \tag{3.119}$$

The rightmost term on the right-hand side may be bounded as we did for (3.118) above, this time using $\|D^2 u^*\|_{C^0(\mathcal{D})}$. We thus have

$$\left\|\partial_j \left(R_\varepsilon - Q_\varepsilon\right)\right\|_{L^2(\mathcal{D})} \le \varepsilon \|\nabla \tau^\varepsilon\|_{L^2(\mathcal{D})} \sum_{i=1}^{d} \|w_i\|_{L^\infty(\mathbb{R}^d)} \|\nabla u^*\|_{C^0(\mathcal{D})}$$

$$+ \|1 - \tau^\varepsilon\|_{L^2(\mathcal{D})} \sum_{i=1}^{d} \|\nabla w_i\|_{L^\infty(\mathbb{R}^d)} \|\nabla u^*\|_{C^0(\mathcal{D})} + O(\varepsilon). \qquad (3.120)$$

We next use (3.116), together with the above bounds on τ^ε, w_i and u^* (here we use the fact that $\nabla w_j \in L^\infty(\mathbb{R}^d)$), and the Cauchy-Schwarz inequality, to prove that

$$\|\nabla (R_\varepsilon - Q_\varepsilon)\|_{L^2(\mathcal{D})} \le C \left|\text{supp}(\nabla \tau^\varepsilon)\right|^{1/2} + C \left|\text{supp}(1 - \tau^\varepsilon)\right|^{1/2} + O(\varepsilon),$$

$$(3.121)$$

where *supp* denotes the support of a function. Given the definition of τ^ε, we have $\text{supp}(\nabla \tau^\varepsilon) \subset \text{supp}(1 - \tau^\varepsilon)$. In addition, the measure of the support of $1 - \tau^\varepsilon$ is of order $\varepsilon |\partial \mathcal{D}|$, where $|\partial \mathcal{D}|$ is the $(d-1)$-dimensional measure of the boundary $\partial \mathcal{D}$. Thus,

$$\|\nabla (R_\varepsilon - Q_\varepsilon)\|_{L^2(\mathcal{D})} \le C\sqrt{\varepsilon}.$$

This inequality and (3.118) imply

$$\|R_\varepsilon - Q_\varepsilon\|_{H^1(\mathcal{D})} \le C\sqrt{\varepsilon}. \qquad (3.122)$$

Moreover,

$$- \text{div} (a_\varepsilon \nabla Q_\varepsilon) = - \text{div} (a_\varepsilon \nabla R_\varepsilon) + \text{div} (a_\varepsilon \nabla (R_\varepsilon - Q_\varepsilon)).$$

Recalling (3.99), estimate (3.122) and that a_ε is bounded, we deduce

$$- \text{div} (a_\varepsilon \nabla Q_\varepsilon) = \varepsilon \, \text{div}(F_\varepsilon) + \sqrt{\varepsilon} \, \text{div}(G_\varepsilon), \quad \|F_\varepsilon\|_{L^2(\mathcal{D})} + \|G_\varepsilon\|_{L^2(\mathcal{D})} \le C,$$

where the constant C does not depend on ε. Since Q_ε vanishes on $\partial \mathcal{D}$, we may multiply this equation by Q_ε and integrate by parts. This gives

$$\int_{\mathcal{D}} a_\varepsilon \nabla Q_\varepsilon \cdot \nabla Q_\varepsilon = \varepsilon \int_{\mathcal{D}} F_\varepsilon \cdot \nabla Q_\varepsilon + \sqrt{\varepsilon} \int_{\mathcal{D}} G_\varepsilon \cdot \nabla Q_\varepsilon \le C\sqrt{\varepsilon} \|\nabla Q_\varepsilon\|_{L^2(\mathcal{D})}.$$

Using the fact that a_ε is coercive, we conclude that

$$\|Q_\varepsilon\|_{H^1(\mathcal{D})} \leq C\sqrt{\varepsilon}. \tag{3.123}$$

We finally collect (3.122) and (3.123), apply the triangle inequality and find

$$\|R_\varepsilon\|_{H^1(\mathcal{D})} \leq C\sqrt{\varepsilon}. \tag{3.124}$$

In passing, we wish to mention the existence of another proof of this result, under similar assumptions. This proof was brought to our attention by Alexei Lozinski. The interested reader may find it in the French version of the present textbook [BL22, Section 3.4.3.4].

We now return to the general case, where u^* and w_j are not as regular as above. We again use the decomposition $R_\varepsilon = R_{1,\varepsilon} + R_{2,\varepsilon}$, where $R_{1,\varepsilon}$ and $R_{2,\varepsilon}$ satisfy (3.106) and (3.107), respectively. Estimate (3.111) is still valid, so we only address the estimate of $\|R_{2,\varepsilon}\|_{H^1(\mathcal{D})}$. To this end, we first prove that

$$\|T_\varepsilon\|_{H^{1/2}(\partial\mathcal{D})} = O\left(\frac{1}{\sqrt{\varepsilon}}\right), \tag{3.125}$$

where, we recall, the definition of the $H^{1/2}$ norm on the $d-1$-dimensional domain $\partial\mathcal{D}$ (we refer to [Bre11] for a general definition) reads as

$$\|h\|_{H^{1/2}(\partial\mathcal{D})}^2 = \|h\|_{L^2(\partial\mathcal{D})}^2 + \iint_{\partial\mathcal{D}\times\partial\mathcal{D}} \frac{|h(x)-h(y)|^2}{|x-y|^d}\,dx\,dy.$$

With a slight abuse of notation, we denote by dx the surface measure on $\partial\mathcal{D}$. Then, using the same bounds as above on u^* and w, we have

$$\|w(./\varepsilon)\|_{H^{1/2}(\partial\mathcal{D})}^2 = \varepsilon^{d-1}\,\|w\|_{L^2(\varepsilon^{-1}\partial\mathcal{D})}^2$$

$$+\varepsilon^{d-2}\iint_{\varepsilon^{-1}\partial\mathcal{D}\times\varepsilon^{-1}\partial\mathcal{D}} \frac{|w(x)-w(y)|^2}{|x-y|^d}\,dx\,dy,$$

$$= \varepsilon^{d-1}\,O\left(\varepsilon^{-(d-1)}\right) + \varepsilon^{d-2}\,O\left(\varepsilon^{-(d-1)}\right),$$

$$= O\left(\varepsilon^{-1}\right),$$

hence

$$\|\nabla u^*\,w(./\varepsilon)\|_{H^{1/2}(\partial\mathcal{D})} = O\left(\frac{1}{\sqrt{\varepsilon}}\right),$$

which proves (3.125), according to the definition of T_ε. In the above argument, we have used that

$$I_\varepsilon = \iint_{\varepsilon^{-1} \partial \mathcal{D} \times \varepsilon^{-1} \partial \mathcal{D}} \frac{|w(x) - w(y)|^2}{|x - y|^d} \, dx \, dy = O\left(\varepsilon^{-(d-1)}\right). \qquad (3.126)$$

Such an estimate is not obvious. We now give some elements of its proof. For simplicity, we only proceed in the case where the set $\partial \mathcal{D}$ is

$$\partial \mathcal{D} = \left\{ x = \underbrace{(x_1, \ldots, x_{d-1}, 0)}_{=x'}, \quad |x'| < 1 \right\},$$

that is, the ball of center 0 and of radius 1 of the hyperplane H defined by $x_d = 0$. Since, in our actual setting, the domain $\mathcal{D}$ is sufficiently regular, it is possible to reduce the question to this particular case using local charts. The above integral then reads as

$$I_\varepsilon = \int_{|x'| < \frac{1}{\varepsilon}} \int_{|y'| < \frac{1}{\varepsilon}} \frac{|w(x', 0) - w(y', 0)|^2}{|x' - y'|^d} \, dx' dy'.$$

If $|x'| < 1/\varepsilon$ and $|y'| < 1/\varepsilon$, then $|x' - y'| < 2/\varepsilon$, and a change of variables gives

$$I_\varepsilon \leq \int_{|x'| < \frac{1}{\varepsilon}} \int_{|z'| < \frac{2}{\varepsilon}} \frac{|w(x', 0) - w(x' + z', 0)|^2}{|z'|^d} \, dz' dx'.$$

We split the integral with respect to z' into two integrals, the first one on the domain $\{|z'| < 1\}$, the second one on the domain $\{1 < |z'| < 2/\varepsilon\}$. We now deal with each of them separately. For the first one, we use a Taylor expansion, writing $w(x' + z', 0) - w(x', 0) = \int_0^1 \nabla w(x' + tz', 0) . z' \, dt$. Inserting this in the integral and applying the Cauchy-Schwarz inequality, we have:

$$\int_{|x'| < \frac{1}{\varepsilon}} \int_{|z'| < 1} \frac{|w(x', 0) - w(x' + z', 0)|^2}{|z'|^d} \, dz' dx'$$

$$\leq \int_{|x'| < \frac{1}{\varepsilon}} \int_{|z'| < 1} \int_0^1 \frac{|\nabla w(x' + tz', 0)|^2 |z'|^2}{|z'|^d} \, dt \, dz' dx'.$$

Using $|x' + tz'| \leq 1 + 1/\varepsilon$, we obtain

$$\int_{|x'|<\frac{1}{\varepsilon}} \int_{|z'|<1} \frac{|w(x',0) - w(x'+z',0)|^2}{|z'|^d} dz' dx'$$

$$\leq \int_{|x''|<1+\frac{1}{\varepsilon}} |\nabla w(x'',0)|^2 dx'' \int_{|z'|<1} |z'|^{2-d} dz'$$

$$\leq C \|\nabla w\|^2_{L^2_{unif}(H)} \, \varepsilon^{-(d-1)} \int_0^1 r^{2-d} r^{d-2} dr,$$

where H is the hyperplane $H = \{x \in \mathbb{R}^d, \quad x_d = 0\}$. The constant C only depends on the dimension d. In the last inequality, we have used that the integral of $|\nabla w|^2$ is bounded from above by its L^1_{unif} norm multiplied by the measure of the domain $\{|x'| \leq 1 + \varepsilon^{-1}\} \subset H$, and we have explicitly computed the integral with respect to z'. The first integral we have considered is thus bounded as follows

$$\int_{|x'|<\frac{1}{\varepsilon}} \int_{|z'|<1} \frac{|w(x',0) - w(x'+z',0)|^2}{|z'|^d} dz' dx' \leq C \|\nabla w\|^2_{L^2_{unif}(H)} \, \varepsilon^{-(d-1)}.$$

$$(3.127)$$

Next, we turn to the integral over $1 < |z'| < 2/\varepsilon$. We simply use that w belongs to L^∞ (recall that, due to the Nash-Moser theory, the corrector is bounded –see our comment at the beginning of Sect. 3.4), and write

$$\int_{|x'|<\frac{1}{\varepsilon}} \int_{1<|z'|<\frac{2}{\varepsilon}} \frac{|w(x',0) - w(x'+z',0)|^2}{|z'|^d} dz' dx'$$

$$\leq 4\|w\|^2_{L^\infty} \int_{|x'|<\frac{1}{\varepsilon}} \int_{1<|z'|<\frac{2}{\varepsilon}} |z'|^{-d} dz' = C\|w\|^2_{L^\infty} \, \varepsilon^{-(d-1)} \int_1^{+\infty} r^{-d} r^{d-2} dr,$$

whence

$$\int_{|x'|<\frac{1}{\varepsilon}} \int_{1<|z'|<\frac{2}{\varepsilon}} \frac{|w(x',0) - w(x'+z',0)|^2}{|z'|^d} dz' dx' \leq C\|w\|^2_{L^\infty} \, \varepsilon^{-(d-1)}.$$

$$(3.128)$$

Collecting (3.127) and (3.128), we have (3.126).   This concludes the proof of (3.125).

Now that we have proved estimate (3.125), we return to Eq. (3.107). Applying Theorem A.4, we know that we can lift the boundary value $T_\varepsilon \in H^{1/2}(\partial\mathcal{D})$ into a function $t_\varepsilon \in H^1(\mathcal{D})$ defined in the domain $\mathcal{D}$, that agrees with T_ε on the boundary, and such that

$$c \, \|T_\varepsilon\|_{H^{1/2}(\partial\mathcal{D})} \leq \|t_\varepsilon\|_{H^1(\mathcal{D})} \leq C \, \|T_\varepsilon\|_{H^{1/2}(\partial\mathcal{D})},$$

where neither of the two constants c and C depends on ε. Actually, only the rightmost inequality will be used below. We now solve

$$\begin{cases} -\operatorname{div} a_\varepsilon \nabla \left(R_{2,\varepsilon} - \varepsilon\, t_\varepsilon\right) = -\varepsilon \operatorname{div} a_\varepsilon \nabla t_\varepsilon & \text{in } \mathcal{D}, \\[2mm] R_{2,\varepsilon} - \varepsilon\, t_\varepsilon \quad\;\; = \quad\;\; 0 & \text{on } \partial\mathcal{D}. \end{cases} \tag{3.129}$$

This is equivalent to (3.107). We note that the right-hand side belongs to $H^{-1}(\mathcal{D})$ and satisfies

$$\|-\operatorname{div} a_\varepsilon \nabla t_\varepsilon\|_{H^{-1}(\mathcal{D})} \leq \|a_\varepsilon\|_{L^\infty(\mathcal{D})}\, \|t_\varepsilon\|_{H^1(\mathcal{D})}.$$

We argue as we did above when we have assumed that the solution vanishes on the boundary: successively multiplying the equation by the solution, integrating by parts and using Theorem A.4, we have

$$\left\|R_{2,\varepsilon} - \varepsilon\, t_\varepsilon\right\|_{H^1(\mathcal{D})} \leq \varepsilon\, \|a_\varepsilon\|_{L^\infty(\mathcal{D})}\, \|T_\varepsilon\|_{H^{1/2}(\partial\mathcal{D})} = O(\sqrt{\varepsilon}).$$

This implies

$$\left\|R_{2,\varepsilon}\right\|_{H^1(\mathcal{D})} = O(\sqrt{\varepsilon}), \tag{3.130}$$

hence, taking into account (3.111), we conclude that

$$\left\|R_\varepsilon\right\|_{H^1(\mathcal{D})} = O(\sqrt{\varepsilon}). \tag{3.131}$$

This estimate gives a rigorous meaning to the two-scale expansion and the associated intuitive arguments. Note, though, that the rate is not as good as the formally expected order $O(\varepsilon)$. In fact, (3.131) cannot be improved in general. Except of course in very special cases, such as the one-dimensional case, or the case we have briefly mentioned above of a periodic domains that "perfectly agrees" with the microscopic period. An inspection of the arguments leading to (3.131) indicates that the order $O(\sqrt{\varepsilon})$ (instead of $O(\varepsilon)$) is due to *boundary effects*. These effects disappear if we can neglect the boundary, either because the norm we use does not see it (the L^2 norm for instance), or because the chosen domain ($\mathcal{D}_0 \subset\subset \mathcal{D}$) eliminates it. The latter situation also happens in dimension 1, where the boundary reduces to a point. This is confirmed by the proof we have performed in the preceding section: staying away from the boundary, the convergence rate is improved, and we recover the order $O(\varepsilon)$.

To conclude this section, let us summarize the regularity assumptions on the coefficient a that we have used in the above arguments.

In Sect. 3.4.3.1, we have only performed preparatory computations. No assumptions is needed there except that a should be coercive and bounded. As always, for a symmetric matrix, (3.1) is sufficient. For a non-symmetric matrix, we also need that $a \in L^{\infty}(\mathbb{R}^d)$.

In Sect. 3.4.3.2, we have essentially used two additional ingredients: on the one hand that $f \in L^2(\mathcal{D})$ implies $u^* \in H^2(\mathcal{D})$, and, on the other hand, that the corrector w is bounded. The latter property holds true with no additional assumption on a. Note however that our proof does not extend to elliptic systems, since it relies on the Nash-Moser theory. On the contrary, the estimate of u^* in H^2 requires that the boundary $\partial \mathcal{D}$ is of class C^2 (see Theorem A.13). It is however possible to circumvent this difficulty by using interior estimates only (Theorem A.12) in order to bound u^* in $H^2(\mathcal{D}_0)$, with $\mathcal{D}_0 \subset\subset \mathcal{D}$. Doing so, we obtain (3.105) if we assume not only that R_ε vanishes on the boundary, but also that it has compact support in $\mathcal{D}$. The periodic setting considered at the end of Sect. 3.4.3.2, in contrast, remains valid without any modification.

In Sect. 3.4.3.3, we also use a C^2 regularity of the boundary, in order to have $u^* \in W^{2,q}$. One could expect, similarly as above, that for interior estimates, such a regularity would not be needed. This is indeed the case, but modifying the proof in this direction requires more elaborate arguments, that we do not wish to develop here. The remaining of the section does not use any extra assumption, apart from (3.1), with the additional L^{∞} bound for a non-symmetric coefficient a. In particular, the fact that the coefficient is Hölder continuous is not required.

Finally, Sect. 3.4.3.4 clearly uses the fact that the coefficient is Hölder continuous. This assumption allows in particular to prove that the gradient of the corrector is bounded. It is unclear whether such an assumption is needed in *full generality*. A possible extension is given in the works [LV00] and [LN03]: they generalize Schauder type estimates to cases in which the coefficient is assumed to be piecewise regular.

3.4.3.5 Convergence in $W^{1,q}$ (and Other "Gradient Norms")

We now aim at estimating the gradient of the residual $R_\varepsilon = u^\varepsilon - u^{\varepsilon,1}$ in L^q norms, for $q \neq 2$. Ideally, we would like to work in the most demanding topology, that is, the case $q = +\infty$. As we shall see, estimates of ∇R_ε in all spaces L^q indeed exist in the literature. Before citing them, and in echo with our remarks at the end of Sect. 3.4.3.4, we immediately add the assumption

$$a \text{ is of regularity } C^{0,\alpha}, \text{ for some } \alpha > 0. \tag{3.132}$$

Such an assumption allows the operator $-\operatorname{div}(a_\varepsilon \nabla\,.)$ to share some of the "nice" properties of the Laplacian... It also allows to prove that, for $\varepsilon > 0$ fixed, the function u^ε, hence R_ε, is sufficiently regular, so that the norms present in (3.134) and (3.135) below exist. Finally, assumption (3.132) is necessary in order to prove that the gradient of the corrector is bounded.

The key result in the periodic case is the following

Proposition 3.6 *Assume that* $a_\varepsilon = a_{per}\left(\frac{\cdot}{\varepsilon}\right)$, *and that the domain* $\mathcal{D}$ *is of class* C^2. *Suppose that* (3.1) *is satisfied. For any interior domain* $\mathcal{D}_0$, *we have the* H^1 *convergence* (3.115), *that is,*

$$\|R_\varepsilon\|_{H^1(\mathcal{D}_0)} \leq C\varepsilon \|f\|_{L^2(\mathcal{D})}, \tag{3.133}$$

provided that $f \in L^2(\mathcal{D})$. *Moreover, if we also assume the regularity* (3.132), *we have the following convergences*

$$\|\nabla R_\varepsilon\|_{L^q(\mathcal{D}_0)} \leq C\varepsilon \|f\|_{L^q(\mathcal{D})}, \tag{3.134}$$

whenever $f \in L^q(\mathcal{D})$, $2 \leq q < +\infty$, *and in the particular case* $q = +\infty$,

$$\|\nabla R_\varepsilon\|_{L^\infty(\mathcal{D}_0)} \leq C\varepsilon \ln(2 + \varepsilon^{-1}) \|f\|_{C^{0,\beta}(\mathcal{D})}, \tag{3.135}$$

whenever $f \in C^{0,\beta}(\mathcal{D})$, $\beta \in]0,1[$. *In the above estimates, all the (possibly different) constants* C *are independent of* ε *and* f.

Remark 3.11 One should not be surprised to see that the case $q = +\infty$ is specific. It is well known that classical regularity estimates of the type $Lu = f \implies \|u\|_{W^{2,q}} \leq \|f\|_{L^q}$ are not valid for $q = +\infty$. A counter-example may be found for instance in [GM12, pages 141–142]. Thus, L^∞ estimates are frequently indirectly obtained via Hölder-type estimates and Sobolev embeddings. In this respect, it is natural that the Hölder norm $\|f\|_{C^{0,\beta}(\mathcal{D})}$ appears in (3.135) instead of the Lebesgue norm $\|f\|_{L^\infty(\mathcal{D})}$. As for the factor $\ln(2 + \varepsilon^{-1})$, it may be seen as a technicality. $\square$

Remark 3.12 All the results stated so far are in fact valid for *systems*. We refer to Remark 3.13 below for more details. $\square$

The proof of these L^q estimates, for any value of q, is particularly difficult. It was originally published in [AL87]. We refer the interested reader to the associated literature (see in particular our adaptation of this proof to the case of defects [BJL20] we will address in Chap. 4). In the present introductory textbook, we will only give the *main ideas* of the proofs and mention the key ingredients, together with their global structure. We will also intuitively motivate the results. In addition, we will primarily be concerned with the L^2 setting, especially because, for the nonperiodic geometries we will consider, and the associated theory of homogenization, the L^2 setting will prove sufficient in many aspects. On the other hand, the interest of the L^q setting mostly lies in the case $q = +\infty$. Such a topology is very precise, at *any* scale, as we will see below.

We could in fact explain the aim of the present section as follows. Our only goal is to make the reader understand why results such as those of Proposition 3.6 are *likely* to hold true. We also aim at providing them with the necessary background to read and understand the articles of the literature. We do not reproduce these proofs, and only give a flavor of them, so that the reader may understand them more easily. As a matter of fact, reading some of these proofs through is an achievement in itself... Nothing unexpected: one cannot hope to prove such powerful results with only basic tools.

Some General Considerations

We again consider the system satisfied by the residual R_ε, that is,

$$
\begin{cases}
- \operatorname{div} (a_\varepsilon \nabla R_\varepsilon) = \varepsilon \operatorname{div} F_\varepsilon & \text{in } \mathcal{D}, \\
\\
\qquad\qquad R_\varepsilon \;\;\; = \;\; \varepsilon\, T_\varepsilon & \text{on } \partial\mathcal{D},
\end{cases}
\tag{3.136}
$$

where F_ε and T_ε are defined by (3.98) and (3.108), respectively.

Let us note that, if we were looking for results for a *given fixed* ε, things would not be so difficult. Then, $W^{k,p}$ estimates immediately follow from classical elliptic regularity theory. Recall indeed that, in Eq. (3.136), the function F_ε is given by (3.98). Thus, the norm of F_ε in a given functional space is (almost) controlled by the norm of the right-hand side f of the original Eq. (1) in the same space. This is due to the specific properties of B and a in the periodic case. Consequently, a simple application of elliptic regularity results, either in the Sobolev spaces (Theorem A.13) or in Hölder spaces (Theorem A.12) would give, respectively, estimates (3.134) and (3.135) of Proposition 3.6, with constants C independent of the right-hand side f. The point here is that we want such results to be valid in the limit $\varepsilon \to 0$. Thus, our arguments need, in some sense, to be *uniform* with respect to ε: we want the constants C in (3.134)–(3.135) to be *independent* of ε. As we will see, this requires (much) more work...

What did we learn from the elements of the proofs we have performed in the preceding sections?

First, we have learnt that, by splitting our problem into a problem with zero boundary data and a problem with zero right-hand side, we are able to distinguish between the contribution from the interior of the domain and the contribution from the boundary. Doing so, we are able to treat each contribution with different tools. It is thus natural to reproduce this strategy here: we define $R_\varepsilon = R_{1,\varepsilon} + R_{2,\varepsilon}$, where $R_{1,\varepsilon}$ and $R_{2,\varepsilon}$ are solution to (3.106) and (3.107), respectively.

We have also learnt that controlling $\nabla R_{2,\varepsilon}$ is possible using a control of $R_{2,\varepsilon}$ itself, owing to the *Caccioppoli inequality*, since $R_{2,\varepsilon}$ is solution to a homogeneous elliptic equation. We expect here that the same type of inequalities will be useful. This will indeed be the case, under a different form.

Finally, we have observed that the arguments we have developed for the operator $- \operatorname{div} (a_\varepsilon \nabla .)$ are mainly inspired by the properties of the Laplacian, or at least of a constant-coefficient elliptic operator. Here again, such considerations will prove important.

The Homogeneous Equation

Assuming that we split $R_\varepsilon = R_{1,\varepsilon} + R_{2,\varepsilon}$ as mentioned above, we start by studying the part (3.107), that is, the solution to the *homogeneous* equation, which has the general form

$$\begin{cases} -\operatorname{div}(a_\varepsilon \nabla v_\varepsilon) = 0 & \text{in } \Omega, \\ \qquad\quad v_\varepsilon \quad\;\; = g & \text{on } \partial\Omega, \end{cases} \tag{3.137}$$

where g is a *fixed* boundary condition. Here Ω denotes a general subdomain of our original domain $\mathcal{D}$, and is, of course, not to be mistaken with the probability space used in some other chapters of the present textbook. The first question we ask is, how to control the solution v_ε, that plays the role of R_ε, in the norms appearing in estimates (3.134)–(3.135). More precisely, we want to estimate $\|\nabla v_\varepsilon\|_{L^q(\Omega)}$ and $\|\nabla v_\varepsilon\|_{C^{0,\beta}(\Omega)}$. These two types of norms, Sobolev on the one hand, and Hölder on the other hand, are likely to play an important part in establishing (3.134)–(3.135). The proof shows that they actually play a role.

As we have pointed out above, a key idea is to first deal with the Laplace operator instead of $-\operatorname{div} a_\varepsilon \nabla$. This is also what we are going to do here, in this (much) more difficult context.

If we bluntly replace the operator $-\operatorname{div} a_\varepsilon \nabla$ by $-\Delta$ in (3.137), then it is classical to prove estimates on $\|\nabla v_\varepsilon\|_{L^\infty(\Omega)}$ and $\|\nabla v_\varepsilon\|_{C^{0,\beta}(\Omega)}$. We now do so.

Since

$$-\Delta v_\varepsilon = 0,$$

and since the Laplacian is a *constant* coefficient operator, we can differentiate this equation. Doing so, we have that each partial derivative $\partial_i v_\varepsilon$, $1 \le i \le d$, satisfies

$$-\Delta(\partial_i v_\varepsilon) = 0. \tag{3.138}$$

We then apply the *Harnack inequality*, under the (weak) form of Corollary A.3. Without loss of generality, we may assume that $0 \in \Omega$. For simplicity, we further suppose that the unit ball $B(0,1)$ centered at the origin is included in the domain Ω. The general case in which $B(0,1)$ may intersect the boundary of Ω can be addressed with similar tools. Corollary A.3 applied to (3.138) (for each index $1 \le i \le d$) gives

$$\sup_{x \in B(0,1/4)} |\nabla v_\varepsilon(x)| \le C \, \|\nabla v_\varepsilon\|_{L^2(B(0,1/2))}, \tag{3.139}$$

where the constant C does not depend on the solution v_ε. The radius $1/4$ that we chose to apply the Harnack inequality is only a technical matter, and this choice will soon become clear below. We next bound the left-hand side from below as follows. Let $0 < \theta < 1/4$ be a fixed radius, and let us apply the Poincaré-Wirtinger on the ball $B(0, \theta)$:

$$\fint_{B(0,\theta)} \left| v_\varepsilon - \fint_{B(0,\theta)} v_\varepsilon \right|^2 \leq C_{PW} \theta^2 \fint_{B(0,\theta)} |\nabla v_\varepsilon|^2, \tag{3.140}$$

where the constant C_{PW} does not depend on v_ε and is related to the Poincaré-Wirtinger inequality on the unit ball, and the coefficient θ^2 accounts for the rescaling of the ball of radius θ. Since $B(0, \theta) \subset B(0, 1/4)$, we can bound the right-hand side of (3.140) from above:

$$\fint_{B(0,\theta)} \left| v_\varepsilon - \fint_{B(0,\theta)} v_\varepsilon \right|^2 \leq C_{PW} \theta^2 \left(\sup_{x \in B(0,1/4)} |\nabla v_\varepsilon(x)| \right)^2.$$

Thus, using (3.139), we have

$$\fint_{B(0,\theta)} \left| v_\varepsilon - \fint_{B(0,\theta)} v_\varepsilon \right|^2 \leq C_{PW} C \theta^2 \fint_{B(0,1/2)} |\nabla v_\varepsilon|^2.$$

Then, we bound the right-hand side using the *Caccioppoli inequality* (we can do so because v_ε solves $-\Delta v_\varepsilon = 0$), and find

$$\fint_{B(0,\theta)} \left| v_\varepsilon - \fint_{B(0,\theta)} v_\varepsilon \right|^2 \leq C \theta^2 \fint_{B(0,1)} |v_\varepsilon|^2. \tag{3.141}$$

As we will explain below (see inequality (3.142) and the subsequent argument), this estimate will allow us to prove a bound in Hölder norm. For now, we would like to see how we can adapt the above reasoning to the operator $- \mathrm{div}(a_\varepsilon \nabla \cdot)$. First, this argument is still valid for the *homogenized* operator associated with Eq. (3.137), that is, the operator $- \mathrm{div}\, a^* \nabla$. Indeed, the main ingredient here is that the Laplace operator, apart from being elliptic, has *constant coefficients,* which is also true for the homogenized operator. This allows to prove (3.139) on the *gradient* of the solution (and not only on the solution itself, which would hold true even for a non-constant coefficient operator). Hence, we have actually obtained (3.141) for the solution v_ε of

$$- \mathrm{div} \left(a^* \nabla v_\varepsilon \right) = 0.$$

Now, we know that, for ε sufficiently small, the original operator $- \mathrm{div}\, a_\varepsilon \nabla$ is close to the homogenized operator $- \mathrm{div}\, a^* \nabla$. Such a property is made rigorous by the homogenization result. Thus, for ε small, estimate (3.141) *should* hold true for our solution v_ε of Eq. (3.137). The argument that proves this intuition is called a *compactness method* and goes by contradiction: one assumes that, even for θ and ε small, (3.141) does not hold and reaches a contradiction. Of course, this argument is detailed in the literature cited above (see [AL87]). Then, a so-called *iteration* method, subsequently followed by a covering technique, allows to prove estimate (3.141) for all positive radii r, and for some exponent ρ:

$$\fint_{B(0,r)} \left| v_\varepsilon - \fint_{B(0,r)} v_\varepsilon \right|^2 \le C r^\rho \fint_{B(0,1)} |v_\varepsilon|^2. \tag{3.142}$$

The final step of the proof consists in noting that the quantity

$$\sup_{r>0} \left(\frac{1}{r^\rho} \fint_{B(0,r)} \left| v - \fint_{B(0,r)} v \right|^2 \right)^{1/2} \tag{3.143}$$

is in fact, according to the so-called *Campanato characterization* of Hölder regularity, *equivalent* to the $C^{0,\beta}$ norm for $\beta = \rho/2$. More precisely, it is equivalent to the $C^{0,\beta}$ *semi*-norm, since it vanishes when v is a constant. We refer to Theorem A.3, the proof of which is given for instance in [Gia83, p 70], or in [GM12, Theorem 5.5]. We have thus described the main steps of the method establishing the *Hölder estimate*

$$[v_\varepsilon]_{C^{0,\beta}(B(0,1/2))} \le C \, \|v_\varepsilon\|_{L^2(B(0,1))}, \tag{3.144}$$

where v_ε is solution to (3.137), C is a constant independent of ε and $[\cdot]_{C^{0,\beta}(B(0,1/2))}$ denotes the $C^{0,\beta}$ semi-norm on the ball $B(0, 1/2)$. So far, we have assumed that the unit ball $B(0, 1)$ was entirely contained in Ω. In the case where this ball intersects the boundary, it is intuitive to think that an estimate such as (3.144) is still valid, with an additional term taking into account the boundary data g on $\partial\Omega$. This would give, for instance,

$$[v_\varepsilon]_{C^{0,\beta}(B(0,1/2))} \le C \, \|v_\varepsilon\|_{L^2(B(0,1))} + C \, [g]_{C^{0,\beta}(B(0,1))}. \tag{3.145}$$

where g is (with a slight abuse of notation) a lifting of the boundary data in the ball $B(0, 1)$.

The first step of the above sketch of proof is the (weak) Harnack inequality (3.139). It allows to control the *gradient* ∇v_ε. Thus, with a similar argument, it is possible to prove the following *Lipschitz estimate*

$$\|\nabla v_\varepsilon\|_{L^\infty(B(0,1/2))} \le C \, \|v_\varepsilon\|_{L^2(B(0,1))}, \tag{3.146}$$

with, here again, a ball $B(0, 1)$ contained in Ω. When this ball intersects $\partial\Omega$, an adaptation similar to (3.145) is also valid. As announced above, inequality (3.146) plays the role of the Caccioppoli inequality that is only valid in the L^2 setting. It should be noted however that the proof of estimate (3.146) is more demanding than that of (3.144). This is not unexpected, since it concerns the *derivative of order one* (that is, the gradient), whereas (3.144) concerns the *derivative of order $\beta < 1$*, with a slight abuse of notation. In order to prove (3.146), we need, as expected, to obtain a good estimate of ∇v_ε, which plays the role of a part of ∇R_ε. Now, the expression of this gradient explicitly involves the corrector (see (3.92)). This point should be kept in mind when we consider nonperiodic problems below.

At this stage, we have made significant progress in the proof of estimates (3.134)–(3.135). Indeed, as we already mentioned, the function v_ε will in the end play the role of R_ε, and we have already proved bounds on the norm $\|v_\varepsilon\|_{L^2} = \|R_\varepsilon\|_{L^2}$ that is present in the right-hand side of (3.144)–(3.146) (recall the H^1 estimate (3.133)).

Although we have accumulated important information, namely estimates (3.144)–(3.146), we have only proceeded in the case of a homogeneous equation, that is, (3.137). We still need to establish estimates on the part $R_{1,\varepsilon}$, that is, for an equation of the type

$$
\begin{cases}
- \operatorname{div} (a_\varepsilon \, \nabla \, v_\varepsilon) = \operatorname{div}(f) & \text{in } \mathcal{D}, \\[2mm]
\hspace{2.5cm} v_\varepsilon \hspace{0.9cm} = \hspace{0.4cm} 0 & \text{on } \partial\mathcal{D},
\end{cases}
\tag{3.147}
$$

instead of (3.137). This is the purpose of the next paragraph.

The Non-homogeneous Equation

Here again, we are going to only give an *outline* of the proof of the desired estimates, for an equation of the type (3.147).

How are we going to extend the above technique of proof for the homogeneous equation to the case of a non-trivial right-hand side? A well-known trick consists in using the *Green function*. We recall that the Green function is defined, in our case (3.106), as the solution to

$$
\begin{cases}
- \operatorname{div}_x (a_\varepsilon \, \nabla_x \, G_\varepsilon(x, y)) = \delta(x - y) & \text{for } x \in \mathcal{D}, \\[2mm]
\hspace{2.5cm} G_\varepsilon(x, y) \hspace{0.9cm} = \hspace{0.4cm} 0 & \text{for } x \in \partial\mathcal{D},
\end{cases}
\tag{3.148}
$$

for each $y \in \mathcal{D}$. The existence and uniqueness of G_ε is proved for instance in [GW82]. In the case of a constant coefficient operator, the Green function satisfies $G_\varepsilon(x, y) = G_\varepsilon(x - y)$, and problem (3.148) may then be reduced to the same problem for a Dirac mass at the origin $y = 0$ (at least if $\mathcal{D} = \mathbb{R}^d$). The solution of (3.147) is then given by a convolution product. In our setting, the coefficient a_ε is not assumed to be constant (and $\mathcal{D} \neq \mathbb{R}^d$), so $G_\varepsilon(x, y)$ is a genuine function of its two variables. It remains that, since the equation is linear, the solution of the non-homogeneous equation is expressed by the *representation formula*

$$
v_\varepsilon(x) = \int G_\varepsilon(x, y) \operatorname{div} f(y) \, dy,
\tag{3.149}
$$

which we do not make precise for now. An estimate of the gradient can then be (formally) obtained by differentiation and integration by parts:

$$\nabla v_\varepsilon(x) = -\int \nabla_x \nabla_y G_\varepsilon(x, y) f(y) \, dy. \tag{3.150}$$

Hence, if we understand the properties of this Green function $G_\varepsilon(x, y)$ and of its gradients $\nabla_x G_\varepsilon(x, y)$, $\nabla_y G_\varepsilon(x, y)$, $\nabla_x \nabla_y G_\varepsilon(x, y)$, and if, here again, we establish that the corresponding estimates are *uniform with respect to* ε, then we will be in position to prove estimates for the solution to a non-homogeneous equation, that is, for functions of the type $R_{1,\varepsilon}$.

Considering the right-hand side of (3.148), we start by observing that it vanishes "quite often"... Indeed, as far as we are interested in establishing an estimate at points $x \neq y$, the Green function $G_\varepsilon(x, y)$ satisfies a homogeneous equation. This observation will prove to be key to show what we need. We may hope that most of the material used to address Eq. (3.137) can carry over to Eq. (3.148). This is indeed the case.

What do we know about the Green function G_ε? We first have general results, that only use the upper and lower bounds (3.1) on the coefficient a_ε in the divergence form operator $-\operatorname{div} a_\varepsilon \nabla$. The first such result is

$$0 \leq G_\varepsilon(x, y) \leq C \frac{1}{|x - y|^{d-2}}, \tag{3.151}$$

for all $x \neq y \in \mathcal{D} \subset \mathbb{R}^d$, $d \geq 3$, for some constant C that depends only on the coercivity constant in (3.1) and on $\|a_\varepsilon\|_{L^\infty} = \|a\|_{L^\infty}$ (hence, in our case, C does not depend on ε!). Such a pointwise estimate is a remarkable fact: it holds not only for the Laplacian, but for *any* divergence form operator with bounded and coercive coefficient. Let us at this stage recall that the Green function of the Laplace operator, at least in the case of the domain $\mathcal{D} = \mathbb{R}^d$, $d \geq 3$, reads as $G(x, y) = \dfrac{C_d}{|x - y|^{d-2}}$, where $C_d = \dfrac{\Gamma(d/2)}{2(d-2)\pi^{d/2}}$, Γ being the Euler function. In the case of a general domain $\mathcal{D}$, the maximum principle allows to prove that the Green function of the Laplace operator behaves similarly to that of the whole space for x close to y, and (if the domain allows) for $|x - y| \to +\infty$. This is the purpose of (3.151).

Another estimate that is valid in the general case is the following:

$$\left\| \nabla_x G_\varepsilon(., y_0) \right\|_{L^{\frac{d}{d-1}, \infty}(\mathcal{D})} + \left\| \nabla_y G_\varepsilon(x_0,) \right\|_{L^{\frac{d}{d-1}, \infty}(\mathcal{D})} \leq C, \tag{3.152}$$

for all $x_0, y_0 \in \mathcal{D} \subset \mathbb{R}^d$, and here again with a constant C depending only on the coercivity constant and on the L^∞ bound of the coefficient. In (3.152), the notation $L^{q,\infty}$, $1 < q < +\infty$, denotes the *Marcinkiewicz*, also called *weak* L^q

space, defined as the set of measurable functions f such that

$$\sup_{s>0} \left(s \text{ meas } \{x \, : \, |f(x)| > s\}^{\frac{1}{q}} \right) < +\infty. \tag{3.153}$$

The associated norm present in (3.152) is defined using the *spherically decreasing rearrangement* of f (we refer for instance to [BL76] for the precise definitions and the associated properties of these functional spaces).

It is again interesting to note, in connection with (3.152), that, in the case of the Laplace operator, $|\nabla_x G(x, y)| \propto \dfrac{1}{|x - y|^{d-1}}$ hence the fact that $\nabla_x G_\varepsilon(., y_0)$ belongs to $L^{\frac{d}{d-1}, \infty}$ is obvious, since

$$\sup_{s>0} \left(s \text{ meas } \left\{ x \, : \, \frac{1}{|x|^{d-1}} > s \right\}^{\frac{d-1}{d}} \right) = \sup_{s>0} \left(s \text{ meas } B\left(0, \left(\frac{1}{s}\right)^{\frac{1}{d-1}}\right)^{\frac{d-1}{d}} \right) < +\infty.$$

The fact that $\nabla_y G_\varepsilon(x_0, .) \in L^{\frac{d}{d-1}, \infty}$ then follows from $\nabla_y G(x, y) = -\nabla_x G(x, y)$.

Estimates (3.151) and (3.152) have been first published in the famous article [GW82]. We also refer to [BLA13], in which the results of [GW82] are revisited, together with some of the results of [AL87] (see also [KLS14]), that we are going to state and use below. We point out that the results of [AL87] also apply to the case of systems, and not only scalar equations, contrary to [GW82] and [BLA13]. We will return to this in Remark 3.13 below. Further results (for higher-order derivatives in particular, and also valid for systems) are proved in [DM95].

Considering (3.151), and in echo to our above remark regarding the right-hand side of (3.148) which vanishes as far as $x \neq y$, we now apply the same strategy as for (3.137) above. We thus derive, in the vein of (3.146), similar estimates on $\nabla_x G_\varepsilon(x, y)$ in function of the L^2-norm of G_ε in a ball of some suitable radius. We next use the estimate (3.151) as follows. Choosing a radius proportional to the distance $|x - y|$, which is possible since we consider x "far from" y, we prove the *pointwise* bound on the gradient:

$$|\nabla_x G_\varepsilon(x, y)| \leq C \frac{1}{|x - y|^{d-1}}, \tag{3.154}$$

for some constant independent of ε, and for all $x \neq y$. It is similarly possible to prove the same estimate on $\nabla_y G_\varepsilon(x, y)$. The function $G_\varepsilon(y, x)$ (note that we have exchanged x and y) is indeed the Green function of the operator $-\text{div}\left(a^T \nabla .\right)$, hence satisfies (3.154). In other words, this estimate holds with ∇_x replaced by ∇_y. Furthermore, we note that in Eq. (3.148), y is seen as a parameter. Hence, we take the derivative with respect to y of this equation, and find that $\nabla_y G_\varepsilon(x, y)$ is solution to (3.148), with $\nabla_y \delta$ instead of the Dirac mass δ. Here again, this distribution vanishes for $x \neq y$, and the same argument as above proves (3.154) for $\nabla_y G_\varepsilon(x, y)$. This method is closely related to the one we have used for an operator with constant coefficients, namely first differentiate the equation and next prove estimates on the gradient. We observe here that the coefficients are indeed constant with respect to y! A similar argument thus applies, giving estimates of the form

$$\left| \nabla_x \nabla_y G_\varepsilon(x, y) \right| \leq C \frac{1}{|x - y|^d}. \tag{3.155}$$

Note, however, that (3.151) is based on fairly general properties of the operator. In sharp contrast, our new estimates (3.154)–(3.155) are consequences of the specific homogenization properties of the operator $- \operatorname{div}_x a_\varepsilon \nabla$ (here, the periodic setting).

Given the above bounds on the Green function, we now use the representation formulas (3.149)–(3.150) of the solution to (3.147) and of its derivatives. Leaving aside (many) technicalities that we do not wish to detail here, we then obtain estimates on ∇v_ε and on the $C^{0,\beta}$-norm of v_ε solution to (3.147). The latter bounds may then be used, together with the estimates on the solution to (3.137), in order to prove bounds on solutions to generic equations of the type

$$\begin{cases} - \operatorname{div} (a_\varepsilon \nabla v_\varepsilon) = f & \text{in } \Omega, \\ v_\varepsilon \quad\quad\quad = g & \text{on } \partial\Omega, \end{cases} \tag{3.156}$$

These results in turn allow to prove (3.134)–(3.135). They can also be used (this is actually a technical step of the proof) to compare the Green function G_ε with the Green function G^* of the homogenized operator $- \operatorname{div}_x a^* \nabla$. Indeed, such a comparison result is obtained upon using a specific (and singular) right-hand side $f(x) = \delta(x - y)$ in Eq. (1). It can thereby be proved that

$$\left| G_\varepsilon(x, y) - G^*(x, y) \right] \leq C \frac{\varepsilon}{|x - y|^{d-1}}, \tag{3.157}$$

and that similar estimates hold on $\nabla_x G_\varepsilon(x, y)$, $\nabla_y G_\varepsilon(x, y)$, $\nabla_x \nabla_y G_\varepsilon(x, y)$ compared to $\nabla_x G^*(x, y)$, $\nabla_y G^*(x, y)$, $\nabla_x \nabla_y G^*(x, y)$, respectively. The latter of these bounds, that is,

$$\left| \nabla_x \nabla_y G_\varepsilon(x, y) - \left(\operatorname{Id} + \nabla w_{per} \left(\frac{y}{\varepsilon} \right) \right) \left(\operatorname{Id} + \nabla w_{per} \left(\frac{x}{\varepsilon} \right) \right) \nabla_x \nabla_y G^*(x, y) \right|$$
$$\leq C \varepsilon \tag{3.158}$$

will be made precise in Chap. 4 (see Eq. (4.53)) and will prove quite useful.

Since the above discussion only outlines the *main steps* of the proof of Proposition 3.6, we refer the reader to the original proof [AL87] for all the details (which are not to be overlooked). We will return to several particular points of this proof in Chap. 4.

3.4.4 Alternative Methods

The so-called *two-scale convergence* method is an alternative method to the techniques presented so far. It has been originally introduced by Gabriel N'Guetseng in [Ngu89], and further formalized by Grégoire Allaire in the famous article [All92]. In its original version, the method was designed for the periodic setting. It has next been extended to more general situations (see for instance [Ngu03a] for nonperiodic deterministic cases and [BMW94] for the stochastic setting).

The main ingredient of this approach is the following

Lemma 3.3 (Two-Scale Convergence) *For any sequence u^ε that is bounded in $H^1(\mathcal{D})$, there exists a function $u_0 \in H^1(\mathcal{D})$ such that, up to extraction, the following classical weak convergence holds:*

$$u^\varepsilon \xrightarrow{\varepsilon \to 0} u_0 \text{ in } H^1(\mathcal{D}).$$

Moreover, there exists a function $u_1 \in L^2(\mathcal{D}, H^1_{\mathrm{per}}(Q))$, such that the following convergence (called two-scale convergence*) holds:*

For any function $\xi \in L^2(\mathcal{D}, C^0_{\mathrm{per}}(Q))$,

$$\int_{\mathcal{D}} \nabla u^\varepsilon(x) \, \xi\left(x, \frac{x}{\varepsilon}\right) dx \xrightarrow{\varepsilon \to 0} \int_{\mathcal{D}} \int_Q (\nabla u_0(x) + \nabla_y u_1(x, y)) \, \xi(x, y) \, dy \, dx.$$

$$(3.159)$$

In Lemma 3.3, given a normed vector space V, $L^2(\mathcal{D}, V)$ denotes the space of V-valued functions f that are square integrable for the norm $\|\cdot\|_V$ of the space V, that is, $\int_{\mathcal{D}} \|f(x)\|_V^2 dx < +\infty$. The associated norm is

$$\|f\|_{L^2(\mathcal{D}, V)} = \left(\int_{\mathcal{D}} \|f(x)\|_V^2 dx \right)^{1/2}.$$

For instance, in the case $V = C^0_{\mathrm{per}}(Q)$ as above, f is a function of $x \in \mathcal{D}$ and $y \in Q$ that is periodic with respect to y, such that $\int_{\mathcal{D}} \sup_{y \in Q} |f(x, y)|^2 dx < +\infty$, and the associated norm is

$$\|f\|_{L^2(\mathcal{D}, C^0_{\mathrm{per}}(Q))} = \left(\int_{\mathcal{D}} \sup_{y \in Q} |f(x, y)|^2 dx \right)^{1/2}.$$

We only outline the proof of this result. First, the fact that u^ε is bounded in $H^1(\mathcal{D})$ implies the first convergence, up to extraction, is classical. Next, for all $\xi \in L^2(\mathcal{D}, C^0_{\text{per}}(Q))$, we have, for any index $1 \leq j \leq d$,

$$\left| \int_{\mathcal{D}} \partial_j u^\varepsilon(x) \xi\left(x, \frac{x}{\varepsilon}\right) dx \right| \leq C \left(\int_{\mathcal{D}} \xi\left(x, \frac{x}{\varepsilon}\right)^2 dx \right)^{1/2}$$

$$\xrightarrow{\varepsilon \to 0} C \|\xi\|_{L^2(\mathcal{D}, L^2_{\text{per}}(Q))}, \qquad (3.160)$$

where the constant C does not depend on ε. The majoration is a simple consequence of the Cauchy-Schwarz inequality, but the convergence needs to be proven (see [All92, Lemma 1.3]). In addition, ∇u^ε is bounded in $L^2(\mathcal{D})$, hence in $L^1(\mathcal{D})$, since the domain $\mathcal{D}$ is bounded. The map $\xi \mapsto \int_{\mathcal{D}} \nabla u^\varepsilon(x) \xi\left(x, \frac{x}{\varepsilon}\right) dx$, which is a linear form on $L^2(\mathcal{D}, C^0_{\text{per}}(Q))$, is thus bounded uniformly with respect to ε. This implies that $\partial_j u^\varepsilon$, considered as an element of $L^2(\mathcal{D}, M_{\text{per}}(Q))$, is a bounded sequence. Here, $M_{\text{per}}(Q)$ denotes the dual space of $C^0_{\text{per}}(Q)$, that is, the set of periodic measures of periodic cell Q. It thus converges weakly (up to an extraction) to $\mu_0 \in L^2(\mathcal{D}, M_{\text{per}}(Q))$. In addition, passing to the limit in the left-hand side of (3.160), we obtain that μ_0 is a continuous linear form on $L^2(\mathcal{D}, L^2_{\text{per}}(Q))$. It can thus be identified as an element χ_j of $L^2(\mathcal{D}, L^2_{\text{per}}(Q))$. Put differently, there exists $\chi_j \in L^2(\mathcal{D}, L^2_{\text{per}}(Q))$ such that

$$\forall \xi \in L^2(\mathcal{D}, C^0_{\text{per}}(Q)),$$

$$\int_{\mathcal{D}} \partial_j u^\varepsilon(x) \xi\left(x, \frac{x}{\varepsilon}\right) dx \xrightarrow[\varepsilon \to 0]{} \int_{\mathcal{D}} \int_Q \chi_j(x, y) \xi(x, y) \, dx \, dy. \qquad (3.161)$$

We want to prove next that the map $(x, y) \mapsto \chi(x, y) = (\chi_1(x, y), \ldots, \chi_d(x, y))$ has the structure of the right-hand side of (3.159). To this end, we first assume that the test function ξ in (3.161) does not depend on y. We deduce, since $\partial_j u^\varepsilon$ converges weakly to $\partial_j u_0$, that

$$\int_Q \chi(x, y) dy = \nabla u_0(x).$$

It follows that

$$\chi(x, y) = \nabla u_0(x) + U(x, y), \qquad \int_Q U(x, y) dy = 0, \qquad (3.162)$$

for some $U \in \left(L^2(\mathcal{D}, L^2_{\text{per}}(Q)) \right)^d$. We may show that U is a gradient with respect to y. We use as test functions some functions $\xi_j(y)$ such that the vector $\Psi = (\xi_j)_{1 \le j \le d}$ is divergence free. Convergence (3.161) then gives

$$\int_{\mathcal{D}} \nabla u^\varepsilon \cdot \Psi \left(\frac{x}{\varepsilon} \right) dx \xrightarrow[\varepsilon \to 0]{} \int_{\mathcal{D}} \nabla u_0(x) \cdot \left(\int_Q \Psi(y) dy \right) dx$$
$$+ \int_{\mathcal{D}} \int_Q U(x, y) \cdot \Psi(y) \, dy \, dx.$$

The div-curl Lemma (Lemma 3.2) allows to pass to the limit in the left-hand side, whence

$$\int_{\mathcal{D}} \nabla u_0 \cdot \left(\int_Q \Psi(y) dy \right) dx = \int_{\mathcal{D}} \nabla u_0(x) \cdot \left(\int_Q \Psi(y) dy \right) dx$$
$$+ \int_{\mathcal{D}} \int_Q U(x, y) \cdot \Psi(y) \, dy \, dx.$$

This implies that, for all $x \in \mathcal{D}$,

$$\int_Q U(x, y) \cdot \Psi(y) \, dy = 0.$$

This being valid for any vector-valued divergence free function Ψ, we apply the de Rham Lemma and infer that $U(x, y) = \nabla_y u_1(x, y)$ for some function $u_1 \in L^2(\mathcal{D}, H^1_{\text{per}}(Q))$.

Although we have used the div-curl Lemma in the above proof, we wish to point out that the original proof in [All92] does not use it. Instead, an integration by parts is used to pass to the limit in the corresponding term. The proof then becomes more delicate. This is why we preferred our approach here.

In the periodic case, $a_\varepsilon = a_{\text{per}} \left(\frac{\cdot}{\varepsilon} \right)$ (with a scalar-valued a_{per} and $f \in L^2(\mathcal{D})$ for simplicity), Lemma 3.3 allows to *prove* that the first terms of the two-scale expansion of Sect. 3.3.1 are correct. Indeed, let us multiply (1) by the function $\varphi_0(x) + \varepsilon \varphi_1 \left(x, \frac{x}{\varepsilon} \right)$, for $\varphi_0 \in H^1(\mathcal{D})$ and $\varphi_1 \in H^1(\mathcal{D}, H^1_{\text{per}}(Q))$ and let us integrate by parts over $\mathcal{D}$. Using the test function $\xi(x, y) = a(y)(\nabla \varphi_0(x) + \nabla_y \varphi_1(x, y))$ in the convergence (3.159), we obtain, in the limit $\varepsilon \to 0$,

$$\int_{\mathcal{D}} \int_Q \left(\nabla u_0(x) + \nabla_y u_1(x, y) \right) a(y) \left(\nabla \varphi_0(x) + \nabla_y \varphi_1(x, y) \right) dy \, dx$$

$$= \int_{\mathcal{D}} f \varphi_0. \qquad (3.163)$$

Taking $\varphi_0 = 0$, we deduce that

$$\int_{\mathcal{D}} \int_Q a(y) \left(\nabla u_0(x) + \nabla_y u_1(x, y) \right) \nabla_y \varphi_1(x, y) \, dy \, dx = 0,$$

holds for all function $\varphi_1 \in H^1(\mathcal{D}, H^1_{\text{per}}(Q))$, hence the equality

$$- \operatorname{div}_y \left[a(y)(\nabla u_0(x) + \nabla_y u_1(x, y)) dy \right] = 0,$$

holds at least in the sense of distributions over Q. Up to some regularity issues, we may then rigorously recover Eq. (3.46). This allows to prove that u_1 is indeed given by (3.47) using the periodic corrector (note that we have proved its existence independently). The first two terms of the expansion are thus the dominant terms. Like the energy method, this approach does not give any convergence rate.

Two-scale convergence theory has been adapted to nonperiodic settings. For the extension to hold, a sufficient condition is that the considered functional space is an algebra equipped with an average. This is for instance the case of almost periodic functions. Other settings are convenient, as we have seen in Chap. 1. We refer the interested reader to the articles [NS11, LNNW09, Ngu06, Ngu04, Ngu03b, Ngu03a]. These works use a notion of algebra (of functions) that is closely related to the interpretation in terms of the stationary ergodic theory of the quasi-periodic and almost periodic cases (see Sect. 4.3.2 below). This also suggests an alternative pathway for the settings we have studied in Sect. 1.4.5. Such ideas go beyond the scope of the present introductory textbook.

To end this chapter, we mention that yet other methods exist, such as for instance the *unfolding method*. Very briefly, this method is based on the definition of an unfolding operator $\mathcal{T}_\varepsilon$ which, to a function $w \in L^2(\mathcal{D})$, associates a function $\mathcal{T}_\varepsilon(w) \in L^2(\mathcal{D} \times Q)$, such that u^ε two-scale converges to $u_0(x, y)$ if and only if $\mathcal{T}_\varepsilon(u^\varepsilon) \longrightarrow u_0$ in $L^2(\mathcal{D} \times Q)$. This allows to replace the two-scale convergence topology by a somewhat more standard topology. The heart of the theory is a compactness lemma of the same type as Lemma 3.3. It ensures that, for any sequence u^ε that is bounded in $H^1(\mathcal{D})$, up to extraction, we have that $\mathcal{T}_\varepsilon(u^\varepsilon)$ converges to some $u_0 \in L^2(\mathcal{D})$, and that $\mathcal{T}_\varepsilon(\nabla u^\varepsilon)$ converges to $\nabla u_0(x) + \nabla_y u_1(x, y)$, for some $u_1 \in L^2(\mathcal{D}, H^1_{\text{per}}(Q))$. We refer to the textook [CDG18] for an exposition of the theory and its applications.

Remark 3.13 (The Case of Systems) Throughout the present textbook, all the convergence rates are given and proved for elliptic *equations*, as opposed to *systems* of equations, in which the unknown u^ε is $\mathbb{R}^m$-valued, for some $m \in \mathbb{N}$. The main reason for this pedagogic choice is that the proofs are considerably simpler in the case of equations. All the results presented in Sect. 3.4 are nevertheless also valid for elliptic systems of PDEs. For instance, Proposition 3.6 is proved in [AL87] for the case of systems.

The method of proof in the case of systems is similar to what we have outlined in this chapter. An important simplification however occurs when we need to use the maximum principle (which in general does not hold for systems) or one of its consequences (the Harnack inequality or the Nash-Moser theory). At each stage where we have used such an argument, one needs to argue using an elliptic regularity estimate in Hölder norm. Such estimates are valid in the case of systems.

In order to give a flavor of these additional difficulties, we refer to the article [KLS14], where, for each proof, the authors indicate the simplifications associated with the scalar-valued case.

Likewise, for a recent presentation of the periodic homogenization theory of systems, we refer to [She18].   □

Chapter 4
Dimension ≥ 2: Some Explicit Cases Beyond the Periodic Setting

Abstract We consider nonperiodic homogenization in dimensions d higher than 2. We first consider the deterministic setting in Sect. 4.1, with "local" defects, by which we mean perturbations in $L^q(\mathbb{R}^d)$, $1 < q < +\infty$, of a periodic background. In Sect. 4.2, we proceed with deterministic cases that cannot be reduced to local defects, as for instance the quasi-periodic and almost periodic cases. We conclude this chapter upon considering the stochastic setting, in Sect. 4.3.

Keywords Nonperiodic homogenization problems · Local defects · Quasi-periodic setting · Almost periodic setting · Random setting · Calderón-Zygmund theory · Concentration-compactness method

At last, we consider *nonperiodic* homogenization problems in dimensions d higher than 2. The preceding chapters may have seemed lengthy. They were however an important (and necessary) preliminary. We have indeed collected there

- some elements of "abstract" homogenization theory, that is, general results that only require minimal assumptions on the considered setting (and on the coefficient a in particular);
- the essence of homogenization theory in the periodic setting for Eq. (1), and
- our first cases of perturbation of the periodic case, for the computation of averages and for one-dimensional homogenization, the latter issue being actually, up to minor modifications, a question of computation of averages (with the notable addition of the "corrector" and how to use it).

We are now going to combine these three basic cornerstones in order to address nonperiodic problems in dimensions $d \geq 2$.

We first consider the deterministic setting in Sect. 4.1, with "local" defects, by which we mean perturbations in $L^q(\mathbb{R}^d)$, $1 < q < +\infty$, of a periodic background. In Sect. 4.2, we proceed with deterministic cases that cannot be reduced to local defects, as for instance the quasi-periodic and almost periodic cases. We conclude this chapter upon considering the stochastic setting, in Sect. 4.3.

© The Author(s), under exclusive license to Springer Nature Switzerland AG 2023
X. Blanc, C. Le Bris, *Homogenization Theory for Multiscale Problems*, MS&A 21,
https://doi.org/10.1007/978-3-031-21833-0_4

4.1   Localized Defects

We first deal with the case of a "localized" perturbation of a periodic geometry: the coefficient $a_\varepsilon = a\left(\frac{\cdot}{\varepsilon}\right)$ is assumed to have the following form:

$$a = a_{per} + \widetilde{a} \quad \text{where } \widetilde{a} \in L^q(\mathbb{R}^d), \quad \text{for some } 1 < q < +\infty. \tag{4.1}$$

This means, in some sense that will be made precise should need be, that the perturbation $\widetilde{a}$ vanishes at infinity. Note that we have excluded the cases $q = 1$ and $q = +\infty$ in (4.1). The case $q = 1$ could be treated with some (after all, irrelevant) technical modifications of the methods we are going to use for $1 < q < +\infty$. In sharp contrast, the case $q = +\infty$ is an important open question, so far as we know. Indeed, in such a case, the fact that the defect vanishes at infinity would only translate into

$$\widetilde{a}(x) \overset{|x| \to +\infty}{\longrightarrow} 0, \tag{4.2}$$

Unfortunately, we are unable to manipulate the space of functions that are bounded and vanish at infinity, even when we assume in addition that these functions are locally regular (Hölder continuous for instance). The case (4.2) is, however, particularly relevant as far as applications are concerned. In the current state of our understanding, we are only able to model the fact that the defect vanishes at infinity using the property $\widetilde{a} \in L^q(\mathbb{R}^d)$, $1 < q < +\infty$. In such a case, (4.2) does not hold as such (some functions in L^q do not tend to 0 at infinity), but as soon as some additional regularity is assumed, for instance when $\widetilde{a}$ is uniformly Hölder continuous, then we recover the condition that $\widetilde{a}$ converges to 0 at infinity. However, this is much less general than the only condition (4.2).

We have seen in Sect. 1.4.2 of Chap. 1 for the computation of averages and in Sect. 2.1.1 of Chap. 2 for one-dimensional homogenization that, whenever $\widetilde{a}$ satisfies (4.1), the homogenized coefficient is unchanged. We may thus expect a similar property here, that is, with obvious notation, $a^* = (a_{per})^*$. Moreover, remembering formula (2.42) in Sect. 2.3 of Chap. 2, we also expect that the corrector w_p, for a fixed $p \in \mathbb{R}^d$, writes $w_p = w_{p,per} + \widetilde{w}_p$ with $\nabla \widetilde{w}_p$ in the same functional space as $\widetilde{a}$, that is, $L^q(\mathbb{R}^d)$. We are going to see, first in the case $q = 2$, then for $1 < q \neq 2 < +\infty$, that both theses properties $a^* = (a_{per})^*$ and $\nabla \widetilde{w}_p \in L^q(\mathbb{R}^d)$, hold true. Then, we will prove that the corrector allows to obtain a strong convergence in H^1, as we have seen in Proposition 3.4. For the latter property, we will give a complete proof in the case $q = 2$, while only a sketch of proof will be provided in the case $1 < q \neq 2 < +\infty$.

We henceforth fix $1 < q < +\infty$ and assume that a is of the form (4.1), where a and a_{per} satisfy the bound and coercivity property (3.1). As before, we assume for simplicity that a and a_{per} are scalar-valued. Most of the arguments we are about to develop are easily adapted to the general case of *matrix-valued* coefficients. However, when this is exceptionally not the case, we will mention it and explain the necessary modifications. We also assume that the coefficient a_{per} is Hölder continuous. We will, in Sect. 4.1.2 (see (4.37)), assume that, likewise, $\widetilde{a}$ is Hölder continuous.

The homogenization problem under study is that of Eq. (1), where the coefficient a satisfies (4.1) (together with the above regularity assumptions), that is the equation

$$- \operatorname{div} \left(\left(a_{per} \left(\frac{\cdot}{\varepsilon} \right) + \widetilde{a} \left(\frac{\cdot}{\varepsilon} \right) \right) \nabla u^{\varepsilon} \right) = f,$$

supplied with homogeneous Dirichlet boundary conditions.

The key issue in the present section is the solvability in a suitable functional space of the corrector equation presumably associated with our homogenization problem. We fix $p \in \mathbb{R}^d$ and study the equation

$$- \operatorname{div} \left((a_{per} + \widetilde{a}) (p + \nabla w_p) \right) = 0. \tag{4.3}$$

This equation is both

- the natural extension to higher dimensions of the one-dimensional equation (2.42) of Sect. 2.3 in Chap. 2, and
- the extension to the nonperiodic case of the periodic equation (3.67) of Chap. 3.

Based on the decomposition (2.42) in Sect. 2.3 of Chap. 2, we look for the solution of (4.3) under the form

$$\begin{cases} w_p = w_{p,per} + \widetilde{w}_p, \\[2mm] w_{p,per} \quad \text{periodic solution to (3.67),} \\[2mm] \nabla \widetilde{w}_p \in L^q(\mathbb{R}^d), \end{cases} \tag{4.4}$$

where $1 < q < +\infty$ is the exponent appearing in $\widetilde{a} \in L^q(\mathbb{R}^d)$ in (4.1).

We are going to present *three* different strategies for proving the existence and uniqueness of the solution to (4.3)–(4.4). The first one is presented in Sect. 4.1.1. It is dedicated to divergence-form equations (4.3) and is specific to the case $q = 2$. It is Hilbertian in nature, and is based on rather elementary methods, such as repeated integrations by parts. The second strategy, in Sect. 4.1.2, applies to *any* exponent $1 < q < +\infty$. It is based on the concentration-compactness method. The strategy of proof is thus much more flexible, and may be adapted to other types of equations. The third strategy, in Sect. 4.1.3, uses classical tools of harmonic analysis, applied to divergence form equations. For the most part, this strategy does not require the coefficient to have a structure such as (4.1), and indeed applies to coefficients much more general than those satisfying (4.1).

Once we have obtained existence and uniqueness of the solution w_p in the space defined by (4.4), we are going to use the corrector for two (related but) different purposes: first it will allow us to derive an explicit expression of the homogenized matrix a^*. We only proceed in the case $q = 2$ and this is the matter of Sect. 4.1.1. The adaptation to the general case $1 < q < +\infty$ is only a matter of minor modifications. Second, we use the corrector to accurately approximate the solution u^ε in the limit of ε vanishing.

4.1.1 The Case of a Defect in $L^2(\mathbb{R}^d)$

Throughout the present section, we fix $q = 2$.

4.1.1.1 Existence (and Uniqueness) of the Corrector

We are going to prove the following result.

Lemma 4.1 *Fix $p \in \mathbb{R}^d$. Assume that a is of the form (4.1) where both a_{per} and a satisfy (3.1), and that $a_{per} \in C^{0,\alpha}$, for some $\alpha > 0$.*

Then, Eq. (4.3) has a unique solution of the form (4.4), up to the addition of a constant. This solution is unique in the class of solutions to (4.3) such that $\nabla w_p \in L^2_{per} + L^2(\mathbb{R}^d)$ if we impose the condition

$$\lim_{R \to +\infty} \frac{1}{|B_R|} \int_{B_R} \nabla w_p = 0, \tag{4.5}$$

where B_R denotes the ball centered at the origin and of radius R.

Important comments on Lemma 4.1 and on its proof will be given below.

Proof of Lemma 4.1

Step 1: Existence Using (3.67), Eq. (4.3) reads as

$$- \operatorname{div}\left((a_{per} + \tilde{a}) \nabla \tilde{w}_p\right) = \operatorname{div}\left(\tilde{a}(p + \nabla w_{p,per})\right) \tag{4.6}$$

and is posed on *the whole space* $\mathbb{R}^d$. We seek for a solution such that $\nabla \tilde{w}_p \in L^2(\mathbb{R}^d)$. Equation (4.6) is linear and can immediately be recast under the following variational form: *find $\tilde{w}_p \in H^1(\mathbb{R}^d)$ such that for all $v \in H^1(\mathbb{R}^d)$, $\mathcal{B}(\tilde{w}_p, v) = \mathcal{L}(v)$, where the bilinear form $\mathcal{B}$ and the linear form $\mathcal{L}$ respectively write*

$$\mathcal{B}(\tilde{w}_p, v) = \int_{\mathbb{R}^d} (a_{per} + \tilde{a}) \nabla \tilde{w}_p . \nabla v$$

and

$$\mathcal{L}(v) = \int_{\mathbb{R}^d} \tilde{a}(p + \nabla w_{p,per}) . \nabla v.$$

Both forms are continuous in $H^1(\mathbb{R}^d)$. Indeed, $\mathcal{B}$ is continuous because $a \in L^\infty(\mathbb{R}^d)$, and $\mathcal{L}$ is continuous because $\widetilde{a}\,(p + \nabla w_{p,per}) \in L^2(\mathbb{R}^d)$ (recall that $\widetilde{a} \in L^2(\mathbb{R}^d)$ and $\nabla w_{p,per} \in L^\infty(\mathbb{R}^d)$, the latter being a consequence of $a_{per} \in C^{0,\alpha}(\mathbb{R}^d)$ and of Theorem A.12, as we have seen in Sect. 3.4.3). Unfortunately, the bilinear form $\mathcal{B}$ is not coercive on $H^1(\mathbb{R}^d)$, which prevents us from applying the Lax-Milgram Lemma. Indeed, no Poincaré-type inequality holds true on the whole space $\mathbb{R}^d$. Such an inequality would imply that $\|\widetilde{w}_p\|_{H^1(\mathbb{R}^d)} \leq C_P \|\nabla \widetilde{w}_p\|_{L^2(\mathbb{R}^d)}$, for some constant $C_P < +\infty$. It is an easy exercise to build a sequence of functions that contradicts this inequality. We will again face similar difficulties in the sequel.

A classical technique to circumvent this difficulty is the so-called *regularization method*, or *approximation method*. We will use it repeatedly in the present textbook (see for instance the stochastic case in Sect. 4.3). We introduce a small parameter $\eta > 0$ and study the equation

$$- \operatorname{div}\left((a_{per} + \widetilde{a})\,\nabla \widetilde{w}_p^\eta\right) + \eta\,\widetilde{w}_p^\eta = \operatorname{div}\left(\widetilde{a}\,(p + \nabla w_{p,per})\right). \tag{4.7}$$

This equation agrees with (4.6) up to the addition of the zero order term $\eta\,\widetilde{w}_p^\eta$ in the left-hand side. The bilinear form associated with Eq. (4.7) reads as

$$\mathcal{B}_\eta(\widetilde{w}_p, v) = \int_{\mathbb{R}^d} (a_{per} + \widetilde{a})\,\nabla \widetilde{w}_p \cdot \nabla v + \eta \int_{\mathbb{R}^d} \widetilde{w}_p\, v.$$

Owing to the coercivity (3.1) of $a_{per} + \widetilde{a}$, and since $\eta > 0$, the bilinear form $\mathcal{B}_\eta$ is, in sharp contrast with $\mathcal{B}$ itself, coercive on $H^1(\mathbb{R}^d)$:

$$\mathcal{B}_\eta(\widetilde{w}_p, \widetilde{w}_p) \geq \min(\mu, \eta)\,\|\widetilde{w}_p\|_{H^1(\mathbb{R}^d)}^2\,,$$

where μ is the coercivity constant of a. In addition, the bilinear form $\mathcal{B}_\eta$ and the linear form $\mathcal{L}$ are continuous on $H^1(\mathbb{R}^d)$, as we have just seen. For $\eta > 0$ fixed, the Lax-Milgram Lemma allows to prove the existence and uniqueness of $\widetilde{w}_p^\eta \in H^1(\mathbb{R}^d)$ solution to the variational formulation associated to (4.7), namely: *for all function* $v \in H^1(\mathbb{R}^d)$,

$$\int_{\mathbb{R}^d} (a_{per} + \widetilde{a})\,\nabla \widetilde{w}_p^\eta \cdot \nabla v + \eta \int_{\mathbb{R}^d} \widetilde{w}_p^\eta\, v = \int_{\mathbb{R}^d} \widetilde{a}\,(p + \nabla w_{p,per}) \cdot \nabla v. \tag{4.8}$$

Using $v = \widetilde{w}_p^\eta$ in (4.8), and successively applying the Cauchy-Schwarz inequality and the Young inequality to bound the right-hand side, we have

$$\left| \int_{\mathbb{R}^d} \widetilde{a}\,(p + \nabla w_{p,per}) \cdot \nabla \widetilde{w}_p \right| \leq \|\widetilde{a}\|_{L^2(\mathbb{R}^d)} \|p + \nabla w_{p,per}\|_{L^\infty(\mathbb{R}^d)} \|\nabla \widetilde{w}_p^\eta\|_{L^2(\mathbb{R}^d)}$$

$$\leq \frac{\mu}{2}\|\nabla \widetilde{w}_p^\eta\|_{L^2(\mathbb{R}^d)}^2 + \frac{1}{2\mu}\|\widetilde{a}\|_{L^2(\mathbb{R}^d)}^2 \|p + \nabla w_{p,per}\|_{L^\infty(\mathbb{R}^d)}^2,$$

where μ is the coercivity constant of a. We then use the coercivity of $\mathcal{B}_\eta$ to bound from below the left-hand side of (4.8), with $v = \widetilde{w}_p^\eta$, and obtain

$$\frac{\mu}{2} \left\| \nabla \widetilde{w}_p^\eta \right\|_{L^2(\mathbb{R}^d)}^2 + \eta \left\| \widetilde{w}_p^\eta \right\|_{L^2(\mathbb{R}^d)}^2 \leq \frac{1}{2\mu} \left\| \widetilde{a} \right\|_{L^2(\mathbb{R}^d)}^2 \left\| p + \nabla w_{p,per} \right\|_{L^\infty(\mathbb{R}^d)}^2 . \tag{4.9}$$

It follows that $\nabla \widetilde{w}_p^\eta$ is bounded in $L^2(\mathbb{R}^d)$, uniformly with respect to η. Hence, up to extraction of a subsequence, we may assume that $\nabla \widetilde{w}_p^\eta$ converges weakly in $L^2(\mathbb{R}^d)$. We denote by $\mathbf{g} \in L^2(\mathbb{R}^d)$ this limit. Since differential operators are continuous with respect to the weak L^2 topology, the equality $\mathrm{curl}\left(\nabla \widetilde{w}_p^\eta\right) = 0$ is satisfied by $\mathbf{g}$. We apply the de Rham Lemma, that we have already used in Chap. 3. It implies that there exists a function $\widetilde{w}_p \in H^1_{loc}(\mathbb{R}^d)$ such that $\mathbf{g} = \nabla \widetilde{w}_p$. Moreover, since $\eta \widetilde{w}_p^\eta = \sqrt{\eta} \left(\sqrt{\eta} \widetilde{w}_p^\eta\right)$ and since $\sqrt{\eta} \left\| \widetilde{w}_p^\eta \right\|_{L^2(\mathbb{R}^d)}$ is bounded (given (4.9)), $\eta \widetilde{w}_p^\eta$ converges strongly to zero in $L^2(\mathbb{R}^d)$. This and the weak convergence $\nabla \widetilde{w}_p^\eta \longrightarrow \nabla \widetilde{w}_p$ allow to pass to the limit in (4.8) and obtain

$$\int_{\mathbb{R}^d} (a_{per} + \widetilde{a}) \, \nabla \widetilde{w}_p . \nabla v \; = \; \int_{\mathbb{R}^d} \widetilde{a} \, (p + \nabla w_{p,per}) . \nabla v, \tag{4.10}$$

for all function $v \in H^1(\mathbb{R}^d)$. This proves the existence of $\widetilde{w}_p \in H^1_{loc}(\mathbb{R}^d)$, solution to (4.6) (in particular) in the sense of distributions. Note that, passing to the weak limit in (4.9), we have the estimate

$$\left\| \nabla \widetilde{w}_p \right\|_{L^2(\mathbb{R}^d)} \leq C \left\| \widetilde{a} \right\|_{L^2(\mathbb{R}^d)} \left\| p + \nabla w_{p,per} \right\|_{L^\infty(\mathbb{R}^d)}.$$

In particular, $\nabla \widetilde{w}_p \in L^2(\mathbb{R}^d)$. Note also that the function w_p that we have defined satisfies condition (4.5) since $\int_Q \nabla w_{p,per} = 0$ on the periodic cell Q, hence

$$\left| \int_{B_R} \nabla w_{p,per} \right| = o\left(|B_R|\right), \text{ while, owing to the Cauchy-Schwarz inequality,}$$

$$\left| \int_{B_R} \nabla \widetilde{w}_p \right| \leq \sqrt{|B_R|} \left(\int_{B_R} |\nabla \widetilde{w}_p|^2 \right)^{1/2}.$$

Thus, $\left| \int_{B_R} \nabla \widetilde{w}_p \right| = o\left(|B_R|\right)$. This concludes the proof of our existence result.

Step 2: Uniqueness We consider two solutions w_p^1 and w_p^2 of (4.3), and assume that each of their respective gradients belongs to $L^2_{per} + L^2(\mathbb{R}^d)$ and satisfies (4.5). We denote by v the difference $v = w_p^1 - w_p^2$, and aim at proving that v is constant (recall that we cannot prove more than uniqueness up to an additive constant, since Eq. (4.3) involves only the gradient ∇w_p, not w_p itself). We know that $\nabla v = \mathbf{g}_{per} + \widetilde{\mathbf{g}}$ with $\mathbf{g}_{per} \in L^2_{per}$ and $\widetilde{\mathbf{g}} \in L^2(\mathbb{R}^d)$. Since $\mathrm{curl}\, \nabla v = 0$, we have $\mathrm{curl}\left(\mathbf{g}_{per} + \widetilde{\mathbf{g}}\right) = 0$. Translating this equality at infinity (in order to do so, we write it in the sense of distributions, replace ∇v by $\nabla v(. + nT)$, for $n \in \mathbb{N}$ and $T \in \mathbb{R}^d$ is a fixed period of a_{per}, and let $n \to +\infty$: the term $\widetilde{\mathbf{g}}$ vanishes while the periodic term $\mathbf{g}_{per}$ remains unchanged), we obtain that curl $\mathbf{g}_{per} = 0$, hence that curl $\widetilde{\mathbf{g}} = 0$. We then apply the de Rham Lemma: each term is a gradient, and we may write $\mathbf{g}_{per} = \nabla v_0$ and $\widetilde{\mathbf{g}} = \nabla \widetilde{v}$, for some v_0 and some $\widetilde{v}$. Concerning v_0, we actually know more. Since w_p^1 and w_p^2 both satisfy condition (4.5), so does v. As above, we apply the Cauchy-Schwarz inequality, showing that the part involving $\widetilde{\mathbf{g}} = \nabla \widetilde{v}$ vanishes. Thus, we are left with the periodic part $\mathbf{g}_{per} = \nabla v_0$. Consequently, we have $\int_Q \mathbf{g}_{per} = 0$. Hence, $\mathbf{g}_{per}$ is not only a gradient: it is the gradient of a periodic function. We denote this function by $v_0 = v_{per}$. So far, we have proved that the difference v of two solutions satisfies

$$- \operatorname{div}((a_{per} + \widetilde{a}) \nabla v) = 0, \tag{4.11}$$

and reads as $v = v_{per} + \widetilde{v}$, where v_{per} is periodic and $\nabla \widetilde{v} \in L^2(\mathbb{R}^d)$. Proving uniqueness amounts to proving that v is constant.

In dimension $d = 1$, this is easy. We can integrate (4.11), finding that $v' = C(a_{per} + \widetilde{a})^{-1}$, for some integration constant C. This may be written as $v' = C a_{per}^{-1} - C \widetilde{a}(a_{per} + \widetilde{a})^{-1} a_{per}^{-1}$, from which we deduce that $v'_{per} = C a_{per}^{-1}$. Now, this function cannot be the derivative of a periodic function, unless $C = 0$ (otherwise the average of v'_{per} does not vanish). Hence, v is constant. In the case $d \geq 2$, as we will see below, the method of proof is an adaptation of a classical technique, used for instance to prove the Caccioppoli inequality. More precisely, the strategy consists in multiplying by a cut-off function and next integrating by parts. This allows to estimate the L^2 norm of the gradient of the solution on a ball by the L^2 norm of the solution on a larger ball. The Caccioppoli inequality is stated in Theorem A.5. For its proof, we refer for instance to [Gia83, Proposition 2.1, pp 76–77].

We first prove that $\nabla v_{per} = 0$. Let χ be a cut-off function such that $\chi = 1$ in the unit ball B_1 centered at $x \in \mathbb{R}^d$ (we do not make this explicit in the sequel), and such that $\chi = 0$ outside the ball B_2 of radius 2 centered at $x = 0$. We assume that χ is smooth, and that its gradient is bounded, say, by 2. We define the rescaled function $\chi_R = \chi(./R)$, and note that $\nabla \chi_R = O(1/R)$ at any point of the annulus $A_{R,2R} = B_R^c \cap B_{2R}$. We multiply (4.11) by the function $\left(v - \fint_{A_{R,2R}} \widetilde{v} \right) \chi_R^2$

where, henceforth, $\fint_D g = \dfrac{1}{|D|} \int_D g$ denotes the average of the function g on the bounded domain D. The choice of this test function may seem strange at first sight. It may be understood as follows: (i) we would like to multiply (4.11) by v in order to obtain bounds on ∇v in L^2 which is our natural framework, (ii) we need to subtract a normalisation constant to v in order to be able to estimate its L^2 norm using a Poincaré-Wirtinger-type inequality, otherwise we have no hope to recover the expression of the gradient norm in both sides of the inequality, (iii) we need a

cut-off function so that all boundary terms in the integration by parts vanish, and
(iv) we need to integrate on large volumes so as to make the cut-off function vanish.
All this will become clear in the following computations. Using an integration by
parts, we obtain

$$\int a|\nabla v|^2 \chi_R^2 = -2 \int a \left(v - \fint_{A_{R,2R}} \tilde{v} \right) \chi_R \nabla v . \nabla \chi_R .$$

We apply the Cauchy-Schwarz inequality to bound the right-hand side. Squaring
and dividing both sides by the integral of $a|\nabla v|^2 \chi_R^2$, we have

$$\int a|\nabla v|^2 \chi_R^2 \leq 4 \int a \left(v - \fint_{A_{R,2R}} \tilde{v} \right)^2 |\nabla \chi_R|^2 .$$

Using the bound $\nabla \chi_R = O(1/R)$ on the annulus $A_{R,2R}$, this implies

$$\int a|\nabla v|^2 \chi_R^2 \leq \frac{C}{R^2} \int_{A_{R,2R}} \left(v - \fint_{A_{R,2R}} \tilde{v} \right)^2 ,$$

for some constant C independent of the radius R. Recalling the decomposition $v = v_{per} + \tilde{v}$, we know that

$$\left(v - \fint_{A_{R,2R}} \tilde{v} \right)^2 \leq 2 v_{per}^2 + 2 \left(\tilde{v} - \fint_{A_{R,2R}} \tilde{v} \right)^2$$

almost everywhere. Hence,

$$\int a|\nabla v|^2 \chi_R^2 \leq 2 \frac{C}{R^2} \int_{A_{R,2R}} v_{per}^2 + 2 \frac{C}{R^2} \int_{A_{R,2R}} \left(\tilde{v} - \fint_{A_{R,2R}} \tilde{v} \right)^2 . \qquad (4.12)$$

Since v_{per} is periodic, the first term of the right-hand side satisfies

$$\int_{A_{R,2R}} v_{per}^2 = O(R^d) . \qquad (4.13)$$

Turning to the second term of the right-hand side, we use the Poincaré-Wirtinger
inequality, that we first apply to an arbitrary function u on the fixed annulus $A_{1,2}$:

$$\int_{A_{1,2}} \left(u - \fint_{A_{1,2}} u \right)^2 \leq C \int_{A_{1,2}} |\nabla u|^2 .$$

Next, a change of variables allows to deduce

$$\int_{A_{R,2R}} \left(u - \fint_{A_{R,2R}} u \right)^2 \le C R^2 \int_{A_{R,2R}} |\nabla u|^2 \tag{4.14}$$

for any arbitrary function u defined on the annulus $A_{R,2R}$. The point here is the scaling law $C R^2$, where C does not depend on R. We apply this to $u = \tilde{v}$, and obtain:

$$\int_{A_{R,2R}} \left(\tilde{v} - \fint_{A_{R,2R}} \tilde{v} \right)^2 \le C R^2 \int_{A_{R,2R}} |\nabla \tilde{v}|^2, \tag{4.15}$$

where $\displaystyle\int_{A_{R,2R}} |\nabla \tilde{v}|^2$ vanishes as $R \longrightarrow +\infty$, because $\nabla \tilde{v} \in L^2(\mathbb{R}^d)$. Inserting (4.13) and (4.15) into (4.12), bounding the left-hand side from below (using that a is coercive), and using that $\chi_R^2 \equiv 1$ in B_R and is everywhere non-negative, we conclude that

$$\int_{B_R} |\nabla v|^2 = O(R^{d-2}) + o(1) = O(R^{d-2}) = o(R^d). \tag{4.16}$$

We use the fact that $v = v_{per} + \tilde{v}$, from which we infer that

$$\int_{B_R} |\nabla v|^2 \ge \frac{1}{2} \int_{B_R} |\nabla v_{per}|^2 - \int_{B_R} |\nabla \tilde{v}|^2 \ge \frac{1}{2} \int_{B_R} |\nabla v_{per}|^2 - \int_{\mathbb{R}^d} |\nabla \tilde{v}|^2,$$

and estimate (4.16) implies *a fortiori* that $\displaystyle\int_{B_R} |\nabla v_{per}|^2 = o(R^d)$. Since ∇v_{per} is periodic, it vanishes identically. Thus, v_{per} is constant.

At this stage, we wish to make the following point: actually, the proof that $\nabla v_{per} = 0$ may be performed in a simpler way: considering Eq. (4.11) in the sense of distributions, it is possible to translate it at infinity. This makes both $\tilde{a}$ and $\nabla \tilde{v}$ vanish, and gives the periodic equation $-\operatorname{div}(a_{per}\nabla v_{per}) = 0$, which, as we already know, implies that $\nabla v_{per} \equiv 0$. We however wanted to give the above proof for pedagogical purpose, and because it is useful in the following.

We next insert $\nabla v_{per} \equiv 0$ into (4.11):

$$- \operatorname{div}((a_{per} + \tilde{a}) \nabla \tilde{v}) = 0. \tag{4.17}$$

We want to prove that (4.17) and the fact that $\nabla \tilde{v} \in L^2(\mathbb{R}^d)$ imply $\nabla \tilde{v} \equiv 0$. This result is well-known, but, as we have just pointed out, we have already proved it! Indeed, the *same* argument as above applies: multiplying (4.17) by the function $\left(\tilde{v} - \fint_{A_{R,2R}} \tilde{v} \right) \chi_R^2$ (note that we have changed v into $\tilde{v}$ in its first occurence!), leads to the estimate analogous to (4.12), that is,

$$\int a|\nabla\widetilde{v}|^2\chi_R^2 \leq \frac{C}{R^2}\int_{A_{R,2R}}\left(\widetilde{v}-\fint_{A_{R,2R}}\widetilde{v}\right)^2.$$

This allows to prove, with the same method, an estimate analogous to (4.16), which now reads as

$$\int_{B_R}|\nabla\widetilde{v}|^2 = o\,(1).$$

Taking the limit $R \to +\infty$, this implies that $\nabla\widetilde{v}$ vanishes identically. We thus have shown that v is a constant, thereby concluding the proof of Lemma 4.1.   □

As announced above, some comments are now in order.

First, we note that we have proved *uniqueness* (up to the addition of a constant) of the corrector w_p, but this result is actually not necessary for the theory of homogenization. In this respect, only the *existence* of a suitable corrector is important. Once we have built *a* corrector, we are able to derive a formula for a^*, that we can use to compute u^*. We can also define the approximation $u^{\varepsilon,1}$, and possibly prove convergence and estimate convergence rates. In the periodic case, uniqueness (up to the addition of a constant) of the corrector is a simple consequence of the Lax-Milgram Lemma. Here, as we have seen in the second step of the proof of Lemma 4.1, uniqueness is not so easy to prove. We nevertheless proved uniqueness because of its *practical* interest. Slightly anticipating on Chap. 5, dedicated to numerical considerations, we indeed point out that designing numerical methods to compute a solution to a partial differential equation such as (4.3), without having first proved uniqueness of this solution, is delicate. We also point out that, on the *methodological* side, proving uniqueness is a first step in the proof of convergence of an algorithm for computing an approximation of the solution.

Second, we have assumed that the coefficient is Hölder continuous: $a_{per} \in C^{0,\alpha}(\mathbb{R}^d)$. This assumption has been used only to prove that $\nabla w_{p,per} \in L^\infty(\mathbb{R}^d)$, which in turn has allowed to show that $\widetilde{a}\,(p + \nabla w_{p,per}) \in L^2(\mathbb{R}^d)$ in the right-hand side of (4.6). As already pointed out in Chap. 3, this assumption may be weakened, since the results of [LN03, LV00] imply that the gradient of the corrector is still bounded if the coefficient is assumed to be piecewise Hölder continuous. Furthermore, since, in the present Sect. 4.1.1, we work in a Hilbert setting (with L^2-type estimates), we could expect that the regularity of the coefficient is irrelevant. Indeed, the coercivity of this coefficient should, in principle, be sufficient for our purpose (in this respect, see the discussion at the end of Sect. 3.4.3.4). This is *more or less* true. For instance, if the function $\widetilde{a}$ is in L^2 and is constant in each cell $k + Q$, $k \in \mathbb{Z}^d$, then without any regularity assumption on a_{per}, we have $\widetilde{a}\,(p + \nabla w_{p,per}) \in L^2(\mathbb{R}^d)$. Indeed, in such a particular case, we note that

$$\left\| \widetilde{a}\,(p + \nabla w_{p,per}) \right\|^2_{L^2(k+Q)} = \left\| \widetilde{a} \right\|^2_{L^2(k+Q)} \left\| p + \nabla w_{p,per} \right\|^2_{L^2(k+Q)}.$$

This allows to bound the associated series in $k \in \mathbb{Z}^d$, since $\left\| p + \nabla w_{p,per} \right\|_{L^2(k+Q)}$ is bounded independently of k (due to periodicity and local L^2 integrability), and the series of general term $\left\| \widetilde{a} \right\|^2_{L^2(k+Q)}$ converges to $\left\| \widetilde{a} \right\|^2_{L^2(\mathbb{R}^d)}$. A similar argument gives the same conclusion under the assumption that $\sum_{k \in \mathbb{Z}^d} \left\| \widetilde{a} \right\|^2_{L^\infty(k+Q)} < +\infty$, which is slightly stronger than $\widetilde{a} \in L^2(\mathbb{R}^d)$. Though very specific, this example indicates that the regularity of the coefficient is not a central issue in the above proof.

In particular, we would like to make the following point, which will be important for the sequel. If we were only interested in solving the equation

$$- \operatorname{div}((a_{per} + \widetilde{a})\,\nabla u) = \operatorname{div} \mathbf{f}, \tag{4.18}$$

posed on the whole space $\mathbb{R}^d$, for $\widetilde{a} \in L^2(\mathbb{R}^d)$ and $\mathbf{f} \in L^2(\mathbb{R}^d)$, then the above *proof* of Lemma 4.1 would apply and allow to conclude that such a solution $u \in L^1_{loc}(\mathbb{R}^d)$ with $\nabla u \in L^2(\mathbb{R}^d)$ exists and is unique (up to the addition of a constant). Such a proof is valid without any assumption of structure or regularity on the coefficient $a_{per} + \widetilde{a}$, apart from the coercivity and the L^∞ bound (3.1). In addition, such a solution satisfies the estimate

$$\left\| \nabla u \right\|_{L^2(\mathbb{R}^d)} \leq C \left\| \mathbf{f} \right\|_{L^2(\mathbb{R}^d)}. \tag{4.19}$$

We leave such an adaptation of the proof of Lemma 4.1 to the reader. We will return to such considerations in the sequel, in Sects. 4.1.2 and 4.1.3, where we will address similar questions in $L^q(\mathbb{R}^d)$, for $q \neq 2$.

We would like to make a third comment, this time about *strict sublinearity*. Lemma 4.1 gives the existence of a corrector w_p in dimension $d \geq 2$. This is the first result of this kind in this textbook in a nonperiodic setting. We therefore think that it is useful to recall the notion of *strict sublinearity* introduced in Sect. 2.3 of Chap. 2 (condition (2.38)). In the present case of a perturbation $\widetilde{a} \in L^2(\mathbb{R}^d)$, we note that the corrector w_p we have built indeed satisfies such a condition, under the assumptions of Lemma 4.1. The periodic part $w_{p,per}$ is of course irrelevant in this respect since it is bounded, according to the Nash-Moser theory (Theorem A.11), and is periodic. We thus only consider $\widetilde{w}_p$.

For a general function such that its gradient belongs to, say, $L^2(\mathbb{R}^d)$ (the extension to $L^q(\mathbb{R}^d)$, $q \neq 2$, follows the same lines), one cannot hope for condition (2.38) to hold true as such. This formulation is a consequence of the regularity of the corrector, and holds true for instance in the one-dimensional case of Chap. 2. Indeed, in this setting, we clearly have $w' \in L^\infty(\mathbb{R})$, hence w is uniformly Lipschitz continuous. On the other hand, *even* a periodic function, which, loosely speaking, is strictly sublinear at infinity, does not necessarily satisfies (2.38). An example may be built with a locally integrable singularity, in each periodic cell. We thus need to generalize (2.38) as follows

$$\varepsilon\, w_p\left(\frac{\cdot}{\varepsilon}\right) \longrightarrow 0 \text{ in } L^1_{loc}(\mathbb{R}^d) \quad \text{as} \quad \varepsilon \to 0. \tag{4.20}$$

It is an easy exercise to check that (2.38) implies (4.20), and that, for instance, any (not necessarily regular) periodic function in $L^1_{loc}(\mathbb{R}^d)$ satisfies (4.20).

We are now in position to remark that a function $\widetilde{w}_p$ the gradient of which is in $L^q(\mathbb{R}^d)$ automatically satisfies (4.20). This may be proved as follows.

First, if $q > d$, then the Morrey inequality (see Theorem A.1) implies that

$$\left|\widetilde{w}_p(x) - \widetilde{w}_p(y)\right| \leq C|x-y|^{1-\frac{d}{q}}, \tag{4.21}$$

where the constant C depends only on $\left\|\nabla\widetilde{w}_p\right\|_{L^q(\mathbb{R}^d)}$, on d and on q. Thus, fixing $y = 0$ and taking the limit $|x| \longrightarrow +\infty$, we have (2.38).

We assume now that $q < d$. We use the Gagliardo-Nirenberg-Sobolev inequality, or more precisely its Corollary A.1: there exists a constant $M \in \mathbb{R}$ such that $v_p = \widetilde{w}_p - M$ satisfies $v_p \in L^{q^*}(\mathbb{R}^d)$, with $\frac{1}{q^*} = \frac{1}{q} - \frac{1}{d}$. Hence, for any ball $B(x_0, 1)$ of radius 1, we have

$$\left\|\varepsilon\widetilde{w}_p\left(\frac{\cdot}{\varepsilon}\right)\right\|_{L^1(B(x_0,1))} \leq \varepsilon M\, |B(x_0, 1)| + \varepsilon\left\|v_p\left(\frac{\cdot}{\varepsilon}\right)\right\|_{L^1(B(x_0,1))}$$

$$= \varepsilon M\, |B(x_0, 1)| + \varepsilon^{d+1}\left\|v_p\right\|_{L^1(B(x_0/\varepsilon, 1/\varepsilon))}.$$

We apply the Hölder inequality to the rightmost term in the above expression and obtain:

$$\varepsilon^{d+1}\left\|v_p\right\|_{L^1(B(x_0/\varepsilon, 1/\varepsilon))} \leq \varepsilon^{d+1}\left|B\left(\frac{x_0}{\varepsilon}, \frac{1}{\varepsilon}\right)\right|^{1-\frac{1}{q^*}}\left\|v_p\right\|_{L^{q^*}(B(x_0/\varepsilon, 1/\varepsilon))}$$

$$\leq C\varepsilon^{d+1-d+\frac{d}{q^*}}\left\|v_p\right\|_{L^{q^*}(\mathbb{R}^d)} = C\varepsilon^{1+\frac{d}{q^*}}\left\|v_p\right\|_{L^{q^*}(\mathbb{R}^d)},$$

which vanishes as $\varepsilon \to 0$. This proves (4.20). We are eventually left with the case $q = d$. We define $\rho \in C^\infty(\mathbb{R}^d)$, such that its support is included in the unit ball $B(0, 1)$ centered at the origin. We also assume that $\rho \geq 0$ and that $\int_{\mathbb{R}^d} \rho = 1$. Let h be defined by $h = \widetilde{w}_p * \rho$. Then $\nabla h = \nabla\widetilde{w}_p * \rho \in L^\infty \cap L^d$. Consequently, h satisfies (4.21) for all $q > d$. On the other hand, the difference $\widetilde{w}_p - h$ satisfies, for almost all $x \in \mathbb{R}^d$,

$$\widetilde{w}_p(x) - h(x) = \widetilde{w}_p(x) \int_{\mathbb{R}^d} \rho(y) dy - \int_{\mathbb{R}^d} \rho(y) \widetilde{w}_p(x - y) dy$$

$$= \int_{B(0,1)} \rho(y) \left(\widetilde{w}_p(x) - \widetilde{w}_p(x - y) \right) dy$$

$$= \int_{B(0,1)} \rho(y) \int_0^1 \nabla \widetilde{w}_p(x - ty) \cdot y \, dt \, dy.$$

We integrate over the ball $B(x_0, 1)$, for $x_0 \in \mathbb{R}^d$ fixed, and obtain

$$\left\| \widetilde{w}_p - h \right\|_{L^1(B(x_0,1))} \leq \int_0^1 \int_{B(0,1)} \left(\int_{B(x_0,1)} |\nabla \widetilde{w}_p(x - ty)| dx \right) \rho(y) dy \, dt.$$

For all $t \in [0, 1]$ and all $y \in B(0, 1)$, $x - ty \in B(x, 1) \subset B(x_0, 2)$, hence

$$\left\| \widetilde{w}_p - h \right\|_{L^1(B(x_0,1))} \leq \int_0^1 \int_{B(0,1)} \left(\int_{B(x_0,2)} |\nabla \widetilde{w}_p(z)| dz \right) \rho(y) dy \, dt$$

$$\leq C \|\nabla \widetilde{w}_p\|_{L^d(B(x_0,2))} \leq C \|\nabla \widetilde{w}_p\|_{L^d(\mathbb{R}^d)},$$

where C depends only on d. We have thus proved that $\widetilde{w}_p - h \in L^1_{unif}(\mathbb{R}^d)$, that is to say, $\widetilde{w}_p = \phi + h$, where $\phi \in L^1_{unif}(\mathbb{R}^d)$, and h satisfies (4.21), whence (4.20). Any function in L^1_{unif} obviously satisfies (4.20). This concludes the proof.

We have so far stated and proved global integrability properties of the gradient $\nabla \widetilde{w}_p$. We actually know more than this. Indeed, $\widetilde{w}_p$ is solution to the corrector equation. As a consequence, it is *regular*. The Nash-Moser theory (Theorem A.11) implies that it is Hölder continuous. We have already used this argument in the periodic case (see Sect. 3.4, page 134). We have also used it above (page 182). It is valid here again because the coefficient $a = a_{per} + \widetilde{a}$ is coercive and bounded. Periodicity is not needed. Since $\widetilde{w}_p$ is regular and $\nabla \widetilde{w}_p \in L^2(\mathbb{R}^d)$, $\widetilde{w}_p$ is strictly sublinear at infinity in the "classical" sense (2.38).

A consequence of the above remark is that the corrector w_p is *in particular* solution to the multidimensional analogue of Eq. (2.39), that is,

$$\begin{cases} - \operatorname{div} \left((a_{per} + \widetilde{a}) (p + \nabla w_p) \right) = 0 \text{ in } \mathbb{R}^d, \\[2mm] \lim_{|x| \to +\infty} \dfrac{w_p(x)}{1 + |x|} = 0. \end{cases} \tag{4.22}$$

Let us note that, on the other hand, we do not know whether w_p is the *only* solution to this equation, up to an additive constant. We have only proved uniqueness under the additional constraint that w_p is the sum of a periodic function and a function the gradient of which is in $L^2(\mathbb{R}^d)$, with the normalization (4.5).

We also point out that, if we directly consider Eq. (4.22), it is possible to prove existence of a solution. Such a proof relies on a different strategy, which is based on a so-called *representation formula* of the solution to (4.3) (or more precisely the solution to the equivalent form (4.6)), that is, an expression of the type

$$\widetilde{w}_p(x) = \int_{\mathbb{R}^d} G(x, y)\, F(y)\, dy,$$

where F denotes a function depending on the right-hand side of (4.6), and G the so-called *Green function* associated to the operator $-\operatorname{div}\left((a_{per} + \widetilde{a})\nabla\,.\right)$ of the left-hand side of (4.6). Such a strategy can in fact be generalized to spaces L^q, $q \neq 2$. The interested reader is referred to [BLL15, Theorem 3.1], where a proof is given that applies in particular to the L^2 setting, in dimension $d \geq 3$. As we will see, the arguments are different from the above proof, which was originally published in [BLL12, Lemma 1]. Green function methods have already been used in Sect. 3.4.3.5, pages 162 ff, in the periodic setting. We will use them again later, for the case of defects (see pages 198 ff).

4.1.1.2 Unchanged Homogenized Coefficient

We are now going to identify the homogenized coefficient a^* associated with the homogenization problem (1)–(4.1). We use the strategy of proof based on the div-curl Lemma, as it was exposed, in the periodic case, in Sect. 3.4.2. The key ingredient is equality (3.84), which we write here as follows

$$\left[a\left(\frac{x}{\varepsilon}\right)\left(p + \nabla w_p\left(\frac{x}{\varepsilon}\right)\right)\right] . \nabla u^\varepsilon = \left[a\left(\frac{x}{\varepsilon}\right)\nabla u^\varepsilon\right] . \left(p + \nabla w_p\left(\frac{x}{\varepsilon}\right)\right).$$
(4.23)

The weak limits stated in (3.82) and (3.83) are still true in the present setting. They read as

$$p + \nabla w_p\left(\frac{x}{\varepsilon}\right) \rightharpoonup p,$$
(4.24)

and

$$a\left(\frac{x}{\varepsilon}\right)\left(p + \nabla w_p\left(\frac{x}{\varepsilon}\right)\right) \rightharpoonup \langle a_{per}\left(p + \nabla w_{p,per}\right)\rangle,$$
(4.25)

respectively. Each L^2 weak convergence (we mean, *local* weak convergence, and we will omit this in the sequel) is already proved for the periodic parts of the coefficient a and of the corrector w_p. We therefore only need to consider the weak

L^2 convergence of $\nabla \widetilde{w}_p \left(\frac{x}{\varepsilon}\right)$, $a \left(\frac{x}{\varepsilon}\right) \nabla \widetilde{w}_p \left(\frac{x}{\varepsilon}\right)$ and $\widetilde{a} \left(\frac{x}{\varepsilon}\right) (p + \nabla w_{p,per}) \left(\frac{x}{\varepsilon}\right)$. The first one is obvious: it is in fact a strong convergence, owing to the equality $\left\| \nabla \widetilde{w}_p \left(\frac{x}{\varepsilon}\right) \right\|_{L^2(\mathbb{R}^d)} = \varepsilon^{d/2} \left\| \nabla \widetilde{w}_p \right\|_{L^2(\mathbb{R}^d)}$, which we have already used several times. This proves (4.24). For the same reason, and since $a \left(\frac{x}{\varepsilon}\right)$ is bounded in L^∞, the product $a \left(\frac{x}{\varepsilon}\right) \nabla \widetilde{w}_p \left(\frac{x}{\varepsilon}\right)$ converges strongly in L^2 to zero. Thus, the second convergence is also strong in L^2. As for the third one, we first note that, still for the same reason, $\widetilde{a} \left(\frac{x}{\varepsilon}\right) p$ converges strongly to zero in L^2. Finally, the term $\widetilde{a} \left(\frac{x}{\varepsilon}\right) \nabla w_{p,per} \left(\frac{x}{\varepsilon}\right)$ is the product of a strongly converging term in L^2 by a weakly converging term in L^2 (to zero). Hence, the product converges weakly in L^1. Now, since $\widetilde{a}$ is in L^∞ and $\nabla w_{p,per}$ is periodic and locally in L^2, the product $\widetilde{a} \left(\frac{x}{\varepsilon}\right) \nabla w_{p,per} \left(\frac{x}{\varepsilon}\right)$ is bounded in L^2. The weak convergence thus holds true in L^2.

We may now apply the same argument as in Sect. 3.4.2, using the div-curl Lemma. To this end, we write Eq. (4.23) under the form (3.85), that is, $\mathbf{f}^\varepsilon \cdot \mathbf{g}^\varepsilon = \mathbf{r}^\varepsilon \cdot \mathbf{s}^\varepsilon$, where only the definitions of $\mathbf{f}^\varepsilon$ and $\mathbf{r}^\varepsilon$ are adapted to the present case. We conclude that the homogenized coefficient a^* associated with our equation with an L^2 defect is equal to the homogenized coefficient of the periodic case, as may be seen in the weak convergence (4.25) above. As expected, and as we have already seen in dimension $d = 1$, a local defect (here in $L^2(\mathbb{R}^d)$) does not change the homogenization properties at the macroscopic scale. The defect will have an impact at the fine scale (as we will shortly see) but not at the macroscopic scale. For now, we keep in mind that

$$a^* = (a_{per})^*. \tag{4.26}$$

4.1.1.3 Using the Corrector

As in the periodic setting studied in Sect. 3.4.3 of Chap. 3, we work in dimension $d = 3$. This allows for simpler arguments, while the generalization to any dimension d is not difficult (see Remark 3.10). We define the approximation

$$u^{\varepsilon,1}(x) = u^*(x) + \varepsilon \sum_{i=1}^{3} \partial_{x_i} u^*(x) \, w_i(x/\varepsilon), \tag{4.27}$$

analogue of (3.90), where w_i is now the corrector defined by (4.3) for $p = e_i$, the i-th vector of the canonical basis, and where u^* is the solution to the homogenized equation. The homogenized coefficient is defined by (4.26). As we have seen above, it reads as

$$[a^*]_{ij} = \text{weak} \lim_{\varepsilon \longrightarrow 0} a(./\varepsilon)\,(\delta_{ij} + \partial_j w_i(./\varepsilon)),$$

$$= \int_Q a_{per}(y)\,(\delta_{ij} + \partial_j w_{i,per}(y))\,dy,$$

for $1 \leq i, j \leq 3$. The computations around (3.92) may be reproduced in the present setting, leading to (3.93), and again give

$$- \text{div}\left(a_\varepsilon\, \nabla(u^\varepsilon - u^{\varepsilon,1})\right) = - \text{div}\left((a^* - a_\varepsilon\,(\text{Id} + \nabla w(./\varepsilon))\,\nabla u^*(x))\right)$$

$$+\varepsilon\, \text{div}\left(a(./\varepsilon)\, w(./\varepsilon)\, \nabla\nabla u^*\right).$$

As in (3.94), we define the matrix-valued function

$$M = a^* - a\,(\text{Id} + \nabla w),$$

that is, $M_{ij} = a_{ij}^* - \sum_{k=1}^{3} a_{ik}\,(\delta_{kj} + \partial_k w_j)$. This function M again satisfies $\text{div}\,M_j = 0$. Applying the same strategy as in Sect. 3.4.3, we next prove that there exists, for all $1 \leq j \leq 3$, a vector-valued function B_j such that $M_j = \text{curl}\,B_j$, which we more concisely write $M = \text{curl}\,B$. Such an existence will allow us to write (3.93) as (3.95), that is,

$$- \text{div}\left(a_\varepsilon\, \nabla(u^\varepsilon - u^{\varepsilon,1})\right) = \varepsilon\, \text{div}\left(B(./\varepsilon) \times \nabla \nabla u^*\right)$$

$$+\varepsilon\, \text{div}\left(a(./\varepsilon)\, w(./\varepsilon)\, \nabla\nabla u^*\right).$$

In order to build the vector-valued function B, we need to impose its structure. The natural structure in Chap. 3 was given by the periodic setting: M was periodic with zero average, so we searched for B a periodic function. Here, M has the form

$$M = a^* - a\,(\text{Id} + \nabla w)$$
$$= \left(a^* - a_{per}\,(\text{Id} + \nabla w_{per})\right) - \left(a_{per}\,\nabla\widetilde{w} + \widetilde{a}\,(\text{Id} + \nabla w)\right)$$
$$=: M_{per} + \widetilde{M}.$$

Moreover, since the homogenized matrix a^* is equal to the homogenized matrix of the periodic setting, we know that the function M_{per} is *exactly* equal to the periodic function denoted by M in Sect. 3.4.3. It reads as $M_{per} = \text{curl}\,B_{per}$. We are now going to build $\widetilde{B}$ such that $\widetilde{M} = \text{curl}\,\widetilde{B}$. As for the periodic setting (see Sect. 3.4.3.1, pages 142 ff), this equality is to be understood columnwise. Put differently, all reduces to solving

$$\operatorname{curl} \mathbf{b} = \mathbf{m},$$

where $\mathbf{m} : \mathbb{R}^3 \longrightarrow \mathbb{R}^3$ is a vector-valued function satisfying

$$\mathbf{m} \in L^2(\mathbb{R}^3)^3, \quad \operatorname{div} \mathbf{m} = 0.$$

Mimicking our strategy of the periodic setting, we are going to solve $-\Delta \mathbf{a} = \mathbf{m}$, and define $\mathbf{b} = \operatorname{curl} \mathbf{a}$. Let us temporarily assume that $\mathbf{m}$ is smooth with compact support. Then, applying the Fourier transform, we have

$$\widehat{\mathbf{a}}(\xi) = \frac{1}{4\pi^2 |\xi|^2} \widehat{\mathbf{m}}(\xi). \tag{4.28}$$

This defines $\widehat{\mathbf{a}} \in L^q(\mathbb{R}^3)$, for all $q < 3/2$. Such an integrability at the origin is indeed obvious since $\widehat{\mathbf{m}}$ is bounded (recall that $\mathbf{m}$ is smooth). On the other hand, $\widehat{\mathbf{m}}$ decays faster than any negative power of $|\xi|$, so the integrability at infinity is also clear. Thus, $\mathbf{a} \in L^\infty(\mathbb{R}^3)$, and $\mathbf{a} \in C^\infty(\mathbb{R}^3)$, since its Fourier transform decays faster than any negative power at infinity). Therefore,

$$\widehat{\mathbf{b}}(\xi) = -\frac{i}{2\pi |\xi|^2} \xi \times \widehat{\mathbf{m}}(\xi)$$

defines a smooth bounded function $\mathbf{b}$. Using that $\operatorname{div} \mathbf{m} = 0$, that is, $\xi . \widehat{\mathbf{m}} = 0$, a straightforward calculation gives that, for all $\xi \neq 0$,

$$2i\pi \xi \times \widehat{\mathbf{b}}(\xi) = \frac{1}{|\xi|^2} \xi \times (\xi \times \widehat{\mathbf{m}}(\xi)) = \widehat{\mathbf{m}} + \frac{1}{|\xi|^2} \underbrace{(\xi . \widehat{\mathbf{m}}(\xi))}_{=0} \xi,$$

that is, $\operatorname{curl} \mathbf{b} = \mathbf{m}$. Since the Fourier transform is an L^2-isometry, we have

$$\|\nabla \mathbf{b}\|_{L^2(\mathbb{R}^3)} = \left\| \frac{1}{|\xi|^2} \xi \otimes (\xi \times \widehat{\mathbf{m}}(\xi)) \right\|_{L^2(\mathbb{R}^3)} \leq \|\widehat{\mathbf{m}}\|_{L^2(\mathbb{R}^3)} = \|\mathbf{m}\|_{L^2(\mathbb{R}^3)} .$$

Hence, the map $\mathbf{m} \longmapsto \nabla \mathbf{b}$, defined on the set of smooth compactly supported functions, is linear and continuous for the L^2 norm. A density argument allows to extend it as a linear continuous map from $L^2(\mathbb{R}^3)$ into itself. We thus have proved existence and uniqueness of $\widetilde{B}$ solution to $\widetilde{M} = \operatorname{curl} \widetilde{B}$, with the continuity estimate

$$\|\nabla \widetilde{B}\|_{L^2(\mathbb{R}^3)} \leq \|\widetilde{M}\|_{L^2(\mathbb{R}^3)} .$$

Slightly anticipating on the sequel, we now mention that it is possible to prove that the map $T : \widetilde{M} \longmapsto \nabla \widetilde{B}$ is not only continuous from L^2 into itself, but that it is also a Calderón-Zygmund operator (see Sect. 4.1.2.1 below, in particular Definition 4.1). This implies that T is continuous from L^q to L^q, for all $q \in]1, +\infty[$. Thus, when $\tilde{a} \in L^q$, we obtain that $\tilde{M} \in L^q$, and that $\nabla \tilde{B} \in L^q$. Consequently, $\tilde{B}$ is strictly sublinear at infinity, in the sense of (2.38), as the corrector w_p was proved to be in Sect. 4.1.1.1.

We also note that (4.28) reads as $\mathbf{a} = \dfrac{1}{4\pi |x|} * \mathbf{m}$, whence

$$\mathbf{b}(x) = \operatorname{curl} \mathbf{a}(x) = \frac{1}{4\pi} \int_{\mathbb{R}^3} \frac{x - y}{|x - y|^3} \times \mathbf{m}(y) dy.$$

Therefore,

$$\tilde{B}(x) = \frac{1}{4\pi} \int_{\mathbb{R}^3} \frac{x - y}{|x - y|^3} \times \tilde{M}(y) dy.$$

This formula is to be understood columnwise, and, for now, is only formal. We are going to *prove* it. We first do so in the case where $\tilde{M}$ is bounded and has compact support. In such a case, we split the above integral into an integral over the ball $B(y, 1)$ and an integral over its complement $B(y, 1)^c$. We obtain

$$\left| \tilde{B}(x) \right| \leq \frac{1}{4\pi} \int_{B(y,1)} \frac{1}{|x - y|^2} \left| \tilde{M}(y) \right| dy + \frac{1}{4\pi} \int_{B(y,1)^c} \frac{1}{|x - y|^2} \left| \tilde{M}(y) \right| dy.$$

The Hölder inequality allows to prove a bound for each term separately:

$$\left| \tilde{B}(x) \right| \leq \frac{1}{4\pi} \left\| \frac{1}{|x|^2} \right\|_{L^1(B(0,1))} \left\| \tilde{M} \right\|_{L^\infty(\mathbb{R}^3)} + \frac{1}{4\pi} \left\| \frac{1}{|x|^2} \right\|_{L^{q'}(B(0,1)^c)} \left\| \tilde{M} \right\|_{L^q(\mathbb{R}^3)}.$$

Since $\tilde{M} \in L^q(\mathbb{R}^3) \cap L^\infty(\mathbb{R}^3)$, with $q < 3$, hence $q' > 3/2$, we have $|x|^{-2} \in L^{q'}(B(0, 1)^c)$. The above inequality thus implies that $\tilde{B} \in L^\infty(\mathbb{R}^3)$. Since the set of bounded compactly supported functions is dense in $L^q(\mathbb{R}^3) \cap L^\infty(\mathbb{R}^3)$, this holds for all $\tilde{M} \in L^q(\mathbb{R}^3) \cap L^\infty(\mathbb{R}^3)$.

A similar proof allows to more generally show that, in all dimensions $d \geq 2$, there exists $\tilde{M} \in L^q(\mathbb{R}^3) \cap L^\infty(\mathbb{R}^3)$, with $q < d$, such that $\tilde{B} \in L^\infty(\mathbb{R}^d)$. The case $q = 2$ is a particular case of this general statement, and may be addressed with simpler tools.

At this stage, we have obtained

$$M = M_{per} + \tilde{M} = \operatorname{curl} B_{per} + \operatorname{curl} \tilde{B} = \operatorname{curl} B,$$

where the function B_{per} is, like $w_{p,per}$, periodic and bounded. The function $\tilde{B}$ (like $\tilde{w}_p$) is strictly sublinear at infinity in the sense of (2.38). It may even be bounded in some cases, depending on the ambient dimension d (here we chose $d = 3$ for simplicity), and on the L^q integrability of the defect $\tilde{a}$. It is actually possible to make precise this strict sublinearity. If the defect is such that $\tilde{a} \in L^q(\mathbb{R}^d)$, $q \neq d$, then, defining

$$v = \min\left(1, \frac{d}{q}\right), \tag{4.29}$$

the functions $\tilde{w}_p$ and $\tilde{B}$ satisfy the estimates

$$\left|\tilde{w}_p(x) - \tilde{w}_p(y)\right| \leq C|x - y|^{1-v}, \quad \left|\tilde{B}(x) - \tilde{B}(y)\right| \leq C|x - y|^{1-v}. \tag{4.30}$$

Moreover, if $q > d$, then $\tilde{w}_p \in L^\infty(\mathbb{R}^d)$ and $\tilde{B} \in L^\infty(\mathbb{R}^d)$. We again point out that the case $q = 2$ is a particular case of the above properties. Note however that the case $q = d$ is excluded in the above arguments. In such a case, since $\tilde{a} \in L^q(\mathbb{R}^d) \cap L^\infty(\mathbb{R}^d)$, $\tilde{a} \in L^r(\mathbb{R}^d)$ for all $r > d$, we may nevertheless apply the above results for these values of the integrability exponent.

We are now in position to reproduce the arguments used in Sect. 3.4.3.2 through Sect. 3.4.3.5. First, if we remove the divergence operators on both sides of Eq. (3.95) we obtain the analogue of (3.96), that is,

$$-a_\varepsilon \nabla(u^\varepsilon - u^{\varepsilon,1}) = \varepsilon B(./\varepsilon) \times \nabla \nabla u^* + \varepsilon a(./\varepsilon) w(./\varepsilon) \nabla \nabla u^*.$$

The difference with Sect. 3.4.3.2 is that, here, the functions B and w may not be bounded. As a consequence, we are no longer able to prove that the terms $\varepsilon B(./\varepsilon)$ and $\varepsilon w(./\varepsilon)$ necessarily behave like $O(\varepsilon)$. Indeed, the contributions of $\tilde{B}$ and $\tilde{w}_p$ give terms that scale at best like ε^v. Thus, $\nabla(u^\varepsilon - u^{\varepsilon,1})$ vanishes at best at the rate $O(\varepsilon^v)$.

As we did in Sect. 3.4.3.2 through 3.4.3.5, we may now progressively make rigorous the original formal arguments. The reasoning is essentially based upon the behaviour of

$$F_\varepsilon = B(./\varepsilon) \times \nabla \nabla u^* + a(./\varepsilon) w(./\varepsilon) \nabla \nabla u^*,$$

defined in (3.98), as $\varepsilon \to 0$. Assuming that all the functions we manipulate vanish on the boundary $\partial \mathcal{D}$, estimate (3.105) of the residual $R_\varepsilon = u^\varepsilon - u^{\varepsilon,1}$ proved in Sect. 3.4.3.2 now becomes

$$\|R_\varepsilon\|_{H^1(\mathcal{D})} = O(\varepsilon^v).$$

Similarly, the interior estimate (3.115) of Sect. 3.4.3.3 may be rigorously proved and reads as

$$\|R_\varepsilon\|_{H^1(\mathcal{D}_0)} = O(\varepsilon^v).$$

We may likewise adapt the arguments of Sect. 3.4.3.5 giving the $W^{1,q}$ and Hölder estimates of Proposition 3.6. These results are obtained upon combining

(i) the behaviour, as $\varepsilon \to 0$ of the functions F_ε and $T_\varepsilon = \sum_{i=1}^{d} \partial_{x_i} u^*(x)\, w_i(x/\varepsilon)$

present in the right-hand side and in the boundary data of Eq. (3.136) satisfied by the residual R_ε, that is,

$$
\begin{cases}
-\,\mathrm{div}\,(a_\varepsilon \nabla R_\varepsilon) = \varepsilon\,\mathrm{div}\,F_\varepsilon & \text{in } \mathcal{D}, \\[2mm]
\qquad\qquad R_\varepsilon \;\;=\;\; \varepsilon\,T_\varepsilon & \text{on } \partial\mathcal{D},
\end{cases}
$$

(ii) and the properties of the operator $-\,\mathrm{div}\,a_\varepsilon \nabla$, which imply estimates (uniformly with respect to ε) on the associated Green function; such estimates, like (3.151) and (3.152), are fairly general, and related only to the bounds (3.1), or may be consequences of the properties of the corrector, as is the case for (3.154)–(3.155) and (3.157)–(3.158).

In the present setting of a periodic structure with defect, the necessary changes in the proof of Proposition 3.6 are clear. In particular, the entire portion of proof regarding the homogeneous equation applies without any modification. In contrast, although this is not obvious in the way we have presented them, the estimates on the decay of the derivatives of the Green function, such as (3.154) and (3.155), critically use the behaviour of the corrector at infinity. Although not conceptually difficult, the necessary adaptations to the case of a nonperiodic corrector are technically delicate. We refer to [BJL20] for a detailed presentation. Instead of developing such adaptations, we prefer to now concentrate on a different question.

4.1.1.4 What if We Use the Periodic Corrector?

Instead of formally deducing (3.96) from (3.95), we are going to perform an alternate, also formal, argument that is more informative.

Let us assume that we are mistaken and that we use the periodic corrector instead of the corrector with defect. More precisely, we use $u^{\varepsilon,1}_{per} = u^* +$

$\varepsilon \sum_{i=1}^{d} \partial_{x_i} u^*\, w_{i,per}(./\varepsilon)$ defined by (3.90), instead of $u^{\varepsilon,1} = u^* + \varepsilon \sum_{i=1}^{d} \partial_{x_i} u^*\, w_i(./\varepsilon)$

defined by (4.27). We thus replace the residual $R_\varepsilon = u^\varepsilon - u^{\varepsilon,1}$ (recall that we have now established its convergence and rate of convergence in various topologies) by $R_{\varepsilon,per} = u^\varepsilon - u^{\varepsilon,1}_{per}$. We are going to compare the equation satisfied by R_ε with the equation satisfied by $R_{\varepsilon,per}$. In view of (3.95), we have

$$
-\,\mathrm{div}\,(a_\varepsilon \nabla R_\varepsilon) = \varepsilon\,\mathrm{div}\left(B(./\varepsilon) \times \nabla \nabla u^*\right) + \varepsilon\,\mathrm{div}\left(a(./\varepsilon)\, w(./\varepsilon) \nabla \nabla u^*\right),
$$
$$
\tag{4.31}
$$

while $R_{\varepsilon, per}$ satisfies

$$- \operatorname{div}\left(a_\varepsilon \nabla R_{\varepsilon, per}\right) = - \operatorname{div}\left(a_\varepsilon \nabla R_\varepsilon\right) - \varepsilon \operatorname{div}\left(a(./\varepsilon)\, \widetilde{w}(./\varepsilon)\, \nabla\nabla u^*\right)$$
$$- \operatorname{div}\left(a(./\varepsilon)\,(\nabla\widetilde{w})(./\varepsilon)\, \nabla u^*(x)\right). \quad (4.32)$$

The comparison of (4.31) with (4.32) illustrates the difference between the case where the defect is taken into account in the corrector and the case where it is not. The rightmost term in (4.32) is formally $O(1)$, while all other terms in the right-hand side of (4.31) and (4.32) are of order $O(\varepsilon)$. Less formally, we may write, directly from the definitions of R_ε and $R_{\varepsilon, per}$,

$$\nabla R_{\varepsilon, per} = \nabla R_\varepsilon - \varepsilon\, \widetilde{w}(./\varepsilon)\, \nabla\nabla u^* - (\nabla\widetilde{w})(./\varepsilon)\, \nabla u^*. \quad (4.33)$$

Using our observations of Sect. 4.1.1.3, we may prove estimates of the type $\|\nabla R_\varepsilon\|_{L^q} = O(\varepsilon^\alpha)$ for some values of $\alpha = \alpha(q) \in]0, 1]$ depending on the L^q norm chosen. If the exponent q is finite, we do not learn anything spectacular: one easily checks that the difference $\nabla R_{\varepsilon, per} - \nabla R_\varepsilon$ behaves (at least) like $O(\varepsilon)$, hence is negligible. Using the "wrong" (that is, periodic) corrector does not deteriorate the quality of approximation. In sharp contrast, the case $q = +\infty$ allows to distinguish between $\nabla R_{\varepsilon, per}$ and ∇R_ε. The rightmost term of (4.33), $(\nabla\widetilde{w})(./\varepsilon)\, \nabla u^*$, is indeed of order $O(1)$ in L^∞ norm, does not vanish as $\varepsilon \to 0$, and dominates in the difference. This is particularly clear if we write equality (4.33) at the microscopic scale, that is, if we change x into $\varepsilon\, x$:

$$\nabla R_{\varepsilon, per}(\varepsilon\, x) = \nabla R_\varepsilon(\varepsilon\, x) - \varepsilon\, \widetilde{w}(x)\, \nabla\nabla u^*(\varepsilon\, x) - (\nabla\widetilde{w})(x)\, \nabla u^*(\varepsilon\, x). \quad (4.34)$$

The rightmost term is of order $O(1)$, and does not vanish as $\varepsilon \to 0$. The approximation of u^ε by $u^{\varepsilon,1}$ at the fine scale is thus of better quality than that of u^ε by $u^{\varepsilon,1}_{per}$.

Some part of this formal argument may be made rigorous. The residual $R_{\varepsilon, per}$ is indeed solution to

$$\begin{cases} - \operatorname{div}\left(a_\varepsilon \nabla R_{\varepsilon, per}\right) = \varepsilon \operatorname{div} F_{\varepsilon, per} & \text{in } \mathcal{D}, \\[2mm] R_{\varepsilon, per} \quad\quad = \quad \varepsilon\, T_{\varepsilon, per} & \text{on } \partial\mathcal{D}. \end{cases} \quad (4.35)$$

This is the analogue of (3.136), where, with obvious notation, $F_{\varepsilon, per}$ and $T_{\varepsilon, per}$ denote the modifications of (3.98)–(3.108) where the "real" corrector w is replaced by its periodic counterpart. As (4.33) shows, we have

$$F_{\varepsilon, per} = F_\varepsilon + a(./\varepsilon)\, \widetilde{w}(./\varepsilon)\, \nabla\nabla u^* + \varepsilon^{-1} a(./\varepsilon)\, (\nabla\widetilde{w})(./\varepsilon)\, \nabla u^*, \quad (4.36)$$

and similar formulas for $T_{\varepsilon, per}$ and T_ε. Equation (4.36) then allows to give a rigorous meaning to our arguments following (4.33). For any $q < +\infty$, the L^q

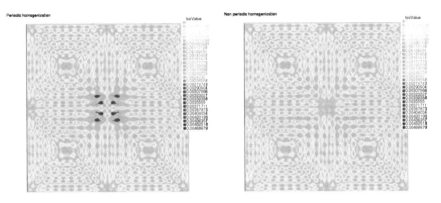

Fig. 4.1 Example of the norm $\left| \nabla u^\varepsilon \left(\varepsilon \cdot \right) - \nabla u^{\varepsilon,1}_{per} \left(\varepsilon \cdot \right) \right|$ (left) and $\left| \nabla u^\varepsilon \left(\varepsilon \cdot \right) - \nabla u^{\varepsilon,1} \left(\varepsilon \cdot \right) \right|$ (right). The computational domain is $(-1, 1)^2$. The accuracy is significantly improved when the nonperiodic corrector is used

norms of $\varepsilon F_\varepsilon$ and of $\varepsilon F_{\varepsilon,per}$ converge to zero at the same rate, although this rate may not be equal to $O(\varepsilon)$, contrary to the periodic case. This common convergence rate has been made precise above. A similar property holds for $T_{\varepsilon,per}$ and T_ε. The proof of Proposition 3.6 hence applies, and we deduce the same convergence rate for $\nabla R_{\varepsilon,per}$ and ∇R_ε. On the other hand, in the L^∞ norm, the situation is different, because such a norm "sees" the microscopic scale, as already mentioned for (4.34). The convergence $\varepsilon F_{\varepsilon,per} \rightarrow 0$ does not hold in such a norm. This is due to the term $a(./\varepsilon) (\nabla \widetilde{w})(./\varepsilon) \nabla u^*$ in (4.36), the L^∞ norm of which is *constant* with respect to ε, while all other terms $\varepsilon F_\varepsilon + \varepsilon a(./\varepsilon) \widetilde{w}(./\varepsilon) \nabla \nabla u^*$, vanish in the L^∞ norm. Indeed, the first term vanishes owing to Proposition 3.6, and the second term vanishes due to the strict sublinearity (4.20) of $\widetilde{w}$. This phenomenon is illustrated by numerical computations in Fig. 4.1.

This does not *prove*, of course, that $\nabla R_{\varepsilon,per}$ does not vanish as $\varepsilon \rightarrow 0$ since $\varepsilon \, \mathrm{div} \left(F_{\varepsilon,per} \right)$ in the right-hand side of (4.35) could tend to 0 even if $\varepsilon F_{\varepsilon,per}$ does not. The above considerations, which constitute a rigorous proof in dimension $d = 1$, however provide a convincing argument that this convergence is unlikely to hold in general. Numerical experiments may also, to some extent, "prove" this fact. Such a numerical study may be found in [BLL12, Section 4].

4.1.2   Case of a Defect in $L^q(\mathbb{R}^d)$, $q \neq 2$

We now consider $\widetilde{a} \in L^q(\mathbb{R}^d)$ with $1 < q < +\infty$ and $q \neq 2$. Of course, the arguments of this section also apply to the case $q = 2$, but our aim is to obtain results valid in the non-Hilbertian case $q \neq 2$.

We are going to concentrate on the proof of existence of a suitable corrector w_p, that is, a solution to (4.4).

Let us temporarily assume that such a corrector exists. We first note that the arguments of Sect. 4.1.1.2 still hold. Thus, the homogenized coefficient a^* is, here again, equal to the coefficient of the periodic setting. The central point of the proof is the div-curl Lemma (see Lemma 3.2 of Sect. 3.4.2). Although we have only proved and stated this result in the L^2 setting, it is an easy exercise to extend it to general exponents q, q' such that $\dfrac{1}{q} + \dfrac{1}{q'} = 1$.

But actually, due to our specific assumptions, the following argument allows to see that it is possible to even apply the L^2 version of Lemma 3.2 directly to the case $q \neq 2$. We first recall that, since we have assumed that the coefficients are Hölder continuous (see (4.37)), using the same argument as in the proof of Lemma 4.1 (in the periodic setting, but this proof only uses the *local* regularity of the coefficient, hence carries over to the present case), we have $\nabla w_p \in L^\infty(\mathbb{R}^d)$. Similarly, because of (4.37), $\tilde{a} \in L^\infty(\mathbb{R}^d)$. If $q < 2$, then a simple application of the Hölder inequality shows that $\tilde{a}$ and $\nabla \tilde{w}_p$ both belong to $L^2(\mathbb{R}^d)$. The argument of Sect. 4.1.1.2 thus applies without any modification. If $q > 2$, we note that the weak convergences we use are *local*. The strong convergences of $\tilde{a}\left(\dfrac{\cdot}{\varepsilon}\right)$ and $\nabla \tilde{w}_p\left(\dfrac{\cdot}{\varepsilon}\right)$ in $L^q(\mathbb{R}^d)$ also hold locally. We thus know, using the Hölder inequality again, that these local convergences hold in L^2. This is sufficient to conclude our proof.

In the same vein, the arguments of Sect. 4.1.1.3 that show the (rate of) convergence of the residual $u^\varepsilon - u^{\varepsilon,1}$ may be generalized to the case $q \neq 2$. Such a generalization is however considerably more intricate. This is indeed already true in the periodic setting (one should compare Sect. 3.4.3.5 to the end of Sect. 3.4.2). We are thus not going to proceed with this generalization here.

We concentrate, as announced, on the existence of w_p. Such an existence result, together with the uniqueness up to the addition of a constant, is a consequence of Proposition 4.1 below. It is indeed sufficient to apply it to the case $r = q, \mathbf{f} = \tilde{a}\,(p + \nabla w_{p,per})$. This function belongs to $L^q(\mathbb{R}^d)$ because $\tilde{a} \in L^q(\mathbb{R}^d)$ and, in view of the Hölder regularity of the coefficients (see (4.37)), $\nabla w_{p,per} \in L^\infty(\mathbb{R}^d)$. We point out that, doing so, we use Proposition 4.1 in a very specific way. Indeed, we only apply the $L^r(\mathbb{R}^d)$ estimate of (4.39) with the exponent r equal to the integrability exponent q of the defect $\tilde{a}$. On the other hand, since q may take any arbitrary value, it is unclear to us whether another strategy exists.

Our central result is thus the following:

Proposition 4.1 *Assume that* (4.1) *holds, and that*

$$\begin{cases} a_{per} \text{ and } \ a_{per} + \tilde{a} \quad \text{satisfy} \quad (3.1) \\[2mm] a_{per},\ \tilde{a} \in C^{0,\alpha}_{\text{unif}}\left(\mathbb{R}^d\right) \quad \text{for some} \quad \alpha > 0. \end{cases} \tag{4.37}$$

Fix $1 < r < +\infty$. *Then, for all* $\mathbf{f} \in L^r(\mathbb{R}^d)$, *there exists a function* $u \in L^1_{loc}(\mathbb{R}^d)$ *such that* $\nabla u \in L^r(\mathbb{R}^d)$, *solution to*

$$- \operatorname{div}(a \nabla u) = \operatorname{div} \mathbf{f} \quad in \ \mathbb{R}^d. \tag{4.38}$$

Such a solution is unique up to the addition of a constant. Moreover, there exists a constant C_r, depending only on the exponent r, on the ambient dimension d and on the coefficient a (in particular, it does not depend on f), such that

$$\|\nabla u\|_{L^r(\mathbb{R}^d)} \leq C_r \|\mathbf{f}\|_{L^r(\mathbb{R}^d)}. \tag{4.39}$$

The central estimate (4.39) of Proposition 4.1 is not unexpected. Such an estimate is clearly a generalization of estimate (4.19) of Sect. 4.1.1.1, the proof of which is actually, as we already pointed out, contained in the proof of Lemma 4.1.

Proposition 4.1 could be mistaken for an indication that any property that holds in the periodic setting is also valid in the presence of a defect. This is not true. Some properties that have been proved in the periodic case have not been extended to the case of defects yet. For instance, the result of [AL89] gives a classification of the solutions to $\operatorname{div}(a_{per} \nabla u) = 0$ having polynomial growth at infinity. To our knowledge, the extension to the case of defects is not known.

In line with this comment, we wish to emphasize the importance of estimate (4.39). It is intuitively clear that, since the equation is *linear*, proving existence and uniqueness of a solution in a suitable functional space is mathematically equivalent to proving an estimate of this type. It is thus important to give some details on the famous results on which the proof of Proposition 4.1, is based. This is the matter of Sect. 4.1.2.1. Compared to the large scope and significance of the results we give there, Proposition 4.1 is no more than an extension, not to say a *corollary*, allowing us to perform homogenization for coefficients of the type (4.1).

Before describing these results, we make the following remark.

Assumption (4.1) rules out the particular case $\tilde{a} \in L^1(\mathbb{R}^d)$. This is to be expected, since L^1 is not reflexive, contrary to the spaces L^r, $1 < r < +\infty$. This may however seem counterintuitive in the present setting. In some sense, a defect $\tilde{a} \in L^1(\mathbb{R}^d)$ indeed decays faster at infinity than a defect $\tilde{a} \in L^q(\mathbb{R}^d)$ for $q > 1$. It should thus be easier to build the corresponding corrector! The point is in fact not to build the corrector, but to show that it lies in the expected functional space. A way to circumvent this difficulty is to prove existence of the corrector in a larger functional space. This goes as follows. Since (4.1) implies that the coefficients are bounded, Hölder inequality readily implies, if we assume that $\tilde{a} \in L^1(\mathbb{R}^d)$, that $\tilde{a} \in L^q(\mathbb{R}^d)$ *for all* $1 < q < +\infty$. We may thus apply Proposition 4.1. Doing so, we prove the existence of a corrector in $\nabla \tilde{w}_p \in L^q(\mathbb{R}^d)$ for all $1 < q < +\infty$, but not in $L^1(\mathbb{R}^d)$. This is sufficient to prove that $a^* = (a_{per})^*$ and to obtain the convergence of the residual $u^\varepsilon - u^{\varepsilon,1}$. Using that $\nabla w_p \in L^\infty(\mathbb{R}^d)$ and the results of [AL91], it is next possible to prove that $\nabla \tilde{w}_p$ belongs to a functional space called *weak-$L^1(\mathbb{R}^d)$*. This space, sometimes denoted by $L^{1,\infty}(\mathbb{R}^d)$, is the set of functions f such that

$$\sup_{\lambda>0}\left(\lambda \text{ measure } \left\{x \in \mathbb{R}^d; \ |f(x)| > \lambda\right\}\right) < +\infty. \tag{4.40}$$

This functional space may be seen as the extension to the case $q = 1$ of the Marcinkiewicz space introduced in (3.153) in Chap. 3. However, (4.40) does not define a norm, but only a quasi-norm, on this space. It remains that this functional space is a complete metric space, when equipped with a suitable distance that we will not make precise here. The reader is referred to [Gra14, Section 1.1 and Theorem 1.4.11] for more details on this subject.

4.1.2.1 Genealogy of a Result

In order to better understand Proposition 4.1, we consider the much simpler situation of the Laplace operator. This is equivalent to assuming that $a = a_{per} + \widetilde{a}$ with $a_{per} \equiv 1$ and $\widetilde{a} \equiv 0$, which is of course consistent with our assumptions (4.1)–(4.37). Equation (4.38) then becomes

$$- \Delta u = \operatorname{div} \mathbf{f} \quad \text{in } \mathbb{R}^d, \tag{4.41}$$

with $\mathbf{f} \in L^r(\mathbb{R}^d)$, and we search for a solution such that $\nabla u \in L^r(\mathbb{R}^d)$. Such a solution will also satisfy (4.39), that is, $\|\nabla u\|_{L^r(\mathbb{R}^d)} \leq C_r \|\mathbf{f}\|_{L^r(\mathbb{R}^d)}$.

We again start with the simplest possible case and next proceed to more difficult situations.

Dimensions "Zero" and 1

If we bluntly remove the divergence operator on both sides of the equation, as we have already done in the preceding chapters, then (4.41) becomes $\nabla u = \mathbf{f}$. The L^r estimate thus becomes obvious. This suggests that the result is likely to hold true in general. Let us now consider Eq. (4.41) *with* the divergence operators on both sides.

In dimension $d = 1$, the result holds true and may be proved easily. Equation (4.41) reads as $-u'' = f'$, which is equivalent to $u' = -f + \text{constant}$. The only integration constant for which $u' \in L^r(\mathbb{R})$ is the constant zero, and we have a unique solution $u(x) = \int_0^x f(t)\,dt$ up to the addition of a constant. This solution satisfies $u' = -f$. Thus, estimate $\|u'\|_{L^r(\mathbb{R})} \leq \|f\|_{L^r(\mathbb{R})}$ holds true and is in fact an equality.

The Radially Symmetric Case

In dimension $d \geq 2$, if f is assumed to be radially symmetric, our argument of the one-dimensional case is easily adapted. Indeed, if $\mathbf{f}(x) = g(r)\,\mathbf{e}_r$, where $r = |x|$ and $\mathbf{e}_r$ is the radial unit vector, then, searching for a radially symmetric solution u, Eq. (4.41) becomes

$$-\partial_r^2 u - \frac{d-1}{r}\partial_r u = \partial_r g + \frac{d-1}{r}g.$$

It is obvious that a particular solution of this equation is $\partial_r u = -g$, that is, $\nabla u = -\mathbf{f}$. On the other hand, the Liouville Theorem (Corollary A.1) implies that the solution u such that $\nabla u \in L^r(\mathbb{R}^d)$ is unique, up to the addition of a constant. We have thus found the unique solution, and the situation is similar to the one-dimensional case.

The above considerations comfort us in our optimism for the general case.

The L^2 Case for the Laplace Operator

In dimensions $d \geq 2$, without any structural assumption such as the radial symmetry considered above, things are less simple. This is due to the fact that (4.41) is equivalent to the condition that $\nabla u + \mathbf{f}$ is a curl, and that curls are not trivial in dimensions $d \geq 2$. If $r = 2$, however, the result may still be proved via at least a couple of elementary methods.

First, an argument using the Fourier transform may be performed. The function $\hat{u}$, Fourier transform (in the sense of tempered distributions) of u solution to Eq. (4.41), satisfies, almost everywhere in $\xi \in \mathbb{R}^d$, $|\xi|^2\,\hat{u}(\xi) = i\sum_{j=1}^{d}\xi_j\,\widehat{\mathbf{f}}_j(\xi)$, where $\widehat{\mathbf{f}}_j$ is of course the Fourier transform of $\mathbf{f}_j$, $1 \leq j \leq d$. It follows that, for all $1 \leq k \leq d$, $\widehat{\partial_k u}(\xi) = i\xi_k\,\hat{u}(\xi) = \sum_{j=1}^{d}\frac{\xi_j\,\xi_k}{|\xi|^2}\widehat{\mathbf{f}}_j(\xi) = \frac{\xi_k}{|\xi|^2}\xi \cdot \widehat{\mathbf{f}}(\xi)$. Using the Cauchy-Schwarz inequality, we deduce that for all $1 \leq k \leq d$, $\left|\widehat{\partial_k u}(\xi)\right|^2 \leq \sum_{j=1}^{d}\left|\widehat{\mathbf{f}}_j(\xi)\right|^2$. This gives exactly the desired estimate $\|\nabla u\|_{L^2(\mathbb{R}^d)} \leq C_2\,\|\mathbf{f}\|_{L^2(\mathbb{R}^d)}$ for $C_2 = \sqrt{d}$. This a priori estimate being proved, it is then easy to modify the above argument in order to prove the existence of a solution, together with the desired estimate.

Another possible proof is provided by the argument we used in the more involved (but L^2) case of Sect. 4.1.1.1: we multiply the equation by $u\chi$, where χ is a cut-off function, integrate by parts in space, and prove the desired estimate.

Before leaving the case $r = 2$, we point out that an L^2 estimate for the solution to (4.41) is intimately linked with the celebrated continuity of the *Riesz operator* in

L^2. This operator (also called *Riesz transform*) is defined, for $1 \leq j \leq d$, by

$$R_j = -i\, \partial_{x_j} (-\Delta)^{-1/2}, \tag{4.42}$$

or, using the Fourier transform, by $\widehat{R_j(f)}\,(\xi) = \dfrac{\xi_j}{|\xi|}\,\hat{f}(\xi)$. A simple computation shows that the operator $\nabla\,(-\Delta)^{-1}\,\mathrm{div}$ that links the gradient ∇u of the solution to (4.41) with the data $\mathbf{f}$ through $\nabla u = \left[\nabla\,(-\Delta)^{-1}\,\mathrm{div}\right]\mathbf{f}$ reads as the "square" of the Riesz operator: $\nabla\,(-\Delta)^{-1}\,\mathrm{div} = R_j\,R_j^*$. We use here the convention of summation over repeated indices, that is,

$$\xi_i\,\widehat{\mathbf{f}_i^e} = \sum_{i}^{d} \xi_i\,\widehat{\mathbf{f}_i^e}. \tag{4.43}$$

The case $r \neq 2$ for the Laplace operator and the Calderón-Zygmund theory
Keeping in mind the above observations on the case $r = 2$, we now turn to the case $r \neq 2$, which is far more complicated. Our presentation is mainly taken from the textbook [Mey90, Chapitres VII-VIII] (in French; see [Mey92] for a translation in English), to which the reader is referred for more details. We first introduce the notion of *Calderón-Zygmund operators*. Consider a linear operator T, defined from the space $\mathcal{D}(\mathbb{R}^d)$ of smooth compactly supported functions, to the space of distributions $\mathcal{D}'(\mathbb{R}^d)$. The *Schwartz kernel Theorem* (see [HÖ3, Theorem 5.2.1, p 128]) implies that there exists a distribution $S \in \mathcal{D}'(\mathbb{R}^d \times \mathbb{R}^d)$, called the *kernel* of T, such that, for all f and all g in $\mathcal{D}(\mathbb{R}^d)$, $\langle T(f), g \rangle = \langle S, g \otimes f \rangle$, where the first bracket is the duality bracket between $\mathcal{D}(\mathbb{R}^d)$ and $\mathcal{D}'(\mathbb{R}^d)$, while the second bracket is the duality bracket between $\mathcal{D}(\mathbb{R}^d \times \mathbb{R}^d)$ and $\mathcal{D}'(\mathbb{R}^d \times \mathbb{R}^d)$. Put differently, and formally using integrals instead of duality brackets (actually, the point here is that such integrals may be ill defined and should be understood as duality brackets, but our aim here is to intuitively understand the meaning of the above assertion), the kernel S is defined by $T(f)(x) = \displaystyle\int_{\mathbb{R}^d} S(x, y)\, f(y)\, dy$ for all function f, hence S satisfies $\displaystyle\int_{\mathbb{R}^d} T(f)(x)\, g(x)\, dx = \int_{\mathbb{R}^d \times \mathbb{R}^d} S(x, y)\, g(x)\, f(y)\, dx\, dy$ for all functions f and g in $\mathcal{D}(\mathbb{R}^d)$. We define the open set

$$D = \left\{ (x, y) \in \mathbb{R}^d \times \mathbb{R}^d \,/\, x \neq y \right\},$$

that is, the product space $\mathbb{R}^d \times \mathbb{R}^d$ without its "diagonal" set, and let

$$K = S_{|D}$$

be the restriction of the kernel S to D. We are now in position to define Calderón-Zygmund operators:

Definition 4.1 (Calderón-Zygmund Operators) The operator T defined above is called a *Calderón-Zygmund operator* if the following conditions are satisfied:

[CZ1]   the function K is locally integrable over D and satisfies

$$\forall (x, y) \in D, \quad |K(x, y)| \leq C_0 \, |x - y|^{-d}, \tag{4.44}$$

   for some constant C_0;
   there exists $\gamma \in]0, 1]$ and $C_1 > 0$ such that
[CZ2]   for $(x, y) \in D$ and $x' \in \mathbb{R}^d$ such that $|x' - x| \leq 1/2 \, |x - y|$,

$$\left| K(x', y) - K(x, y) \right| \leq C_1 \, \left| x' - x \right|^{\gamma} \, |x - y|^{-d-\gamma}, \tag{4.45}$$

[CZ3]   for $(x, y) \in D$ and $y' \in \mathbb{R}^d$ such that $|y' - y| \leq 1/2 \, |x - y|$,

$$\left| K(x, y') - K(x, y) \right| \leq C_1 \, \left| y' - y \right|^{\gamma} \, |x - y|^{-d-\gamma}, \tag{4.46}$$

[CZ4]   the operator T may be extended as a continuous linear operator from $L^2(\mathbb{R}^d)$ into itself.

Note that, in the case $\gamma = 1$, conditions (4.45) and (4.46) are equivalent to

$$\left| \partial_{x_j} K \right| + \left| \partial_{y_j} K \right| \leq C_1 \, |x - y|^{-d-1}, \quad 1 \leq j \leq d. \tag{4.47}$$

The Riesz operator (4.42) is a prototypical example of a Calderón-Zygmund operator. This is also the case for the operator

$$T_{Lap} = \nabla (-\Delta)^{-1} \operatorname{div} \tag{4.48}$$

that we consider here. The operator T_{Lap} is a *convolution operator*, which means that its kernel K is of the form $K(x, y) = k(x - y)$. This is due to the fact that the operator T_{Lap} is associated with a linear partial differential equation with *constant coefficients*, that is, an equation of the type (4.41), which is translation invariant. The interest of the Calderón-Zygmund theory is that it addresses operators that *are not* necessarily convolution operators. This includes in particular operators associated with non-constant coefficient equations, as for instance (4.38).

For now, let us consider the kernel of the operator $\nabla (-\Delta)^{-1} \operatorname{div}$. The Green function of the Laplace operator over $\mathbb{R}^d$ is well known, and, in dimension $d \geq 3$ (we omit dimensions $d = 1$ and $d = 2$ for simplicity) satisfies, up to multiplication by a constant,

$$G(x, y) \propto |x - y|^{-d+2} \quad \text{for } (x, y) \in D. \tag{4.49}$$

Computing $\partial_{x_i} \partial_{y_j} |x - y|^{-d+2}$, we find that the kernel of the operator $\nabla (-\Delta)^{-1}$ div reads as (up to multiplication by a constant)

$$K(x, y) \propto (x_i - y_i)(x_j - y_j) |x - y|^{-d-2} \quad \text{for } (x, y) \in D, \quad (4.50)$$

where we have used a compact notation: in (4.50), it is understood that the kernel associated to the i-th component of the operator involves linear combinations of terms of the type $(x_i - y_i)(x_j - y_j) |x - y|^{-d-2}$. A precise calculation of these terms is not needed here. The point is the *homogeneity* of the terms involved. It is clear that a kernel of the type (4.50) satisfies condition (4.44). In addition, differentiating (4.50), we see that condition (4.47) is satisfied too. Since we already know that the operator T_{Lap} is continuous from $L^2(\mathbb{R}^d)$ into itself, it is a Calderón-Zygmund operator.

The property we are particularly interested in, in the theory of Calderón-Zygmund operators, is the following: *a Calderón-Zygmund operator may be extended into a linear continuous operator from $L^r(\mathbb{R}^d)$ into itself, for all $1 < r < \infty$.* Applying this property to T_{Lap}, we find estimate (4.39). We are not going to prove this property here. We only mention that the method of proof consists in first establishing that a Calderón-Zygmund operator may be extended into a continuous operator from $L^1(\mathbb{R}^d)$ into *weak-L^1*, defined by (4.40). We recall that this space is larger than $L^1(\mathbb{R}^d)$ since, for any $f \in L^1(\mathbb{R}^d)$,

$$\lambda \text{ measure } \left\{ x \in \mathbb{R}^d; \ |f(x)| > \lambda \right\} \leq \int_{\{x \in \mathbb{R}^d; \ |f(x)| > \lambda\}} |f| \leq \int_{\mathbb{R}^d} |f|.$$

A few elementary properties of this space have been stated in the paragraph following (4.40). At this stage of the proof, Calderón and Zygmund introduced the celebrated *Calderón-Zygmund decomposition* of an L^1 function. This decomposition consists in writing the considered function as a sum of a function in $L^1 \cap L^2$ and a series of oscillating localized terms. In contrast, if the range of the operator is to be included in $L^1(\mathbb{R}^d)$, the operator has to be defined on a strict subspace of $L^1(\mathbb{R}^d)$, called the *Hardy space* $\mathbb{H}^1(\mathbb{R}^d)$. In the same vein, we note that Calderón-Zygmund operators are in general not continuous from $\mathbb{H}^1(\mathbb{R}^d)$ into itself. More precisely, it is possible to describe Calderón-Zygmund operators that are continuous from $\mathbb{H}^1(\mathbb{R}^d)$ into itself by the so-called $T(1) = 0$ criterion. Similar properties hold for the space $L^\infty(\mathbb{R}^d)$: a Calderón-Zygmund operator is not continuous from this space into itself. In order to address this issue, one needs to consider spaces that are larger than $L^\infty(\mathbb{R}^d)$ itself, such as BMO (the space of functions of "Bounded Mean Oscillation"), which contains for instance the function $\log |x|$. We do not wish to develop these questions further, and refer to [Mey92] for more details.

Applying the so-called *interpolation theory* between $r = 1$ and $r = 2$, together with the fact that the considered operator is continuous from $L^1(\mathbb{R}^d)$ into *weak-L^1* and from $L^2(\mathbb{R}^d)$ into $L^2(\mathbb{R}^d)$, we obtain an extension of this operator from $L^r(\mathbb{R}^d)$ into itself, for all $1 < r \leq 2$. Next, a duality argument applied to the adjoint operator (which is a Calderón-Zygmund operator too), allows to build an extension as an operator from $L^r(\mathbb{R}^d)$ into itself, for all $2 \leq r < \infty$. It should be noted that the preceding sentences summarize very important and difficult results in harmonic analysis. As we already said, the interested reader may consult the enlightening course of Yves Meyer in [Mey90] or [Mey92], with, as a preliminary introduction to harmonic analysis, the reference textbook [Ste93].

In the context of non-constant coefficient operators, our purpose is the continuity of the operator from L^r into itself. The conclusion of the above considerations is that, provided the operator satisfies (4.44) through (4.46), proving that it is continuous from L^r into itself, for all $1 < r < +\infty$, reduces to proving that it is continuous from L^2 into itself. This is a *tremendous* simplification. Except that showing continuity from L^2 into itself may prove more difficult than expected. We proved it easily in the case of the operator T_{Lap}, but, in full generality, things are more delicate. The celebrated *T(1) Theorem* of David and Journé answers this question by giving a necessary and sufficient condition for the L^2 continuity (see its exact statement in [DJ84, Theorem 1]). We will not pursue in this direction, but the reader is highly encouraged to read this reference, in order to better understand the link between these questions and what we study here, in the framework of homogenization. Still, we keep in mind the following fact: before proving continuity in L^r, $1 < r < +\infty$, it is essential to first prove it in the case $r = 2$. This is consistent with our approach in the preceding chapter and at the beginning of the present chapter.

The Case $r \neq 2$ for the Periodic Operator

As we said above, the Calderón-Zygmund has been developed to address the question of non-constant coefficient operators. This is precisely the case we are interested in, and the question is already relevant in the periodic setting. In such a case, we have already proved continuity in L^2, and we could obtain continuity in L^r, provided we show that the considered operator is of Calderón-Zygmund type. In this respect, it is sufficient to study the properties of the associated kernel, as we have done when proving (3.151), (3.154) and (3.155). With a little more work, these estimates indeed imply properties (4.44) through (4.46). This allows to show that the operator is a Calderón-Zygmund operator. Historically, the proof given in the article [AL91] did not follow this strategy: L^r continuity has been proved directly there, using the properties of the associated kernel. Note that in [AL91], in addition to being periodic, the coefficient is assumed to be Hölder continuous. We now give the main ideas of this article.

In order to prove continuity in $L^r(\mathbb{R}^d)$ of the operator

$$T_{per} = \nabla \left(-\operatorname{div}\left(a_{per} \nabla.\right)\right)^{-1} \operatorname{div}, \tag{4.51}$$

we study its kernel, as we have done above for the operator T_{Lap}. This kernel involves second-order derivatives of the type $\partial_{x_i} \partial_{y_j} G^{per}(x, y)$ of the Green function $G^{per}(x, y)$, which plays the same role as the Green function $G(x, y)$ of the Laplace operator, defined in (4.49). Pointwise estimates on $G^{per}(x, y)$ and its derivatives are in fact *included* in homogenization theory for Eq. (1) with a periodic coefficient a_{per}, when precise convergence rates such as those of Proposition 3.6 are considered. We are going to give a schematic view of this fact. For complete proofs, we refer to the original articles [AL87] and [AL91]. A pedagogical presentation may also be consulted in the course [Pra16].

The comprehensive study of the (rate of) convergence in ε in various norms of the homogenization approximation of the solution u^ε to (1) for a periodic coefficient $a = a_{per}$ has been initiated in [AL87]. In this work (and many subsequent ones), pointwise estimates of the Green function $G_\varepsilon(x, y)$ of the operator $L_{per} = -\operatorname{div}\left(a_{per}\left(\frac{\cdot}{\varepsilon}\right)\nabla\cdot\right)$ are key steps in the proof of the main results. Since $G_\varepsilon(x, y)$ is solution of the equation with the (very peculiar) right-hand side $\delta(x - y)$, we should be able to understand how this solution $G_\varepsilon(x, y)$ converges, in some sense to be made precise, to the homogenized equation, involving the differential operator

$$L_* = -\operatorname{div}\left(a^*\nabla\cdot\right), \tag{4.52}$$

that is, a Green function of the same kind as the Green function of the Laplace operator. We have mentioned above pointwise estimates on $G_\varepsilon(x, y)$ and on its derivatives ((3.151)–(3.154)–(3.155)), together with its belonging to suitable functional spaces (in the vein of (3.152)). We have also mentioned, in (3.157) and (3.158), estimates on the distance between $G_\varepsilon(x, y)$ and its homogenized limit. We anticipated there that we would make precise this estimate on the cross derivative $\nabla_x \nabla_y G_\varepsilon(x, y)$. We state it now:

$$\left|\partial_{x_i}\partial_{y_j} G_\varepsilon(x, y) - \left(\operatorname{Id} + \nabla w_{per}\left(\frac{y}{\varepsilon}\right)\right)\left(\operatorname{Id} + \nabla w_{per}\left(\frac{x}{\varepsilon}\right)\right)\partial_{x_i}\partial_{y_j} |x - y|^{-d+2}\right|$$
$$\leq C\varepsilon^\alpha, \tag{4.53}$$

for ε small enough and $|x - y|$ bounded from above and from below. Note that, in (4.53), we recognize an expansion similar to $u^{\varepsilon,1}$. Intuitively (actually, this can be made rigorous), estimate (4.53) implies

$$\left|\partial_{x_i}\partial_{y_j} G_{per}(x, y) - \left(\operatorname{Id} + \nabla w_{per}(y)\right)\left(\operatorname{Id} + \nabla w_{per}(x)\right)\partial_{x_i}\partial_{y_j} |x - y|^{-d+2}\right|$$
$$\leq C |x - y|^{-d-\alpha}, \tag{4.54}$$

for some $\alpha > 0$ and for large values of $|x - y|$. Studying the continuity of the operator $T_{per} = \nabla L_{per}^{-1} \operatorname{div}$ from $L^r(\mathbb{R}^d)$ into itself reduces, using estimate (4.54), to studying the continuity of the operator $T_* = \nabla L_*^{-1} \operatorname{div}$. Such a question has been addressed above, since the homogenized matrix a^* is constant, and since the properties of this operator are similar to those of the operator T_{Lap}. To be more precise, one directly shows that the kernel $\partial_{x_i} \partial_{y_j} G_{per}(x, y)$ of the operator T_{per} satisfies estimates that ensure that T_{per} satisfies the same properties as T_{Lap}. When $|x - y|$ is at a finite distance from the origin, the kernel of the operator satisfies suitable properties, in view of elliptic regularity, since the coefficient a_{per} is Hölder continuous. On the other hand, for $|x - y|$ far away from the origin, the kernel is close to, and has the suitable properties of, the kernel of T_{Lap}. We refer the reader to the article [AL91] for a rigorous version of these formal arguments.

We have so far obtained

– the continuity result in L^r, $1 < r < +\infty$, for the operator with periodic coefficient $a = a_{per}$ and
– the continuity result in L^2 for the operator with perturbed coefficient $a = a_{per} + \tilde{a}$.

Our last task is now to prove the result in L^r, $1 < r < +\infty$, $r \neq 2$, for the operator with defect. This is the aim of Proposition 4.1, the proof of which is given in the following section.

4.1.2.2   Proof of Proposition 4.1

The purpose of this section is the proof of Proposition 4.1, which has originally been published in [BLL18].

As we have just pointed out, Proposition 4.1 is in some sense the combination of a result in L^r (for all $1 < r < +\infty$) for the operator with periodic coefficients, and a result in L^2 for the operator with defect. This intuition will become clear in the proof below: arguments from both of these elementary building blocks will be used.

Another intuitive interpretation of the proof is the following: *if the result holds true for some reference operator, then it holds true for a suitable perturbation of this operator.* Here, the reference operator is the periodic operator. It could be a simpler operator such as the Laplace operator, or it could be a more complicated operator, for which the result would have been proved beforehand and independently. In the proof below, we add a perturbation that disappears in some sense at infinity, and prove that the result of the periodic setting still holds.

When the defect $\tilde{a}$ is omitted, the coefficient a is periodic and estimate (4.39) has been established in the article [AL91]. We have summarized the corresponding proof in Sect. 4.1.2.1. When $\tilde{a} \neq 0$, we note that, since $\tilde{a} \in L^q(\mathbb{R}^d) \cap C_{unif}^{0,\alpha}(\mathbb{R}^d)$, $\tilde{a}(x) \overset{|x| \to \infty}{\longrightarrow} 0$. The operator $-\operatorname{div}(a \nabla.)$ is thus intuitively close to the periodic operator $-\operatorname{div}(a_{per} \nabla.)$ at infinity, and estimate (4.39) is expected to hold for functions the support of which is far from the origin. On the other hand, locally, at finite distance from the origin, estimate (4.39) is a consequence of elliptic regularity (see Theorem A.13) and of the fact that it holds in the case $q = 2$, as we saw in

Sect. 4.1.1.1. Combining these two observations, we expect to be able to prove the result. The rigorous proof of Proposition 4.1 exactly implements this strategy.

We first address the case of an exponent $2 \leq r < +\infty$. The case $1 < r \leq 2$ will be treated afterwards, using a duality argument.

Preliminary

The method we use is known as a *continuation* (or *connexity*) argument. We define the coefficient $a_t = a_{per} + t\tilde{a}$ for $0 \leq t \leq 1$. When $t = 0$, the results of Proposition 4.1 are known, since the coefficient is periodic. We want to prove them for $t = 1$. We introduce the following property $\mathcal{P}$:

A coefficient a that satisfies, for some $1 \leq q < +\infty$, assumptions (4.1)–(4.37), is said to satisfy property $\mathcal{P}$ if the statements of Proposition 4.1 hold true for Eq. (4.38) with coefficient a.

We define the interval

$$\mathcal{I} = \{t \in [0, 1] \,/\, \forall s \in [0, t], \text{ property } \mathcal{P} \text{ holds for } a_s\}. \tag{4.55}$$

We intend to successively prove that $\mathcal{I}$ is not empty, then that it is open and closed (both notions being understood relatively to the closed interval $[0, 1]$), which will show, by connexity, that $\mathcal{I} = [0, 1]$, thus the result claimed for $t = 1$.

The Interval $\mathcal{I}$ Is not Empty and Is Open

The fact that $\mathcal{I} \neq \emptyset$ is obvious, since, owing to the results in the periodic setting (proved in [AL91, Theorem A] and briefly recalled above) we have $0 \in \mathcal{I}$. The property $u \in L^1_{loc}$ is a consequence of elliptic regularity theory and of the fact that $\mathbf{f} \in L^1_{loc}(\mathbb{R}^d)$. This property actually immediately carries over to all $t \in [0, 1]$ as soon as we know that there is a solution in the following argument. Thus, we will not mention it anymore.

We next show that $\mathcal{I}$ is open (relatively to the interval $[0, 1]$). For this purpose, we suppose that $t \in \mathcal{I}$ and wish to prove property $\mathcal{P}$ on $[t, t + \varepsilon[$ for some $\varepsilon > 0$ such that $t + \varepsilon \leq 1$. In order to solve, for $\mathbf{f} \in L^r(\mathbb{R}^d)$

$$- \operatorname{div}\left((a_t + \varepsilon\,\tilde{a})\,\nabla u\right) = \operatorname{div} \mathbf{f},$$

we write it as follows:

$$\nabla u = T_t\left(\varepsilon\,\tilde{a}\,\nabla u + \mathbf{f}\right), \tag{4.56}$$

where $T_t = \nabla\left(-\operatorname{div}\left(a_t\,\nabla.\right)\right)^{-1}\operatorname{div}$ (with a notation coming from (4.51) and similar formulae) is the linear map that to each $\mathbf{f} \in L^r(\mathbb{R})$ associates the gradient $\nabla u \in L^r(\mathbb{R}^d)$ of the solution u to $-\operatorname{div}(a_t\nabla u) = \operatorname{div}\mathbf{f}$. Such a map exists since $\mathcal{P}$ is satisfied for $a = a_t$. It is moreover continuous from L^r to L^r, with norm C_r. In addition, if $C_r \varepsilon \|\tilde{a}\|_{L^\infty(\mathbb{R}^d)} < 1$, the map $v \mapsto T_t (\varepsilon \tilde{a} v + \mathbf{f})$ is a contraction in L^r. We may thus apply the Banach fixed point Theorem to this map: equation $\mathbf{v} = T_t(\varepsilon \tilde{a} \mathbf{v} + \mathbf{f})$ has a unique solution in $L^r(\mathbb{R}^d)$. Such a solution is a gradient by construction, since it lies in the range of T_t: $\mathbf{v} = \nabla u$, where u is solution to (4.56). It satisfies estimate (4.39) with a constant $C_r \left(1 - C_r \varepsilon \|\tilde{a}\|_{L^\infty(\mathbb{R}^d)}\right)^{-1}$ instead of C_r. This proves our claim.

The Interval $\mathcal{I}$ Is Closed if the Continuity Constants Are Bounded

We now show, and this is the key point of the proof, that the interval $\mathcal{I}$ is *closed*. We consider a sequence $t_n \in \mathcal{I}$, $t_n \leq t$, $t_n \longrightarrow t$ as $n \longrightarrow +\infty$. For all $n \in \mathbb{N}$, we know that, for any $\mathbf{f} \in L^r(\mathbb{R}^d)$, there exists a solution u (unique up to the addition of a constant) with $\nabla u \in L^r(\mathbb{R}^d)$ of equation

$$- \operatorname{div}\left(a_{t_n} \nabla u\right) = \operatorname{div} \mathbf{f} \quad \text{in } \mathbb{R}^d,$$

and that this solution satisfies

$$\|\nabla u\|_{L^r(\mathbb{R}^d)} \leq C_n \|\mathbf{f}\|_{L^r(\mathbb{R}^d)} \tag{4.57}$$

for some constant C_n depending on n but not on $\mathbf{f}$ nor on u. We want to show the same properties for the limit t of the sequence (t_n).

We first temporarily admit that the sequence of constants (C_n) in (4.57) is uniformly bounded in n and conclude. For $\mathbf{f} \in L^r(\mathbb{R}^d)$ fixed, we consider the sequence of solutions u^n of

$$- \operatorname{div}\left(a_{t_n} \nabla u^n\right) = \operatorname{div} \mathbf{f} \quad \text{in } \mathbb{R}^d,$$

which we may write as

$$- \operatorname{div}\left(a_t \nabla u^n\right) = \operatorname{div}\left(\mathbf{f} + (a_{t_n} - a_t) \nabla u^n\right).$$

The sequence of gradients ∇u^n is bounded in $L^r(\mathbb{R}^d)$, and therefore converges weakly in this space (up to extraction of a subsequence) to some function which we know, by construction, is a gradient. We denote it by ∇u. Since the sequence $a_{t_n} - a_t$ converges strongly in L^∞, the weak convergence of ∇u^n to ∇u allows to pass to the limit in the above equation. Thus, u is solution to $- \operatorname{div}(a_t \nabla u) = \operatorname{div} \mathbf{f}$. On the other hand, since the sequence of solutions u^n satisfies for each $n \in \mathbb{N}$ the L^r estimate, because the constants C_n are bounded and because the norm is lower semi-continuous for the weak L^r topology, we obtain the estimate on ∇u. There remains to prove uniqueness, up to the addition of a constant, in the *general* class of solutions u such that $\nabla u \in L^r$. Put differently, since the equation is linear, we wish to prove that the solution u to

$$- \operatorname{div}(a_t \nabla u) = 0 \quad \text{in } \mathbb{R}^d, \tag{4.58}$$

such that $\nabla u \in L^r(\mathbb{R}^d)$ is a constant. We note that (4.58) also reads as

$$- \operatorname{div}(a_{per} \nabla u) = t \operatorname{div}(\widetilde{a} \nabla u). \tag{4.59}$$

We are going to use a *bootstrap* argument. This consists in repeatedly applying an argument, in such a way that at each step, we recycle in the equation a stronger property. In the right-hand side of the equation, owing to the Hölder inequality, we have $\widetilde{a} \nabla u \in L^{r_1}(\mathbb{R}^d)$ for $\dfrac{1}{r_1} = \dfrac{1}{r} + \dfrac{1}{q}$. We use here the fact that $\widetilde{a} \in L^q(\mathbb{R}^d)$, hence, in some sense, as announced at the beginning of the present Sect. 4.1.2.2, vanishes at infinity, thereby improving the integrability of the product $\widetilde{a} \nabla u$ (note that we will again use this argument in the remainder of the proof). We next apply the results of [AL89], that state that any solution to $- \operatorname{div}(a_{per} \nabla v) = 0$ is constant when it grows slowlier than a polynomial at infinity. In particular, the solution to $- \operatorname{div}(a_{per} \nabla u) = \operatorname{div}(f)$ is unique if $\nabla u \in L^r$. This property is satisfied here in (4.59), hence u is the unique solution, up to the addition of a constant. This implies that $\nabla u \in L^{r_1}(\mathbb{R}^d)$. The exponent r_1 is strictly smaller than r. We have thus improved the integrability at infinity. Recall indeed that, intuitively (and formally), a function in $L^s(\mathbb{R}^d)$ decays faster at infinity as the exponent s decreases. Then, we can repeat the argument with the exponent r_1 instead of r. Doing so, we build by induction a sequence of exponents r_n such that $\nabla u \in L^{r_n}(\mathbb{R}^d)$ and $\dfrac{1}{r_n} = \dfrac{1}{r_{n-1}} + \dfrac{1}{q}$, that is, $\dfrac{1}{r_n} = \dfrac{1}{r} + \dfrac{n}{q}$. Recall that we have assumed that $r \geq 2$, hence $\dfrac{1}{r} \leq \dfrac{1}{2}$. If, in addition, $q \geq 2$, that is, $\dfrac{1}{q} \leq \dfrac{1}{2}$, we may choose n sufficiently large to make $\dfrac{1}{r_n}$ large, that is, r_n so small as to have $1 \leq r_n \leq 2$. On the other hand, if $q < 2$, then since $\widetilde{a}$ is in $L^\infty(\mathbb{R}^d)$, it belongs to $L^2(\mathbb{R}^d)$ owing to the Hölder inequality, and we may return to the preceding case. In both cases, we have, for some $n \in \mathbb{N}$, $\nabla u \in L^{r_n}(\mathbb{R}^d)$ with $1 \leq r_n \leq 2$. By assumption, $\nabla u \in L^r(\mathbb{R}^d)$ with $r \geq 2$. Thus, applying the Hölder inequality again, $\nabla u \in L^2(\mathbb{R}^d)$, since $\dfrac{1}{2} = \dfrac{\alpha}{r} + \dfrac{1-\alpha}{r_n}$ for some suitable $0 \leq \alpha \leq 1$. We thus have a solution u of (4.58) the gradient of which is not only in $L^r(\mathbb{R}^d)$ but also in $L^2(\mathbb{R}^d)$. The results we have collected so far allow to conclude that $\nabla u \equiv 0$: we simply apply estimate (4.19), for which we have mentioned that the integrability of $\widetilde{a}$ is not relevant. Uniqueness up to the addition of a constant is thus proved. Note that, as announced above, we used here a crucial ingredient from the L^2 setting to deal with the case L^r... Note also that the key point of the argument is a uniqueness result for the linear homogeneous equation (4.58). Such a result is known as the *Liouville Theorem: if $Lu = 0$ in $\mathbb{R}^d$, and u belongs to a suitable class of functions, then $u = 0$ (or is a constant)*. We will return to such questions in the sequel. See for instance the comments at the end of the present section, on page 211.

Recall that we have so far *assumed* that the sequence of constants C_n in (4.57) is bounded, and that we now need to prove this fact. Recall also that we have assumed $r \geq 2$, and that in the end, we will need to address the case $r \leq 2$.

The Continuity Constants Are Bounded: A Simplified Version of the Concentration-Compactness Method

In order to prove our claim, we argue by contradiction. We therefore assume that there exist sequences $\mathbf{f}^n \in L^r(\mathbb{R}^d)$ and u^n with $\nabla u^n \in L^r(\mathbb{R}^d)$ such that

$$- \operatorname{div}\left(a_{t_n} \nabla u^n\right) = \operatorname{div} \mathbf{f}^n \quad \text{in } \mathbb{R}^d, \tag{4.60}$$

$$\left\|\nabla u^n\right\|_{L^r(\mathbb{R}^d)} = 1, \tag{4.61}$$

for all $n \in \mathbb{N}$, and

$$\left\|\mathbf{f}^n\right\|_{L^r(\mathbb{R}^d)} \overset{n \longrightarrow +\infty}{\longrightarrow} 0, \tag{4.62}$$

We readily note that (4.60) also reads as

$$- \operatorname{div}\left(a_t \nabla u^n\right) = \operatorname{div}\left(\mathbf{f}^n + (a_t - a_{t_n}) \nabla u^n\right),$$

where, as $n \longrightarrow +\infty$, the term $(a_t - a_{t_n}) \nabla u^n$ converges to zero in $L^r(\mathbb{R}^d)$ since $a_t - a_{t_n}$ converges to zero in $L^\infty(\mathbb{R}^d)$ and ∇u^n is bounded in $L^r(\mathbb{R}^d)$. Without loss of generality, we may thus replace $\mathbf{f}^n + (a_t - a_{t_n}) \nabla u^n$ by $\mathbf{f}^n$. We therefore assume that

$$- \operatorname{div}\left(a_t \nabla u^n\right) = \operatorname{div} \mathbf{f}^n \tag{4.63}$$

instead of (4.60), while keeping (4.61) and (4.62) unchanged. Clearly, (4.63) is simpler than (4.60), since the operator no longer depends on n. We now concentrate on the sequence ∇u^n. The argument we now give is typical of the *concentration-compactness method,* in a simplified version. We have already mentioned this method at the beginning of Sect. 3.4.2, then only for a terminology issue. We now apply it to our specific setting, and conclude the proof. We will then give some comments on this method.

We claim that

$$\exists \eta > 0, \quad \exists 0 < R < +\infty, \quad \forall n \in \mathbb{N}, \quad \left\|\nabla u^n\right\|_{L^r(B_R)} \geq \eta > 0, \tag{4.64}$$

where B_R denotes the ball of radius R centered at the origin.

We again argue by contradiction (we recall that the main argument we are conducting here is itself also an argument by contradiction) and assume that, contrary to (4.64),

$$\forall 0 < R < +\infty, \quad \|\nabla u^n\|_{L^r(B_R)} \xrightarrow{n \to +\infty} 0. \tag{4.65}$$

Since the coefficient $\widetilde{a}$ is uniformly continuous (recall that it is indeed uniformly Hölder continuous), and belongs to $L^r(\mathbb{R}^d)$, it vanishes at infinity. Thus, for all $\delta > 0$, there exists a sufficiently large radius R such that

$$\|\widetilde{a}\|_{L^\infty(B_R^c)} \leq \delta, \tag{4.66}$$

where B_R^c is the complement of the ball B_R. We then write

$$\|\widetilde{a}\,\nabla u^n\|_{L^r(\mathbb{R}^d)}^r = \int_{B_R} |\widetilde{a}\,\nabla u^n|^r + \int_{B_R^c} |\widetilde{a}\,\nabla u^n|^r$$

$$\leq \|\widetilde{a}\|_{L^\infty(\mathbb{R}^d)}^r \|\nabla u^n\|_{L^r(B_R)}^r + \|\widetilde{a}\|_{L^\infty(B_R^c)}^r \underbrace{\|\nabla u^n\|_{L^r(\mathbb{R}^d)}^r}_{=1}$$

Thus, owing to (4.66),

$$\|\widetilde{a}\,\nabla u^n\|_{L^r(\mathbb{R}^d)}^r \leq \|\widetilde{a}\|_{L^\infty(\mathbb{R}^d)}^r \|\nabla u^n\|_{L^r(B_R)}^r + \delta^r.$$

For R fixed, the convergence (4.65) allows to choose n_0 such that, for all $n \geq n_0$, the first term of the right-hand side is bounded from above by δ^r. Thus,

$$\forall n \geq n_0, \quad \|\widetilde{a}\,\nabla u^n\|_{L^r(\mathbb{R}^d)}^r \leq 2\delta^r.$$

This holds for any $\delta > 0$. It follows that $\widetilde{a}\,\nabla u^n$ converges strongly to zero in $L^r(\mathbb{R}^d)$. We insert (4.62) and this convergence into (4.63), which we then rewrite as

$$-\operatorname{div}\left(a_{per}\,\nabla u^n\right) = \operatorname{div}\left(\mathbf{f}^n + t\,\widetilde{a}\,\nabla u^n\right).$$

We next deduce, using the continuity in $L^r(\mathbb{R}^d)$ of the *periodic* operator T_{per} defined in (4.51), that ∇u^n converges strongly to zero in $L^r(\mathbb{R}^d)$. This contradicts (4.61) and concludes the proof of (4.64). The reader should note that, here again, we have used (in a rigorous way) the intuition explained at the beginning of Sect. 4.1.2.2: far from the origin, the operator is similar to the periodic operator because the defect vanishes.

The bound (4.61) implies that, up to the extraction of a subsequence that we do not make explicit, the sequence ∇u^n weakly converges in $L^r(\mathbb{R}^d)$ to the gradient ∇u

of some function u. Passing to the limit in the equation in the sense of distributions yields

$$- \operatorname{div}(a_t \nabla u) = 0. \tag{4.67}$$

Unfortunately, we cannot pass to the weak limit in the lower bound (4.64). We thus have no additional information on u. At this stage, u could vanish identically, and Eq. (4.67) would read as $0 = 0$! In order to circumvent this difficulty, we show, and this is a key point, that the convergence of ∇u^n to ∇u is in fact *strong* in $L^r_{loc}(\mathbb{R}^d)$. This is not true in general, but here we are going to use that u_n and u are both solutions to elliptic equations with similar properties. This will allow us to establish strong convergence. Our argument below has a zigzag pattern. We use the weak convergence of gradients to prove strong convergence of the functions. This in turn allows to prove strong convergence of the gradients, due to elliptic regularity.

The Rellich Theorem implies that, up to extraction of a subsequence we again omit, the sequence u^n converges strongly to u in $L^r_{loc}(\mathbb{R}^d)$. This is actually true up to the addition of a constant (possibly depending on n). But we may include this constant in u^n for each n and still denote the limit by u. We next use this convergence to prove that the sequence of gradients converges strongly. We start from

$$- \operatorname{div}\left(a_t \left(\nabla u^n - \nabla u\right)\right) = \operatorname{div} \mathbf{f}^n.$$

We multiply this equation by $(u^n - u)\chi_R$, where χ_R is a "usual" cut-off function satisfying $\chi_R = 1$ in B_R, $\chi_R = 0$ in B^c_{R+1}, and $\chi_R \geq 0$ everywhere. Integrating by parts, we obtain

$$\int \chi_R \left[a_t \nabla(u^n - u)\right] \cdot \nabla(u^n - u) = - \int \left[a_t \nabla(u^n - u)\right] \cdot (\nabla \chi_R)(u^n - u)$$

$$- \int \chi_R \nabla(u^n - u) \cdot \mathbf{f}^n - \int (u^n - u)\mathbf{f}^n \cdot \nabla \chi_R.$$

We may now pass to the limit in each of the three terms of the right hand side. The terms $\nabla(u^n - u)$ converge only weakly in L^r, but the terms involving $u^n - u$ and $\mathbf{f}^n$ converge strongly in L^r, at least locally. Since we have assumed $r \geq 2$, these convergence *a fortiori* hold true locally in L^2. Therefore, each scalar product converges: the right-hand side vanishes as $n \to +\infty$, for a fixed value of R. Owing to the form of χ_R and to the coercivity of a_t, we obtain the strong convergence $\nabla u^n \longrightarrow \nabla u$ in $L^2(B_R)$. We are *almost* done: the convergence is strong, but only in L^2, and not in L^r, as we have claimed. To conclude, we apply the classical interior regularity result (see for example [GM12, Theorem 7.2])

$$\|\nabla v\|_{L^r(B_{R/2})} \leq C(R) \left(\|\nabla v\|_{L^2(B_R)} + \|\mathbf{f}\|_{L^r(B_R)}\right), \tag{4.68}$$

for all $0 < R < +\infty$, with a constant $C(R)$ depending only on R, and any solution v of $-\operatorname{div}(a_t \nabla v) = \operatorname{div} \mathbf{f}$. In order to briefly explain this estimate, let us mention that classical techniques of elliptic PDEs allow to reduce the question to the case of an elliptic operator with constant coefficients. For such an operator (think for instance about the Laplace operator), arguments *a la Caccioppoli* give (4.68), without the term $\|\nabla v\|_{L^2}$. This term comes out when we return to the non-constant coefficient operator. Applying (4.68) to $v = u^n - u$ and $\mathbf{f} = \mathbf{f}^n$, and using (4.62), we obtain the *strong* convergence of ∇u^n to ∇u in $L^r(B_R)$. Using this strong convergence, we may pass to the limit in the lower bound (4.64). We thus know that ∇u does not vanish identically. However, u is a solution to (4.67). We are back to a situation similar to (4.58). With the same method, we may thus conclude that $\nabla u = 0$ and reach a contradiction. This ends the proof in the case $2 \leq r < +\infty$.

The Adjoint Case

We now turn to the case $1 < r < 2$. At first sight, this case should be easier than the case $r \geq 2$, since integrability is better at infinity. We argue by duality. The results of Proposition 4.1 have been established for $2 \leq r < +\infty$, so we may apply them to the adjoint of the operator $-\operatorname{div}(a \nabla.)$, that is, the operator $-\operatorname{div}(a^T \nabla.)$. Indeed, if the coefficient a satisfies the assumptions of Proposition 4.1, so does its tranpose matrix a^T. We fix $\mathbf{f} \in L^r(\mathbb{R}^d)$, with $1 < r \leq 2$, and note that $2 \leq r' < +\infty$ where r' is the conjugate exponent of r, that is, $\dfrac{1}{r} + \dfrac{1}{r'} = 1$. To any $\mathbb{R}^d$-valued function $\mathbf{g} \in L^{r'}(\mathbb{R}^d)$, we can associate the unique (up to the addition of a constant) solution v such that $\nabla v \in L^{r'}(\mathbb{R}^d)$, of $-\operatorname{div}(a^T \nabla v) = \operatorname{div} \mathbf{g}$. Its gradient ∇v is a linear continuous function of $\mathbf{g}$ in $L^{r'}(\mathbb{R}^d)$, in view of our proof for the case of an exponent larger than 2. We may thus define the linear form $L_{\mathbf{f}}$ by $L_{\mathbf{f}}(\mathbf{g}) = \displaystyle\int_{\mathbb{R}^d} \mathbf{f} \cdot \nabla v$ on $L^{r'}(\mathbb{R}^d)$. Applying Proposition 4.1, we have

$$\left| L_{\mathbf{f}}(\mathbf{g}) = \int_{\mathbb{R}^d} \mathbf{f} \cdot \nabla v \right| \leq \|\mathbf{f}\|_{L^r(\mathbb{R}^d)} \|\nabla v\|_{L^{r'}(\mathbb{R}^d)}$$

$$\leq C_{r'} \|\mathbf{f}\|_{L^r(\mathbb{R}^d)} \|\mathbf{g}\|_{L^{r'}(\mathbb{R}^d)}. \tag{4.69}$$

This shows that the linear form $L_{\mathbf{f}}$ is continuous on $L^{r'}(\mathbb{R}^d)$. Hence there exists $U \in L^r(\mathbb{R}^d)$ such that

$$L_{\mathbf{f}}(\mathbf{g}) = \int_{\mathbb{R}^d} \mathbf{f} \cdot \nabla v = \int_{\mathbb{R}^d} \mathbf{g} \cdot U,$$

and we read on estimate (4.69) that

$$\|U\|_{L^r(\mathbb{R}^d)} \leq C_{r'} \|\mathbf{f}\|_{L^r(\mathbb{R}^d)}.$$

We now identify U more precisely. Assuming that $\mathbf{g}$ also satisfies div $\mathbf{g} = 0$ in $\mathbb{R}^d$, we have $- \operatorname{div}(a^T \nabla v) = \operatorname{div} \mathbf{g} = 0$ with $\nabla v \in L^{r'}(\mathbb{R}^d)$, and thus, by estimate (4.39), $\nabla v \equiv 0$. For such $\mathbf{g}$, we have $L_\mathbf{f}(\mathbf{g}) = \int_{\mathbb{R}^d} \mathbf{f} . \nabla v = 0$, hence $0 = \int_{\mathbb{R}^d} \mathbf{g} . U$. This holds for all $\mathbf{g} \in L^{r'}(\mathbb{R}^d)$ such that $\operatorname{div}(\mathbf{g}) = 0$. As a consequence, there exists u such that $U = \nabla u$, hence $\nabla u \in L^r(\mathbb{R}^d)$ with

$$\|\nabla u\|_{L^r(\mathbb{R}^d)} \leq C_{r'} \, \|f\|_{L^r(\mathbb{R}^d)} \, .$$

We finally show that u satisfies $- \operatorname{div}(a \nabla u) = \operatorname{div} \mathbf{f}$. To this end, we consider the specific case where $v \in \mathcal{D}(\mathbb{R}^d)$ and set $\mathbf{g} = -a^T \nabla v$, so that $- \operatorname{div}(a^T \nabla v) = \operatorname{div} \mathbf{g}$. Applying the above, we have $\int_{\mathbb{R}^d} \mathbf{f} . \nabla v = \int_{\mathbb{R}^d} \mathbf{g} . \nabla u$. The left-hand side is the duality product $-\langle \operatorname{div} \mathbf{f}, v \rangle_{\mathcal{D}'(\mathbb{R}^d), \mathcal{D}(\mathbb{R}^d)}$, while, by definition, the right-hand side reads as

$$- \int_{\mathbb{R}^d} a^T \nabla v . \nabla u = - \int_{\mathbb{R}^d} \nabla v . a \nabla u = \langle \operatorname{div}(a \nabla u), v \rangle_{\mathcal{D}'(\mathbb{R}^d), \mathcal{D}(\mathbb{R}^d)}.$$

Since this holds true for all $v \in \mathcal{D}(\mathbb{R}^d)$, it follows that $- \operatorname{div}(a \nabla u) = \operatorname{div} \mathbf{f}$, which concludes the proof.   □

A Few Remarks About the Proof and the Concentration-Compactness Method
As we said in the course of the proof of Proposition 4.1, we give here a few more details on the *principle* of the above proof and in particular on the *concentration-compactness method*.

In retrospect, the reader may note how versatile the above proof is. This is confirmed by the fact that the *same* strategy of proof applies to several cases of different coefficients and different types of equations. We do not wish to give the corresponding details here (see [BLL18, BLL19]), but it is good to bear this fact in mind.

In the end, what really mattered? First, the *structure* of the coefficient a: it is the sum of a periodic coefficient a_{per} (the unperturbed environment) and a function $\widetilde{a}$ (the defect, which vanishes at infinity) belonging to L^q. In some loose sense, $\widetilde{a}$ vanishes at infinity, intuitively because it is integrable, and this vanishing is made rigorous by the additional Hölder regularity assumption. The latter indeed implies that $\widetilde{a}(x) \longrightarrow 0$ at infinity. The structure of the coefficient a has been used twice. First, it has been used before Proposition 4.1, where it has allowed us to reduce the existence and uniqueness of the corrector solution of (4.3), given our knowledge of the periodic corrector, to the existence of uniqueness in $L^r(\mathbb{R}^d)$ of a solution to an equation of the type (4.38). It has been used a second time when we split the problem of Proposition 4.1 into two parts:

(i) the *local* part, in which, for a large variety of coefficients a (provided they are bounded, coercive, and possibly regular when fine norms are considered) solving a linear equation of the type (1) and proving a continuity estimate of the solution with respect to the data is *simple*, and relies on usual methods of elliptic partial differential equations;

(ii) the part *far from the origin*, which is expected to be more delicate, and for which we may hope to rely on well-known "reference" cases, such as, in our specific setting, the periodic coefficient that we obtain asymptotically since the defect $\tilde{a}$ vanishes at infinity.

What are part **(i)** and part **(ii)** respectively useful for? The answer lies in the behaviour of the sequence considered in the contradiction argument. If something is likely to go wrong (here the desired estimate; in some other situations, the minimization of some functional), then we consider a typical pathological sequence. Obviously, in the type of problem we study here, the difficult part is part **(ii)**, at infinity. Thus, if a pathology occurs, it is related to some sequence of functions escaping at infinity. The concentration-compactness method then comes into play. The problem considered in Proposition 4.1 belongs to the class of *locally compact* problems, in the sense that such a problem would be compact (hence "easy") if it were set in a bounded domain (here, this corresponds to part **(i)**!). The concentration-compactness method not only applies to such questions, but also to different types of problems, where the issue is local and related to *concentration* problems (typically the appearance of a Dirac mass). In locally compact problems, *compactness* of the sequence and its suitable behaviour are guaranteed as long as we have established that no mass can escape at infinity. The key point of the above proof is to scrutinize the pathological sequence ∇u_n, coming from the contradiction argument (4.60)–(4.62), and understand where the major portion of the mass of ∇u_n goes. This is estimated by the norm in (4.61). If this mass entirely escapes at infinity, as would imply (4.65), then it only essentially interact with the periodic coefficient (because $\tilde{a}$ vanishes at infinity). But then, the situation is covered by the results for the periodic setting. Consequently, a part of the mass must stay at finite distance of the origin, as we see in (4.64). Using compactness, we then conclude that this part of the sequence converges to the solution of a Liouville problem (Eq. (4.67)), which leads to a contradiction.

4.1.3   A Proof of a Different Nature

In this section, we present an alternative proof of an L^r continuity result of the same type, apart from some subtleties that are irrelevant here, as Proposition 4.1. Like the proof given in Sect. 4.1.2, the present proof applies to all spaces L^r, $1 < r < +\infty$. It has however four important differences with the proof of Sect. 4.1.2, which we now list:

(i) it applies to divergence-form equations only, while, as we have mentioned above, the proof of Sect. 4.1.2 is versatile and allows to address different forms of equations;

(ii) it does not use the *structure* (4.1) of the coefficient a, and is thus valid for any coercive and regular coefficient (note, however, that the *application* of the existence and continuity result for equation $-\operatorname{div} a \nabla u = \operatorname{div} \mathbf{f}$ in the proof of the existence of a corrector solution to $-\operatorname{div} (a (p + \nabla w)) = 0$ does use the structure of the coefficient);

(iii) it also applies to some nonlinear equations provided they are in divergence form, as for instance the *quasi-linear* equations of the type $-\operatorname{div} \left(a |\nabla u|^{s-1} \nabla u \right) = \operatorname{div} \mathbf{f}$, with $s \geq 1$;

(iv) it uses ingredients from harmonic analysis, which are elementary but more elaborate than those used so far, and which the reader may find interesting to get familiar with.

The central result we are going to prove is the following.

Lemma 4.2 *Consider a coefficient a satisfying the bounds (3.1), and such that $a \in C_{\mathrm{unif}}^{0,\alpha} (\mathbb{R}^d)$ for some $\alpha > 0$. Fix $2 \leq r < +\infty$. Then, for any $\mathbf{f} \in L^2 \cap L^r (\mathbb{R}^d)$ and any $u \in L_{loc}^1 (\mathbb{R}^d)$ such that $\nabla u \in L^2 (\mathbb{R}^d)$, solution to (4.38), we have $\nabla u \in L^r (\mathbb{R}^d)$ and estimate (4.39) holds true, for some constant C_r depending only on the exponent r, the dimension d and the coefficient a.*

We shall give the proof of Lemma 4.2 below. We first indicate how this result can be used to recover the results of Proposition 4.1. If we compare the statement of Lemma 4.2 to that of Proposition 4.1, we see that, in Lemma 4.2, we need to assume the following points that are not contained in the assumptions of Proposition 4.1: (a) the right-hand side $\mathbf{f}$ is not only in $L^r (\mathbb{R}^d)$ but in $L^2 \cap L^r (\mathbb{R}^d)$, (b) a solution with gradient in L^2 exists (so that one may next establish that the gradient belongs to the expected functional space L^r and satisfies the continuity estimate (4.39)) and (c) the exponent r is larger than or equal to 2.

In order to recover Proposition 4.1 from Lemma 4.2, we thus need to address each of these additional assumptions. As we will see, we may succeed in doing so... or we may fail, depending on the situation. Let us indeed recall that the assumptions we use here are weaker than those of Proposition 4.1, so we may expect weaker results.

We start with the following result, which is a corollary of the *method of proof* of Lemma 4.1. This lemma proves that everything goes well for the particular exponent $r = 2$. As we will see below, this exponent plays a crucial role in the statement of Lemma 4.2 below. We have mentioned this briefly in the comments following the proof of Lemma 4.1, when discussing estimate (4.19). Using this new result, we may then only assume that $\mathbf{f} \in L^r (\mathbb{R}^d)$ instead of $\mathbf{f} \in L^2 \cap L^r (\mathbb{R}^d)$, and still prove the existence of the considered solution. This will be detailed below.

Lemma 4.3 *We assume that the coefficient a is bounded and coercive in the sense of (3.1), but we no longer assume, as in Lemma 4.1, the particular form (4.1), nor the Hölder continuity of the coefficient (we actually do not even assume continuity).*

Then, for all $\mathbf{f} \in L^2(\mathbb{R}^d)$, *there exists* $u \in L^1_{loc}(\mathbb{R}^d)$, *such that* $\nabla u \in L^2(\mathbb{R}^d)$, *solution to*

$$- \operatorname{div}(a \nabla u) = \operatorname{div} \mathbf{f} \quad in \ \mathbb{R}^d. \tag{4.70}$$

Such a solution is unique up to the addition of a constant. In addition, there exists a constant C_2, *depending only on the dimension d and on the coefficient a (thus independent of* $\mathbf{f}$), *such that*

$$\|\nabla u\|_{L^2(\mathbb{R}^d)} \leq C_2 \|\mathbf{f}\|_{L^2(\mathbb{R}^d)}. \tag{4.71}$$

Proof of Lemma 4.3 The proof is, as announced above, a replication of that of Lemma 4.1. It is actually rather the other way around: the proof of Lemma 4.1 is an *adaptation* of the (simpler) proof of Lemma 4.3, which we are about to give now. We proceed in this order only for pedagogical purposes.

In order to prove the existence of a solution, we use the same method as in Step 1 of the proof of Lemma 4.1: we regularize Eq. (4.70) by adding a term of order zero as in (4.7) and consider, for all $\eta > 0$,

$$- \operatorname{div}\left(a \nabla u^{\eta}\right) + \eta u^{\eta} = \operatorname{div} \mathbf{f}. \tag{4.72}$$

The presence of $\eta > 0$ allows to restore coercivity in the space $H^1(\mathbb{R}^d)$ of the bilinear form associated with the left-hand side of the variational formulation. The Lax-Milgram Lemma thus allows to conclude that a solution $u^{\eta} \in H^1(\mathbb{R}^d)$ exists and is unique. As in the proof of Lemma 4.1, we have

$$\left\|\nabla u^{\eta}\right\|^2_{L^2(\mathbb{R}^d)} + \eta \left\|u^{\eta}\right\|^2_{L^2(\mathbb{R}^d)} \leq C \|\mathbf{f}\|^2_{L^2(\mathbb{R}^d)},$$

for some constant C that is *independent* of the parameter η. Passing to the weak limit (up to extraction of a subsequence), we obtain a solution u of (4.70) (or, more precisely, of its variational formulation, which implies (4.70) in the sense of distributions, and even in a strong sense given our assumptions), such that $\nabla u \in L^2(\mathbb{R}^d)$ and that estimate (4.71) holds true.

As for uniqueness, we mimic Step 2 of the proof of Lemma 4.1. The difference v between two solutions satisfies $\nabla v \in L^2(\mathbb{R}^d)$ and $-\operatorname{div}(a \nabla v) = 0$. We use again a smooth cut-off function χ, with $\nabla \chi$ bounded by 2 and such that $\chi = 1$ in the ball B_1 of radius 1 centered at some point $x \in \mathbb{R}^d$, and $\chi = 0$ outside the ball B_2 of radius 2 centered at the same point x. We define its rescaled function $\chi_R = \chi(./R)$, which satisfies $\nabla \chi_R = O(1/R)$ at each point of the annulus $A_{R,2R} = B_R^c \cap B_{2R}$. We multiply the above equation by $\left(v - \fint_{A_{R,2R}} v\right) \chi_R^2$. Integrating by parts, we obtain

$$\int a|\nabla v|^2 \chi_R^2 = 2 \int a \left(v - \fint_{A_{R,2R}} v\right) \chi_R \nabla v. \nabla \chi_R.$$

The argument of the proof of Lemma 4.1 giving (4.12) applies and yields

$$\int a|\nabla v|^2 \chi_R^2 \leq \frac{C}{R^2} \int_{A_{R,2R}} \left(v - \fint_{A_{R,2R}} v \right)^2.$$

We estimate the right-hand side as in (4.14) and obtain, successively using the coercivity of the coefficient and the fact that $\chi_R^2 \equiv 1$ in B_R,

$$\int_{B_R} |\nabla v|^2 \leq C \int_{A_{R,2R}} |\nabla v|^2.$$

By assumption, we know that $\nabla v \in L^2(\mathbb{R}^d)$ (this is a key assumption as compared to Lemma 4.1, which allows us to proceed without the structure assumption (4.1) on the coefficient a). The right-hand side thus vanishes as $R \to +\infty$. We therefore conclude that $\nabla v \equiv 0$, and that uniqueness holds true, up to the addition of a constant. This ends the proof of Lemma 4.3.   □

Using Lemma 4.3, we are now in position to modify the statement of Lemma 4.2, replacing the sentence *"for any* $\mathbf{f} \in L^2 \cap L^r(\mathbb{R}^d)$ *and any* $u \in L^1_{loc}(\mathbb{R}^d)$ *such that* $\nabla u \in L^2(\mathbb{R}^d)$*, solution to (4.38), we have* $\nabla u \in L^r(\mathbb{R}^d)$ *and..."* by the sentence *"for any* $\underline{\mathbf{f} \in L^r(\mathbb{R}^d)}$ *(and not only* $\mathbf{f} \in L^2 \cap L^r(\mathbb{R}^d)$*),* *there exists* $u \in L^1_{loc}(\mathbb{R}^d)$ *such that* $\nabla u \in L^r(\mathbb{R}^d)$*, solution to (4.38), and such that..."*

In order to prove this, it is sufficient to fix $\mathbf{f} \in L^r(\mathbb{R}^d)$, and build a sequence $\mathbf{f}_n \in L^2 \cap L^r(\mathbb{R}^d)$, $n \in \mathbb{N}$, converging to $\mathbf{f}$ in $L^r(\mathbb{R}^d)$ as $n \to +\infty$. For each n, since $\mathbf{f}_n \in L^2(\mathbb{R}^d)$, we may apply Lemma 4.3 and prove the existence of a solution u_n to Eq. $-\operatorname{div}(a \nabla u_n) = \operatorname{div} \mathbf{f}_n$, such that $\nabla u_n \in L^2(\mathbb{R}^d)$. For each n, we then apply Lemma 4.2, proving that the solution u_n satisfies (4.39), hence that ∇u_n is bounded in $L^r(\mathbb{R}^d)$. Up to extraction of a subsequence, we may pass to the weak limit in this space, and obtain a function in L^r that is a gradient (because differential operators are continuous with respect to the weak topology), that we denote by $\nabla u \in L^r(\mathbb{R}^d)$. Using the convergence $\mathbf{f}_n \to \mathbf{f}$ in $L^r(\mathbb{R}^d)$, we infer from (4.39) that this weak limit ∇u satisfies $\|\nabla u\|_{L^r(\mathbb{R}^d)} \leq C_r \|\mathbf{f}\|_{L^r(\mathbb{R}^d)}$. As announced, we have just proved the *existence* of a solution with its gradient in $L^r(\mathbb{R}^d)$ that satisfies the estimate. The points (a) and (b) of page 212 above are thus addressed for what concerns existence of a solution. The reader should note that, actually, the extension we have performed does not depend on the assumption $r \geq 2$: it also would hold true if we had Lemma 4.2 for $r \leq 2$. This latter point is also related to item (c).

The next question we need to address is uniqueness of such a solution with gradient in $L^r(\mathbb{R}^d)$. Things are here more delicate. In our linear setting, uniqueness amounts to proving that if $\nabla v \in L^r(\mathbb{R}^d)$ and $-\operatorname{div}(a \nabla v) = 0$, then $\nabla v \equiv 0$. Under the general assumptions that the coefficient a is only bounded and coercive, this does not hold true. We may indeed recall here the following classical example, borrowed from [Mey63]: in dimension $d = 2$, let

$$a(x) = \begin{pmatrix} 1 - (1 - \mu^2)\frac{x_1^2}{|x|^2} & (1 - \mu^2)\frac{x_1 x_2}{|x|^2} \\ (1 - \mu^2)\frac{x_1 x_2}{|x|^2} & 1 - (1 - \mu^2)\frac{x_2^2}{|x|^2} \end{pmatrix},$$

where $\mu \in]0, 1[$ is a fixed parameter. A simple computation shows that the eigenvalues of $a(x)$ are 1 and μ^2, hence $a(x)$ is uniformly coercive. Moreover, the function

$$u(x) = x_1 \left(x_1^2 + x_2^2\right)^{\frac{\mu-1}{2}} = x_1|x|^{\mu-1},$$

satisfies $-\operatorname{div}(a\nabla u) = 0$. Note first that this function is strictly sublinear at infinity and locally bounded. Moreover, its gradient reads as

$$\nabla u(x) = \begin{pmatrix} |x|^{\mu-1} + (\mu - 1)x_1^2|x|^{\mu-3} \\ (\mu - 1)x_1 x_2|x|^{\mu-3} \end{pmatrix} = |x|^{\mu-3}\begin{pmatrix} \mu x_1^2 + x_2^2 \\ (\mu - 1)x_1 x_2 \end{pmatrix}. \qquad (4.73)$$

Hence, $|\nabla u(x)| \leq 2|x|^{\mu-1}$, which implies that $\nabla u \in L^r(\mathbb{R}^2 \setminus B(0, 1))$, for all $r > \frac{d}{1-\mu} = \frac{2}{1-\mu}$. Of course, we do not have $\nabla u \in L^r(\mathbb{R}^2)$, since the behaviour at the origin does not allow for such an integrability. Actually, $|\nabla u|^r$ is integrable over B_1 if and only if $r < \frac{d}{1-\mu} = \frac{2}{1-\mu}$. A slight modification of our solution is then in order so that it becomes smooth at the origin. To this end, we use the matrix

$$a_\chi(x) = \chi(x)a(x) + (1 - \chi(x))\begin{pmatrix} 1 & 0 \\ 0 & 1 \end{pmatrix},$$

where χ is a smooth cut-off function, such that $\chi(x) = 1$ if $|x| > 2$, and $\chi(x) = 0$ if $|x| < 1$. We define $v = \chi u$, and compute

$$\operatorname{div}\left(a_\chi(x)\nabla v(x)\right) = \operatorname{div}((1 - \chi)\chi\nabla u) + \operatorname{div}((1 - \chi)u\nabla\chi)$$
$$+ \operatorname{div}(\chi ua\nabla\chi) + 2\chi\nabla\chi . (a\nabla u).$$

We denote by ϕ the right-hand side of this equation:

$$\phi = \operatorname{div}((1 - \chi)\chi\nabla u) + \operatorname{div}((1 - \chi)u\nabla\chi) + \operatorname{div}(\chi ua\nabla\chi) + 2\chi\nabla\chi . (a\nabla u).$$

The functions $(1 - \chi)\chi$ and $\nabla\chi$ are compactly supported, and their support does not contain the origin. In addition, a and u are smooth in $\mathbb{R}^2 \setminus \{0\}$. As a consequence, ϕ is smooth and compactly supported. Next, we solve

$$-\operatorname{div}(a_\chi\nabla w) = \phi.$$

We see that $v + w$ solves $\text{div}(a_\chi \nabla(v + w)) = 0$. We are next going to prove that $\nabla v + \nabla w \not\equiv 0$, and that $\nabla v + \nabla w \in L^r(\mathbb{R}^2)$, for all $r > \dfrac{2}{1 - \mu}$.

We first show that $\nabla v + \nabla w \not\equiv 0$. To this end, we prove that $\nabla w \in L^r(\mathbb{R}^2)$ for all $r > d/(d - 1) = 2$. Since $|\nabla u| \geq \mu |x|^{\mu - 1}$, the behaviour at infinity of ∇v implies $\nabla v \notin L^r(\mathbb{R}^2)$ for $r \leq \dfrac{2}{1 - \mu}$, hence $\nabla v + \nabla w \not\equiv 0$. In order to establish that $\nabla w \in L^r(\mathbb{R}^2)$, we write

$$\nabla w(x) = \int_{\mathbb{R}^d} \nabla_x G_{a_\chi}(x, y)\phi(y)dy, \tag{4.74}$$

where G_{a_χ} is the Green function of the operator $-\text{div}(a_\chi \nabla)$. This function satisfies (3.152). Using the Hölder inequality in Marcinkiewicz spaces (see [BS88, page 220]), and the fact that ϕ is smooth and compactly supported, we deduce that $\nabla w \in L^\infty(\mathbb{R}^2)$. Thus, to prove that $|\nabla w|^r$ is integrable, we only need to address the behaviour at infinity. We note that $\text{supp}(\phi) \subset B_2$, where B_2 is the ball of radius 2 centered at the origin. We use (4.74), the Hölder inequality (between L^r and $L^{r'}$) and the fact that $\text{supp}(\phi) \subset B_2$, to obtain

$$\int_{B_4^c} |\nabla w(x)|^r dx \leq \|\phi\|_{L^{r'}(\mathbb{R}^2)} \int_{B_4^c} \int_{B_2} |\nabla_x G_{a_\chi}(x, y)|^r dy \, dx.$$

In the integral on the right-hand side, x and y are such that $|x - y| > 2$. We thus have $|\nabla_x G_{a_\chi}(x, y)| \leq C_0$, for some constant C_0 independent of x and y. We then split the integral as follows:

$$\int_{B_4^c} |\nabla w(x)|^r dx$$

$$\leq \|\phi\|_{L^{r'}(\mathbb{R}^2)} \int_{B_2} \left(\sum_{n \geq 0} \int_{C_0 2^{-n-1} < |\nabla_x G_{a_\chi}(x,y)| \leq C_0 2^{-n}} |\nabla_x G_{a_\chi}(x, y)|^r dx \right) dy$$

$$\leq \|\phi\|_{L^{r'}(\mathbb{R}^2)} \int_{B_2} \left(\sum_{n \geq 0} \left| \left\{ |\nabla_x G_{a_\chi}(\cdot, y)| > C_0 2^{-n-1} \right\} \right| C_0^r 2^{-nr} \right) dy.$$

The definition (3.153) of the space $L^{2,\infty}$ and the property (3.152) satisfied by G_{a_χ} then give

$$\int_{B_4^c} |\nabla w(x)|^r dx$$

$$\leq \|\phi\|_{L^{r'}(\mathbb{R}^2)} \sup_{y \in B_2} \|\nabla_x G_{a_\chi}(\cdot, y)\|_{L^{2,\infty}(\mathbb{R}^2)} \sum_{n \geq 0} |B_2| C_0^2 2^{2n+2} C_0^r 2^{-nr}$$

$$\leq C \sum_{n \geq 0} 2^{(2-r)n}.$$

The sum in the right-hand side is finite whenever $r > 2$. This shows that $\nabla w \in L^r(\mathbb{R}^2)$, for $r > 2$, hence that $\nabla v + \nabla w \neq 0$.

We next prove that $\nabla v + \nabla w \in L^r(\mathbb{R}^2)$ for $r > \dfrac{2}{1-\mu}$. We already know that $\nabla w \in L^r$. On the other hand, ∇v is a bounded function and, in view of the definition of χ, its behaviour at infinity is that of ∇u. Hence, $\nabla v \in L^r(\mathbb{R}^2)$. We have thus proved that

$$\text{div}(a_\chi \nabla(v + w)) = 0, \quad \nabla(v + w) \in L^r(\mathbb{R}^d), \quad \forall r > \frac{2}{1-\mu}, \quad \nabla(v + w) \neq 0.$$

$$(4.75)$$

Of course, since $\mu \in]0, 1[$, the exponent r satisfies $r > 2$, which is expected since we know that if $r = 2$, Lemma 4.3 implies uniqueness. In contrast, for any exponent $r > 2$, we now know that we do not have uniqueness, unless we impose some additional assumptions (such as periodicity for instance).

The last point we address is item (c) of page 212, that is, the exponents $r < 2$. We fix $1 \leq r < 2$, $\mathbf{f} \in L^2 \cap L^r(\mathbb{R}^d)$, $u \in L^1_{loc}(\mathbb{R}^d)$, such that $\nabla u \in L^2(\mathbb{R}^d)$, solution of (4.38) (more precisely, solution of the corresponding variational formulation). We wish to prove that $\nabla u \in L^r(\mathbb{R}^d)$ and that estimate (4.39) holds true.

Let $\mathbf{g} \in L^{r'}(\mathbb{R}^d)$, where r' is the conjugate exponent of r, so that $r' \geq 2$. Considering a sequence $\mathbf{g}_n \in L^2 \cap L^{r'}(\mathbb{R}^d)$, $n \in \mathbb{N}$ that converges to $\mathbf{g}$ in $L^{r'}(\mathbb{R}^d)$ as $n \to +\infty$, we may apply the above argument and prove the existence of $\nabla v \in L^{r'}(\mathbb{R}^d)$, satisfying

$$\|\nabla v\|_{L^{r'}(\mathbb{R}^d)} \leq C_{r'} \|\mathbf{g}\|_{L^{r'}(\mathbb{R}^d)} \tag{4.76}$$

and solution to $-\text{div}(a^T \nabla v) = \text{div}\,\mathbf{g}$ (actually, here again, of its variational formulation).

The $L^r(\mathbb{R}^d)$ norm of ∇u is given by

$$\|\nabla u\|_{L^r(\mathbb{R}^d)} = \sup_{\mathbf{g} \in L^{r'}(\mathbb{R}^d),\ \mathbf{g} \neq 0} \frac{\left|\int_{\mathbb{R}^d} \nabla u \cdot \mathbf{g}\right|}{\|\mathbf{g}\|_{L^{r'}(\mathbb{R}^d)}}. \tag{4.77}$$

Successively using the variational formulations of equations $-\operatorname{div}(a\,\nabla u) = \operatorname{div}\mathbf{f}$ and $-\operatorname{div}(a^T\,\nabla v) = \operatorname{div}\mathbf{g}$, we have

$$\int_{\mathbb{R}^d} \nabla u \cdot \mathbf{g} = \int_{\mathbb{R}^d} \nabla v \cdot \mathbf{f}. \tag{4.78}$$

We insert (4.76) and (4.78) into (4.77), and obtain

$$\|\nabla u\|_{L^r(\mathbb{R}^d)} \leq C_{r'} \sup_{\mathbf{g} \in L^{r'}(\mathbb{R}^d),\ \mathbf{g} \neq 0} \frac{\left|\int_{\mathbb{R}^d} \nabla v \cdot \mathbf{f}\right|}{\|\nabla v\|_{L^{r'}(\mathbb{R}^d)}} \leq C_{r'} \|\mathbf{f}\|_{L^p(\mathbb{R}^d)}.$$
$$\tag{4.79}$$

This proves the desired result: $\nabla u \in L^r(\mathbb{R}^d)$ and estimate (4.39) holds true.

We also wish to point out that, as already mentioned, we may apply the above extensions to the case $r < 2$: there *exists* a solution satisfying $\nabla u \in L^r(\mathbb{R}^d)$ and estimate (4.39). Concerning uniqueness, we may proceed as follows: $\nabla u \in L^r(\mathbb{R}^d)$, $r < 2 \leq d$, hence, applying the Gagliardo-Nirenberg-Sobolev inequality (see Corollary A.1), there exists a constant $M \in \mathbb{R}$ such that $u - M \in L^{r^*}(\mathbb{R}^d)$, where $\frac{1}{r^*} = \frac{1}{r} - \frac{1}{d}$. Since $\operatorname{div}(a\nabla u) = 0$, a consequence of the Harnack inequality (see Corollary A.4) is that $u(x)$, if it is not a constant, behaves like a positive power of $|x|$ at infinity, which is a contradiction. Hence, $\nabla u = 0$. Note that, actually, this argument is valid for any $r < d$. Hence, in the case $r \geq 2$ above, we have a uniqueness result if $2 \leq r < d$. This is not in contradiction with the counter-example (4.73), since we have imposed $d = 2$ there.

We now mention an alternative proof of *uniqueness,* under a weaker assumption than that of Lemma 4.3, that is, $\nabla u \in L^2(\mathbb{R}^d)$. It is stated as follows:

Lemma 4.4 *Let a be a bounded and coercive matrix, that is, satisfying* (3.1). *Assume that $u \in L^1_{loc}(\mathbb{R}^d)$ is a solution to*

$$-\operatorname{div}(a\nabla u) = 0. \tag{4.80}$$

If $\nabla u \in L^r(\mathbb{R}^d)$ with $2 \leq r < d$, then u is constant.

Recall that we have a counter-example to uniqueness (see (4.75) and the discussion above), in the case $r > d = 2$.

Proof of Lemma 4.4 We first apply Corollary A.1, that is a consequence of the Gagliardo-Nirenberg-Sobolev inequality (Theorem A.2). We know that there exists a constant $K \in \mathbb{R}$ such that

$$u - K \in L^{r^*}(\mathbb{R}^d), \quad r^* = \frac{rd}{d-r}.$$

Without loss of generality, we may assume that $K = 0$. We define

$$q = \frac{r(d-2)}{d-r},$$

and we thus have $\nabla |u|^{q/2} \in L^2(\mathbb{R}^d)$. Indeed, in the sense of distributions,

$$\nabla |u|^{q/2} = \frac{q}{2} |u|^{q/2-1} \nabla u, \tag{4.81}$$

with $|u|^{q/2-1} \in L^{\frac{r^*}{q/2-1}}(\mathbb{R}^d)$ and $\nabla u \in L^r(\mathbb{R}^d)$, hence, in view of the Hölder inequality, $\nabla |u|^{q/2} \in L^s(\mathbb{R}^d)$, with

$$\frac{1}{s} = \frac{q/2-1}{r^*} + \frac{1}{r} = \left(\frac{r(d-2)}{2(d-r)} - 1 \right) \frac{d-r}{rd} + \frac{1}{r} = \frac{d-2}{2d} - \frac{d-r}{rd} + \frac{1}{r}$$

$$= \frac{dr - 2r - 2d + 2r + 2d}{2rd} = \frac{1}{2}.$$

We then define a cut-off function χ that satisfies:

$$\chi \in C^\infty(\mathbb{R}^d), \quad \chi = 1 \text{ in } B_1, \quad \chi = 0 \text{ in } B_2^c, \quad 0 \le \chi \le 1,$$

set $\chi_R(x) = \chi(x/R)$, use $\chi_R u |u|^{q-2}$ as a test function in the variational formulation of (4.80), and obtain

$$(q-1) \int_{\mathbb{R}^d} |u|^{q-2} \chi_R^2 \, a \nabla u \cdot \nabla u = -2 \int_{\mathbb{R}^d} u |u|^{q-2} \chi_R \, a \nabla u \cdot \nabla \chi_R.$$

Owing to the coercivity (3.1) of the matrix a, this implies

$$(q-1) \int_{\mathbb{R}^d} |u|^{q-2} \chi_R^2 |\nabla u|^2 \le \frac{2M}{\mu} \int_{\mathbb{R}^d} |\nabla u \cdot \nabla \chi_R| \, |u|^{q-1} \chi_R.$$

We use (4.81) in the left-hand side, which reads as $\dfrac{q-1}{(q/2)^2} \displaystyle\int_{\mathbb{R}^d} \left| \nabla |u|^{q/2} \right|^2 \chi_R^2$. Proceeding similarly for the right-hand side, and still using (4.81), we have

$$\frac{q-1}{(q/2)^2} \int_{\mathbb{R}^d} \left| \nabla |u|^{q/2} \right|^2 \chi_R^2 \le \frac{2M}{\mu} \frac{2}{q} \int_{\mathbb{R}^d} \left| \nabla |u|^{q/2} \cdot \nabla \chi_R \right| |u|^{q/2} \chi_R.$$

The Cauchy-Schwarz inequality applied to the right-hand side then gives

$$\int_{\mathbb{R}^d} \left| \nabla |u|^{q/2} \right|^2 \chi_R^2 \leq \frac{Mq}{\mu(q-1)} \left(\int_{\mathbb{R}^d} \left| \nabla |u|^{q/2} \right|^2 \chi_R^2 \right)^{1/2} \left(\int_{\mathbb{R}^d} |u|^q |\nabla \chi_R|^2 \right)^{1/2},$$

which implies

$$\int_{\mathbb{R}^d} \left| \nabla |u|^{q/2} \right|^2 \chi_R^2 \leq \left(\frac{Mq}{\mu(q-1)} \right)^2 \int_{\mathbb{R}^d} |u|^q |\nabla \chi_R|^2 \leq \frac{C}{R^2} \int_{B_{2R} \setminus B_R} |u|^q,$$

in view of $|\nabla \chi_R| \leq \|\nabla \chi\|_{L^\infty}/R$. The constant C in the above estimate depends only on d, μ, M, q, χ. We then apply the Hölder inequality in $L^{d/2}$ and $L^{d/(d-2)}$ and obtain

$$\int_{\mathbb{R}^d} \left| \nabla |u|^{q/2} \right|^2 \chi_R^2 \leq \frac{C}{R^2} \left(\int_{B_{2R} \setminus B_R} |u|^{\frac{dq}{d-2}} \right)^{1-\frac{2}{d}} \left(R^d \right)^{\frac{2}{d}} = C \left(\int_{B_{2R} \setminus B_R} |u|^{r^*} \right)^{1-\frac{2}{d}},$$

where the constant C, which may be different from the preceding one, depends only on d, μ, M, q, χ. Since $u \in L^{r^*}(\mathbb{R}^d)$, taking the limit $R \to +\infty$ in this inequality, we thus have $\nabla |u|^{q/2} = 0$, so u is a constant. $\square$

All these extensions being proved, we now give the proof of Lemma 4.2.

Proof of Lemma 4.2 As mentioned above, the proof of Lemma 4.2 that we now give carries over to nonlinear settings. It has indeed been originally published in [Iwa83, Theorem 2] specifically for nonlinear equations, but we present here a simplified version for the linear case.

Moreover, for pedagogical purposes, we first prove the result in the case $a \equiv 1$, that is, for equation

$$- \Delta u = \operatorname{div} \mathbf{f}, \tag{4.82}$$

and we will indicate at the end of that proof the necessary modifications to address the case (4.38). This is actually in line with the proof of [Iwa83], which considers (nonlinear) translation invariant equations.

We also mention that we give the proof in the case $r > 2$, because the case $r = 2$ has been already dealt with in Lemma 4.3.

We fix $x_0 \in \mathbb{R}^d$, $R > 0$, a ball B_R centered at the origin such that $x_0 \in B_{R/2}$, and consider the equation

$$\begin{cases} - \Delta w = 0 & \text{in } B_R, \\ \\ w = u & \text{on } \partial B_R. \end{cases} \tag{4.83}$$

Our first step is a classical "Hilbertian" argument which we have already used before in the present textbook. Multiplying Eq. (4.83) by $w - u$ and integrating, we find $\int_{B_R} \nabla w \cdot \nabla(w - u) = 0$. The same procedure applied to (4.82) gives $\int_{B_R} \nabla u \cdot \nabla(w - u) = -\int_{B_R} \mathbf{f} \cdot \nabla(w - u)$. We then take the difference between these equalities, and find

$$\int_{B_R} |\nabla(w - u)|^2 = \int_{B_R} \mathbf{f} \cdot \nabla(w - u).$$

A simple application of the Cauchy-Schwarz inequality leads to the estimate

$$\|\nabla(w - u)\|_{L^2(B_R)} \leq \|\mathbf{f}\|_{L^2(B_R)}. \tag{4.84}$$

In a second step, we use the following inequality:

$$\forall A, B, c \in \mathbb{R}, \ \forall \beta > 0, \quad |A^2 - c| \leq |B^2 - c| + \frac{\beta + 1}{\beta}(A - B)^2 + \beta A^2. \tag{4.85}$$

We first use (4.85) with $A = |\nabla u(x)|$, $B = |\nabla w(x)|$, $c = |\nabla u(x)|^2 + |\nabla w(x)|^2$, $\beta = 1$, and integrate over the ball $B_{R/2}(x_0)$:

$$\int_{B_{R/2}(x_0)} |\nabla w|^2 \leq \int_{B_{R/2}(x_0)} |\nabla u(x)|^2 \, dx + 2 \int_{B_{R/2}(x_0)} \big||\nabla u(x)| - |\nabla w(x)|\big|^2 \, dx$$

$$+ \int_{B_{R/2}(x_0)} |\nabla u(x)|^2 \, dx$$

$$\leq 2 \int_{B_R} \left(|\nabla u|^2 + |\mathbf{f}|^2\right), \tag{4.86}$$

where, for the second integral of the right-hand side, we have used the triangle inequality $||\nabla u| - |\nabla w|| \leq |\nabla(u - w)|$ and the bound (4.84) since $x_0 \in B_{R/2}$, hence $B_{R/2}(x_0) \subset B_R$.

We next use (4.85) with $A = |\nabla u(x)|$, $B = |\nabla w(x)|$, $c = |\nabla w(x_0)|^2$ and β to be chosen later. We integrate on the ball $B_\rho(x_0)$ for $\rho \leq r/2$ with $r \leq R$ to be fixed below (hence *a fortiori* $B_\rho(x_0) \subset B_R$), and find

$$\int_{B_\rho(x_0)} \big||\nabla u(x)|^2 - |\nabla w(x_0)|^2\big| \, dx \leq \int_{B_\rho(x_0)} \big||\nabla w(x)|^2 - |\nabla w(x_0)|^2\big| \, dx$$

$$+ \frac{\beta + 1}{\beta} \int_{B_\rho(x_0)} \big||\nabla u(x)| - |\nabla w(x)|\big|^2 \, dx$$

$$+ \beta \int_{B_\rho(x_0)} |\nabla u(x)|^2 \, dx$$

$$\leq \int_{B_\rho(x_0)} \left| |\nabla w(x)|^2 - |\nabla w(x_0)|^2 \right| \, dx$$

$$+ \frac{\beta + 1}{\beta} \int_{B_R} |\mathbf{f}|^2 + \beta \int_{B_\rho(x_0)} |\nabla u|^2,$$

$$(4.87)$$

with the same argument as above for the second integral on the right-hand side.

We recall the following elliptic regularity results: *if the function w is solution to $-\Delta w = 0$ in a domain D, then w satisfies the following estimates*

$$\sup_{x \in B_r(x_0)} |\nabla w(x)| \leq C \left(\fint_{B_r(x_0)} |\nabla w|^2 \right)^{1/2}, \qquad (4.88)$$

$$\sup_{x \in B_\rho(x_0)} |\nabla w(x) - \nabla w(x_0)| \leq C \left(\frac{\rho}{r}\right)^\alpha \left(\fint_{B_r(x_0)} |\nabla w|^2 \right)^{1/2} \qquad (4.89)$$

for some constant C and some exponent $\alpha > 0$ depending only on the dimension, and where the radii $\rho > 0$ and $r > 0$ are such that $\rho \leq r/2$ and $B_r(x_0) \subset D$. We recall that, for any measurable set B of finite measure $|B|$, and any function f that is integrable over B, the notation $\fint_B f$ stands for the mean value of f over B, that is,

$$\fint_B f = \frac{1}{|B|} \int_B f(x) dx.$$

Estimate (4.88) is in fact Corollary A.3 applied to the first derivatives of w, which are harmonic. The second estimate (4.89) may be seen as a consequence of the Nash-Moser regularity results (see Theorem A.11), here again applied to ∇w.

For $x \in B_\rho(x_0)$, $\rho \leq r/2$ and $D = B_R$, we use the regularity results (4.88) and (4.89), finding

$$\left| |\nabla w(x)|^2 - |\nabla w(x_0)|^2 \right| \leq |\nabla w(x) - \nabla w(x_0)| \, (|\nabla w(x)| + |\nabla w(x_0)|)$$

$$\leq C \left(\frac{\rho}{r}\right)^\alpha \fint_{B_r} |\nabla w|^2. \qquad (4.90)$$

We choose $r = R/2$ and use inequality (4.86) to bound the right-hand side from above. We obtain

$$\left| |\nabla w(x)|^2 - |\nabla w(x_0)|^2 \right| \leq C \left(\frac{\rho}{R} \right)^\alpha \fint_{B_R} \left(|\nabla u|^2 + |\mathbf{f}|^2 \right). \tag{4.91}$$

Next, we insert this estimate into the right-hand side of (4.87), which gives

$$\int_{B_\rho(x_0)} \left| |\nabla u(x)|^2 - |\nabla w(x_0)|^2 \right| dx \leq C \left(\frac{\rho}{R} \right)^\alpha \rho^d \fint_{B_R} \left(|\nabla u|^2 + |\mathbf{f}|^2 \right)$$
$$+ \frac{\beta+1}{\beta} \int_{B_R} |\mathbf{f}|^2 + \beta \int_{B_\rho(x_0)} |\nabla u|^2.$$

This implies

$$\fint_{B_\rho(x_0)} \left| |\nabla u(x)|^2 - |\nabla w(x_0)|^2 \right| dx \leq C \left(\frac{\rho}{R} \right)^\alpha \fint_{B_R} |\nabla u|^2 + \beta \fint_{B_\rho(x_0)} |\nabla u|^2$$
$$+ \left(\frac{\beta+1}{\beta} \left(\frac{\rho}{R} \right)^{-d} + C \left(\frac{\rho}{R} \right)^\alpha \right) \fint_{B_R} |\mathbf{f}|^2. \tag{4.92}$$

We impose $\beta \leq 1$, so that $\beta + 1 \leq 2$, we take $R = \rho \beta^{-1/\alpha}$, and choose β sufficiently small, so that $\rho \leq r/2 \leq R/4$ is satisfied (recall that this assumption was used to establish (4.91)). Estimate (4.92) then reads as

$$\fint_{B_\rho(x_0)} \left| |\nabla u(x)|^2 - |\nabla w(x_0)|^2 \right| dx \leq \beta \fint_{B_R} |\nabla u|^2 + \beta \fint_{B_\rho(x_0)} |\nabla u|^2$$
$$+ \frac{1}{\beta^{1+d/\alpha}} \fint_{B_R} |\mathbf{f}|^2, \tag{4.93}$$

up to some irrelevant multiplicative constants that do not depend on ρ, x_0, β, and which we omit. Independently of this, a simple argument allows to prove, using the triangle inequality, that, for any domain $B \subset \mathbb{R}^d$, for any integrable function g and any constant $c \in \mathbb{R}$,

$$\fint_B \left| g(x) - \fint_B g(y) \, dy \right| dx \leq 2 \fint_B |g(x) - c| \, dx. \tag{4.94}$$

We apply this to $g = |\nabla u|^2$, $c = |\nabla w(x_0)|^2$ and $B = B_\rho(x_0)$. Using (4.93), this gives

$$\fint_{B_\rho(x_0)} \left| |\nabla u(x)|^2 - \fint_{B_\rho(x_0)} |\nabla u|^2 \right| dx \leq \beta \fint_{B_R} |\nabla u|^2 + \beta \fint_{B_\rho(x_0)} |\nabla u|^2$$
$$+ \frac{1}{\beta^{1+d/\alpha}} \fint_{B_R} |\mathbf{f}|^2, \tag{4.95}$$

here again up to irrelevant multiplicative constants. As announced above, we are now going to use some notions from harmonic analysis. For $f \in L^1_{loc}(\mathbb{R}^d)$, we introduce the following two functions defined for $x \in \mathbb{R}^d$ by

$$M[f](x) = \sup_{r>0} \fint_{B_r(x)} |f|(y)\, dy, \tag{4.96}$$

$$f^\sharp(x) = \sup_{r>0} \fint_{B_r(x)} \left| f(y) - \fint_{B_r(x)} f(z)\, dz \right| dy. \tag{4.97}$$

These functions are called *Hardy-Littlewood maximal function* and the *Fefferman and Stein maximal operator* (or *sharp maximal function*), respectively. The reader is referred to [Ste93, Chapter I] and [Ste93, Chapter IV] for these notions.

The right-hand side of (4.95) is first bounded using $M\left[|\nabla u|^2\right]$ and $M\left[|\mathbf{f}|^2\right]$ defined by (4.96):

$$\fint_{B_\rho(x_0)} \left| |\nabla u(x)|^2 - \fint_{B_\rho(x_0)} |\nabla u|^2 \right| dx$$
$$\leq 2\beta\, M\left[|\nabla u|^2\right](x_0) + \frac{1}{\beta^{1+d/\alpha}}\, M\left[|\mathbf{f}|^2\right](x_0).$$

Since this holds for any ball $B_\rho(x_0)$, we can take the supremum of the left-hand side, this time using (4.97):

$$\left(|\nabla u|^2\right)^\sharp(x_0) \leq 2\beta\, M\left[|\nabla u|^2\right](x_0) + \frac{1}{\beta^{1+d/\alpha}}\, M\left[|\mathbf{f}|^2\right](x_0). \tag{4.98}$$

Inequality (4.98) will allow us to conclude shortly. But let us pause and consider it in some more details. In order to understand it, we proceed as usual in this textbook, simplifying it upon omitting the operators $M[.]$ and $.^\sharp$. Then it formally reads as

$$|\nabla u|^2(x_0) \leq 2\beta\, |\nabla u|^2(x_0) + \frac{1}{\beta^{1+d/\alpha}}\, |\mathbf{f}|^2(x_0).$$

which implies, taking β sufficiently small,

$$|\nabla u|^2 (x_0) \leq C |\mathbf{f}|^2 (x_0). \tag{4.99}$$

Should it hold true, this inequality (4.99) would be a pointwise estimate. This would be an extremely powerful result. Indeed, it would allow to prove that $\nabla u \in L^q(\mathbb{R}^d)$, for all $1 \leq q \leq +\infty$, provided $\mathbf{f} \in L^q(\mathbb{R}^d)$. It would also trivially imply the desired estimate (4.39). Unfortunately, we are not allowed to eliminate $M[.]$ and $.^\sharp$ in (4.98). We are nevertheless able to now conclude that (4.39) holds true.

To this end, we recall the following two estimates of harmonic analysis. They relate the norm of a function g to the norms of the functions $M[g]$ and $g^\sharp$. First, we have the so-called *Stein inequality*

$$\|g\|_{L^q(\mathbb{R}^d)} \leq \|M[g]\|_{L^q(\mathbb{R}^d)} \leq C_q^S \|g\|_{L^q(\mathbb{R}^d)}, \tag{4.100}$$

valid for all $1 < q < +\infty$, with an explicit constant C_q^S, depending only on the exponent q and on the dimension d. Second, we have

$$\|M[g]\|_{L^q(\mathbb{R}^d)} \leq C_q^{FS} \|g^\sharp\|_{L^q(\mathbb{R}^d)}, \tag{4.101}$$

called the *Fefferman-Stein inequality*. It holds for any $1 \leq q < +\infty$, with an explicit constant C_q^{FS} that depends only on the exponent q and on the dimension d. These inequalities are stated and proved in [Ste93, Chapter I, Theorem 1, p 13] and [Ste93, Chapter IV, Theorem 2,p 148].

We are going to use these inequality for $q = \dfrac{r}{2}$. We first return to (4.98) and take the $L^{r/2}(\mathbb{R}^d)$ norm of both sides and find:

$$\left\| \left(|\nabla u|^2\right)^\sharp \right\|_{L^{r/2}(\mathbb{R}^d)} \leq 2\beta \left\| M\left[|\nabla u|^2\right] \right\|_{L^{r/2}(\mathbb{R}^d)} + \frac{1}{\beta^{1+d/\alpha}} \left\| M\left[|\mathbf{f}|^2\right] \right\|_{L^{r/2}(\mathbb{R}^d)}.$$

We use the Fefferman-Stein inequality (4.101) to bound from below the left-hand side:

$$\left\| M\left[|\nabla u|^2\right] \right\|_{L^{r/2}(\mathbb{R}^d)} \leq 2 C_{r/2}^{FS} \beta \left\| M\left[|\nabla u|^2\right] \right\|_{L^{r/2}(\mathbb{R}^d)}$$

$$+ \frac{1}{\beta^{1+d/\alpha}} C_{r/2}^{FS} \left\| M\left[|\mathbf{f}|^2\right] \right\|_{L^{r/2}(\mathbb{R}^d)},$$

and we choose β small enough, that is, such that $2 C_{r/2}^{FS} \beta < 1$, in order to obtain

$$\left\| M\left[|\nabla u|^2\right] \right\|_{L^{r/2}(\mathbb{R}^d)} \leq C \left\| M\left[|\mathbf{f}|^2\right] \right\|_{L^{r/2}(\mathbb{R}^d)},$$

for some irrelevant constant C that does not depend on **f**. We then apply the Stein inequality (4.100), which states that the norms $\|.\|_{L^{r/2}(\mathbb{R}^d)}$ and $\|M[.]\|_{L^{r/2}(\mathbb{R}^d)}$ are equivalent. This allows us to conclude that estimate (4.39) holds true. As announced, we have circumvented the difficulty that inequality (4.99) is not valid in general. We have nevertheless reached the desired conclusion, without proving this pointwise inequality (for a good reason indeed: except in dimension 1, it does not hold in general).

As announced, we now indicate the minor modifications in order to adapt the above arguments to the case of Eq. (4.38) with a general coefficient satisfying the assumptions of Lemma 4.2. Since (4.82) now reads as (4.38), Eq. (4.83) is changed into $-\text{div}\,(a\,\nabla w) = 0$. The integration by parts can nevertheless be performed without any modification, and the coercivity of coefficient a is enough to prove estimate (4.84). The arguments following this inequality only consist in inserting pointwise estimates deduced from the general estimate (4.85) and carry through as such. It is only when we use the regularity results (4.88) and (4.89) that we need to modify our arguments. We point out that these results still hold, provided the coefficient a is of class $C^{0,\alpha}$, owing to elliptic regularity results (Theorem A.12). The end of the proof using maximal and sharp functions is then valid without any modification. This concludes the proof of Lemma 4.2.   □

4.2   Other Explicit Deterministic Cases

In the explicit cases we consider in this section, the point is not to show that a homogenized limit exists. Indeed, since we assume that the coefficients are bounded and coercive, such an existence is a simple consequence of the abstract theory and of Proposition 3.1.

The key issue is to *explicitly determine* the homogenized matrix a^*. To this end, we first show that a corrector exists. Once such an existence is proved, we are then in position to express the matrix a^* as a function of (averages of) the corrector. Here, average is to be understood in a suitable sense: periodic average in the periodic setting, expectation in the random setting, etc... The key argument at this stage is that of Sect. 3.4.2, based on the div-curl Lemma, that is, Lemma 3.2. We do not reproduce it in the present Sect. 4.2, especially since we will explicitly revisit it in the stationary setting of Sect. 4.3.3.

The second key issue is to make precise the convergence of the oscillating solution u^ε to the homogenized solution u^* using an approximation $u^{\varepsilon,1}$ involving the corrector and a two-scale expansion (with, possibly, a truncation). This would allow to prove convergence rates in various topologies. Such questions are, given our pedagogical purpose, too delicate for the settings we consider here.

We thus only concentrate on the existence of a corrector, and leave aside all its implications.

4.2.1 Non-local Defects

We now return to settings, in dimensions $d \geq 2$, where the defects are not localized and do not vanish at infinity, in any L^q, $1 \leq q \leq +\infty$. An example of such a setting, where the defects become rare at infinity, has been studied in dimension one in Sect. 1.4.3. We only cite here the central result and refer the interested reader to [Gou22] for more details.

We consider a coefficient a of the form (4.1), that is,

$$a = a_{per} + \tilde{a}, \tag{4.102}$$

but instead of imposing that $\tilde{a} \in L^q(\mathbb{R}^d)$, we assume here that $\tilde{a}$ is of the form

$$\tilde{a}(x) = \sum_{z \in \mathcal{G}} \varphi(x - z), \tag{4.103}$$

where the set $\mathcal{G}$ is a discrete set of points of $\mathbb{R}^d$ such that all the points are exponentially far away from each other as their distance to the origin increases. An example of such a set is given in Fig. 4.2, together with the associated Voronoi

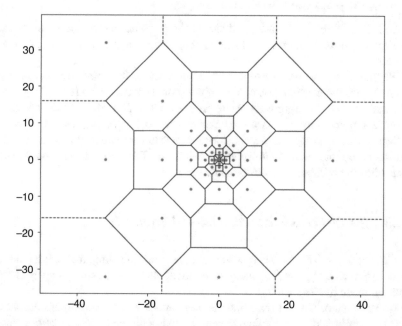

Fig. 4.2 An example of point defects that are "rare" at infinity. Copyright ©2022 American Institute of Mathematical Sciences. Reprinted with permission. All rights reserved (reproduced from [Gou22, Figure 2])

cells. Here, the function φ is assumed compactly supported and of class C^∞. It is of course possible to generalize the prototypical form (4.103) by taking the closure (for the L^2_{unif} for instance, or any suitable uniform norm) of the algebra generated by such functions. This gives a Banach space, denoted by $\mathcal{B}$, in which the theory can be developed. The results of [Gou22] show that, if $\tilde{a} \in \mathcal{B}$, and under the usual assumptions of coercivity, upper bound and Hölder regularity of the coefficient, there exists a corrector w_p of the form

$$w_p = w_{p,per} + \tilde{w}_p,$$

where $w_{p,per}$ is the periodic corrector and $\nabla \tilde{w}_p \in \mathcal{B}$. In addition, w_p is unique (up to the addition of a constant) in this class and it is possible to use this corrector to obtain convergence results similar to those established when the defects are localized.

As in the one-dimensional case addressed in Sect. 1.4.3, the point is to understand the behaviour of the coefficient at infinity. Indeed, for a sequence $a(x + x_n)$, where $(x_n)_{n \in \mathbb{N}}$ is a sequence going to infinity, two different situations may occur:

1. the distance between x_n and the set G tends to infinity, hence $a(x+x_n)$ converges to $a_{per}(x)$ as $n \to +\infty$;
2. up to extraction, the sequence x_n remains close to G, hence $a(x + x_n)$ converges to $a_{per}(x) + \varphi(x)$ up to translation.

In the former case, the situation reduces to the periodic setting. In the latter case, the situation resembles that of a local defect case. In both cases, we know how to proceed.

We thus expect to be able to conclude. In a sense, the above strategy of proof that identifies the possible behaviours of the sequence $a(x + x_n)$ for $|x_n| \longrightarrow +\infty$ is reminiscent of the arguments of the proof of Proposition 4.1 and the comments we have subsequently given on the concentration-compactness method at the end of Sect. 4.1.2. Here again, we consider the possible *problem at infinity*, or more precisely the possible *problems* (plural!) at infinity. The study of each possible case constitutes the backbone of the complete proof.

4.2.2 Quasi-Periodicity and Almost-Periodicity

In this section, we address the question of quasi-periodic homogenization (and to a lesser extent almost periodic homogenization) as such. Let us straight away explain what we mean by "as such".

We know since Chap. 1 that, like the periodic setting, the quasi-periodic setting may be interpreted as a particular case of the stationary ergodic setting, upon a suitable choice of the probability space (that is, a periodic structure in higher dimensions with the normalized Lebesgue measure, see Sect. 1.6.3). On the other hand, the quasi-periodic case is also a particular case of the almost periodic setting. The latter may itself be interpreted as a particular stationary ergodic setting, here again with a suitable choice of the probability space (the Bohr compactification $\mathbb{G}$, with the associated Haar measure, see, here again, Sect. 1.6.3).

One may thus think that studying the stationary ergodic case is sufficient to both solve the quasi-periodic and the almost periodic homogenization problems.

This is where a subtlety comes in. What do we mean by "homogenization"? This actually contains several questions: proving the existence of a homogenized limit, computing this limit and the associated homogenized coefficients, establishing approximation results such as the two-scale expansion, estimating convergence rates in various norms, etc... All these questions are variations of the same issue, but each of them may require different tools. Consider the first three questions: for studying them, no corrector is needed. It should indeed be noted that in Proposition 3.1, such an existence is not required to prove the existence of a homogenized limit. In order to proceed further and identify the homogenized coefficient and make precise the approximation of the solution, we recall from Propositions 3.4 and 3.5 that "almost" solutions to the corrector problem (in the sense of (3.56) and (3.57)) are sufficient to conclude. In order to obtain further results, it is possible to prove, as in the stochastic setting, the existence of an actual solution to the corrector problem. Now, establishing this existence *almost surely* is sufficient to solve many of the relevant issues, as we will see in Sect. 4.3. We conclude from these arguments that proving that a corrector exists is a more demanding issue than the question of homogenization itself, which gives more information on the underlying mathematical phenomena.

We may think, however, that if we prove existence (and uniqueness in some adequate sense) of a suitable corrector in the stationary ergodic setting –this is actually what we are going to do in Proposition 4.3 of Sect. 4.3.1– then, we will be able to deduce the same result for the periodic, quasi-periodic, and almost periodic settings. Things are not that simple. We will see in Sect. 4.3.2 that, though the periodic corrector may be recovered from the stationary setting, the question is more subtle in the quasi-periodic and almost periodic settings: in both situations, the stochastic formalism, although it allows to prove homogenization, does not allow to recover a "real" corrector.

We also wish to make the following point about the quasi-periodic setting. A quasi-periodic coefficient is in particular almost periodic. We may thus consider it as such, and try to solve the question of existence of a corrector in such a setting. If we follow this route, then, even if we indeed build such a corrector, all we know is that it has an almost periodic gradient, perhaps is itself almost periodic, but certainly not that it has a quasi-periodic gradient.

On the other hand, in the one-dimensional case, $1+w'$ is a multiple of a^{-1}, hence, is quasi-periodic, with the same quasi-periods as the coefficient a. This suggests that, intuitively, in the quasi-periodic setting, under some suitable assumptions to be determined, it should be possible to prove that there exists a corrector with a quasi-periodic gradient, with the same quasi-periods as the coefficient a. This also suggests that, in the almost-periodic setting, a corrector with an almost-periodic gradient should exist. Does such a result hold true? To the best of our knowledge,

there exists no general direct (that is, purely deterministic) proof of the existence
of an almost periodic corrector. Under additional assumptions, it is possible to
prove the existence of such a corrector (which is actually bounded), but such proofs
are difficult and go beyond the scope of the present textbook. Note that most of
them reproduce the steps of the proof in the stochastic case. We refer to the article
[AGK16] for more details. We will give further comments on this question in
Sect. 4.2.2.2 below. To some extent (and this will be discussed further in Sect. 4.3.2),
the stationary setting does not help.

We only wish to study the first question: can we solve the corrector problem
when the coefficient is quasi-periodic, and what are the properties of the solution?
We will admit that, once such a corrector is built, homogenization theory follows,
with all the properties (correction, convergence rate, etc...) we saw in the periodic
setting, upon adapting the corresponding proofs. It is actually a good exercise for
the reader to return to such proofs and check what carries over to the quasi-periodic
case, what is modified, and what no longer holds.

The following two somewhat contradictory comments usefully complement the
above discussion.

The first comment is slightly provocative. It is unclear to us whether the
quasi-periodic setting (and the almost periodic setting likewise) has any *practical*
interest. There certainly exists *some* situations in which a quasi-periodic geometry
is naturally present. This is the case for instance when one considers an interface
between two periodic media with incommensurable periods (think of a periodic cell
equal to the unit cube in one half-space, and a periodic cell equal to the cube of
length $\sqrt{2}$ on the other half-space). In materials science, this is the situation of a
twin boundary. Another example from materials science is the case of *quasicrystals*.
Such nonperiodic crystalline structures have been discovered in the 1980s, and
may be modelled by quasi-periodic functions (or more generally by almost periodic
functions, see [Sen95]). These situations are nevertheless relatively rare compared
to the large variety of cases in the engineering sciences.

On the other hand, until periodic geometries with defects were first considered in
homogenization, the quasi-periodic setting and the almost periodic setting were two
useful frameworks to show variations (and generalizations) of the periodic setting
in which many but *not all* properties are preserved. This is all the more true since
these are two *deterministic* settings. The random setting is far more difficult and not
so well understood. In addition, it does not give any information on some specific
issues of the deterministic setting. The above remarks are reason enough to consider
quasi-periodic and almost-periodic homogenization.

4.2.2.1   Quasi-Periodicity

In order to fix the ideas, and for simplicity (this is a restriction neither for the
arguments we are about to use, nor for the results we are about to prove), we choose
the case of a (scalar-valued) coefficient $a(x, y)$ that is quasi-periodic in dimension 2,

and is the trace of a periodic 4-dimensional function, in the following sense

$$a(x, y) = A_{per}(x, x, y, y), \tag{4.104}$$

where $A_{per} = A_{per}(x_1, x_2, x_3, x_4)$ is a regular function, defined on the space $\mathbb{R}^4$, periodic of period S with respect to variables x_1 and x_3, and periodic of period T with respect to variables x_2 and x_4, with of course $T/S \notin \mathbb{Q}$. Note indeed that if $T/S \in \mathbb{Q}$, the coefficient $a(x, y)$ is itself periodic, and is uninteresting. As we will see below, we are going to use the convention that capital letters indicate mathematical objects associated with the "augmented" space $\mathbb{R}^4$, while lowercase letters indicate mathematical objects associated with the original ambient space $\mathbb{R}^2$. Hence the notation A_{per}, which does not necessarily imply that this function is matrix-valued. The present setting is of course a very specific case, given the wide variety of possible situations within the quasi-periodic setting. This example will nevertheless allow us to capture the main phenomenon and to illustrate a general method.

We of course assume that the coefficient A_{per} satisfies the coercivity and bounds (3.1). We might expect to be able to deal with A_{per} of class $C^{0,\alpha}$, but it turns out that our specific arguments will lead us to impose higher regularity: we will need to assume that the coefficient is of class C^k for some k large enough (the size of k will be fixed depending on the ambient dimension and of the quotient T/S, in the course of the proof).

Our aim in this section is to prove the following result:

Proposition 4.2 (Existence of a Corrector, Quasi-Periodic Setting) *Consider, in dimension $d = 2$, the corrector equation*

$$\begin{cases} -\operatorname{div}\left(a\left(p + \nabla w_p\right)\right) = 0 \text{ in } \mathbb{R}^d, \\[2mm] \lim_{|x| \to +\infty} \dfrac{w_p(x)}{1 + |x|} = 0, \end{cases} \tag{4.105}$$

where the scalar-valued coefficient a is quasi-periodic, coercive and bounded in the sense of (3.1), and reads as (4.104) with a coefficient A_{per} of class $C^k\left(\mathbb{R}^4\right)$ for k sufficiently large (we will see in the proof that $k \geq 3$ is a sufficient condition).

Then, Eq. ((4.105)-1) has a solution w_p, regular and strictly sublinear at infinity, in the sense of ((4.105)-2). Its gradient is quasi-periodic, and has the same form as (4.104), that is, it is the trace of a 4-dimensional gradient, with the same period as the coefficient A_{per}, the trace of which is a. This gradient is bounded, its average vanishes, and has bounded derivatives. This solution w_p, called the corrector (associated with the vector p), is unique up to the addition of a constant.

If in addition we assume that the ratio of the periods S and T of A_{per} is not a Liouville number, in the sense of Definition 1.3 (and if necessary assuming further regularity on the coefficient A_{per}), then the corrector w_p is itself quasi-periodic, bounded, and has the form (4.104).

Since the coefficient a reads as (4.104), a first natural attempt to prove Proposition 4.2, is to write Eq. ((4.105)-1) as an equation in dimension $2d = 4$. In this equation, the coefficient becomes A_{per} and is periodic. We may thus be in position to use the same methods as in the periodic case, at the beginning of Sect. 3.4 of Chap. 3. Let us briefly explore this path. As we will see, "something like this" is going to give the result in the end.

We first introduce some notation.

We use lower case letters, like $w_p(x, y)$, for functions defined on $\mathbb{R}^2$, and we use upper case letters, like $W_p(x_1, x_2, x_3, x_4)$, for functions defined on $\mathbb{R}^4$. The fixed vector $p = (p_x, p_y) \in \mathbb{R}^2$ in Eq. ((4.105)-1) is similarly lifted as the vector $P = (p_x, p_x, p_y, p_y) \in \mathbb{R}^4$.

We likewise keep the notation ∇ and div for the gradient and divergence operators in dimension $d = 2$, and use D and DIV for the gradient and divergence operators in dimension $2d = 4$. We denote by $\overline{D}$ the operator $\overline{D} = (\partial_{\frac{x_1+x_2}{2}}, \partial_{\frac{x_3+x_4}{2}})$ acting on functions $W(x_1, x_2, x_3, x_4)$ defined on $\mathbb{R}^4$. For this, we use the system of coordinates $\left(\frac{x_1+x_2}{2}, \frac{x_1-x_2}{2}\right)$, so that, in the original coordinates (x_1, x_2), we have $\partial_{\frac{x_1+x_2}{2}} = \partial_{x_1} + \partial_{x_2}$ and $\partial_{\frac{x_1-x_2}{2}} = \partial_{x_1} - \partial_{x_2}$. The adjoint operator of $\overline{D}$ is denoted by $-\overline{\text{DIV}}$.

Proof of Proposition 4.2 If we understand Eq. ((4.105)-1) as the trace in $(x_1, x_2, x_3, x_4) = (x, x, y, y)$ of an equation posed in the space $\mathbb{R}^4$, and if we introduce the unknown function $W_{p,per}$ on $\mathbb{R}^4$ (with the secret hope that this function is periodic) such that we expect $w_p(x, y) = W_{p,per}(x, x, y, y)$, then Eq. ((4.105)-1) is lifted into

$$- \overline{\text{DIV}} \left(A_{per} \left(p + \overline{D}W_{p,per}\right)\right) = 0. \tag{4.106}$$

This equation *looks like* the corrector equation in dimension 4, but it is different. The *actual* corrector equation would imply the differential operators D and DIV instead of $\overline{D}$ and $\overline{\text{DIV}}$. Let us nevertheless see whether we can easily solve Eq. (4.106). The difficulty is the following. Although we started from an equation that is elliptic on $\mathbb{R}^2$, Eq. (4.106) is not elliptic on $\mathbb{R}^4$, because "some components" are missing in the differential operator. Think for example about the case $a = 1$ and the Laplace equation $-(\partial_{xx} + \partial_{yy})w_p = 0$ set on the space $\mathbb{R}^4$.

It is nevertheless a good strategy. The idea of lifting a quasi-periodic equation into a periodic equation in higher dimensions will indeed allow us to prove the desired result. Such a method was originally introduced by Sergei Kozlov (one of the major contributors to homogenization theory and an author of the reference textbook [ZKO94]). He used it for a slightly different purpose in [Koz79] (see also Sect. 4.2.2.2), but the idea is his. The argument we now give has been developed in detail and in a more general setting in [BLL15, Sections 5.3 & 5.4]. The technique is sometimes called "dimension doubling".

In order to address Eq. (4.106), we first restore the ellipticity of the equation. To this end, we add a small term, which will in the end vanish, but which temporarily reinstates ellipticity in the equation. More precisely, we consider on $\mathbb{R}^4$

$$- \overline{\mathrm{DIV}} \left(A_{per} \left(p + \overline{D} W_{p,per,\eta} \right) \right) - \eta \, \mathrm{DIV} \, D \, W_{p,per,\eta} + \eta \, W_{p,per,\eta} = 0.$$

(4.107)

where $-\mathrm{DIV} \, D$ is the *genuine* Laplace operator in dimension 4. It is perhaps more illuminating to write this equation as follows

$$- \overline{\mathrm{DIV}} \left(A_{per} \, \overline{D} W_{p,per,\eta} \right) - \eta \, \Delta \, W_{p,per,\eta} + \eta \, W_{p,per,\eta}$$
$$= \overline{\mathrm{DIV}} \left(A_{per} \, p \right), \quad (4.108)$$

where $-\Delta$ is the *4-dimensional* Laplace operator, and $-\overline{\mathrm{DIV}} \left(A_{per} \, \overline{D} \cdot \right)$ is a non-negative operator, since A_{per} is coercive. We thus have a standard elliptic equation, in which the coefficients and the right-hand side are both periodic. The theory of periodic elliptic problems, which we have already used in Sect. 3.4 of Chap. 3, allows to prove that there exists unique periodic solution $W_{p,per,\eta} \in H^1_{per}(Y)$ of the associated variational formulation. Here, $Y = \left] -\dfrac{1}{2}, \dfrac{1}{2} \right[^4$ is the unit cube of $\mathbb{R}^4$. This solution $W_{p,per,\eta}$ is (at least) solution to Eq. (4.108) in the sense of distributions, and satisfies the estimate

$$\int_Y A_{per} \left| \overline{D} \, W_{p,per,\eta} \right|^2 + \eta \int_Y \left| D \, W_{p,per,\eta} \right|^2 + \eta \int_Y \left| W_{p,per,\eta} \right|^2 \le C,$$

(4.109)

where C is a constant independent of $\eta > 0$. It is a simple exercise to prove (4.109), starting from (4.108), multiplying by $W_{p,per,\eta}$, integrating by parts and applying the Cauchy-Schwarz inequality. The details are left to the reader.

Owing to estimate (4.109), which is uniform with respect to η, we know that, up to extraction, the sequence $\overline{D} \, W_{p,per,\eta}$ converges weakly in $L^2(Y)$ to some function belonging to $L^2(Y)$. This limit reads as $\overline{D} \, W_p$ for some function W_p (the property of being a –partial– gradient passes to the weak limit), which is periodic and has zero average (such properties also pass to the limit). We also know that $\sqrt{\eta} \, D \, W_{p,per,\eta}$ and $\sqrt{\eta} \, W_{p,per,\eta}$ are bounded in $L^2(Y)$, hence $\eta \, D \, W_{p,per,\eta}$ and $\eta \, W_{p,per,\eta}$ tend to zero strongly in $L^2(Y)$. These convergences allow us to pass to the limit in (4.107), or more precisely in the variational formulation of (4.108), and to obtain a solution W_p, at least in the sense of distributions, to (4.106). Now, in order to obtain ((4.105)-1) from (4.106), we need to take the trace on the plane $x_1 = x_2 = x$, $x_3 = x_4 = y$. In order to do so, we need W_p to be sufficiently regular. We do not have a priori such an information. It is thus not possible, at this stage, to prove the existence of a solution to ((4.105)-1), let alone to show that this solution is quasi-periodic.

Recall now that we have assumed that the coefficient A_{per} is as regular as required. We may thus apply the operator D to Eq. (4.108). Doing so, we see that $DW_{p,per,\eta}$ satisfies an equation of the same type, with a right-hand side that we do not make precise, but which in particular involves $D A_{per}$. Using the same argument as above, we may thus prove estimate (4.109) with $DW_{p,per,\eta}$ instead of $W_{p,per,\eta}$. Consequently, repeating the above argument, we have that $D \overline{D} W_{p,per,\eta}$ converges weakly in $L^2(Y)$. Again applying the operator D, the same convergence holds for $D^2 \overline{D} W_{p,per,\eta}$. We repeat this process k times and prove weak convergence for $D \overline{D} W_{p,per,\eta}$, ..., $D^k \overline{D} W_{p,per,\eta}$. We choose $2k > 4$, that is, $k \geq 3$, so that the Sobolev space $H^k_{per}(Y)$ is continuously embedded into the space of continuous functions of $\mathbb{R}^4$. We thereby obtain that the limit $\overline{D} W_p$ is continuous. Such a property allows to take the trace on the plane $(x_1 = x_2 = x, x_3 = x_4 = y)$ of the function $\overline{D} W_p$ and of Eq. (4.106). We thus obtain a function ∇w_p which is quasi-periodic with the same quasi-periods S and T as the coefficient, and solution to Eq. ((4.105)-1). Since ∇w_p is the trace of $\overline{D} W_p$, which is periodic and continuous, hence bounded, we deduce that ∇w_p is (continuous and) bounded. Moreover, since $\overline{D} W_p$ has zero average (as a periodic function), it holds that ∇w_p has zero average as a quasi-periodic function, in view of Proposition 1.2. This trivially implies the strict sublinearity ((4.105)-2). Of course, if A_{per} is assumed to be smooth, so is ∇w_p.

The above arguments prove the existence of a solution, but give no information on uniqueness. To prove uniqueness, we consider two regular solutions of (4.105), with quasi-periodic gradients. Their difference, denoted by u, has a quasi-periodic gradient, and is strictly sublinear at infinity (about this point, see the end of Sect. 1.5.1), and is solution to $-\operatorname{div}(a \nabla u) = 0$. We multiply this equation by u, integrate over a ball B_R, and use that the coefficient a is coercive and bounded. We obtain

$$\frac{1}{|B_R|} \int_{B_R} |\nabla u|^2 \leq C \frac{1}{|B_R|} \int_{\partial B_R} |\nabla u| \, |u|,$$

for some constant C independent of R. We know that the coefficient a is of class $C^{0,\alpha}$ for $\alpha > 0$ (actually, it is smooth, but we only need Hölder regularity here). Thus, applying the Nash-Moser Theorem (Theorem A.11)) we prove that ∇u is bounded. Thus,

$$\frac{1}{|B_R|} \int_{B_R} |\nabla u|^2 \leq C \frac{1}{|B_R|} \int_{\partial B_R} |u|,$$

for some (new) constant C independent of R. Since u is strictly sublinear at infinity, the right-hand side vanishes as $R \to +\infty$. Hence, the average of the quasi-periodic non-negative function $|\nabla u|^2$ vanishes. This implies that u is a constant.

It should be noted that we did not prove uniqueness of the solution to (4.105) in general. We only proved uniqueness of the regular solution with a quasi-periodic gradient.

The last point of the proof of Proposition 4.2 concerns the corrector w_p itself, as opposed to its gradient. We now assume that T/S is not a Liouville number (see Definition 1.3). We apply the Gårding inequality (proved in Chap. 1, Lemma 1.2). Since the derivatives of $\overline{D} W_{p,per,\eta}$ are square integrable (this is obvious for the first derivatives, while, for higher derivatives, one may need to assume further regularity for the coefficient a), we thus have a bound on $W_{p,per,\eta} - \dfrac{1}{|Y|} \displaystyle\int_Y W_{p,per,\eta}$ in $L^2(Y)$. When passing to the limit for the sequence $\overline{D} W_{p,per,\eta}$, we may thus pass to the limit for the sequence of periodic functions $W_{p,per,\eta} - \dfrac{1}{|Y|} \displaystyle\int_Y W_{p,per,\eta}$. We may of course assume that the limit is still equal to W_p, since constants vanish when applying the gradient operator. The function W_p is thus periodic and continuous and bounded. We finally take the trace to obtain that w_p is itself quasi-periodic and bounded.   □

4.2.2.2   Almost Periodicity

The article [Koz79] by Sergei Kozlov is rightfully considered as a major contribution to the theory of almost periodic homogenization. The author proves there that a homogenized limit exists. Under suitable assumptions, a convergence rate is proved. The existence of an almost periodic corrector (or of a corrector with almost periodic gradient) is however not established. The proof of homogenization in [Koz79] uses a density argument, building on the fact that the set of trigonometric polynomials (see Definition 1.2) is dense in the set almost periodic functions. It is thus sufficient to prove the existence of a corrector when the coefficient is a trigonometric polynomial. In addition, proving, in such a case, that the gradient of the corrector is almost periodic (and not, as in Sect. 4.2.2.1, quasi-periodic) is sufficient. Thus, [Koz79, Theorem1], establishes that, if the coefficient is a trigonometric polynomial, then there exists a corrector that is strictly sublinear at infinity, with an almost periodic gradient. As we already mentioned in Sect. 4.2.2.1, the *method of proof* (by increasing the dimension) is however much more general. It has allowed us to prove Proposition 4.2! On the other hand, the proof of [Koz79, Theorem1] does not rely on a regularization of the higher dimensional operator, but rather on an approximation of the partial differential equation by a finite dimensional problem (a Galerkin approximation, as in numerical analysis), and gives, by the above mentioned density argument, a corrector the gradient of which is almost periodic, and which is strictly sublinear at infinity. We think that it is not necessary to reproduce it in this introductory textbook, since we have already mentioned the main ingredients in the previous section.

Note that, in particular, the technique of [Koz79] allows for an approximation of the homogenized matrix a^* by a limit process. A convergence rate can then be obtained under suitable assumptions. As one may expect, such conditions are, as above, of *diophantine* type on the commensurability of the "periods". All this allows to proceed with homogenization in an almost periodic setting, without resorting to its interpretation as a particular case of the stochastic setting, as we mention in Sect. 4.3.2 below.

In summary, no uniqueness result is stated for whichever corrector constructed in [Koz79], while some arguments are given therein that suggest existence of a corrector in a general almost-periodic setting might not hold true. Counter examples are given of some "generic" partial differential equations with almost periodic coefficients that do not admit almost periodic solutions.

Finally, note that the article [AGK16] extends the work [Koz79] by focusing on the existence of a *bounded* corrector, with some additional assumptions on the almost periodic coefficient. As we already said, the proof is of a different nature, and uses the stochastic setting.

4.2.3 Back to Algebras of Functions Applied to Homogenization

In this section, we return to the notion of algebra of functions introduced in Sects. 1.4.4 and 1.4.5. We saw there that our assumptions (H1)–(H2)–(H3') allow to construct functional spaces in which any function has an average (see (1.43)–(1.44)), hence allowing for homogenization in "zero-dimension". In addition, the considerations of Chap. 2 show that the one-dimensional case also reduces to the question of existence of averages. It is thus possible to proceed with homogenization in this setting in dimension 1 too. Note that assumption (H3') is absolutely necessary for our purpose. To realize this, assume that the coefficient a writes

$$a(x) = 1 + \sum_{k \in \mathbb{Z}} \varphi(x - k - Z_k), \tag{4.110}$$

where φ has compact support and the sequence $(Z_k)_{k \in \mathbb{Z}}$ is bounded (say, $|Z_k| < 1$ to fix the ideas). It is then possible to build a particular sequence $(Z_k)_{k \in \mathbb{Z}}$ such that $1/a$ does not have an average, while a remains bounded from above and away from 0. To be more precise, if $\varphi = \mathbf{1}_{[0,1[}$, a simple computation shows that, on the one hand,

$$\left\langle \frac{1}{1 + \sum \varphi(\cdot - k)} \right\rangle = \frac{1}{2},$$

since the function inside the average is in fact constant and equal to $1/2$, and, on the other hand, if θ_k is a suitably chosen sequence (see below), then

$$\left\langle \frac{1}{1 + \sum \varphi(\cdot - k - \theta_k)} \right\rangle = \int_0^1 \int_0^1 \int_0^1 \frac{1}{1 + \varphi(x - y) + \varphi(x + 1 - z)} dx\,dy\,dz$$

$$= \frac{7}{9}. \qquad (4.111)$$

The latter equality is based on the fact that, for any continuous and compactly supported function ϕ of two variables, and for all $x \in \mathbb{R}$,

$$\lim_{N \to +\infty} \frac{1}{2N} \sum_{k=-N}^{N} \phi(x - \theta_k, x + 1 - \theta_{k-1}) = \int_0^1 \int_0^1 \phi(x - y, x + 1 - z)dy\,dz.$$

The fact that there exists a sequence θ_k satisfying such a property is not obvious, and we refer for instance to [Fra63, Theorem 15] for a possible construction. We may then build a function a of the form (4.110), with the sequence Z_k defined by

$$Z_k = \begin{cases} \theta_k & \text{if } 2^{2n} \leq |k| < 2^{2n+1}, \quad n \in \mathbb{N}, \\ 0 & \text{if } 2^{2n+1} \leq |k| < 2^{2n+2}, \quad n \in \mathbb{N}. \end{cases}$$

Using an adaptation of the computations that allowed to obtain (4.111), it is then possible to prove that

$$\lim_{n \to +\infty} \frac{1}{2^{2n+1}} \int_{-2^{2n}}^{2^{2n}} \frac{1}{a}(x)dx = \frac{1}{2}, \qquad (4.112)$$

while

$$\lim_{n \to +\infty} \frac{1}{2^{2n+2}} \int_{-2^{2n+1}}^{2^{2n+1}} \frac{1}{a}(x)dx = \frac{7}{9}, \qquad (4.113)$$

which implies that the average of the function $1/a$ does not exist.

As a consequence, a strictly sublinear corrector does not exist either. Such a corrector would indeed read as $1 + w' = C/a$, for some constant C. It would thus satisfy, on the one hand,

$$1 + \frac{1}{2^{2n+1}} \left(w(2^{2n}) - w(-2^{2n}) \right) = C \frac{1}{2^{2n+1}} \int_{-2^{2n}}^{2^{2n}} \frac{1}{a}(x)dx$$

and, on the other hand,

$$1 + \frac{1}{2^{2n+2}} \left(w(2^{2n+1}) - w(-2^{2n+1}) \right) = C \frac{1}{2^{2n+2}} \int_{-2^{2n+1}}^{2^{2n+1}} \frac{1}{a}(x)dx.$$

The strict sublinearity of w and the convergences (4.112) and (4.113) would then imply $C = 2$ and $C = 9/7$, respectively, which is contradictory. Note that this example does not satisfy assumption (H3). Indeed, (H3) implies, as we saw, that $1/a$ has an average. Since (H3') implies (H3), it does not satisfy (H3'). The above considerations show the large variety of possible cases, including cases in which averages of functions do not exist. Our assumptions (H1)–(H2)–(H3') rule out at least this case, but we do not know, in the present state of our understanding, whether they are sufficient to develop a theory of homogenization.

In higher dimensions, it is of course possible to extend the construction of the functional spaces $\mathcal{A}^{(s,p)}$. While the general theory of homogenization allows to prove that a homogenized coefficient exists (simply apply Proposition 3.1), it is unclear how to obtain explicit formulas to compute such a coefficient. A first step in this direction would be to solve the corrector equation in the associated functional space $\mathcal{A}^{(2)}$ (for instance). This could be formalized as follows: for $p \in \mathbb{R}^d$ fixed, find $w_p \in L^1_{loc}(\mathbb{R}^d)$ satisfying

$$\begin{cases} - \operatorname{div} \left(a \left(\nabla w_p + p \right) \right) = 0, \\ \nabla w_p \in \mathcal{A}^{(2)}, \\ \langle \nabla w_p \rangle = 0. \end{cases} \tag{4.114}$$

Should such a solution exist, it would be strictly sublinear at infinity. This is implied by the third line of (4.114) and ensured by the structure of $\mathcal{A}^{(2)}$. The proof of this result is easy, and we leave it to the reader. In addition, this corrector would allow to define the homogenized coefficient as the average

$$\left[a^* \right]_{ij} = \langle a \left(e_j + \nabla w_{e_j} \right) \cdot e_i \rangle,$$

in the spirit of the settings we studied so far (periodic, quasi-periodic, etc...) Such a strategy, could it be performed, would use the structure of algebra and associated properties, first to solve system (4.114), and then to establish the above formula for a^*.

Carrying out such an agenda seems for now out of reach. It remains that the above considerations have proved useful to motivate research efforts and construct new settings suitable for a homogenization theory and inspired by the notion of algebra introduced in Sect. 1.4.5, although not as general.

4.3 The Stochastic Setting

This section is dedicated to the homogenization of Eq. (1) with a stationary oscillating coefficient, in the ergodic setting. We thus consider equation

$$
\begin{cases}
-\operatorname{div}\left(a\left(\dfrac{x}{\varepsilon}, \omega\right) \nabla u^{\varepsilon}(x, \omega)\right) = f(x) \text{ in } \mathcal{D} \subset \mathbb{R}^d, \\[2mm]
u^{\varepsilon} = 0 & \text{on } \partial\mathcal{D}.
\end{cases}
\tag{4.115}
$$

The reader should note that, in (4.115), the right-hand side and the boundary condition are deterministic. Equation (4.115) is of course the generalization to higher dimensions of Eq. (2.52) studied in Sect. 2.4 of Chap. 2. In order to set this equation in a mathematically rigorous framework, we reproduce the formalization of the stationary ergodic setting presented in Sect. 1.6 of Chap. 1 (see in particular Sect. 1.6.2). The only difference is that we work in higher dimensions, but this generalization is straightforward.

4.3.1 Existence of a Corrector

We saw in Sect. 2.4 of Chap. 2 that, at least in dimension 1, we are able to solve the stationary corrector equation associated to the homogenization of Eq. (4.115). Aiming at generalizing Sect. 2.4, we introduce the natural extension of (2.59) and (2.60) to higher dimensions: for $p \in \mathbb{R}^d$ fixed, we introduce the so-called *stationary corrector problem*:

$$
\begin{cases}
-\operatorname{div}(a(x, \omega)(p + \nabla w_p(x, \omega))) = 0 \text{ in } \mathbb{R}^d, \\[2mm]
\nabla w_p(x, \omega) \text{ stationary}, \\[2mm]
\mathbb{E}\displaystyle\int_Q \nabla w_p(x, \omega)\, dx = 0.
\end{cases}
\tag{4.116}
$$

In (4.116), the notion of stationarity is to be understood as the generalization to higher dimensions of (1.74), that is, a function $v = v(x, \omega)$ is *stationary* if

$$
\forall k \in \mathbb{Z}^d, \quad v(x + k, \omega) = v(x, \tau_k \omega)
\tag{4.117}
$$

almost surely in $\omega \in \Omega$ and for almost all $x \in \mathbb{R}^d$. Our goal is to prove that this problem has a solution w_p. The best we can hope for is to prove existence of a solution *almost surely*, then to show that this solution is unique up to the addition of a *random* variable. Recall indeed that if $X(\omega)$ is any random variable, then $w_p(x, \omega) + X(\omega)$ has the same gradient as $w_p(x, \omega)$, hence is solution to (4.116).

We are going to prove the following result:

Proposition 4.3 (Existence of a Corrector, Stationary Setting) *There exists a solution $w_p(x, \omega)$ to problem* (4.116), *in the following sense: Eq.* ((4.116)-1) *is, almost surely, satisfied in the sense of distributions $\mathcal{D}'\left(\mathbb{R}^d\right)$. In addition, the function $\nabla w_p(x, \omega)$ is, as stated in* ((4.116)-2), *stationary, and belongs to the space $\left(L^2\left(\Omega, L^2(Q)\right)\right)^d$. The condition* ((4.116)-3) *is satisfied, hence the solution w_p is almost surely strictly sublinear at infinity. It is unique up to the addition of a random variable.*

Note that we implicitly used in Proposition 4.3 that, if ∇w_p is stationary with zero average, then w_p is strictly sublinear at infinity. Actually, we only proved this fact in dimension 1 (see the end of Sect. 1.6.4), but such a proof is easily generalized to higher dimensions.

Equations (2.59) and (2.60) of Chap. 2 taught us *which* corrector equation we should write. It is now the proof of existence of a corrector in the periodic setting, given at he beginning of Sect. 3.4 of Chap. 3, that gives us a hint on how to prove that such a corrector exists, in the sense of Proposition 4.3. In this proof (the reader is referred to Eqs. (3.67) through (3.73)), we have introduced a suitable functional space (actually, the space $H^1_{per,0}$ defined by (3.72)) and written down the variational formulation of the equation (see (3.71)). We have then studied this formulation using the Lax-Milgram Lemma. We are going to proceed similarly. Only the technical details are more involved, since we need to deal with the space variable x, and with the additional variable ω, parameter of the family of problems (4.116).

Apart from these technical details, the essential additional issue in the stationary problem (4.116) with respect to the periodic problem (3.68) is a question of *coercivity*. The operator $-\operatorname{div} a \nabla$, in the periodic case, can be made elliptic, despite the fact that there is no zero-order term, by considering the space $H^1_{per,0}$ defined by (3.72). This is an important consequence of the Poincaré-Wirtinger inequality, already used several times: if v_{per} is a periodic function with zero average $\int_Q v_{per} = 0$, we have

$$\int_Q |v_{per}|^2 \le C \int_Q |\nabla v_{per}|^2,$$

where C is a constant independent of v_{per}. The bilinear form $\int_Q a_{per} \nabla v_{per} \nabla w_{per}$ associated with the operator $-\operatorname{div} a_{per} \nabla$ is thus coercive for the H^1 norm that equips the space $H^1_{per,0}$. Unfortunately, as we have seen in Sect. 1.6.4, page 59, no inequality of the type (4.14) holds in the space of stationary functions, even when imposing that the average $\mathbb{E}\left(\int v(x, \omega) \, dx\right)$ vanishes.

To circumvent this difficulty, several strategies may be developed. We are going to argue using a *regularization* of problem (4.116), that is, an additional zero order term that will eventually vanish and restores the coercivity of the bilinear form.

More precisely, we consider, as an approximation of (4.116), the problem, for $p \in \mathbb{R}^d$ and a regularization constant $\eta > 0$,

$$\begin{cases} -\operatorname{div}\left(a(x, \omega)\left(p + \nabla w_{p,\eta}(x, \omega)\right)\right) + \eta\, w_{p,\eta}(x, \omega) = 0 \text{ in } \mathbb{R}^d, \\ \\ w_{p,\eta}(x, \omega) \text{ stationary.} \end{cases} \tag{4.118}$$

In (4.118), the property "$w_{p,\eta}$ stationary" means that $w_{p,\eta}$ satisfies (4.117). Note that, besides introducing the zero order term $+\eta\, w_{p,\eta}$, we have dropped the zero average condition. Note also that we now look for a function $w_{p,\eta}$ as a stationary function itself, contrary to (4.116), in which only the gradient ∇w_p is expected to be stationary.

The regularized problem (4.118) may be studied using classical tools. We first define the following functional space

$$H = \left\{ v \in L^2\left(\Omega, L^2(Q)\right) \,/\, v \text{ stationary and } \nabla_x v \in \left(L^2\left(\Omega, L^2(Q)\right)\right)^d \right\}. \tag{4.119}$$

Here again, v stationary means that v satisfies (4.117). We equip this space with the following norm:

$$\|v\|_H^2 = \mathbb{E}\left(\int_Q |v(x, \omega)|^2 \, dx + \int_Q |\nabla v(x, \omega)|^2 \, dx \right), \tag{4.120}$$

so that H is a *Hilbert space*. We leave the proof of this fact to the reader. It is not difficult and is a good opportunity to become more familiar with the notion of stationarity. The subset of stationary functions that are regular with respect to x is dense in H. The norm (4.120) may also be expressed as a *large volume average*:

$$\|v\|_H^2 = \text{p.s.} \lim_{R \to +\infty} \frac{1}{|B_R|} \int_{B_R} |v(x, \omega)|^2 + |\nabla v(x, \omega)|^2 \, dx, \tag{4.121}$$

where B_R denotes the ball of radius R centered at the origin. In this formula, B_R could be replaced by any *suitable* sequence of domains tending to $\mathbb{R}^d$. We do not make precise the notion of a "suitable" sequence of domains. Such considerations are clearly beyond the scope of the present textbook. We only mention that any sequence of balls or cubes the size of which go to infinity is "suitable". Property (4.121) is of course a direct consequence (and a generalization to higher dimensions) of the ergodicity property stated in Theorem 1.5, Eq. (1.78) of Chap. 1. The proof is again left to the reader. A possible strategy is to cover the ball B_R by cubes $k + Q$ for appropriate values of $k \in \mathbb{Z}^d$, and apply (1.78) to the stationary function $f(x, \omega) = |v(x, \omega)|^2 + |\nabla v(x, \omega)|^2$. The fact that the sequence of domains is "suitable" allows to control the remainder of the integral on the part of the domain left uncovered by the union of cubes.

The following property will be used in the proof of Proposition 4.3 below: for any $v \in H$, we have

$$\mathbb{E}\left(\int_Q \nabla v(x, \omega)\, dx\right) = 0. \tag{4.122}$$

This equality would be easy to prove by an integration by parts (or, more precisely, the *Green formula*), if v were a periodic function. The proof for stationary functions uses the same argument. In dimension 1 for instance, we have, successively integrating by part, using the continuity and the stationarity of v,

$$\mathbb{E}\left(\int_0^1 v'(x, \omega)\, dx\right) = \mathbb{E}\left(v(1, \omega) - v(0, \omega)\right) = \mathbb{E}(v(1, \omega)) - \mathbb{E}(v(0, \omega)) = 0,$$

at least if v is assumed to be regular. A density argument then allows to prove (4.122) for all $v \in H$. The proof is more involved in dimensions $d \geq 2$, but is based on the same arguments, namely the Green formula and a density argument.

We are now in position to give the

Proof of Proposition 4.3 We reproduce the proof of [PV81, Theorem 2].

We first consider the variational formulation of the regularized problem (4.118) in the Hilbert space (4.119) equipped with the norm (4.120): *find $w_{p,\eta} \in H$ such that, for all $v \in H$,*

$$\mathbb{E}\int_Q \left(a(x, \omega)\, \nabla w_{p,\eta}(x, \omega) \,.\, \nabla v(x, \omega) \,+\, \eta\, w_{p,\eta}(x, \omega)\, v(x, \omega)\right)\, dx$$

$$= -\mathbb{E}\int_Q a(x, \omega)\, p \,.\, \nabla v(x, \omega)\, dx. \tag{4.123}$$

This variational formulation has the "standard" form $\mathcal{B}(w_{p,\eta}, v) = \mathcal{L}(v)$, where $\mathcal{B}(w_{p,\eta}, v)$ is the bilinear form of the left-hand side of (4.123), and $\mathcal{L}(v)$ is the linear form of the right-hand side. These forms are obviously continuous, and the key point is that $\mathcal{B}$ is coercive, since

$$\mathcal{B}(v, v) = \mathbb{E}\int_Q \left(a(x, \omega)\, |\nabla v(x, \omega)|^2 \,+\, \eta\, |v(x, \omega)|^2\right)\, dx$$

is equivalent to the norm (4.120) defined on H. The Lax-Milgram Lemma thus applies: there exists a unique solution $w_{p,\eta}$ to the above variational formulation. It is possible to show that this implies that $w_{p,\eta}$ is almost surely a solution to ((4.118)-1) in the sense of distributions. This equation is however not our main concern, so we skip this point (the proof of which, in any event, follows a pattern we have already seen in Chap. 3). We note that, using $v = w_{p,\eta}$ as a test function in (4.123), and using the Cauchy-Schwarz inequality to bound the right-hand side, we have the estimate

$$\mathbb{E} \int_Q |\nabla w_{p,\eta}(x,\omega)|^2 \, dx + \eta \, \mathbb{E} \int_Q |w_{p,\eta}(x,\omega)|^2 \, dx \leq C, \qquad (4.124)$$

for some C independent of η. Inequality (4.124) implies that, as $\eta \to 0$, the sequence $\nabla w_{p,\eta}$ converges weakly, up to extraction, in $\left(L^2 \left(\Omega, L^2(Q) \right) \right)^d$ to some $T \in \left(L^2 \left(\Omega, L^2(Q) \right) \right)^d$. We next prove that T is a gradient. To this end, we assume that the ambient dimension is $d = 3$, as we have done in Sect. 3.4.3.1. The generalization to any values of d is not difficult (see Remark 3.10). We fix $\varphi \in (C^\infty(Q))^d$, compactly supported in Q, and $X \in L^2(\Omega)$. Then

$$\mathbb{E} \left(X \int_Q T \cdot \operatorname{curl} \varphi \right) = \lim_{\eta \to 0} \mathbb{E} \left(X \int_Q \nabla w_{p,\eta} \cdot \operatorname{curl} \varphi \right) = 0, \qquad (4.125)$$

owing to the following integration by parts:

$$\int_Q \nabla w_{p,\eta} \cdot \operatorname{curl} \varphi = - \int_Q w_{p,\eta} \operatorname{div} (\operatorname{curl} \varphi) = 0,$$

where the boundary terms vanish because φ has compact support in Q. We deduce from (4.125) that, almost surely, $\int_Q T \cdot \operatorname{curl} \varphi = 0$. This holds for any function $\varphi \in (C^\infty(Q))^d$ with compact support, so the de Rham Lemma, which we have already used in Sect. 3.4.3.1, implies that there exists $w_p \in L^2(\Omega, H^1(Q))$ such that $T = \nabla w_p$. We have thus proved that $\nabla w_{p,\eta}$ converges to ∇w_p weakly in $\left(L^2 \left(\Omega, L^2(Q) \right) \right)^d$. In addition, using (4.124) again, since the sequence $\sqrt{\eta} \, w_{p,\eta}$ is bounded in $L^2 \left(\Omega, L^2(Q) \right)$, the sequence $\eta \, w_{p,\eta}$ converges strongly to zero in this space. We may thus pass to the limit in the variational formulation (4.123) and obtain the existence of a function w_p such that $\nabla w_p \in \left(L^2 \left(\Omega, L^2(Q) \right) \right)^d$ and, for all $v \in H$,

$$\mathbb{E} \int_Q a(x,\omega) \nabla w_p(x,\omega) \cdot \nabla v(x,\omega) \, dx = - \mathbb{E} \int_Q a(x,\omega) \, p \cdot \nabla v(x,\omega) \, dx. \qquad (4.126)$$

We now prove two additional properties satisfied by ∇w_p. First, ∇w_p is stationary (hence, condition ((4.116)-2) holds true). Indeed, any $f(x,\omega)$ that is a weak limit in $L^2 \left(\Omega, L^2(Q) \right)$ of a sequence stationary functions $f_n(x,\omega)$ is itself stationary. To see this, we note that, for any function $\varphi \in L^2(\Omega)$, any function $\chi \in L^2(Q)$, any $k \in \mathbb{Z}^d$, and any $n \in \mathbb{N}$,

$$\mathbb{E} \int_Q f_n(x + k, \omega) \, \varphi(\omega) \, \chi(x) \, dx \; = \; \mathbb{E} \int_Q f_n(x, \tau_k \omega) \, \varphi(\omega) \, \chi(x) \, dx,$$

since f_n is stationary. Passing to the limit $n \to +\infty$, we obtain a similar equality for f, namely,

$$\mathbb{E} \int_Q f(x + k, \omega) \, \varphi(\omega) \, \chi(x) \, dx \; = \; \mathbb{E} \int_Q f(x, \tau_k \omega) \, \varphi(\omega) \, \chi(x) \, dx.$$

This property holds for any $\varphi \in L^2(\Omega)$ and any $\chi \in L^2(Q)$. Since the vector space spanned by products of such functions is dense in $L^2\left(\Omega, L^2(Q)\right)$, we infer that $f(x + k, \omega) = f(x, \tau_k \omega)$ in $L^2\left(\Omega, L^2(Q)\right)$, that is, f is stationary. We apply the same technique, with $\varphi \equiv 1$, $\chi \equiv 1$, to prove that $\mathbb{E}\left(\int_Q \nabla w_{p,\eta}(x, \omega) \, dx\right) = 0$, in view of (4.122) and of the fact that $w_{p,\eta}$ is stationary. This gives condition ((4.116)-3) for ∇w_p.

The final key point is to prove that the variational formulation (4.126) implies Eq. ((4.116)-1) in the sense of distributions (actually, our proof will show that the variational formulation (4.123) implies Eq. ((4.118)-1)). Such an implication is straightforward once we have proved the following: if $F(y, \omega)$ is a stationary function in $L^2\left(\Omega, L^2(Q)\right)$ and if $\mathbb{E} \int_Q F(x, \omega) \Phi(x, \omega) \, dx \; = \; 0$ for any stationary function $\Phi(x, \omega)$ in $L^2\left(\Omega, L^2(Q)\right)$, then $F(y, \omega) = 0$ almost surely and for almost all x. Such a property would be obvious if we could use $\Phi = F$ in the above equality. Such a strategy is however not possible here since F (that is, w_p above) does not belong to the space of test functions: ∇w_p is stationary, but a priori w_p is not. We thus need a *constructive* and flexible proof, which we now give. For $\varphi \in L^2(\Omega)$, $\chi \in \mathcal{D}\left(\mathbb{R}^d\right)$ we introduce the function

$$\Phi(x, \omega) = \sum_{k \in \mathbb{Z}^d} \chi(x + k) \, \varphi(\tau_{-k} \omega), \tag{4.127}$$

which is well defined, since the sum $\displaystyle\sum_{k \in \mathbb{Z}^d}$ contains only a finite number of terms for a fixed $x \in \mathbb{R}^d$, and is stationary, since, for $j \in \mathbb{Z}^d$,

$$\begin{aligned}
\Phi(x + j, \omega) &= \sum_{k \in \mathbb{Z}^d} \chi(x + k + j) \, \varphi(\tau_{-k} \, \omega) \\
&= \sum_{k' \in \mathbb{Z}^d} \chi(x + k') \, \varphi(\tau_{-k'} (\tau_j \, \omega)) \\
&= \Phi(x, \tau_j \, \omega).
\end{aligned}$$

We use this function Φ as a test function in $\mathbb{E} \int_Q F(x, \omega) \Phi(x, \omega)\,dx = 0$. We obtain

$$\mathbb{E} \int_Q F(x, \omega) \Phi(x, \omega)\,dx = \mathbb{E} \int_Q F(x, \omega) \sum_{k \in \mathbb{Z}^d} \chi(x + k)\, \varphi(\tau_{-k}\omega)\,dx$$

$$= \mathbb{E} \int_Q \sum_{k \in \mathbb{Z}^d} F(x + k, \tau_{-k}\omega) \chi(x + k)\,dx\, \varphi(\tau_{-k}\omega),$$

$$= \sum_{k \in \mathbb{Z}^d} \mathbb{E} \int_{Q+k} F(x, \tau_{-k}\omega) \chi(x)\,dx\, \varphi(\tau_{-k}\omega),$$

where we have used that $F(x, \omega)$ is stationary and that $\displaystyle\sum_{k \in \mathbb{Z}^d}$ contains only a finite number of non-vanishing terms, which allows to exchange the sum and the integral. The group action τ being measure invariant, we have, for all $k \in \mathbb{Z}^d$,

$$\mathbb{E} \int_{Q+k} F(x, \tau_{-k}\omega) \chi(x)\,dx\, \varphi(\tau_{-k}\omega) = \mathbb{E} \int_{Q+k} F(x, \omega) \chi(x)\,dx\, \varphi(\omega),$$

and we thus obtain

$$\mathbb{E} \int_Q F(x, \omega) \Phi(x, \omega)\,dx = \sum_{k \in \mathbb{Z}^d} \mathbb{E} \int_{Q+k} F(x, \omega) \chi(x)\,dx\, \varphi(\omega),$$

$$= \mathbb{E} \int_{\mathbb{R}^d} F(x, \omega) \chi(x)\,dx\, \varphi(\omega).$$

Since this quantity vanishes for all function $\varphi \in L^2(\Omega)$, we have $\int_{\mathbb{R}^d} F(x, \omega) \chi(x)\,dx = 0$ almost surely. Such a property holds true for all $\chi \in \mathcal{D}(\mathbb{R}^d)$. Hence, almost surely, $F(x, \omega) = 0$ in the sense of distributions in $x \in \mathbb{R}^d$, almost everywhere and in $L^2(\Omega, L^2(Q))$. The latter properties are all equivalent to one another since we have assumed $F \in L^2(\Omega, L^2(Q))$.

We now return to Eq. (4.126). Using $v(x, \omega) = \Phi(x, \omega)$ (defined by (4.127)) as a test function, and applying the above technique (which amounts to applying the above with $F(x, \omega)$ successively being equal to the components of $a(x, \omega)(p + \nabla w_p(x, \omega)))$, we deduce that

$$\int_{\mathbb{R}^d} a(x, \omega) \nabla w_p(x, \omega) \cdot \nabla \chi(x)\,dx = - \int_{\mathbb{R}^d} a(x, \omega)\, p \cdot \nabla \chi(x)\,dx,$$

hence that Eq. ((4.116)-1) holds in the sense of distributions.

It remains to prove uniqueness of the solution. Since the equation is linear, this amounts to proving that, if $p = 0$, the solution w_0 to (4.126) with ∇w_0 stationary satisfies $\nabla w_0 = 0$. Similarly to what we have seen above, using $v = w_0$ as a test function would yield the result. But it is not possible since w_0 is not necessarily stationary. We thus solve the following problem

$$- \Delta \psi_j^\gamma + \gamma \psi_j^\gamma = \partial_j w_0,$$

where $j \in \{1, \ldots, d\}$ and $\gamma > 0$ is a regularization parameter that will eventually vanish. This problem is of course to be understood in the sense of its variational formulation in the set of stationary functions. More precisely, we define

$$V = \left\{ v \in L^2 \left(\Omega, H^1_{loc}(\mathbb{R}^d) \right), \quad v \text{ stationary} \right\},$$

and look for $\psi_j^\gamma \in V$, such that

$$\forall v \in V, \quad \mathbb{E} \int_Q \nabla \psi_j^\gamma \cdot \nabla v + \gamma \, \mathbb{E} \int_Q \psi_j^\gamma v = \mathbb{E} \int_Q \partial_j w_0 v.$$

Applying the Lax-Milgram Lemma in the Hilbert space V proves that this problem has a unique solution $\psi_j^\gamma \in V$. Of course, $\nabla \psi_j^\gamma$ is stationary. We next define the stationary function

$$v^\gamma(x, \omega) = \sum_{j=1}^d \partial_j \psi_j^\gamma(x, \omega),$$

which we may use as a test function in (4.126) with $p = 0$. This gives

$$\mathbb{E} \int_Q a(x, \omega) \nabla w_0(x, \omega) \cdot \nabla v^\gamma(x, \omega) dx = 0. \tag{4.128}$$

We next prove that ∇v^γ converges to ∇w_0 as $\gamma \to 0$. To this end, we note that $(-\Delta + \gamma) v^\gamma = -\Delta w_0$, hence,

$$\mathbb{E} \int_Q |\nabla v^\gamma|^2 + \gamma \, \mathbb{E} \int_Q (v^\gamma)^2 \leq \left(\mathbb{E} \int_Q |\nabla w_0|^2 \right)^{1/2} \left(\mathbb{E} \int_Q |\nabla v^\gamma|^2 \right)^{1/2}.$$

This proves that ∇v^γ is bounded in $L^2(\Omega \times Q)$, and that $\gamma v^\gamma \longrightarrow 0$ in $L^2(\Omega \times Q)$. We may thus pass to the limit in (4.128), and obtain $\nabla w_0 = 0$. □

Several remarks are in order.

First, Proposition 4.3 states that the gradient of the corrector w_p is stationary, but all we know is that w_p itself is strictly sublinear at infinity. The latter fact is a consequence of the fact that ∇w_p is stationary and has zero average (we have proved this property in dimension 1 in Sect. 1.6.4, page 60, but the proof may be easily extended to higher dimensions). The stationarity of w_p itself is not guaranteed in full generality. In fact, we have established in Chap. 2, Sect. 2.4, the explicit formula (2.61) for the one-dimensional corrector, that is, $w(x, \omega) = -x + a^* \int_0^x a^{-1}(y, \omega)\, dy$, which apart from very special situations, is not stationary, as we have seen in Chap. 1, pages 57 ff.

On the other hand, in dimensions $d \geq 3$, it has been recently proved in [GO17] that the corrector is itself stationary, *provided* an additional spectral gap condition is satisfied. Loosely speaking, this spectral gap condition amounts to a weak version of the Poincaré inequality. More precisely, the spectral gap condition is a property of the probabilistic setting in which the considered equations are posed. We refer for instance to [BGL14, Chapitre 4] for an extensive presentation of this setting and related questions. The spectral gap condition may be stated as follows: there exists a constant C such that, for any function f for which the left-hand side makes sense,

$$\int_{\mathbb{R}^d} f^2 \, d\mu - \left(\int_{\mathbb{R}^d} f \, d\mu \right)^2 \leq C\, D(f), \qquad (4.129)$$

where $D(f)$ is the *Dirichlet form* associated to the considered setting. In our specific setting, this form reads as $\int_{\mathbb{R}^d} |\nabla f|^2 \, d\mu$ and the measure μ is the probability measure defining the "usual" expectation value, in the stationary setting. Note that the left-hand side of (4.129) also writes $\int_{\mathbb{R}^d} \left(f - \int_{\mathbb{R}^d} f \, d\mu \right)^2$. In the periodic setting interpreted as a particular case of the stationary setting (we will see this in detail in Sect. 4.3.2 below), inequality (4.129) coincides with the Poincaré-Wirtinger inequality. This particular case allows for a simple explanation of the terminology "spectral gap". The constant in the Poincaré-Wirtinger inequality is indeed equal to the inverse of the difference between the first two eigenvalues of the operator $-\Delta$ defined on H^1_{per}. The fact that this constant is well-defined is equivalent to the spectrum of this operator having a "gap". As a side remark, let us also mention that the spectral gap condition plays an important part in the study of long-time behaviour of solutions to parabolic equations. We will mention some related questions at the end of Sect. 6.3.3. The spectral gap condition implies ergodicity, but is more demanding than the latter.

When a spectral gap condition holds true, it can be proved, as we announced above, that, in dimensions $d \geq 3$, the corrector w_p defined above is *itself* stationary (in addition to having a stationary gradient). This result has been proved recently in [GO17]. This may be intuitively understood by noting that the bound (4.124) on $\nabla w_{p,\eta}$ implies essentially a similar bound on $w_{p,\eta}$ itself, uniformly with respect to η. This allows to pass to the limit on the corrector itself in the space of stationary functions.

Another, more natural, remark is the following.

Remark 4.1 We have worked above in the *discrete* stationary framework defined by (4.117). We have already pointed out that such a setting is more natural in the study of perturbed periodic geometries. An analogous result to Proposition 4.3 is valid and easily stated in the *continuous* stationary setting. The proof, which we omit, is a simple adaptation of the method used above: first one regularizes the problem by a zero-order term, then writes the variational formulation (remove the integrals $\int_Q \ldots dx$ in (4.123)), then passes to the limit. The only minor difficulty arises when proving that the solution of the variational formulation is solution to the equation in the sense of distributions: one needs to replace (4.127) by its continuous stationary version $\Phi(x, \omega) = \int_{\mathbb{R}^d} \chi(x + y)\,\varphi(\tau_{-y}\,\omega)\,dy$. $\square$

We have obtained, with Proposition 4.3, the almost sure existence of a corrector the gradient of which is stationary. We are now going to study the consequence of this on the particular stationary settings we focus on in this textbook.

4.3.2   Back to the Particular Settings

Application to the Periodic Setting
First, as we pointed out in Sect. 1.6.3 of Chap. 1, the periodic setting may be identified as a particular stationary case. The argument of Chap. 1 has been given in dimension 1, but it is straightforward to adapt it to any dimension d. Our aim here is thus not to prove the existence of a periodic corrector, since we have already proved this, but to recover such an existence from the random setting, as a way to better understand the latter setting.

Let us first identify the periodic setting as a particular case of the *continuous* stationary setting. This is the classical identification given in homogenization textbooks. As in Sect. 1.6.3 of Chap. 1, we use the torus as the probability space, the Lebesgue measure as the probability measure, and the translation $\tau_x \omega = x + \omega$ as the group action.

We then consider a periodic coefficient $a_{per}(x)$ and define $a(x, \omega) = a_{per}(x + \omega)$, almost surely in ω and almost everywhere in $x \in \mathbb{R}^d$. Such a coefficient is stationary, since

$$a(x + y, \omega) = a_{per}(x + y + \omega) = a(x, \tau_y \omega),$$

almost surely in ω and almost everywhere in $x \in \mathbb{R}^d$ and $y \in \mathbb{R}^d$. Applying the variant of Proposition 4.3 adapted to the continuous stationary setting gives the almost sure existence of a corrector $w_p(x, \omega)$, the gradient of which is stationary, and such that $\mathbb{E} \nabla w_p(x, \omega) = 0$ for almost all $x \in \mathbb{R}^d$. We know, by construction, with the same argument as in Sect. 1.6.3 and by stationarity, that the function $\nabla w_p(x, \omega)$ satisfies $\nabla w_p(x', \omega') = \nabla w_p(x, \omega)$ whenever $x' + \omega' = x + \omega \bmod(1)$. Hence, $\nabla w_p(x, \omega)$ is a periodic function of $x + \omega$. It is a gradient with respect to x (because its curl identically vanishes) and we denote it by $\nabla w_{p,per}(x + \omega)$. Note that our notation with a subscript *per* anticipates periodicity. At this stage, we do not know yet that $w_{p,per}$ is periodic of the variable $x + \omega$. We nevertheless know that its gradient, as a periodic function of the variable $x + \omega$, is of zero average. Indeed, condition $\mathbb{E} \nabla w_p(x, \omega) = 0$ reads, since Ω is the torus and the probability measure is the Lebesgue measure, as

$$\int_{\mathbb{T}^d} \nabla w_p(x + \omega)\, d\omega = 0,$$

for almost all $x \in \mathbb{R}^d$. This is equivalent to the fact that the average of the periodic function $\nabla w_{p,per}$ vanishes. The latter is thus the gradient of some periodic function, which confirms our notation $w_{p,per}$. Note that we may make this function unique by assuming that its average vanishes. We then observe that the set of all ω for which we have proved existence of a corrector is of full measure. This allows us to choose at least a point ω_0 in this set and define $w_{p,per}(x + \omega_0) = w_p(x, \omega_0)$. This function $w_{p,per}(x + \omega_0)$ is solution to the stationary corrector equation with coefficient $a(x, \omega_0) = a_{per}(x + \omega_0)$. We have thus obtained a solution to the periodic corrector equation. It suffices to change $x + \omega_0$ into x, that is, "read" the equation at a translated point, in order to recover the usual formulation.

It is alternatively possible to identify the periodic setting as a particular case of the *discrete* stationary setting. We then define the random coefficient $a(x, \omega) = a_{per}(x)$, almost surely, which is stationary in the discrete sense, since $a(x + k, \omega) = a_{per}(x + k) = a_{per}(x) = a(x, \tau_k \omega)$ for all $k \in \mathbb{Z}^d$. Applying Proposition 4.3, we obtain the almost sure existence of a corrector $w_p(x, \omega)$, the gradient of which is stationary in the discrete sense, with zero average, that is,

$$\mathbb{E} \int_Q \nabla w_p(x, \omega)\, dx = 0.$$

We then define $w_{p,per}(x) = \mathbb{E}\, w_p(x, \omega)$, again using a notation that anticipates on our expected conclusion. Owing to the discrete stationarity of $\nabla w_p(x, \omega)$, we know that $\nabla w_{p,per}$ is periodic. In addition, its average vanishes, by application of the Fubini Lemma and the zero average condition of $\nabla w_p(x, \omega)$. Thus, $w_{p,per}$ itself is periodic. We also know, in view of the fact that the coefficient a_{per} is deterministic, hence the expectation value commutes with the differential operator $-\operatorname{div}(a_{per} \nabla \,.)$, that $w_{p,per}$ is solution to $-\operatorname{div}(a_{per}(p + \nabla w_{p,per})) = 0$. We conclude that $w_{p,per}$ defined this way is a periodic corrector, and is thus, up to an additive constant, *the* periodic corrector.

We may actually prove that, almost surely, $w_p(x, \omega)$ (and not only its expectation) is equal, up to the addition of *random* constant, to the periodic corrector. Indeed, $w_{p,per}$ is *in particular* a stationary function in the discrete sense. Hence its gradient is stationary. Furthermore, $w_{p,per}$ solves the variational formulation

$$\mathbb{E} \int_Q a_{per}(x) \, \nabla w_{p,per}(x) \cdot \nabla v(x, \omega) \, dx \;=\; -\mathbb{E} \int_Q a_{per}(x) \, p \cdot \nabla v(x, \omega) \, dx,$$

for all stationary functions $v(x, \omega)$. This is due to the fact that, here again, we may exchange expectation value and differential operators, and recover the variational formulation of the periodic corrector problem. We next apply the uniqueness result of the stationary corrector, established in [PV81] and reproduced above in the proof of Proposition 4.3, in order to conclude that $\nabla w_{p,per}(x) = \nabla w_p(x, \omega)$, almost surely and almost everywhere, hence that $w_{p,per}(x) - w_p(x, \omega)$ is a random variable independent of x.

Application to the Quasi-Periodic Setting
We have seen in Sect. 1.6.3 of Chap. 1 that the quasi-periodic case may be identified as a particular case of the continuous stationary setting. It is in fact a simple adaptation of the identification of the periodic case as a stationary setting, given in Chap. 1 and recalled above.

This suggests a route to possibly deduce existence of a quasi-periodic corrector (or at least a corrector the gradient of which is quasi-periodic) using the almost sure existence of a corrector $w_p(x, \omega)$, the gradient of which is stationary and satisfies $\mathbb{E} \nabla w_p(x, \omega) = 0$ for almost all $x \in \mathbb{R}^d$, established by the analogue of Proposition 4.3 in the continuous stationary setting. The modification required from the periodic to the quasi-periodic setting will however deeply modify the situation we have exposed in the previous paragraph. The strategy will eventually fail.

The one-dimensional case is sufficient to understand the obstruction. In this case, we know, from Sect. 2.4 of Chap. 2, that there exists a corrector the derivative of which is quasi-periodic. Can we recover this corrector from the stationary setting? As we will see, in order to do so, we need to "cheat".

We choose a coefficient $a_{q-per}(x)$ that is quasi-periodic in the real variable $x \in \mathbb{R}$, and assume that this coefficient is continuous. This coefficient is the trace of a regular coefficient in some higher dimension. For simplicity, and since this does not change our arguments, we assume that $a_{q-per}(x)$ is the trace of a periodic coefficient of dimension $d = 2$, and even assume

$$a_{q-per}(x) = a_{per}\left(x, \frac{x}{\sqrt{2}}\right),$$

where $a_{per}(x, y)$ is a regular coefficient that is periodic with the periodic cell $Q = [0, 1]^2$, and of course coercive. The quasi-periods of the coefficient $a_{q-per}(x)$ are thus 1 and $\sqrt{2}$. The stationary setting associated with this quasi-periodic situation is obtained by taking as probability space the unit cube $\Omega = Q = [0, 1]^2$, as probability measure the Lebesgue measure, and as the action $\tau_x \omega = \omega + \left(x, \dfrac{x}{\sqrt{2}}\right) = \left(\omega_1 + x, \omega_2 + \dfrac{x}{\sqrt{2}}\right)$ for almost all $\omega = (\omega_1, \omega_2) \in \Omega$ and

$x \in \mathbb{R}$. The fact that $\sqrt{2}$ is irrational implies that this setting is ergodic. We then define the function $a(x, \omega) = a_{per}\left(\omega_1 + x, \omega_2 + \dfrac{x}{\sqrt{2}}\right)$. By definition, this coefficient is stationary (and regular, bounded, coercive). We know from the results of the stationary setting that there exists almost surely a solution to the corrector equation (2.59), namely

$$-\frac{d}{dx}\left(a(x, \omega)\left(1 + \frac{dw}{dx}(x, \omega)\right)\right) = 0.$$

In addition, w' is stationary and $\mathbb{E}w' = 0$. Elliptic regularity (in the present one-dimensional setting, it is actually sufficient to solve explicitly the equation to obtain such regularity) shows that the function w' is regular. The question is now whether this equation implies the quasi-periodic corrector equation, and whether w' is quasi-periodic. At this stage, we face an unexpected issue... Let $\omega \in \Omega$ be such that the equation is satisfied at ω. Given our experience in the periodic setting, we expect to find the quasi-periodic corrector equation *translated* at the point ω. In order to obtain that w' is quasi-periodic, we need to recognize this function as the trace of a periodic regular function. We know that w' is stationary, hence reads as $w'(x, \omega) = W_{per}\left(\omega_1 + x, \omega_2 + \frac{x}{\sqrt{2}}\right)$, for some periodic function W_{per} belonging to $L^2(\Omega)$. While we have seen above that the function $x \mapsto w'(x, \omega)$ is (almost surely) regular in the variable x, this does not imply that the function $(x, y) \mapsto W_{per}\left(\omega_1 + x, \omega_2 + \frac{y}{\sqrt{2}}\right)$ is (almost surely) regular in the *two-dimensional* variable $(x, y) \in \mathbb{R}^2$. As a consequence, taking the trace $x = y$ is not allowed *a priori*, and we cannot conclude on the quasi-periodic character of w'... Our only option is to "cheat" and use the specificity of the one-dimensional setting: we may explicitly solve the equation, that is, write $a(x, \omega)(1 + w'(x, \omega)) = c(\omega)$ for some random variable $c(\omega)$. We then prove that this random variable is trivial (we have already used such an argument in Chap. 2, page 84) due to stationarity and ergodicity. We thus have an explicit expression of w' as a function of $a(x, \omega)$. This formula clearly implies that w' is quasi-periodic and solves the equation *"surely"*!

In contrast, in the case of dimensions $d \geq 2$, we cannot argue as we just have. No explicit formula is known for the solution of the corrector equation and we can deduce from the stationary setting neither the corrector equation "surely", nor the fact that the gradient of the solution is quasi-periodic. This however does not prevent to develop a homogenization theory, provided, as we mentioned above, we make precise what we mean by "homogenization theory". These considerations *a posteriori* validate the relevance of Sect. 4.2.2.1.

Application to the Almost Periodic Setting

We finally turn to the almost periodic setting.

Here again, we know since Chap. 1 that the almost periodic setting can be identified as a special case or the continuous stationary setting. The probability space is then the Bohr compactification $\mathbb{G}$ and the probability measure is the associated Haar measure. The group action is the "translation" $\tau_x \omega = x + \omega$, where the addition law $+$ is the extension of the usual addition on $\mathbb{R}^d$ to the Bohr compactification.

The question we examine is, as above, the possibility to deduce the existence of an almost periodic corrector (or at least a corrector with an almost periodic gradient) using the almost sure existence of a corrector $w_p(x, \omega)$ in this stationary setting. For the same reasons as for the quasi-periodic setting, the strategy will fail.

In order to better understand the issue, though, let us be more precise about the identification of the almost periodic setting as a particular case of the stationary setting.

The first option is to consider *continuous* almost periodic functions, namely almost periodic functions in the sense of Bohr (see Definition 1.4), that is, as we saw in Chap. 1, the closure for the uniform norm of the set of trigonometric polynomials over $\mathbb{R}^d$ (see Theorem 1.2). The theory recalled in Chap. 1 (see in particular Theorem 1.3) allows to identify this space with the set $C(\mathbb{G})$ of continuous functions on the Bohr compactification $\mathbb{G}$. The continuous almost periodic coefficient $a_{p-per}(x)$ may thus be considered as a stationary coefficient $a(x, \omega)$. We know from the theory of the stationary setting that the corrector obtained almost surely solves the corrector equation. We then assume that the coefficient $a_{p-per}(x)$ is not only continuous, but Hölder continuous $C^{0,\alpha}$, for some $\alpha > 0$. Using elliptic regularity estimates (in line with what we have mentioned in the previous paragraph), we infer that, loosely speaking (but all this may be proved rigorously), for any ω such that the equation holds, $\nabla w_p(x, \omega)$ is a continuous function of the variable $x \in \mathbb{R}^d$. Put differently, the abstract function of $L^2(\mathbb{G})$

$$x \longmapsto \nabla W_p(x + \omega),$$

which is naturally associated with $\nabla w_p(x, \omega)$, is, almost surely in ω, continuous in x. However, this *does not imply* that this function ∇W_p is continuous on $\mathbb{G}$. Indeed, the real-valued directions x are not *all the directions* in $\mathbb{G}$, since we know, from Chap. 1, that $\mathbb{R}^d$ is of zero Haar measure in the Bohr compactification (see page 48). Consequently, ∇W_p does not belong to $C(\mathbb{G})$, and we cannot identify it with an almost periodic function.

An alternate option is to consider almost periodic functions that are not necessarily continuous, that is, more precisely, to work in the space of almost periodic functions in the sense of Besicovitch (see Definition 1.6). This space is the closure, for the semi-norm $\langle |f|^p \rangle^{1/p}$, $1 \leq p < +\infty$, of the set of trigonometric polynomials. We know that this space may be identified with the quotient space $L^p(\mathbb{G})$ where the null element is the subset of functions the L^p norm of which vanishes. In this quotient space, the functions f and $f + \varphi$ where φ is, for instance, a compactly supported function, belong to the same equivalence class, hence are considered identical. Applying the existence result of the stationary theory yields the existence of a suitable *equivalence class*, that is, the existence of a solution the gradient of which is stationary, and which is defined up to the addition of a function with zero average norm $\langle | \cdot |^p \rangle^{1/p}$. It also yields a corrector equation holding almost surely in ω and in the sense of distributions in x. In summary, we have, in the equivalence class solution, (a) an element that is an almost periodic (in the sense of Besicovitch) gradient up to the addition of a "negligible" function, and (b) an element that is, almost surely, the gradient of a solution to the corrector equation, here again up to the addition of a negligible function. But we do not necessarily find, almost surely, a function the gradient of which is almost periodic that solves the corrector equation. Let alone the fact that, should such a function exist, it would not necessarily solve the equation better than almost surely.

Of course, in both of the above attempts, we can consider the particular case of dimension 1. We then have, independently, an explicit expression for the solution of the corrector equation and know that its derivative is almost periodic. But in neither of the above arguments, we were able to recover this solution using the stationary theory.

4.3.3 Convergence to the Homogenized Problem

We return to the general stationary setting of Sect. 4.3.1.

Now that we have a corrector with stationary gradient, we are in position to apply the same strategy as in Chap. 3, when we identified an explicit expression for the homogenized coefficient in the periodic setting.

We consider the case of discrete stationarity, bearing in mind that all the arguments may be easily adapted to the notion of continuous stationarity.

First, we apply step by step the arguments of the beginning of Sect. 3.1.

We introduce the variational formulation of Eq. (4.116). For a right-hand side $f \in L^2(\mathcal{D})$, (actually, $f \in H^{-1}(\mathcal{D})$ would be sufficient) it writes: *find* $u^\varepsilon \in L^2\left(\Omega, H_0^1(\mathcal{D})\right)$ *such that, for all function* $v \in L^2\left(\Omega, H_0^1(\mathcal{D})\right)$,

$$\mathbb{E} \int_{\mathcal{D}} a_\varepsilon(x, \omega) \nabla u^\varepsilon(x, \omega) . \nabla v(x, \omega) \, dx = \mathbb{E} \langle f, v(., \omega) \rangle_{H^{-1}(\mathcal{D}), H_0^1(\mathcal{D})}.$$

$$(4.130)$$

This is equivalent, in view of the functional spaces we use for u^ε and v, to find almost surely (that is, $\mathbb{P}$-almost everywhere in $\omega \in \Omega$), a variational formulation of the type (3.2) in the space variable x. The formulation (4.130) is formally obtained from (4.116) by multiplying by $v(x, \omega)$, integrating over $\mathcal{D}$, integrating by parts in x, and taking the expectation value. Conversely, using the test function $v(x, \omega) = \theta(x)\phi(\omega)$ where $\theta \in C^\infty(\mathbb{R}^d)$ has compact support and $\phi \in L^2(\Omega)$, and using the fact that the vector space spanned by such functions is dense, we recover (3.2).

Since the coefficient $a(x, \omega)$ is assumed to almost surely satisfy condition (3.1), we may use the Lax-Milgram Lemma to prove existence and uniqueness of a solution u^ε. Then, proceeding as in Sect. 3.1, we obtain

$$\mathbb{E}\int_{\mathcal{D}}\left|\nabla u^\varepsilon(x)\right|^2\,dx \leq C\,\mathbb{E}\left[\|f\|_{H^{-1}(\mathcal{D})}\,\|u^\varepsilon\|_{H_0^1(\mathcal{D})}\right],$$

$$\leq C\,\|f\|_{H^{-1}(\mathcal{D})}\left(\mathbb{E}\left[\|u^\varepsilon\|^2_{H_0^1(\mathcal{D})}\right]\right)^{1/2},$$

where the constant C does not depend on ε, and where we have used the Cauchy-Schwarz inequality in $L^2(\Omega)$. Applying the Poincaré inequality, we deduce that

$$u^\varepsilon \quad \text{is bounded in } L^2\left(\Omega, H_0^1\left(\mathcal{D}\right)\right) \quad \text{uniformly in } \varepsilon. \tag{4.131}$$

Up to an extraction that we omit, the sequence u^ε converges weakly in $L^2\left(\Omega, H_0^1\left(\mathcal{D}\right)\right)$ to some function u^* (this limit is *a priori* random!). The sequence $a_\varepsilon \nabla u^\varepsilon$ similarly converges weakly in $L^2\left(\Omega, L^2\left(\mathcal{D}\right)^d\right)$, to some vector-valued (random) function r^*.

We now reproduce the argument of the periodic setting of Sect. 3.4.2, or equivalently, of the periodic case with defect in L^q of Sect. 4.1.1.2.

Equation (4.116) being linear, we may pass to the weak limit. Doing so, we obtain, as in (3.80), that $-\operatorname{div} r^* = f$. On the other hand, for all $p \in \mathbb{R}^d$, the properties of the corrector w_p and the ergodic theorem (Theorem 1.5 in dimension 1, the generalization to higher dimensions being straightforward) imply the convergence (3.82), that is,

$$p + \nabla w_p\left(\frac{x}{\varepsilon}, \omega\right) \longrightarrow p.$$

in $L^2\left(\Omega, L^2(\mathcal{D})\right)$, while, in the same space,

$$a\left(\frac{x}{\varepsilon}, \omega\right)\left(p + \nabla w_p\left(\frac{x}{\varepsilon}, \omega\right)\right) \longrightarrow \mathbb{E}\int_Q a\left(p + \nabla w_p\right).$$

As in (3.83), we define the constant matrix $(a^*)^T$ by

$$a\left(\frac{x}{\varepsilon}, \omega\right)\left(p + \nabla w_p\left(\frac{x}{\varepsilon}, \omega\right)\right) \longrightarrow \mathbb{E}\int_Q a\left(p + \nabla w_p\right) = (a^*)^T\,p, \tag{4.132}$$

and we consider the scalar quantity

$$\left[a\left(\frac{x}{\varepsilon}, \omega\right)\left(p + \nabla w_p\left(\frac{x}{\varepsilon}, \omega\right)\right)\right] \cdot \nabla u^\varepsilon = \left[a\left(\frac{x}{\varepsilon}, \omega\right)\nabla u^\varepsilon\right] \cdot \left(p + \nabla w_p\left(\frac{x}{\varepsilon}, \omega\right)\right),$$

analogue of (3.84). We write this quantity under the form $\mathbf{f}^\varepsilon \cdot \mathbf{g}^\varepsilon = \mathbf{r}^\varepsilon \cdot \mathbf{s}^\varepsilon$, and apply the argument that, in (3.85), has allowed us to pass to the limit in each product, based on the div-curl Lemma, namely Lemma 3.2. We obtain $\left[(a^*)^T\,p\right] \cdot \nabla u^* = r^* \cdot p$, that is,

$$[a^* \nabla u^*] \cdot p = r^* \cdot p.$$

Since this holds true for any $p \in \mathbb{R}^d$, we conclude that $r^* = a^* \nabla u^*$, hence, passing to the limit in Eq. (4.115), that $-\text{div}(a^* \nabla u^*) = f$ for the homogenized matrix a^* defined by (4.132).

We have thus proved the following result

Proposition 4.4 (Homogenization in the Discrete Stationary Case) *The solution $u^\varepsilon(x, \omega)$ of problem (4.115) converges weakly in the space $L^2 (\Omega, H_0^1 (\mathcal{D}))$, as $\varepsilon \to 0$, to $u^*(x)$, a deterministic function, solution to*

$$\begin{cases} -\text{div}(a^* \nabla u^*(x)) = f(x) \text{ in } \mathcal{D} \subset \mathbb{R}^d, \\ \\ u^* = 0 \qquad\qquad\qquad \text{on } \partial\mathcal{D}, \end{cases} \qquad (4.133)$$

where the deterministic matrix a^* is defined by

$$[a^*]_{ij} = \mathbb{E} \int_Q a \left(e_j + \nabla w_{e_j}\right) e_i, \qquad (4.134)$$

where w_{e_j} is solution, for $p = e_j$, of the stationary corrector equation (4.116), the existence of which is given by Proposition 4.3. We also have convergence of the fluxes:

$$a \left(\frac{x}{\varepsilon}, \omega\right) \nabla u^\varepsilon(x, \omega) \longrightarrow a^* \nabla u^* \qquad (4.135)$$

in $L^2(\Omega \times \mathcal{D})$.

The reader could consider arguing as follows. When "ω is fixed" (which is an informal way to say that what follows is true *almost surely*), problem (4.115) is a "general" homogenization problem. Given the bounds assumed on the coefficient a, such a problem may be addressed using the general homogenization theory of Sect. 3.1.1, namely using Proposition 3.1. The argument does provide existence of a homogenized limit, but no information on the homogenized matrix. The latter matrix is a priori a *random* matrix $a^*(\omega)$, that depends on the extraction, since the subsequence the existence of which is established by Proposition 3.1 depends on ω. The proof provided above for Proposition 4.4 is much more informative. Considering the *family* of problems (4.115), as opposed to separately considering each problem (4.115) for each $\omega \in \Omega$, we obtain that the homogenized coefficient is *deterministic*. It is the transversal nature of the stationary ergodic property that intertwines the problems for different values of ω (think of the operator τ) and makes the limit deterministic.

In order to recover the deterministic character of the homogenized limit using the above argument "ω by ω", an additional argument is needed. One such argument uses oscillatory test functions (as indicated in Sect. 3.4.2). Doing so, one proves the following *almost sure* homogenization result (as opposed to the weak convergence result stated in Proposition 4.4)

Proposition 4.5 *Under the assumptions of Proposition 4.4, and with the same notation, we have the following almost sure convergences:*

$$u^\varepsilon \rightharpoonup u^* \text{ in } H^1(\mathcal{D}), \tag{4.136}$$

$$u^\varepsilon \longrightarrow u^* \text{ in } L^2(\mathcal{D}), \tag{4.137}$$

and

$$a\left(\frac{\cdot}{\varepsilon}\right)\nabla u^\varepsilon \rightharpoonup a^*\nabla u^* \text{ in } L^2(\mathcal{D}). \tag{4.138}$$

Remark 4.2 Propositions 4.4 and 4.5 carry over to the *continuous* stationary setting. The only modification is that the homogenized matrix a^* becomes $\left[a^*\right]_{ij} = \mathbb{E}\left(a\left(e_j + \nabla w_{e_j}\right)e_i\right)$, while Proposition 4.3 is replaced by Remark 4.1. □

It should be emphasized that the above results do not claim any strong convergence, even in the space $L^2(\Omega \times \mathcal{D})$. Proving such a result requires some additional work. We indeed do not have compactness in the variable ω, contrary to the variable x, in which the Rellich Theorem gives compactness. Strong convergence results in $L^2(\Omega \times \mathcal{D})$ have been proved for instance in [PV81, Theorem 3] (for an elliptic operator with a zero-order term, of the form $L_\varepsilon u = -\operatorname{div}(a_\varepsilon \nabla u) + \alpha u, \alpha > 0$) and in [Koz80, Proposition 1].

The question of strong convergence in $L^2(\Omega, H_0^1(\mathcal{D}))$ also arises. It holds when using a correction, as in the periodic setting (see formula (3.55)). Here again, such results are proved in [PV81, Theorem 3] (still with a regularized operator), and [Koz80, Proposition 1].

For recent directions of research and results in homogenization theory for linear elliptic equations in the stationary setting (and the related settings), we also refer to the advanced textbook [AKM19] and, as examples of research articles, [AGK16, GO17]. The connections with a Calderón-Zygmund type theory developed for random operators, analogous to that we discussed for the periodic and perturbed settings in the earlier sections of the present chapter, are, in particular, the topic of [AD16, GNO20]. The methods used in all those works are substantially different from the techniques we use here. Developing further such methods would go beyond the scope of the present introductory textbook.

Chapter 5
Numerical Approaches

Abstract The present chapter is an introduction to computational approaches for multiscale problems such as (1). We successively discuss herein three different categories of methods. The first method is a direct application of homogenization theory. The second method, multiscale approaches, uses homogenization theory as a guideline to design an efficient computational approach. The third method is designed as a way to address random settings that are close to nice deterministic (say, periodic) settings.

Keywords Finite element method · Multiscale method · MsFEM · HMM · Variance reduction · Stochastic Homogenization

The present chapter is an introduction to computational approaches for multiscale problems such as (1). We successively discuss herein three different categories of methods.

In Sect. 5.1, we put in action the approach developed theoretically in Chaps. 1 through 4. We consider (1), assuming that the coefficient a_ε is such that this problem admits a homogenized limit, of the form (3.8), that is amenable to numerical computations. We have to approximate numerically the corrector functions w_p associated to the problem, next the homogenized coefficient a^* appearing in Eq. (3.8) and finally construct a two-scale approximation $u^{\varepsilon,1}$ of the exact solution u^ε to (1). We complete this agenda in Sects. 5.1.2 through 5.1.5, after recalling some elementary facts and tools in Sect. 5.1.1. We in particular proceed with the setting of periodic problems in Sect. 5.1.3, periodic problems perturbed by defects in Sect. 5.1.4 and stochastic problems in Sect. 5.1.5.

We present in Sect. 5.2 a somewhat different, both more modern and more general strategy, that of *multiscale numerical approaches*. Homogenization theory is then only a guideline and not any more a strict agenda. It provides the intuition for the introduction of novel computational approaches. The mathematical justification for the approaches, possibly based on arguments from homogenization theory, is not a prerequisite for using the approaches in practice. Put differently, the implementation does not require the analytical developments of the theory. The numerical analysis of the approach is, however, welcome, with a view to first certifying the approach using an error estimation, and second suitably calibrating the various discretization at play in the implementation. We proceed with one prototypical example of such a multiscale approach, the Multiscale Finite Element Method, abbreviated in $MsFEM$. We present its one-dimensional version in Sect. 5.2.1, and next its two-dimensional version in Sect. 5.2.2. The latter presentation applies *mutatis mutandis* to all dimensions higher than or equal to 2. We then briefly discuss in Sect. 5.2.3 the most famous alternative to $MsFEM$ approaches, namely the HMM approach (Heterogeneous Multiscale Method). To conclude Sect. 5.2, Sect. 5.2.4 presents a variant, based on a different standpoint, just as yet another example of the variety of possible approaches and Sect. 5.2.5 discusses some general questions, in particular related to practical issues.

The third and last section of the chapter, Sect. 5.3, is motivated by the following claim, which will be substantiated in Sect. 5.1.5: the *random* multiscale setting is extremely expensive computationally. In the modelling phase, striking a balance between a periodic setting that is easy to approach numerically but too idealistic and remote from reality, and the random setting that does embrace many practical cases but is often prohibitively expensive computationally is a practical challenge with high risk and high profit. We present, in Sects. 5.3.1 and 5.3.2 respectively, two examples of possible modelling and computational strategies, in a framework that we call *"weakly" stochastic*, to encode the fact that this setting is actually a small perturbation, in a sense to be made precise, of an otherwise periodic setting.

Let us conclude this introduction by emphasizing that a comprehensive study of computational approaches for multiscale problems would require an entire monograph, not to say *several volumes*. First of all, we would need to extensively cover the material needed to understand in depth the existing numerical approaches for classical (as opposed to multiscale) problems. For this matter, we refer the reader to the reference books [Qua18, QV08, BS08, EG04], or one of their many possible alternatives. If we may express a preference, though, we strongly recommend the former one of the list, namely [Qua18]. To our readers who also can read French, we would like to recommend the three books [Ber01, Jol90, RBD98], for a somewhat unusual, and extremely didactic, presentation of the finite elements methods we will use in the sequel, along with some detailed implementation aspects. We also wish to observe that, in spite of the existence of hundreds of research articles discussing multiscale computational approaches, we are only aware of one textbook, [EH09], that specifically studies (a family of) such approaches.

5.1 The Classical Approach put in Action

Throughout this section and as announced above, we consider the classical homogenization approach and we put it in action computationally. Starting from the multiscale problem (1)–(2), we assume the necessary conditions on the coefficient a so that a homogenized coefficient a^* exists *and* is amenable to numerical computations. Put differently, we assume that we have at hand a "formula" that yields a^*, and not only that a^* is defined by the limit (3.60) in Proposition 3.5, possibly up to an extraction.

We have to first approximate the corrector function, next compute the homogenized coefficient which will enable us to solve the homogenized equation, and eventually construct the first two terms of the two-scale expansion. This program is that announced in the items **(i)–(ii)–(iii)–(iv)** of page 124 in Chap. 3.

The prototypical situation we consider is the periodic setting, where the coefficient a^* reads as (3.54) in function of the periodic corrector function solution to (3.67) and is next inserted in the homogenized equation (3.8). We devote Sect. 5.1.3 to this program. The reader, however, is already well aware that the periodic setting is not the only setting to put the strategy in action. We will see two other settings in Sects. 5.1.4 and 5.1.5. The former section considers perturbations of the periodic settings, while the latter considers the stochastic setting and, as we will see, is far more demanding in terms of computational efforts. To begin with, we recall in Sect. 5.1.1 some basic elements on discretization techniques. We readily apply what we present in Sect. 5.1.1 to the specific case of the homogenized equation in Sect. 5.1.2.

5.1.1 Solution Procedure for an Elliptic Boundary-Value Problem Posed on a Bounded Domain

For the reader's convenience, we briefly recall in this section some elementary facts about the discretization, using the finite element method, of an elliptic boundary-value problem posed on a bounded domain. The reader familiar with the approach may skip the section and directly proceed to Sect. 5.1.2 where we apply it to the homogenized problem.

More specifically, we assume that the ambient dimension is $d \geq 2$. The specific case of dimension $d = 1$ has been discussed in Sect. 2.1.2 in Chap. 2. We choose as model problem the following problem

$$\begin{cases} -\operatorname{div}(a(x)\,\nabla u(x)) = f(x) & \text{in } \mathcal{D}, \\ \qquad\quad u(x) \qquad\quad = 0 & \text{on } \partial\mathcal{D}, \end{cases} \tag{5.1}$$

where $f \in L^2(\mathcal{D})$. We notice that, when $a = a_\varepsilon$, our original problem (1)–(2) takes this form. Similarly, when $a = a^*$, the problem coincides with the homogenized problem (3.8). On the other hand, the corrector problem (3.67) may also be put in the form (5.1) *provided* we slightly modify it in order to account for the *periodic* boundary conditions. We will discuss this later.

We now describe the main steps of the discretization of (5.1): variational formulation, construction of a regular and quasi-uniform mesh of the domain, introduction of a finite element approximation space, derivation of the algebraic system, solution procedure for the system, numerical analysis that certifies the approach and provides the suitable error estimation. Our brief exposition must not be mistaken with a genuine numerical analysis course on the topic. We refer the reader to the textbooks mentioned in the introductory section of this chapter and more specifically here to [Qua18, Chapter 4], [QV08, Part I], [BS08, Chapters 3 & 5].

The variational formulation of (5.1) is the two-dimensional variant of the one-dimensional formulation (2.10), that is: *Find* $u \in H_0^1(\mathcal{D})$ *such that, for all functions* $v \in H_0^1(\mathcal{D})$,

$$\int_{\mathcal{D}} a(x)\, \nabla u(x) \cdot \nabla v(x)\, dx = \int_{\mathcal{D}} f(x)\, v(x)\, dx. \tag{5.2}$$

In the one-dimensional setting, the interval $[0, 1]$ has been divided into N subintervals $\left[\dfrac{i}{N}, \dfrac{i+1}{N}\right]$, $i = 0, \ldots, N-1$, each of length $H = \dfrac{1}{N}$. From dimension $d = 2$ up, we similarly need to introduce a *mesh*. The simplest case, which is the only one we are going to consider, is a mesh consisting of triangles over the two-dimensional domain $\mathcal{D}$. Also for simplicity, we assume that $\mathcal{D}$ is indeed a polygonal domain, such that the triangular mesh we construct entirely and exactly covers $\mathcal{D}$. The exterior edges of the triangles contiguous to the boundary $\partial \mathcal{D}$ therefore exactly coincide with $\partial \mathcal{D}$. In all other cases, some technicalities appear: meshing in dimension $d = 3$ is a highly complex process, quadrangular two-dimensional mesh elements can cause subtleties, and when the domain itself is not a polygon, the quality of approximation of its boundary needs to be assessed. We carefully stay away from all these types of technicalities here.

Similarly, we disregard all the *practical* difficulties related to the construction of a "good" finite element mesh. This question keeps many researchers busy for their entire professional lifetime, occupies dozens of bookshelves in libraries, and is addressed using millions of lines of software. For the issues we focus our attention on, all we need to know is that a suitable mesh has been constructed following the rules of the art, using the best software available. In dimension $d = 2$ and for polygonal domains, this is a very decent assumption. Anyone can have access to a freeware that can provide this service.

We now denote by $\mathcal{T}_H$ the union of triangles τ_k the mesh consists of. We assume that the mesh has all the suitable properties in terms of *regularity* so that the classical numerical analysis results for the convergence of finite element methods on that particular mesh hold true. In short, and somewhat vaguely stated, we assume that the triangles are all of similar a shape, so that their edges are essentially of size H. We refer the reader to the bibliography, say e.g. [Qua18, Chapter 4, Section 4.5], where a detailed formalization of this may be found.

For brevity, we only mention here the following two notions. First, the mesh is said *regular* when the following property is satisfied, uniformly in H: for all the elements (that is, for us, all the triangles) τ_k of the mesh,

$$\frac{\text{diam}\,(\tau_k)}{\rho(\tau_k)} \leq C,$$

for a constant C that is independent of k and H, and where diam (τ_k) denotes the diameter of the element τ_k and $\rho(\tau_k)$ the radius of the circle inscribed in that element. This property guarantees that the *interpolation* operators on the nodes of the mesh have suitable approximation properties. Second, the mesh is said to be *quasi-uniform* when, in addition to the above regularity property, it holds

$$\rho(\tau_k) \geq c\,H,$$

for all k, again for a constant c that is independent of k and H. In a regular mesh, the elements cannot be arbitrarily deformed, neither too squeezed not too dilated. They can however be arbitrarily small. If in addition the mesh is quasi-uniform, then they can no longer be arbitrarily small. All elements have the typical size of the mesh.

   We next define the space V_H as the finite-dimensional space spanned by the $N-1$ continuous, piecewise affine functions, vanishing on the boundary of the domain, that are defined as follows

$$\chi_i(x) = \begin{cases} 1 & \text{at the node } i \text{ of the mesh} \\ \text{affine on each of the triangles sharing this node as a vertex} & \\ 0 & \text{on all the other triangles} \end{cases} \qquad (5.3)$$

for $i = 1, \ldots, N-1$ (here we denote by $N-1$ the number of *internal* nodes, that is those nodes that do not lie on the boundary), or, equivalently,

$$\chi_i(x) = \begin{cases} 1 & \text{at the node } i \text{ of the mesh} \\ \text{affine on each triangle} & \\ 0 & \text{at all other nodes of the mesh.} \end{cases}$$

These functions, indexed by the vertices of the triangles, are the multi-dimensional version of the one-dimensional "hat" functions defined in (2.11). They are termed as the $\mathbb{P}_1$ *finite elements* associated to the triangular mesh $\mathcal{T}_H$. They are also called the *nodal basis* of the finite element method considered. It is to be noted that N is here *of the order of* H^{-2}, since the two-dimensional domain $\mathcal{D}$ has been subdivided into triangles of size H. The specific, precise relation relating N to H depends upon the details of the mesh. For the sequel, and for the practice, only the order of magnitude of N with respect to H matters. Figure 5.1 summarizes pictorially the above construction.

Fig. 5.1 A $\mathbb{P}_1$ finite element on a two-dimensional triangular mesh: the function has value 1 at the (red) node considered and 0 at all other (blue) nodes. It is affine on all the triangles that share this node as a vertex

We are now in position to state the discrete variational formulation associated to (5.2): *Find $u_H \in V_H$ such that, for any function $v_H \in V_H$,*

$$\int_{\mathcal{D}} a(x) \nabla u_H(x) . \nabla v_H(x) \, dx = \int_{\mathcal{D}} f(x) v_H(x) \, dx. \qquad (5.4)$$

This formulation is of course analogous to (2.12).

The space V_H constructed above and next used in (5.4) is a finite-dimensional subspace of the infinite dimensional functional space $H_0^1(\mathcal{D})$ appearing in the variational formulation (5.2). One speaks of an *internal approximation*. The general approach consisting in approximating the variational formulation by an analogous formulation in finite dimension is called a *Galerkin approximation*. The terminology often assumes (although in some peculiar cases this can be misleading, as we will see later on) that the space V_H of the *discrete* variational formulation is a finite-dimensional *subspace* of the infinite dimensional space considered in the variational formulation.

As in Sect. 2.1.2, we notice that a typical function in V_H reads as $v_H(x) = \sum_{j=1}^{N-1} (v_H)_j \chi_j(x)$ for arbitrary scalar coefficients $(v_H)_j$ and that the solution is searched for under the form $u_H(x) = \sum_{j=1}^{N-1} (u_H)_j \chi_j(x)$. Formulation (5.4) therefore equivalently reads as

$$\sum_{j=1}^{N-1} \left[\int_{\mathcal{D}} a(x) \nabla \chi_j(x) . \nabla \chi_i(x) \, dx \right] (u_H)_j = \int_{\mathcal{D}} f(x) \chi_i(x) \, dx, \qquad (5.5)$$

for all $i = 1, \ldots, N-1$. This expression (5.5) in turn reads as the linear algebraic system

$$[A] [u_H] = [f_H]. \qquad (5.6)$$

This algebraic equation exactly agrees with the algebraic formulation obtained in (2.14) in the one-dimensional setting. The matrix $[A]$ and the column vectors $[u_H]$ et $[f_H]$ are respectively defined for $1 \le i, j \le N$, by

$$[A]_{ij} = \int_{\mathcal{D}} a(x) \nabla \chi_j(x) . \nabla \chi_i(x) \, dx,$$

$$[u_H]_j = (u_H)_j, \quad [f_H]_j = \int_{\mathcal{D}} f(x) \chi_j(x) \, dx. \qquad (5.7)$$

The reader may wish to notice, in passing, that we do not bring up the issue of the *numbering* of the nodes of the mesh. The issue is of considerable practical importance, though. It indeed affects the structure of the matrix $[A]$, and hence the methods to employ in order to solve the algebraic system (5.6). On the other hand, this issue does not affect at all the questions we examine in this section. Likewise, and again following our policy to focus our attention elsewhere, we do not discuss in details the actual computation of the entries of the matrix $[A]$ and the column vector $[f_H]$:

(i) for the matrix $[A]$, called the *stiffness* or the *rigidity* matrix, the techniques at play for computing the entries are *assembling techniques*; they adequately browse the nodes of the mesh, progressively storing the contribution of each finite element considered to the entries of the matrix $[A]$,

(ii) for the right-hand side (and also more generally for the computation of each entry of the matrix $[A]$), numerical *quadrature formulae* are at play. In the particular case when the function f is a linear combination of the finite elements χ_k, the integrals on the right-hand side may be computed analytically, in terms of the elementary integrals $\int_{\mathcal{D}} \chi_i(x)\,\chi_j(x)\,dx$. In the general case, however, the integrals involving the function f must be approximated numerically by quadrature formulae (which would be exact if f were a certain polynomial function of fixed order but are only approximations otherwise).

The procedure to solve the algebraic equation (5.6) is formally identical to that for (2.14). The only practical difficulty is the *size* of the algebraic system itself. For an identical meshsize H (that is, for an identical expected accuracy), the number N of degrees of freedom (number of scalar unknowns quantities) is of the order of H^{-2} and not any longer H^{-1}. As H decays, the numerical approaches to solve the algebraic system (5.6) have thus to be carefully chosen. Direct methods to invert the matrix itself, such as the Gauss elimination or the LU factorization, are often ineffective because they require too large a memory size. They are outperformed by iterative methods, such as the conjugate gradient method, *GMRES*, or more generally the *Lanczos algorithm*. For a comprehensive description and analysis of all such approaches, we refer to [Saa03] among a huge literature on the topic.

The numerical analysis of the finite element method in dimension $d \geq 2$ follows the same pattern as that in dimension $d = 1$ which we have recalled in Sect. 2.1.2. The key ingredient is Céa's Lemma, which bounds the numerical error by the approximation error and which we have already seen in its one-dimensional version (2.21) in Chap. 2:

$$\|u - u_H\|_{H^1(\mathcal{D})} \leq C \inf_{v_H \in V_H} \|u - v_H\|_{H^1(\mathcal{D})}, \qquad (5.8)$$

As in Chap. 2, the right-hand side is estimated using the approximation properties of the space V_H. For the specific finite element space $\mathbb{P}_1$ we have chosen here, the following generalization of (2.22) holds, using the properties assumed on the mesh:

$$\inf_{v_H \in V_H} \|v - v_H\|_{H^1(\mathcal{D})} \leq C\,H\,\|v\|_{H^2(\mathcal{D})}, \qquad (5.9)$$

for any arbitrary function $v \in H^2(\mathcal{D})$ and a constant C independent of both v and H. For a sufficiently regular coefficient a (say, e.g., $a \in W^{1,\infty}(\mathcal{D})$), by virtue of the elliptic regularity Theorem A.14, we know that the solution u to (5.1) is H^2. Applying (5.9) to $v = u$, we thus obtain from (5.8)–(5.9),

$$\|u - u_H\|_{H^1(\mathcal{D})} \leq C H \|u\|_{H^2(\mathcal{D})} . \tag{5.10}$$

This estimation is called an *a priori error estimate*. The terminology a priori refers to the fact that the numerical solution u_H needs not be already computed so that we have an estimation of its accuracy with respect to the exact solution as the meshsize H varies. It is important to note that, since the constant C in (5.10) is not explicitly known (it can be in turn estimated numerically, but accurately approximating it is difficult in general and in any event requires an additional possibly expensive computation—see [LK10] for an example of a work in this direction), only the scaling law of the error in function of the meshsize H matters in (5.10). The practically relevant information is that if we half the meshsize H then we may expect that the error is also (at worst) halfed. The reader should not overlook this rough information. It is extremely valuable for the practice.

Estimate (5.10) being called an a priori estimate, it is no surprise that there also exists *a posteriori error estimates*. The latter estimate consists in bounding the numerical error by a quantity that provides additional useful information on the accuracy achieved but also depends upon the actual value of the numerical solution u_H and therefore can only be known *after* the computation (and in fact at the price of an additional computational workload, almost as expensive as the original computation of u_H). An a posteriori estimate also involves unknown constants. Such constants, however, may factor out if *ratios* of accuracy are considered. This explains why a posteriori estimators are typically useful for estimating the *local* accuracy and indicate regions of the computational domain $\mathcal{D}$ where this accuracy must be improved as compared to other regions. This is instrumental, for instance, in *adaptive mesh refinement* techniques. A posteriori error estimation is a world of its own. For what specifically concerns multiscale problems, it is still research in progress.

5.1.2   Application to the Homogenized Problem

Applying the finite element method recalled in the previous section to the homogenized problem (3.8), that is

$$\begin{cases} - \operatorname{div}(a^* \, \nabla u^*(x)) = f(x) & \text{in } \mathcal{D}, \\ \qquad\quad u^*(x) \quad\;\; = 0 & \text{on } \partial\mathcal{D}, \end{cases} \tag{5.11}$$

is straightforward. Provided the coefficient a^* is assumed sufficiently regular, which will be the case throughout this textbook since a^* is most often a constant, and provided the domain $\mathcal{D}$ is also regular, the solution u^* is H^2 Sobolev regular and the accuracy of the $\mathbb{P}_1$ approximation on a regular mesh of size H is given by

$$\left\| u^* - (u^*)_H \right\|_{H^1(\mathcal{D})} \leq C H \left\| u^* \right\|_{H^2(\mathcal{D})}. \tag{5.12}$$

In all fairness, we have to acknowledge that the assumption of regularity of the domain $\mathcal{D}$ is not satisfied in all the cases we address. Indeed, we have assumed in the previous section that the domain is polygonal, which only yields a Lipschitz regularity and not more. This does not allow, in full generality, to apply the classical elliptic regularity results such as Theorem A.14 and indeed obtain (5.12). Such an H^2 regularity result actually turns out to also hold true for *two-dimensional* open polygonal domains: we refer to [Gri11, Theorem 4.3.1.4] for the Laplace operator $-\Delta$ and mention that the result may be generalized without difficulty to any elliptic operator with constant coefficients. It is also possible to obtain such a result for *convex* polygonal domains in any dimension (a way to do this is to follow the proof of [Gri11, Theorem 3.2.1.2], and extract the estimate from these arguments, but going into such a proof would take us too far away from our point.) We are thus going to "sneak" this additional assumption of convexity of the domain $\mathcal{D}$ in our mathematical setting. It is not restrictive practically for the type of problems we consider. The key feature of (5.12) is that the meshsize H necessary to reach a given accuracy on u^* is independent of ε, the characteristic small lengthscale within the original problem (1)–(2). Heuristically, the numerical approach shares this common feature with the theory: the small lengthscale ε has disappeared macroscopically.

The reader has of course to bear in mind that u^* *is not* u^ε. It only looks like it, up to terms depending upon ε. The rates of convergence that we have established in Chaps. 3 and 4 formalize and quantify this. If the ultimate goal is to approach u^ε and not only u^*, estimation (5.12) is not the end of the day.

Going further than (5.12) requires either to construct an explicit numerical approximation using the two-scale expansion of u^ε truncated at a certain order, which is essentially following the "historical" line of thought which we describe throughout this Sect. 5.1, or to switch to a completely different computational strategy, which we will address in Sect. 5.2.

But before we get to that, we have to compute the coefficient a^* of (5.11) explicitly. For all the practical settings we present in this textbook, this means first computing the corrector function...

5.1.3 Application to the Computation of the Periodic Corrector Function $w_{p,per}$, and next of a^*

We now compute the periodic corrector function $w_{e_j,per}$ in order to next evaluate the homogenized coefficient a^* via (3.54) in the periodic setting, that is,

$$a_{ij}^* = \int_Q \left(a_{per}(y)\,(e_j + \nabla_y w_{e_j,per}(y))\right) . e_i \, dy.$$

The periodic corrector function $w_{p,per}$, for $p \in \mathbb{R}^d$ fixed, is, we recall, the periodic solution, unique up to an additive constant, to Eq. (3.67), that is,

$$- \operatorname{div}\left(a_{per}(y)\,(p + \nabla w_{p,per}(y))\right) = 0.$$

We have established the variational formulation (3.71) of such a periodic equation in Sect. 3.4 of Chap. 3. In the particular case $z_{per} = w_{p,per}$, $h_{per}(y) = \operatorname{div}_y(a_{per}(y)\,p)$, the formulation reads as: *Find $w_{p,per} \in H_{per,0}^1$ such that, for all $v_{per} \in H_{per,0}^1$,*

$$\int_Q a_{per}(y)\,\nabla w_{p,per}(y) . \nabla v_{per}(y)\, dy = - \int_Q a_{per}(y)\, p . \nabla v_{per}(y)\, dy,$$

$$(5.13)$$

where the functional space

$$H_{per,0}^1 = \left\{ v_{per} \in H^1(Q);\ v_{per} \text{ periodic},\ \int_Q v_{per}(y)\, dy = 0 \right\}$$

is defined in (3.72). This formulation can essentially be addressed using the approach recalled in Sect. 5.1.1. The only two differences are (a) that the right-hand side is a function of ∇v_{per} and not of v_{per} itself, but since it remains a continuous linear form on $v_{per} \in H^1$, this does not introduce any considerable complications, and more importantly (b) that the functional space $H_{per,0}^1$ is not straightforward to approximate. We now discuss the latter issue.

We first introduce a regular mesh of the periodic cell Q. The meshsize is deliberately denoted by h and not H, in order to emphasize

(i) that h is in nature associated to a microscopic problem, namely the corrector problem (5.13), whereas, in sharp contrast, H is associated to the macroscopic homogenized problem (5.11);
(ii) that the choice of h, different from H, is dictated by specific considerations of accuracy that will be addressed below.

We now address the choice of functional space $V_{per,h}$ for the approximation of problem (5.13). Our goal is evidently to obtain a formulation such as (2.12), which would here look like *Find $w_{p,per,h} \in V_{per,h}$ such that, for all function $v_{per,h} \in V_{per,h}$,*

$$\int_Q a_{per}(y) \, \nabla w_{p,per,h}(y) \cdot \nabla v_{per,h}(y) \, dy$$

$$= -\int_Q a_{per}(y) \, p \cdot \nabla v_{per,h}(y) \, dy, \qquad (5.14)$$

for a well chosen functional space $V_{per,h}$. We expect this space to be a subspace of $H^1_{per,0}$ defined in (3.72).

The starting point to define $V_{per,h}$ is a $\mathbb{P}_1$ finite element space V_h constructed from the mesh of the periodic cell Q. The mesh should comply with the periodicity constraint. Its trace on two opposite sides of Q have to be identical. We may then encode periodicity by imposing that any given function of $V_{per,h}$ takes the same value on two "identical" vertices on two opposite sides of the cell. Put differently, the degrees of freedom on a given side are fused with those on the opposite side. The space so constructed then allows to approximate a periodic function $v_{per} \in H^1(Q)$ in (3.72).

We notice, to illustrate our construction, that, in dimension $d = 1$, it amounts to extend our definition (2.11) of the basis functions χ_i of Chap. 2 to the two particular indices $i = 0$ and $i = N$ upon setting

$$\chi_0(x) = \chi_N(x) = \begin{cases} 1 & \text{in} \quad x = 0 \text{ and } x = 1, \\ 0 & \text{if} \quad x \in \left[\dfrac{1}{N}, 1 - \dfrac{1}{N}\right], \\ \text{affine} & \text{if} \quad x \in \left[0, \dfrac{1}{N}\right], \\ \text{affine} & \text{if} \quad x \in \left[1 - \dfrac{1}{N}, 1\right]. \end{cases}$$

It is straightforward to figure out the similar construction for the square $Q = [0, 1]^2$ in dimension $d = 2$, or the cube $[0, 1]^3$ in dimension $d = 3$. From an algebraic viewpoint, that is within (5.6)–(5.7), the approach amounts to only adding *half* the number of nodes of the boundary to the size of the matrix $[A]$.

On the other hand, the constraint $\int_Q v_{per}(y) \, dy = 0$ in (3.72) is a *global* constraint and is therefore slightly more delicate to implement practically. It affects the *sum* of coefficients, that is, proceeding with our one-dimensional illustration, the sum $\displaystyle\sum_{j=0}^{N} (v_H)_j$. Although ensuring this condition is not an issue in principle, one has to be cautious in practice. We can either add one line and one column to the matrix, all entries of the last line being equal to 1. Or we can eliminate one degree of freedom, expressing it in terms of all the other ones. The latter option amounts to eliminating one line and one column from the matrix $[A]$. Up to irrelevant details, the last line of the new matrix will have entries that are mostly 1. Such algebraic operations, either way, couple all the columns of the matrix together and therefore break down the possibly nice structure of the original finite element matrix. For $\mathbb{P}_1$ finite elements, e.g., such matrices are indeed tridiagonal, or block-tridiagonal. This property makes them particularly amenable to dedicated algorithms for solving the associated linear algebraic system. Breaking the structure restricts the portfolio of possible algorithms and eventually endangers the efficiency of the procedure.

One alternative practical option is to observe that the purpose of the scalar condition $\int_Q v_{per}(y)\,dy = 0$ is only to reinstate uniqueness of the corrector function $w_{p,per}$ and is not a concern for the *existence* of such a function. The practitioner may therefore wish to disregard this condition and only forcefully fix a value of *one* degree of freedom within the computation, for instance the value zero at a specific node along the boundary. In effect, this eliminates the node from the problem. The local modification of the matrix that follows only marginally affects the structure of that matrix and essentially preserves all its properties. An alternative option is to modify the equation solved by adding a penalty term for the condition

$$\int_Q v_{per}(y)\,dy = 0.$$

Actually, one might even proceed one step further in the simplification. Provided an iterative method is employed to solve the linear algebraic system (as opposed to a direct method which would compute the inverse matrix $[A]^{-1}$), one may even solve the system directly on the space

$$V_{per,h} = \{v_{per,h} \in V_h;\ v_{per,h} \text{ periodic}\}.$$

and discard the condition $\int_Q v_{per,h}(y)\,dy = 0$. It then amounts to uniquely identifying $\nabla v_{per,h}$ instead of $v_{per,h}$ itself, which is only defined up to the addition of a constant. Fixing the initial guess for the iterative algorithm implicitly selects such a constant and removes the degeneracy. Since only the gradient $\nabla v_{per,h}$ is in fact relevant, this does not affect the computations that follow.

It is important to note that, in any event, the particular choice of implementation of the condition is irrelevant for the numerical analysis of the approach. It would only matter for the study of the iterative algorithm used to solve the linear system, which is not our concern here. Irrespective of the implementation, we may thus proceed assuming that the approximation (5.14) of problem (5.13) is posed on the space

$$V_{per,h} = \left\{v_{per} \in V_h(Q);\ v_{per} \text{ periodic},\ \int_Q v_{per}(y)\,dy = 0\right\}. \tag{5.15}$$

The corresponding variational formulation is (5.14)–(5.15).

The basic elements of numerical analysis recalled in Sect. 5.1.1 readily apply to the problem at hand, regardless of the fact that our recollection there was focused on a boundary value problem with homogeneous Dirichlet conditions while the problem here is *periodic* and includes a constraint. The analysis is indeed based on the structure of the variational formulation and essentially relies upon both the Lax-Milgram Lemma and Céa's Lemma. The linearity of the problem and the fact that the discretization space is included in the functional space used for the original variational formulation are the only two properties that matter. Both properties are preserved in (5.14)–(5.15).

In practice, the meshsize h will be chosen in order to reach the expected accuracy on the corrector function $w_{p,per,h}$, or, more precisely, on its gradient $\nabla w_{p,per,h}$.

The next step, when the corrector problem is solved for each of the canonical directions $p = e_i$, $i = 1, \ldots, d$ (in an overwhelming proportion of the practical cases, $d = 2$ or 3), consists in a simple integration over the periodic cell to compute the homogenized matrix a^*. In (3.54), that is,

$$a_{ij}^* = \int_Q \left(a_{per}(y)\,(e_j + \nabla w_{e_j,per}(y))\right).e_i\,dy,$$

the function $w_{e_j,per}$ is known in function of its expansion on the finite elements χ_k, where k varies over the nodes of the mesh. The above integral may therefore be in turn expanded into elementary integrals of the type

$$\int_Q a_{per}(y)\,\nabla\chi_k(y).e_i\,dy.$$

The latter integrals have actually already been evaluated when assembling the right-hand side of the discretization of (5.13). When the coefficient a_{per} is known analytically, and when its expression allows for an explicit expression of the integral, no quadrature formula is necessary and the integration is exact. In the periodic case, the determination of the gradients $\nabla w_{e_j,per}$ is thus essentially the only computational workload.

As soon as the entries of the matrix a^* have been computed, the matrix is inserted in the homogenized problem. The latter problem is solved numerically as explained in the previous section. To obtain an H^1 approximation of u^ε, the approximated correctors are again used in the truncated two-scale expansion $u^{\varepsilon,1}$ (following (3.90) of Chap. 3).

To conclude this section on the periodic setting, we now list the various sources of numerical errors appearing in our sequence of computations. What error do we indeed commit if we replace the exact solution u^ε to the original problem (1)–(2) by

$$u^{\varepsilon,1,h,H}(x) = u_H^*(x) + \varepsilon \sum_{i=1}^d \partial_{x_i} u_H^*(x)\,w_{e_i,per,h}\left(\frac{x}{\varepsilon}\right) \qquad (5.16)$$

which is the numerical approximation of its truncated two-scale expansion (3.90), that is

$$u_{per}^{\varepsilon,1}(x) = u^*(x) + \varepsilon \sum_{i=1}^{d} \partial_{x_i} u^*(x)\, w_{e_i,per}(x/\varepsilon)? \tag{5.17}$$

The list of the successive errors committed is as follows:

(E1) we have approximated $w_{e_i,per}$ by $w_{e_i,per,h}$: the approximation error is controlled by the a priori error estimate (5.10) regarding (5.13) *versus* (5.14), that is (provided the coefficient a_{per} is sufficiently regular so that $w_{e_i,per,h}$ is H^2 Sobolev regular)

$$\left\| w_{e_i,per} - w_{e_i,per,h} \right\|_{H^1(Q)} \leq C\,h\, \left\| w_{e_i,per} \right\|_{H^2(Q)}.$$

In principle, we would have to add to this error the error arising from the fact that, when assembling the rigidity matrix of the corrector problem, its entries have been computed using a quadrature formula, as mentioned on page 263, item **(ii)**;

(E2) we have approximated a^*, that is (3.54), using

$$[a_h^*]_{ij} = \int_Q \left(a_{per}(y)\,(\nabla w_{e_j,per,h}(y) + e_j) \right) . e_i \, dy. \tag{5.18}$$

As already noticed, the error only depends on the approximation error (E1), provided a_{per} is known explicitly and its particular expression allows for an analytic integration in (5.18) once $\nabla w_{e_j,per,h}$ has been computed. Otherwise, an additional error owing to the quadrature formulae possibly employed to evaluate (5.18) is to be accounted for. Note that the matrices and vectors assembled when solving the cell problem may be used here to simplify the computation of (5.18). Generally, the error $|a^* - a_h^*|$ is at most of order h. Such an error is essentially irrelevant. In sharp contrast, this error will deserve a special attention when we approach nonperiodic settings later on.

(E3) we have approximated u^* by u_H^*: our discussion in Sect. 5.1.2 suggests that the error is controlled by the estimation (5.12),

$$\left\| u^* - (u^*)_H \right\|_{H^1(\mathcal{D})} \leq C\,H\, \left\| u^* \right\|_{H^2(\mathcal{D})}.$$

We should, however, not forget that (5.12) only estimates the difference between the exact solution to (5.11) and its approximation. Now, since we did not have direct access to a^* we had to approximate it using a_h^*, which we have inserted into the homogenized problem. The additional error affecting the outcome may be evaluated (for instance using the Strang Lemma [Cia78, Theorem 4.1.1, page 186]) by estimating how both the exact solution u^* to (5.11) and the numerical solution vary when the coefficient in the respective equation is modified. We will not proceed with the specific evaluation of this error, but is possible to carefully conduct such an analysis;

(E4) using the previous observations, we may estimate the error between $u^{\varepsilon,1,h,H}$ and $u^{\varepsilon,1}$, in function of h, H, and ε. We finally need to estimate that between $u^{\varepsilon,1}$ and u^{ε}, which only depends upon ε and is independent of the discretization parameters h and H. This error has been analyzed in our theoretical developments of Chap. 3.

Given the above list (E1) through (E4), it is immediate to realize (although we will not show the details of the proof of our claim) that the global error can be estimated using a bound of the type

$$\left(\sum_{\tau_k \in \mathcal{T}} \left\| u^{\varepsilon} - u^{\varepsilon,1,h,H} \right\|_{H^1(\tau_k)}^2 \right)^{1/2} = O\left(\varepsilon + H + h\right), \tag{5.19}$$

where each of the parameters ε, h, H may appear with a certain power, possibly different from 1. For instance, a term in $\sqrt{\varepsilon}$ is expected, given (3.124). In practice, ε is, in general, the smallest one of these three parameters and H (often of the same order of magnitude as h) is the largest one. The reader may also notice that the norm appearing on the left-hand side of (5.19) is a so-called "broken H^1 norm", which resembles $\| \cdot \|_{H^1(\mathcal{D})}$. The consideration of such a broken norm owes to the fact that the functions $\partial_{x_i} u_H^*(x)$ appearing in $u^{\varepsilon,1,h,H}$ within (5.19) are not *globally* H^1 throughout the domain $\mathcal{D}$.

5.1.4 Application to the Nonperiodic Setting

We now revisit in the (deterministic) *nonperiodic* setting the work performed in the periodic setting in the previous section. We focus our attention on the new features of the problem, leaving aside what is a mere repetition of the considerations of the periodic setting. Moreover, we concentrate ourselves on the two following specific cases:

(a) a localized, say L^2, defect embedded in an otherwise periodic structure, that is the setting we have considered from the theoretical perspective in Sect. 4.1.1 of Chap. 4, and

(b) a quasi-periodic structure, as studied theoretically in Sect. 4.2.2.1 of that same chapter.

We will encounter our first serious additional difficulty with respect to the periodic setting upon considering setting (a). In sharp contrast with Sect. 5.1.3, *the corrector problem is an equation posed on the entire ambient space $\mathbb{R}^d$* (for our exposition we take $d = 2$) *and not any longer on a bounded subdomain.* This difficulty will also be present in setting (b), but yet another difficulty will then add up: the homogenized coefficient a^* is not an average over a bounded domain (such as the periodic cell Q) but is to be understood as *the asymptotic limit over large volumes*

$$[a^*]_{ij} = \lim_{R \to +\infty} R^{-d} \int_{[-R/2, R/2]^d} \left(a(y)\left(\nabla w_{e_j}(y) + e_j\right)\right) . e_i \, dy, \qquad (5.20)$$

as a consequence of the weak convergence (3.60). It turns out that, because the local defect vanishes at infinity, this additional difficulty does not arise in setting **(a)** and a^* is then identical to the periodic homogenized coefficient, itself computable "as usual" on the periodic cell itself. We will return to this later.

In both settings **(a)** and **(b)**, we are going to successively describe how to numerically approximate the solution to an equation posed on the entire space. We will next compute averages over large volumes.

5.1.4.1   A Localized, L^2, Defect Embedded in an Otherwise Periodic Structure

We recall that the setting we consider here has been introduced and studied theoretically in Sect. 4.1.1 of Chap. 4. The coefficient there is assumed to be of the form (4.1) for $q = 2$, that is, $a = a_{per} + \tilde{a}$ for some a_{per} that is a periodic, Hölder continuous function of class $C^{0,\alpha}$, for a certain exponent $\alpha > 0$, and $\tilde{a} \in L^2(\mathbb{R}^d) \cap L^\infty(\mathbb{R}^d)$. We have established in Lemma 4.1, that there exists a corrector function w_p, unique up to the addition of a constant and solution to (4.3), that is,

$$- \operatorname{div}\left((a_{per} + \tilde{a})(p + \nabla w_p)\right) = 0, \quad \text{on } \mathbb{R}^d,$$

and that reads as the sum $w_p = w_{p,per} + \tilde{w}_p$ in (4.4). In this sum, $w_{p,per}$ is the "usual" periodic corrector, solution to (3.67) (the numerical approximation of which we have described in Sect. 5.1.3), and $\nabla \tilde{w}_p \in L^2(\mathbb{R}^d)$. We have then mentioned in Sect. 4.1, page 172, that the homogenized coefficient a^* is identical to the periodic homogenized coefficient a^*_{per}.   The latter fact holds true because in the weak convergence (3.60) of the periodic setting, the terms where $\tilde{a}\left(\dfrac{x}{\varepsilon}\right)$ or $\nabla \tilde{w}_p\left(\dfrac{x}{\varepsilon}\right)$ appear all vanish since both functions strongly converge to zero in $L^2(\mathbb{R}^d)$ as $\varepsilon \to 0$. This is also seen on the limit over large volumes (5.20) which we recalled above.

Since all the other computational tasks may be completed as in Sect. 5.1.3, the one and only novelty of the setting is thus the solution procedure for the corrector problem (4.3), posed over the whole space $\mathbb{R}^d$. To this end, it is useful to first rewrite (4.3) under the specific form we have introduced for the proof of Lemma 4.1, that is (4.6)

$$- \operatorname{div}\left((a_{per} + \tilde{a})\nabla \tilde{w}_p\right) = \operatorname{div}\left(\tilde{a}\,(p + \nabla w_{p,per})\right), \quad \text{on } \mathbb{R}^d.$$

This equation being of the general form

$$- \operatorname{div}(a\,\nabla u) = \operatorname{div}\mathbf{f}, \quad \text{on } \mathbb{R}^d, \tag{5.21}$$

(a form that reminds us of Eq. (4.38) and that is genuinely related to the theoretical questions we have considered in Sect. 4.1), we now need to understand how to numerically approximate the solution u to (5.21).

The underlying question is that of the numerical solution of problems posed on unbounded domains. Here, as is the case for many of the computational issues we too briefly consider in this chapter, we only scrape the surface of things. To start with, we observe that if we henceforth restrict ourselves to finite element techniques, we have to *truncate* the computational domain. Considering an infinite mesh is indeed unpractical. It would lead to an algebraic system of infinite size. It is an intuitive idea to restrict Eq. (5.21) to a "sufficiently large" bounded domain, say, to keep things simple, $[-R/2, R/2]^d$ with R large. Put differently, we choose to approximate u by u_R solution to

$$- \operatorname{div}(a\,\nabla u_R) = \operatorname{div}\mathbf{f}, \quad \text{in } [-R/2, R/2]^d. \tag{5.22}$$

The information we sorely miss is *the boundary condition we have to complement* (5.22) *with*. The obvious ideal condition is $u_R = u$ where u is the exact solution to (5.21), in which case u_R would perfectly agree with u throughout $[-R/2, R/2]^d$. This guess is unrealistic, since precisely we are *looking for u*... For such a problem, there actually exist many sophisticated techniques (along the lines of the so-called *transparent boundary conditions*) to circumvent the difficulty. But we will avoid opening this Pandora box. We will adopt here a simple, pragmatic solution. The truncated domain $[-R/2, R/2]^d$ is in practice expected to be large. The differential operator on the left-hand side of (5.22) being a diffusion operator, it is expected that the solution u_R only loosely depends of the actual boundary condition we set. At least this must be true in a large subdomain, say $[-\sqrt{R/2}, \sqrt{R/2}]^d$. This expectation may indeed be formalized mathematically, but we skip this fact. We therefore choose

$$u = 0 \quad \text{on the boundary of } [-R/2, R/2]^d. \tag{5.23}$$

However crude this boundary condition is, it turns out to perform rather well. We could equally well have imposed *periodic* boundary conditions, or homogeneous Neumann boundary conditions, that is, $\nabla u \cdot \mathbf{n} = 0$, thereby capitalizing on the fact that u actually denotes $\widetilde{w}_p$ and that ∇w_p is expected to belong to $L^2\left(\mathbb{R}^d\right)$. All these options are perfectly valid. Ours is only one option among many.

We will not discuss here the theoretical issue that arises from the previous description, that is to estimate the additional numerical error specifically due to the approximation of u by u_R, or, in our original notation, that of $\widetilde{w}_p$ by $\widetilde{w}_{p,R}$.

Suffice it to mention here that, whatever this error is, we may reduce it by the following trick. This "trick" is of course not a trick, literally. It is actually based on a detailed theoretical and numerical analysis and it provides us with a perfect transition toward the new questions that we are about to investigate in the quasi-periodic and next in the stochastic settings. Rather than (5.22), we may consider as approximation to problem (5.21):

$$- \operatorname{div} (a \, \nabla \widetilde{u}_R) + \eta_R \, \widetilde{u}_R = \operatorname{div} \mathbf{f}, \quad \text{in } [-R/2, \, R/2]^d, \qquad (5.24)$$

for a coefficient $\eta_R > 0$ typically related to R by a scaling law of the type $\eta_R \approx \dfrac{1}{R^2}$. The explicit relation between η_R and R may be made precise. We supply (5.24) with the boundary conditions we originally selected for (5.22). The intuition behind the addition of the zero order term "$+ \eta_R \, \widetilde{u}_R$" is the wish to damp the spurious effects owing to the imperfect choice of the boundary condition (5.23). It can be readily realized in dimension $d = 3$, where things are a bit more explicit. We take $a \equiv 1$ in (5.21) for simplicity of exposition. As the reader may know, the Green function associated to the operator $-\Delta + \eta_R$ reads as the so-called *Yukawa potential*

$$G_{\eta_R}(x) \, \propto \, \frac{1}{|x|} \, \exp\left(-\sqrt{\eta_R} \, |x|\right). \qquad (5.25)$$

up to an irrelevant multiplicative factor. As expected, this function converges, in the limit $\eta_R \to 0$, to the Green function of the Laplacian operator in dimension $d = 3$, that is $\dfrac{1}{|x|}$, again up to an irrelevant multiplicative factor. The exponential decay within (5.25) makes the solution to (5.22) even more insensitive to the choice of a poor boundary condition. The error between $\widetilde{u}_R$ solution to (5.24) and u_R solution to (5.22) (or directly u solution to (5.21)) may be estimated. The parameter η_R may then be calibrated in function of the size R of the truncated domain (and possibly some other discretization parameters in the problem) so that the approximation yields the desired accuracy.

Applying the above techniques allows to solve (4.3) numerically. We refer the reader to Fig. 4.1 of Chap. 4 which shows how much a defect affects the quality of the approximation of u^ε.

It is now time to address our second nonperiodic setting, the quasiperiodic setting.

5.1.4.2   The Quasiperiodic Setting

We revisit here the quasiperiodic setting which we have studied theoretically in Sect. 4.2.2.1 of Chap. 4. As in that section, we choose to work with a particular quasiperiodic coefficient that essentially embeds all the difficulty of the general quasiperiodic setting. We thereby spare the reader unnecessary technicalities. In addition, we work in dimension $d = 2$. Our quasiperiodic coefficient is scalar-valued, coercive and bounded in the sense of (3.1), and sufficiently regular. It reads as (4.104), that is

$$a(x, y) = a_{per}(x, x, y, y),$$

where $a_{per} = a_{per}(x_1, x_2, x_3, x_4)$ is a regular function defined on $\mathbb{R}^4$, which is S-periodic in the variables x_1 and x_3, and T-periodic in the variables x_2 and x_4. The ratio T/S is of course assumed irrational, so that the function a is *not* periodic. We have established in Proposition 4.2 that, under these conditions, there exists a corrector function, unique up to the addition of a constant and the gradient of which is a quasiperiodic function, that is solution to Eq. (4.105), that is

$$- \operatorname{div}\left(a\,(p + \nabla w_p)\right) = 0 \quad \text{in } \mathbb{R}^d, \tag{5.26}$$

$$\langle \nabla w_p \rangle = 0. \tag{5.27}$$

The gradient ∇w_p shares the structure (4.104) of the coefficient a.   It has quasiperiods S and T.

The first question we need to address from the numerical standpoint is the approximation of problem (5.26), posed, as in the previous section, on the entire space $\mathbb{R}^d$ (for, here, $d = 2$). We again have to suitably truncate the problem on a supposedly sufficiently large, bounded domain such as $[-R/2, R/2]^2$ with R large. The boundary condition to be set on the boundary of that domain is again not straightforward to guess. For lack of a better option, the condition can be a periodic condition, or a Dirichlet homogenous boundary condition. If the ratio T/S *were* rational (which it is not) and if R were a common integer multiple to both T and S, then the condition of periodicity would be correct. On the other hand, one may also add a zero-order term as in (5.24). We will make this choice below.

As compared to the previous section, the significant new difficulty is the following. *Even if* we were able to *exactly* solve the corrector equation (5.26), which we are not, and therefore exactly determine w_p, (or, more practically, even when we are able to efficiently and accurately approximate the solution to (5.26)), the next question to be addressed is the computation of the average (5.20), that is,

$$\left[a^*\right]_{ij} = \lim_{R \to +\infty} R^{-d} \int_{[-R/2, R/2]^d} \left(a(y)\,(\nabla w_{e_j}(y) + e_j)\right) . e_i\, dy.$$

This average is intrinsically *not* equal to an average over any bounded domain, in contrast to the two examples considered above of the periodic setting and the localized defect. We *must* consider the limit of large volumes $R \to +\infty$. To better understand how to best proceed, we have to temporarily return to dimension $d = 1$ and to some questions regarding averages that we have already approached in Chap. 1. We are in particular going to revisit, this time from a computational perspective, the setting of quasiperiodic functions of Sect. 1.5.1. The questions we address are, as will be shortly seen, closely related to questions in *signal processing*.

Consider a quasiperiodic, real valued function b defined on the real line. We know from Proposition 1.2 that b has an average (actually equal to that of the periodic function defined on a higher dimensional space function from which b was defined taking a trace). We also know, from Proposition 1.3, that this average is obtained as the limit

$$\langle b \rangle = \lim_{R \to +\infty} R^{-1} \int_{-R/2}^{R/2} b(x)\,dx. \tag{5.28}$$

The rate of convergence in (5.28) is, at best and this corresponds to the particular periodic case, $\dfrac{1}{R}$. Such a rate of convergence is considered slow for the practice and it is worth trying to improve it. In our computational problem of interest, the reason is that, at a given expected accuracy, one could then choose a truncated computational domain $[-R/2, R/2]^2$ of smaller size R for the numerical resolution of (5.26). This will save computational time. Or, the other way around: for a given size R, improving the accuracy in (5.20) will improve the accuracy on the homogenized matrix $[a^*]$.

We borrow from signal processing the following useful idea to accelerate the above convergence. It has been specifically introduced in the context of homogenization in the work [BLB10], based on a similar endeavour in the context of molecular dynamics for the computation of long-time ergodic averages in [CCC05]. We first write the right-hand side of (5.28) as the average

$$\langle b \rangle_{R^{-1}1_R} = R^{-1} \int_{-R/2}^{R/2} b(x)\,dx, \tag{5.29}$$

where $\langle b \rangle_\varphi = \displaystyle\int_{-R/2}^{R/2} b(x)\varphi(x)dx$, for any non-negative function φ, of integral 1, and supported in the interval $[-R/2, R/2]$ (here, $\varphi = R^{-1}1_R$, where 1_R denotes the characteristic function of that interval). We next introduce, for $k \in \mathbb{N}^*$ fixed, the function

$$\begin{cases} \varphi \in C^k\left(\left[-\dfrac{1}{2}, \dfrac{1}{2}\right]\right), \quad \varphi \geq 0, \quad \displaystyle\int_{-\frac{1}{2}}^{\frac{1}{2}} \varphi = 1, \\[2ex] \varphi = 0 \quad \text{in } \left]-\infty, -\dfrac{1}{2}\right] \cup \left[\dfrac{1}{2}, +\infty\right[, \\[2ex] \forall 0 \leq j \leq k-1, \quad \dfrac{d^j}{dx^j}\varphi\left(-\dfrac{1}{2}\right) = \dfrac{d^j}{dx^j}\varphi\left(\dfrac{1}{2}\right) = 0 \end{cases} \tag{5.30}$$

which we call a *filter function*. We denote by $\varphi_R(.) = R^{-1} \varphi \left(\frac{\cdot}{R} \right)$ and we consider (using the same notation as in (5.29))

$$\langle b \rangle_{\varphi_R} = \int_{-R/2}^{R/2} b(x)\,\varphi_R(x)\,dx. \tag{5.31}$$

We hope, and this will indeed be the case, that replacing (5.29) with (5.31) allows to accelerate the convergence (5.28). We more precisely aim at obtaining

$$\langle b \rangle - \lim_{R \to +\infty} \int_{-R/2}^{R/2} b(x)\,\varphi_R(x)\,dx = O\left(R^{-k}\right),$$

for some $k \geq 1$. In order to understand why this acceleration holds under suitable circumstances, we scrutinize the case when the function b is not only quasiperiodic but is indeed *periodic*. To fix the idea, say it is of period 1. This particular case is of course only useful for our pedagogic purpose. In practice, when the function is periodic, this average is straightforward to compute, taking R as the period. But let us *pretend*, for the good sake, that we know that b is periodic but that we do not know its period. Then we have to embrace the limit as $R \to \infty$. To some extent, this is not so far from a certain practice. We could imagine practical modelling situations where assuming periodicity sounds reasonable but the period still remains inaccessible.

We thus assume that b is periodic (and, as usual for us, bounded). The rate of convergence (5.29) is straightforward to establish (and we indeed did so in (1.16) in Chap. 1):

$$\langle b \rangle - \lim_{R \to +\infty} R^{-1} \int_{-R/2}^{R/2} b(x)\,dx = O\left(R^{-1}\right).$$

On the other hand, we may calculate (5.31) using the Fourier transform

$$\int_{-R/2}^{R/2} b(x)\,\varphi_R(x)\,dx = \int_{\mathbb{R}} b\,\varphi_R = \int_{\mathbb{R}} \widehat{b}\,\widehat{\varphi_R} = \sum_{j \in \mathbb{Z}} \widehat{b}(j)\,\overline{\widehat{\varphi}(j\,R)},$$

whence

$$\langle b \rangle - \int_{-R/2}^{R/2} b(x)\,\varphi_R(x)\,dx = -\sum_{j \in \mathbb{Z} \setminus \{0\}} \widehat{b}(j)\,\overline{\widehat{\varphi}(j\,R)},$$

since $\widehat{b}(0) = \langle b \rangle$ and $\widehat{\varphi}(0) = 1$. We recall that

$$\widehat{\varphi^{(k)}}(\xi) = (2i\pi\xi)^k\,\widehat{\varphi}(\xi).$$

In addition, $\varphi \in C^k\left(\left[-\frac{1}{2}, \frac{1}{2}\right]\right)$ identically vanishes outside $\left[-\frac{1}{2}, \frac{1}{2}\right]$ so that all its derivatives up to order $k-1$ vanish in $\pm\frac{1}{2}$. It follows that $\varphi^{(k)} \in L^1(\mathbb{R})$ and that its Fourier transform is bounded. Consequently, for all $\xi \in \mathbb{R}\backslash\{0\}$,

$$|\widehat{\varphi}(\xi)| \leq \frac{1}{(2\pi)^k}\left\|\widehat{\varphi^{(k)}}\right\|_{L^\infty(\mathbb{R})}|\xi|^{-k} \leq \frac{1}{(2\pi)^k}\left\|\varphi^{(k)}\right\|_{L^1(\mathbb{R})}|\xi|^{-k} \leq C|\xi|^{-k},$$

where the constant C is independent of ξ. We deduce from the above observation that

$$\left|\sum_{j\in\mathbb{Z}\backslash\{0\}} \widehat{b}(j)\,\overline{\widehat{\varphi}(j\,R)}\right| \leq C\,R^{-k}\sum_{j\in\mathbb{Z}\backslash\{0\}}\left|\widehat{b}(j)\right||j|^{-k}$$

$$\leq C\,R^{-k}\,\|b\|_{L^2([-\frac{1}{2},\frac{1}{2}])}\left(\sum_{j\in\mathbb{Z}\backslash\{0\}}|j|^{-2k}\right)^{\frac{1}{2}}$$

$$= O\left(R^{-k}\right),$$

where the intermediate estimation is obtained using the discrete Cauchy-Schwarz inequality and upon observing that $\sum_{j\in\mathbb{Z}}\left|\widehat{b}(j)\right|^2$ is equal to the L^2 norm of the function b, since the Fourier transform is an isometry on that functional space. This establishes that the acceleration claimed in (5.31) indeed occurs as k grows. Of course, a similar proof holds in any ambient dimension d, the filter function being then defined as $\varphi_R(.) = R^{-d}\,\varphi\left(\frac{\cdot}{R}\right)$ with

$$\begin{cases} \varphi \in C^k\left(\left[-\frac{1}{2}, \frac{1}{2}\right]^d\right), \quad \varphi \geq 0, \quad \int_{[-\frac{1}{2},\frac{1}{2}]^d}\varphi = 1, \\[2mm] \varphi = 0 \quad \text{in } \mathbb{R}^d\backslash\left[-\frac{1}{2}, \frac{1}{2}\right]^d, \\[2mm] \forall 0 \leq j \leq k-1, \; D^j\varphi = 0 \quad \text{on } \partial\left(\left[-\frac{1}{2}, \frac{1}{2}\right]^d\right), \end{cases} \qquad (5.32)$$

instead of (5.30). In the above expression, $D^j\varphi$ denotes the tensor of the derivatives of order j of φ: it is the gradient for $j = 1$, the Hessian matrix for $j = 2$, the third order tensor containing all the third order derivatives for $j = 3$, and so on and so forth. In the quasiperiodic setting, it is likewise possible to establish the acceleration. The proof essentially follows the same pattern. We refer the reader to [BLB10, Proposition 4]. In that case, however, it is necessary for the proof to be valid and the result to hold to add the following assumption on the quasiperiods $T_1, \ldots, T_m$ of the coefficient (in the sense of Definition 1.1):

$$\exists C > 0, \quad \forall \eta > 0, \quad \forall p \in \mathbb{Z}^m, \quad \text{such that} \quad \left| \sum_{j=1}^{m} \frac{p_j}{T_j} \right| \leq \eta, \quad \sum_{j=1}^{m} |p_j| \geq \frac{C}{\eta}.$$

We wish to thank our colleague Sonia Fliss for pointing out to us that we had overlooked this additional assumption in our original publication [BLB10]. The above additional assumption, called a *Diophantine condition*, is also present, for similar reasons, in [CCC05].

We know at this stage how to accelerate the computation of the average of a quasiperiodic function. We still need to apply this procedure to the specific average (5.20). Although the integrand therein is in theory a quasiperiodic function, it has been, in practice, approximated on a truncated domain of size R. This may introduce a spurious effect in our acceleration technique, as we will now see. Assume indeed that we apply the acceleration technique upon blindly replacing (5.20) by

$$[a^*]_{ij} = \lim_{R \to +\infty} \int_{[-R/2, R/2]^d} \left(a(y) \left(\nabla w_{e_j}(y) + e_j \right) \right) . e_i \, \varphi_R(y) \, dy, \qquad (5.33)$$

where the notation is self-explanatory. Then we are doomed. . . .

To understand the difficulty, let us again temporarily return, without any loss of generality, to the one-dimensional setting, and consider the corrector problem for a quasiperiodic coefficient $a_{\text{q-per}}$. This problem is easy to solve, as we know since Chap. 2. It reads as (see e.g. (2.39))

$$-\frac{d}{dy}\left(a_{\text{q-per}}(y) \left(1 + \frac{d}{dy} w_{\text{q-per}}(y) \right) \right) = 0,$$

supplied with the condition of strict sublinearity at infinity. We could solve this problem explicitly. Let us however pretend again that we do not know this possibility and truncate the problem on the interval $[-R/2, R/2]$ for a supposedly large R and, as suggested above, for periodic boundary conditions:

$$\begin{cases} -\dfrac{d}{dy}\left(a_{\text{q-per}}(y) \left(1 + \dfrac{d}{dy} w_{\text{q-per}, R}(y) \right) \right) = 0 \text{ in } [-R/2, R/2], \\[2mm] w_{\text{q-per}, R}(-R/2) = w_{\text{q-per}, R}(R/2). \end{cases} \qquad (5.34)$$

We readily obtain

$$1 + w'_{\text{q-per},R}(y) = \left(R^{-1} \int_{-R/2}^{R/2} a_{\text{q-per}}^{-1} \right)^{-1} a_{\text{q-per}}(y)^{-1}.$$

If we now use this approximated expression for the corrector in the filtered average formula (5.31), this yields

$$a_R^* = \int_{-R/2}^{R/2} a_{\text{q-per}} \left(1 + w'_{\text{q-per},R} \right) \varphi_R$$

$$= \left(R^{-1} \int_{-R/2}^{R/2} a_{\text{q-per}}^{-1} \right)^{-1}. \tag{5.35}$$

as an approximate value for the homogenized coefficient! Put differently, the filter has completely disappeared from the computation. The approximation is identical to the classical one. All our efforts have been useless.

The key observation that now saves us is the following: we should introduce the filter *also in the corrector problem itself.* Let us indeed replace (5.34) with

$$\begin{cases} -\dfrac{d}{dy} \left[\varphi_R(y) \left(a_{\text{q-per}}(y) \left(1 + \dfrac{d}{dy} \tilde{w}_{\text{q-per},R}(y) \right) + \lambda_R \right) \right] = 0 \text{ in } [-R/2, R/2], \\ \\ \displaystyle\int_{-R/2}^{R/2} \varphi_R \left(\tilde{w}_{\text{q-per},R} \right)' = 0. \end{cases} \tag{5.36}$$

The well-posedness of problem (5.36) is not evident at first sight. One way to proceed is to work in the functional space $H^1([-R/2, R/2], \varphi_R(x)dx)$ for which the measure $\varphi_R(x)dx$ replaces the Lebesgue measure. The associated variational problem reads as: find $(w, \lambda) \in H^1([-R/2, R/2], \varphi_R(x)dx) \times \mathbb{R}$ such that

$$\forall (v, \mu) \in H^1([-R/2, R/2], \varphi_R(x)dx) \times \mathbb{R},$$

$$\int_{-R/2}^{R/2} a_{\text{q-per}} v' w' \varphi_R + \int_{-R/2}^{R/2} \lambda v' \varphi_R - \int_{-R/2}^{R/2} \mu w' \varphi_R = - \int_{-R/2}^{R/2} a_{\text{q-per}} v' \varphi_R.$$

We may then apply the Lax-Milgram Lemma in the Hilbert space

$$H^1([-R/2, R/2], \varphi_R(x)dx) \times \mathbb{R}.$$

To this end, we need first to prove a Poincaré-Wirtinger type inequality in that space. We refer to [BLB10, Proposition 2] for the details of its proof. The reader may then easily check that the solution of the above variational problem is indeed a solution in the distribution sense to (5.36).

We observe that the constant λ_R in the first line of (5.36) appears as the Lagrange multiplier associated to the constraint within the second line. In the absence of a filter, this constraint is equivalent to the periodicity of the corrector imposed in the second line of (5.34). Solving (5.36) into

$$1 + (\widetilde{w}_{q-\text{per},R})'(y) = \left(R^{-1} \int_{-R/2}^{R/2} a_{q-\text{per}}^{-1} \varphi_R \right)^{-1} a_{q-\text{per}}(y)^{-1}$$

therefore provides us with the following new approximation of the homogenized coefficient:

$$\begin{aligned}
(\widetilde{a}_R^*)^{-1} &= \int_{-R/2}^{R/2} a_{q-\text{per}} \left(1 + (\widetilde{w}_{q-\text{per},R})' \right) \varphi_R \\
&= \left(R^{-1} \int_{-R/2}^{R/2} a_{q-\text{per}}^{-1} \varphi_R \right)^{-1}.
\end{aligned}$$

Replacing (5.34) with (5.36) has reinstated the filter in the average approximating a^*. The expected acceleration can therefore occur.

The "full" filtering technique (by this we mean the filtering technique both applied to the corrector equation and the average) introduced in dimension $d = 1$ formally carries over to all dimensions $d \geq 2$. The filtered truncated corrector problem (5.36) now reads as

$$\left\{ \begin{aligned}
&-\text{div}\left[\varphi_R(y) \left(a_{q-\text{per}}(y) \left(e_j + \nabla \widetilde{w}_{e_j,R}(y) \right) + \lambda_R \right) \right] = 0 \text{ in } [-R/2, R/2]^d, \\
&\int_{[-R/2,R/2]^d} \varphi_R \, \nabla \widetilde{w}_{e_j,R} = 0,
\end{aligned} \right.$$

$$(5.37)$$

for a filter function defined in (5.32) and a Lagrange multiplier λ_R that is now a vector in $\mathbb{R}^d$. In (5.37), e_j denotes the j-th canonical basis vector of $\mathbb{R}^d$ and the solution $\widetilde{w}_{e_j,R}$ is searched for in $H^1\left([-R/2, R/2]^d, \varphi_R(x)dx \right)$. As was the case for (5.36), we refer to [BLB10, Proposition 2] for the well-posedness of (5.37). The filtered average is computed using (5.33). No theoretical analysis of the approach is known in dimensions $d \geq 2$. The numerical experiments conducted however show that the acceleration theoretically established at all orders k in dimension $d = 1$ actually saturates at the second order $k = 2$ in dimensions $d \geq 2$. Since a rate of convergence in $O(R^{-2})$ clearly outperforms a rate $O(R^{-1})$, the approach is nevertheless interesting. The gain is illustrated on Fig. 5.2, on a periodic case. "On average", the method using filtering is the better one, although a miracle occurs for the unfiltered approach when the size of the truncated domain occasionally

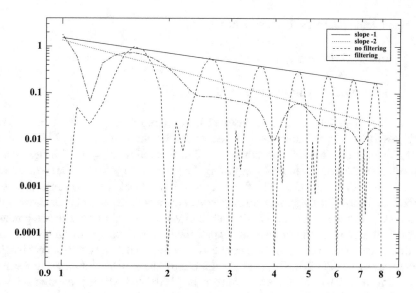

Fig. 5.2 (reproduced from [BLB10, Figure 1].) Filtering method on a periodic case. Horizontally, R denotes the size of the truncated computational domain used to define the approximated corrector problem. The error between the computed homogenized coefficient and its exact value is shown in function of R. Both coordinates are shown in log-scale. The exact value is computed using the same numerical approach for the corrector problem on the unit cell, and a simple integration over that cell to get the average. The dashed line is obtained with the method that does not use filtering. The chain line shows the result with filtering. Copyright ©2010 American Institute of Mathematical Sciences. Reprinted with permission. All rights reserved

coincides with a multiple of the period. On the other hand, say when $R = N + 1/2$, for N integer, the filtered approach is accurate at the order R^{-2}, instead of R^{-1} for the unfiltered approach. We refer to [BLB10] for numerical results showing a similar gain in actual quasiperiodic settings.

Although the gain of only one order in the rate of convergence, from R^{-1} to R^{-2}, may be deemed suboptimal from a certain idealistic perspective, the fact that the insertion of the filter barely modifies the computational workload is a clear advantage of the approach. On the other hand, the application of the approach to the almost periodic setting is limited by the fact that, and this is already observed in dimension $d = 1$, the speed-up in $O(R^{-k})$ is generically not realized. For the above reasons, a far better approach, motivated by Blanc and Le Bris [BLB10], has been introduced by Antoine Gloria in [Glo11] (see also [GH16]). The approach is reminiscent of a theoretical idea we have already repeatedly used in our theoretical developments.

The underlying idea of this alternative approach is to replace the filtering (5.37) of the truncated corrector problem by the approximation approach already employed say in Chap. 4 for the proof of Lemma 4.1 (Eq. (4.7)), or that of Proposition 4.3 (Eq. (4.118)). The exact same idea has lately been applied in (5.24), this time for computational purposes. It consists in substituting

$$
\begin{cases}
-\operatorname{div}\left(a\left(1+\nabla w_R\right)\right)+\eta_R\, w_R = 0 & \text{in } [-R/2,\, R/2]^d, \\[2ex]
w_R \quad \text{periodic on the boundary } [-R/2,\, R/2]^d,
\end{cases}
\tag{5.38}
$$

for (5.37). The solution is next inserted in a filtered average of the type (5.33), this average being not necessarily taken on the whole domain $[-R/2, R/2]^d$ but possibly only on a well chosen subdomain. The theoretical analysis of the approach is well documented. It allows for adjusting the parameter $\eta_R > 0$ so that a speed-up in R^{-4} is essentially achieved. We refer to [GH16, Glo11] for more details. As mentioned above, the approach also applies to the stochastic setting, which we will address in the next section. Other contributions in the vein of the developments presented here aim at improving the approximation of the corrector in the various settings. We wish to mention, in particular the coupling with an exterior problem in the series of works [CELS15, CEL20a, CEL20b], the approximation using a wave equation in [AR16], a parabolic problem in [AAP19, AAP21b], another elliptic equation [AAP21a], etc.

5.1.5   The Random Setting

We devote this section to the computational approximation of the random problem

$$
\begin{cases}
-\operatorname{div}\left(a\left(\dfrac{x}{\varepsilon},\omega\right)\nabla u^\varepsilon(x,\omega)\right) = f(x) \text{ in } \mathcal{D}\subset\mathbb{R}^d, \\[2ex]
u^\varepsilon(\cdot,\omega) = 0 & \text{on } \partial\mathcal{D},
\end{cases}
$$

introduced in Eq. (4.115) of Sect. 4.3 in Chap. 4.

The practical program to achieve the numerical approximation in the random setting is essentially similar to that completed in Sect. 5.1.4.2 above in the quasiperiodic setting. We need, in a first stage, to numerically approximate the solution to the corrector problem (4.116)

$$
\begin{cases}
-\operatorname{div}(a(x,\omega)\,(p+\nabla w_p(x,\omega))) = 0 \text{ in } \mathbb{R}^d, \\[2ex]
\nabla w_p(x,\omega) \text{ stationary}, \\[2ex]
\mathbb{E}\displaystyle\int_Q \nabla w_p(x,\omega)\,dx = 0,
\end{cases}
\tag{5.39}
$$

where, we recall, Q denotes the unit cube of $\mathbb{R}^d$. The equation is posed on the entire space $\mathbb{R}^d$. It has been studied theoretically in Proposition 4.3 of Sect. 4.3.1.

In a second stage, we next have to approximate the homogenized coefficient (4.134), that is,

$$[a^*]_{ij} = \mathbb{E} \int_Q a \left(e_j + \nabla w_{e_j}\right) . e_i. \tag{5.40}$$

The third and final stage is that of the numerical solution to the homogenized problem

$$\begin{cases} - \operatorname{div}(a^* \nabla u^*(x)) = f(x) \text{ in } \mathcal{D} \subset \mathbb{R}^d, \\ \\ u^* = 0 \qquad\qquad\qquad \text{on } \partial\mathcal{D}, \end{cases}$$

derived as (4.133) in Proposition 4.4 of Sect. 4.3.3. The latter stage proceeds with the exact same techniques as those of Sect. 5.1.2, since the random character originally present in (4.115) has completely disappeared in (4.133). Although the specific random setting we choose is that of discrete stationarity (thus the integral on Q within (5.39) and (5.40)), our discussion would carry over in its entirety to the continuous stationary setting, up to minor modifications.

The novelty as compared to Sect. 5.1.4.2 is of course the dependence of the problem with respect to the random variable $\omega \in \Omega$, in addition to that with respect to the already present space variable $x \in \mathbb{R}^d$. Our *discretization* of the problem will therefore involve not only the now usual discretization of the problem using a truncation over a bounded domain along with a finite element type approach to solve it, but also the consideration of a sample of *realizations* of the random variables at play.

Put differently, we have

(i) to consider the corrector problem (4.116)—or more precisely its approximated version on a truncated domain—, for a finite, *statistical sample* $\{\omega_1, \ldots, \omega_M\}$, of size $M \in \mathbb{N}$, of realizations of the random field $a(x, \omega_m)$, $1 \leq m \leq M$, and to approximate numerically its solution using a well established technique such as one employed in the previous section, and

(ii) next to approximate the expectation (4.134) using an *empirical mean*

$$\frac{1}{M} \sum_{m=1}^{M} \int_Q a(., \omega_m) \left(e_j + \nabla w_{e_j}(., \omega_m)\right) . e_i \tag{5.41}$$

as prescribed by the *Monte-Carlo method* introduced in Chap. 1.

Several issues arise. They may be classified into two different categories:

(a) *(Influence of the truncation in x):* from a theoretical standpoint, if we had truncated problem (4.116) on a bounded domain, then, *even if* we were able to exactly solve the truncated problem (which we are not, so there exists an error due to the discretization in x) and *even if* we were able to achieve this for all $\omega \in \Omega$ (in the sense, almost surely) which again we are not since we may only consider a finite sample of realizations, *and if* correspondingly we were able to exactly compute the expectation value (4.134) in step **(ii)**, then would the homogenized coefficient evaluated converge to the exact homogenized coefficient, in the limit of truncated domains covering the entire space $\mathbb{R}^d$? Can we, for a fixed truncation, estimate the error committed?

In a simpler version inherited from our work in the previous sections (see for instance (5.34)), the question may be reformulated as follows. Define, almost surely, the *truncated* stochastic corrector problem as

$$
\begin{cases}
- \operatorname{div}(a(x, \omega)\,(p + \nabla w_{p,R}(x, \omega))) = 0 \text{ in } [-R/2,\, R/2]^d, \\[2mm]
w_{p,R}(x, \omega) = 0, \text{ or periodic, or } \ldots \text{ on the boundary } \partial\left([-R/2,\, R/2]^d\right).
\end{cases}
\tag{5.42}
$$

Is it then true that, almost surely,

$$
\lim_{R \to +\infty} a_R^* = a^*
\tag{5.43}
$$

where

$$
[a_R^*]_{ij} = R^{-d} \int_{[-R/2, R/2]^d} a\left(e_j + \nabla w_{e_j, R}\right) . e_i,
\tag{5.44}
$$

and if yes, what is the rate of convergence with respect to R? May we additionally find numerical techniques to speed up the convergence?

The right-hand side of (4.134) is indeed nothing else in its original definition than the weak limit defined in (3.60) and thus, as in (5.20), the limit of large volumes of the quantity considered.

In view of the bounds available on the various quantities involved, the almost sure limit (5.43) next implies the same limit in various other senses, for instance in the L^2 sense, that is in quadratic mean.

(b) *(Somewhat symmetrically to* **(a)**, *influence of the truncation in ω):* from a theoretical standpoint again, if we assume we are able to exactly solve in x the corrector problem and if we approach the expectation value

$$
\mathbb{E} \int_Q a\left(e_j + \nabla w_{e_j}\right) . e_i
$$

on the right-hand side (4.134) by the empirical mean

$$\frac{1}{M} \sum_{m=1}^{M} \int_Q a(.,\omega_m) \left(e_j + \nabla w_{e_j}(.,\omega_m)\right) . e_i$$

defined in (5.41), at which rate with respect to M does the estimator converge to the exact expectation value? Do there exist techniques to speed up the convergence?

Let us emphasize, with a pedagogical view, that the above two categories of questions are generic in the applications of Monte-Carlo approaches. The two categories respectively correspond to the decomposition of the error into the so-called *systematic error* and *statistical error*. In dimension $d = 1$, this decomposition explicitly reads as follows. The error in the computation of

$$(a^*)^{-1} = \text{a.s.} \lim_{R \to +\infty} R^{-1} \int_{[-R/2,R/2]} a^{-1}(x,\omega)\, dx \qquad (5.45)$$

may be split as:

$$\text{a.s.} \lim_{R \to +\infty} \left[(a^*)^{-1} - \mathbb{E}\left(R^{-1} \int_{[-R/2,R/2]} a^{-1}(x,\omega)\, dx \right) \right]$$

$$+\text{a.s.} \lim_{R \to +\infty} \left[\mathbb{E}\left(R^{-1} \int_{[-R/2,R/2]} a^{-1}(x,\omega)\, dx \right) - R^{-1} \int_{[-R/2,R/2]} a^{-1}(x,\omega)dx \right]$$

that is, denoting by $(a^*)^{-1}(R,\omega) = \left(R^{-1} \int_{[-R/2,R/2]} a^{-1}(x,\omega)\, dx \right),$

$$\text{a.s.} \lim_{R \to +\infty} \left[(a^*)^{-1} - \mathbb{E}\left((a^*)^{-1}(R,\omega) \right) \right] \qquad (5.46)$$

$$+\text{a.s.} \lim_{R \to +\infty} \left[\mathbb{E}\left((a^*)^{-1}(R,\omega) \right) - (a^*)^{-1}(R,\omega) \right]. \qquad (5.47)$$

The first term (5.46) is the systematic error. It is the difference of two deterministic terms. The second term (5.47) is the statistical error. It is the difference between a random variable and its expectation value.

It is to be noted that, in practice, one does not use only $M = 1$ realization ω. The sample contains M such realizations. It does not affect our discussion at all. We may indeed repeat the above decomposition for the approximation of $(a^*)^{-1}$ as

$$\text{a.s.} \lim_{R \to +\infty} \left[(a^*)^{-1} - \mathbb{E} \left(R^{-1} \int_{[-R/2, R/2]} a^{-1}(x, \omega) \, dx \right) \right]$$

$$+ \text{a.s.} \lim_{R \to +\infty} \left[\mathbb{E} \left(R^{-1} \int_{[-R/2, R/2]} a^{-1}(x, \omega) \, dx \right) \right.$$

$$\left. - \frac{1}{M} \sum_{m=1}^{M} R^{-1} \int_{[-R/2, R/2]} a^{-1}(x, \omega_m) \, dx \right] \qquad (5.48)$$

where the quantity

$$\frac{1}{M} \sum_{m=1}^{M} R^{-1} \int_{[-R/2, R/2]} a^{-1}(x, \omega_m) \, dx$$

has the exact same expectation value for all values of M.

The systematic error (5.46) only depends on the accuracy of our approximation of the corrector problem on a bounded domain. Not bringing up the issue of elaborated techniques such as the truncation or filtering techniques we have mentioned in Sect. 5.1.4.2, it is classical to consider that, at worst,

$$\text{systematic error} \approx \frac{1}{R}. \qquad (5.49)$$

The reader is familiar with the statistical error (5.48), which we have introduced as early as in Chap. 1. For independent realizations ω_i, the rate of convergence is given by the Central Limit Theorem. The prefactor (which we omit now but will be the specific topic of interest in Sect. 5.1.5.2 below) depends on the variance of the random variable considered:

$$\text{statistical error} \approx \frac{1}{\sqrt{M}}. \qquad (5.50)$$

Note, however, that in the specific case considered, the random variable

$$R^{-1} \int_{[-R/2, R/2]} a^{-1}(x, \omega) \, dx$$

we approach the expectation value of is itself *parameterized* by R. This dependency upon R affects the variance, thus the prefactor appearing in (5.50). For instance, formally for $R = +\infty$, the random variable is actually deterministic, so that its variance vanishes. So does the statistical error (5.50)!

In any event, the take away message from the above discussion is that the error essentially decays as

$$\text{error} \approx \frac{1}{R} + \frac{1}{\sqrt{M}}, \tag{5.51}$$

where, *implicitly*, the parameter R is also involved in the latter term. Under suitable assumptions, its impact on the estimation may be studied and explicitly quantified. We will see such a case in the next section.

We are now going to successively reply, in Sect. 5.1.5.1, to the questions of the first category **(a)** and in Sect. 5.1.5.2 to those of the second category **(b)**.

5.1.5.1 Convergence

The first general proof of the almost- sure convergence (5.43) has appeared in [BP04], a seminal article that generated many works in its wake. We briefly summarize here its main ideas.

The One Dimensional Setting
To start with, let us consider, as we always do, the one-dimensional setting introduced in Sect. 2.4. The corrector problem

$$\begin{cases} -\dfrac{d}{dy}\left(a(y,\omega)\left(1 + \dfrac{d}{dy}w(y,\omega)\right)\right) = 0 \\ \dfrac{dw}{dy} \text{ stationary,} \\ \mathbb{E}\displaystyle\int_Q \dfrac{dw}{dy} = 0, \end{cases}$$

is approximated by the truncated problem

$$\begin{cases} -\dfrac{d}{dy}\left(a(y,\omega)\left(1 + \dfrac{d}{dy}w_R(y,\omega)\right)\right) = 0 \text{ in } [-R/2, R/2], \\ \\ w_R(-R/2,\omega) = w_R(R/2,\omega) \end{cases} \tag{5.52}$$

exactly as we did for the quasiperiodic problem (5.34). Its solution reads as

$$1 + (w_R)'(x,\omega) = a_R^*(\omega)\, a^{-1}(x,\omega),$$

where the integration constant $a_R^*(\omega)$ is fixed by the periodic boundary conditions in (5.52). Integrating the above expression over the interval $[-R/2, R/2]$, we indeed obtain

$$\left(a_R^*(\omega)\right)^{-1} = R^{-1} \int_{-R/2}^{R/2} (a(x,\omega))^{-1} \, dx. \tag{5.53}$$

This expression, actually identical to (5.44) when the latter is made specific to the one-dimensional setting, immediately provides the approximation of the homogenized coefficient, defined by

$$\left(a^*\right)^{-1} = \mathbb{E} \int_{-1/2}^{1/2} (a(x,\omega))^{-1} \, dx. \tag{5.54}$$

Since Sect. 1.6 of Chap. 1, the reader already knows the (positive) answer to the question *Does the coefficient $a_R^*(\omega)$ defined by (5.53) almost surely converge to the coefficient a^* defined in (5.54)?*, The function $(a(x,\omega))^{-1}$ is a stationary function and this property which in fact defines its average, is definitely true.

The reply to the question *At what rate does the convergence occur?* is also known, from the same section: *It depends. . . .*

We indeed recall the variety of situations encoded in the stationary setting: periodicity, quasiperiodicity, almost periodicity, i.i.d. random variables, etc. We have provided in Chap. 1 the example (1.85) of the function

$$a(x,\omega) = 1 + \sum_{k \in \mathbb{Z}} X_k(\omega) \mathbf{1}_Q(x-k), \tag{5.55}$$

where $Q = [0,1]$ and the sequence X_k is i.i.d. We have then established in (1.89) that the rate of the almost sure convergence

$$\int_0^1 \left(a\left(\frac{x}{\varepsilon},\omega\right)\right)^{-1} dx \xrightarrow{\varepsilon \to 0} \mathbb{E}\left(\int_0^1 (a(.,\omega))^{-1}\right) \tag{5.56}$$

is $\sqrt{\varepsilon}$. This amounts to saying that the rate of convergence of (5.53) to (5.54) is $\dfrac{1}{\sqrt{R}}$.

This elementary setting actually also provides a pedagogic and illuminating testbed for the decomposition of the error into systematic and statistical errors, as introduced in (5.46) and (5.47) respectively.

The decomposition of the error committed for the computation of the coefficient $(a^*)^{-1}$ is indeed straightforward. Taking the expectation value of both sides of (5.53) yields

$$\mathbb{E}\left(\left(a_R^*(\omega)\right)^{-1}\right) = \mathbb{E}\left(R^{-1} \int_{-R/2}^{R/2} (a(x,\omega))^{-1} \, dx\right)$$

$$= R^{-1} \int_{-R/2}^{R/2} \mathbb{E}\left((a(x,\omega))^{-1}\right) \, dx.$$

Since the function $\mathbb{E}\left((a(x,\omega))^{-1}\right)$ is 1-periodic, the error between the latter quantity and the average of this function, that is

$$\int_{-1/2}^{1/2} \mathbb{E}\left((a(x,\omega))^{-1}\right) dx = \mathbb{E}\int_{-1/2}^{1/2}(a(x,\omega))^{-1} dx = (a^*)^{-1}$$

is at most of the order of $\dfrac{1}{R}$. We have known this fact since Chap. 1. This is the systematic error. It can even identically vanish if we have been sufficiently smart to pick an integer value for R! The statistical error adds up to the systematic error. For the function a considered, it scales as $\dfrac{1}{\sqrt{R}}$. One should bear in mind, as recalled above, that the latter error sensitively depends upon the assumptions on a.

The reader may notice, however, that our choice of discussing the error estimation on $(a^*)^{-1}$ is a bit biased. To some extent, we have cheated. We have indeed unfairly exploited the fact that we know that, in the one-dimensional setting, everything is more conveniently expressed in terms of the *inverse* coefficients $(a)^{-1}$, instead of a. An honest, and therefore more informative and representative analysis is to be conducted on a^* itself. The explicit form of a, such as e.g. (5.55), is then necessary. We leave this exercise to the reader. After some tedious calculations, the conclusions will be similar. We had rather focus our attention to dimensions $d \geq 2$ and address the general difficulty.

The Multidimensional Setting
From $d = 2$ up, the corrector problem to solve in the random setting is problem (4.116), that is,

$$\begin{cases} -\operatorname{div}(a(x,\omega)\,(p + \nabla w_p(x,\omega))) = 0 \text{ in } \mathbb{R}^d, \\[2mm] \nabla w_p(x,\omega) \text{ stationary}, \\[2mm] \mathbb{E}\int_Q \nabla w_p(x,\omega)\,dx = 0. \end{cases}$$

It is typically approximated by the truncated problem (5.42), namely

$$\begin{cases} -\operatorname{div}(a(x,\omega)\,(p + \nabla w_{p,R}(x,\omega))) = 0 \text{ in } [-R/2, R/2]^d, \\[2mm] w_{p,R}(x,\omega) = 0, \text{ or periodic or } \dots \text{ on the boundary } \partial\left([-R/2, R/2]^d\right), \end{cases}$$

or a modification thereof, such as the insertion of a zero order term, a filter function, etc., as mentioned in Sect. 5.1.4.2. We have just seen a one-dimensional version in (5.52). An approximation

$$\left[a_R^*(\omega)\right]_{ij} = R^{-d} \int_{[-R/2, R/2]^d} a\left(e_j + \nabla w_{e_j,R}\right) . e_i \tag{5.57}$$

of the homogenized coefficient $\left[a^*\right]_{ij}$ is inferred from the solution to (5.42). The relevant question is then that of the convergence (5.43) of the random variable $\left[a_R^*(\omega)\right]_{ij}$ to the exact coefficient, as the truncation parameter R grows to infinity.

We cannot emphasize enough the following observation regarding the definition (5.57) of the approximated homogenized coefficient. This observation may have escaped the reader's notice already in the analogous general formula (5.43) and, in dimension 1, in the formulae (5.45) and (5.53). Although we are going to establish that, *asymptotically* in $R = +\infty$, the coefficient a_R^* agrees with the exact *deterministic* value a^*, the coefficient a_R^* is, for all $R < +\infty$, a *random variable*. As long as $R < +\infty$, the truncation has destroyed the ergodicity property and transformed a deterministic quantity into a random one.

The proof of the almost sure convergence (5.43) completed in [BP04] is actually rather elementary. To fix the ideas, say we consider homogeneous boundary conditions in (5.42). Alternative boundary conditions could be treated equally easily. Up to the change

$$\overline{w_{p,R}}(x, \omega) = R^{-1} w_{p,R}(Rx, \omega)$$

of unknown function, (5.42) reads as

$$\begin{cases} -\operatorname{div}(a(Rx, \omega)(p + \nabla \overline{w_{p,R}}(x, \omega))) = 0 \text{ in } [-1/2, 1/2]^d, \\ \\ \overline{w_{p,R}}(x, \omega) = 0 \quad \text{on} \quad \partial\left([-1/2, 1/2]^d\right). \end{cases} \tag{5.58}$$

Since the coefficient $a(Rx, \omega)$ has the form $a\left(\dfrac{x}{\varepsilon}, \omega\right)$ when $\varepsilon = R^{-1}$, the left-hand side of (5.58) is a differential operator with an oscillatory stationary coefficient. The limit $\varepsilon \to 0$, that is $R \to +\infty$, of this problem is obtained by homogenization. Proposition 4.4, however, does not straightforwardly apply to (5.58). First, yet another change of unknown function is needed, namely

$$\chi_{p,R}(x, \omega) = \overline{w_{p,R}}(x, \omega) + p . x, \tag{5.59}$$

so that the problem reads as

$$\begin{cases} -\operatorname{div}(a(Rx, \omega) \nabla \chi_{p,R}(x, \omega)) = 0 \text{ in } [-1/2, 1/2]^d, \\ \\ \chi_{p,R}(x, \omega) = p . x \quad \text{on} \quad \partial\left([-1/2, 1/2]^d\right). \end{cases} \tag{5.60}$$

Second, we notice that Proposition 4.4 is stated with homogeneous Dirichlet boundary conditions. We claim, and the reader may easily verify this fact, that its result
is equally valid for *non-homogeneous* boundary conditions, provided those are not
oscillatory (that is, here, independent of R). Applying this variant of the proposition,
we obtain that, respectively, $\chi_{p,R}(x, \omega) \longrightarrow \chi^*$ in $L^2\left(\Omega, H^1\left([-1/2, 1/2]^d\right)\right)$ and
$a(Rx, \omega)\nabla\chi_{p,R}(x, \omega) \longrightarrow a^*\nabla\chi^*$ in $L^2\left(\Omega \times [-1/2, 1/2]^d\right)$, where χ^* is the
solution to the homogenized problem

$$\begin{cases} -\operatorname{div}(a^*\nabla\chi^*) = 0 \text{ in } [-1/2, 1/2]^d, \\[2mm] \chi^*(x) = p \cdot x \quad \text{on} \quad \partial\left([-1/2, 1/2]^d\right). \end{cases}$$

Since a^* is a constant matrix, the solution is $\chi^*(x) = p \cdot x$. The above convergences
thus read as

$$\overline{w_{p,R}}(x, \omega) \xrightarrow[R \to +\infty]{} 0 \quad \text{in} \quad L^2\left(\Omega, H^1\left([-1/2, 1/2]^d\right)\right),$$

$$a(Rx, \omega)\left(p + \nabla\overline{w_{p,R}}(x, \omega)\right) \xrightarrow[R \to +\infty]{} a^*p \quad \text{in} \quad L^2\left(\Omega \times [-1/2, 1/2]^d\right).$$

In addition, Proposition 4.5 also implies the following almost sure convergence:

$$a(Rx, \omega)\left(p + \nabla\overline{w_{p,R}}(x, \omega)\right) \xrightarrow[R \to +\infty]{} a^*p \quad \text{in} \quad L^2\left([-1/2, 1/2]^d\right).$$

The latter convergence in turn implies, for all $1 \leq i, j \leq d$,

$$\lim_{R \to +\infty} \int_{[-1/2, 1/2]^d} a(Rx, \omega)\left(e_j + \nabla\overline{w_{e_j,R}}\right) \cdot e_i = \left[a^*\right]_{ij}. \tag{5.61}$$

This limit, once rescaled, exactly expresses the convergence of the coefficient (5.57),
that is (5.43). Since the limit obtained is obviously unique, all the limits above, all
taken after a suitable extraction, in fact do not depend on the specific extraction.

The next question is that of the rate of convergence (5.43).

On the Rate of Convergence and How to Improve It
As in dimension 1, it is possible to study the rate of convergence with respect to R
in (5.43). We have already recalled that, given the variety of situations contained in
the stationary setting, the rate is sensitive to the specific assumptions made on the
random coefficient $a(.\, , \omega)$ and may largely vary. The so-called *mixing* assumptions
are often in order. Even under the suitable assumptions, the theoretical analysis may
be quite technical, as we saw in Chap. 1. We will therefore not proceed with this
analysis. We refer to the bibliography, and, to start with, to [BP04].

Irrespective of the theoretical analysis we have just skipped, the practical question we have asked at the beginning of this Sect. 5.1.5 remains: *Do there exist techniques to speed up the convergence?* We actually already know some reply. We may indeed *combine* (5.42) and (5.38) so that we define the truncated problem as

$$
\begin{cases}
-\operatorname{div}\left(a(x,\omega)\,(p+\nabla w_{p,R}(x,\omega))\right) + \eta_R\, w_{p,R}(x,\omega) = 0 \\
\qquad\qquad\qquad\qquad\qquad\qquad \text{in } [-R/2, R/2]^d, \\[2ex]
w_{p,R}(x,\omega) = 0 \quad \text{on the boundary } \partial\left([-R/2, R/2]^d\right),
\end{cases}
\tag{5.62}
$$

for some suitably chose parameter η_R. The works [GH16, Glo11] provide the necessary analysis to establish that, combining this with an appropriate filtering of the average, it is indeed possible to accelerate the convergence.

We prefer to focus, in the upcoming Sect. 5.1.5.2, on the other component of the error, namely the *statistical* error.

The first reason is that, in the literature, the latter component of the error has not been paid so much attention as the former. And this by far. The reader may however consider, and there would be some truth in this, that it is not reason enough to devote time to the study of the statistical error. We fortunately have a much better, second reason, which is more relevant and pragmatic. In practice, the size R of the computational domain that one can afford to solve the truncated corrector problem is *very* limited. The workload is indeed heavy. The domain must first be meshed. Next the problem is solved numerically, and this needs to be done repeatedly, for the many realizations M of the statistical sample. Even solving *one* such corrector problem could be considered expensive computationally, particularly in ambient dimension $d = 3$, and more critically when the problem is not an equation but a system of equations as it could be for instance the case in elasticity modelling. The task of solving multiple such problems is therefore not to be taken lightly. As the workload grows exponentially in R, the practice is therefore to keep R as small as possible. A compromise has to be found, between the representativity of the domain and the computational cost.

Because R cannot be taken as large as one would dream of, the regime in which the approach is put in action is remote from the asymptotic regime $R \to +\infty$. We surely do not learn much from the theoretical limit $R \approx +\infty$ when in practice $R = 4$. Such theoretical studies are definitely useful to guarantee the mathematical correctness of the approach and possibly to introduce novel approximation methods and improve some features of the existing discretization techniques. At a fixed computational cost, they suggest approaches that may be much profitable, in terms of the systematic error. They however must be complemented with an analysis of the statistical error, which better fits the actuality of the computations performed.

One could object to the previous discussion that, also in the context of Monte-Carlo methods, the number M of realizations that can in practice be considered is not as large as wished. Therefore the Central Limit Theorem could appear too idealized to be practically relevant. We even came across many practical works where practitioners only consider *one* realization. We will not pass any judgment on such works. Suffice it to say they exist. It however remains that increasing the number M of realizations considered

- is often less expensive in terms of computational workload (increasing R increases exponentially the size of the linear algebraic systems to solve, and thus correspondingly the memory requirement, short of implementing dedicated approaches of high performance computing),
- only linearly affects the computational cost, in contrast to the size R that acts nonlinearly,
- is intrinsically and easily amenable to parallel computation, since the exact same procedure has to be repeated, independently, for all realizations ω_m.

For the above reasons, we consider it relatively simpler to increase M, thereby better approaching the asymptotic regime. The mathematical analysis of the limit $M \to \infty$, where the variance of the random objects involved plays a crucial role, is thus our focus.

Before we get to it, we recall that we have noticed in the one-dimensional context studied above that the statistical error is generically larger than the systematic error. The errors are typically respectively of sizes $\dfrac{1}{\sqrt{R}}$ and $\dfrac{1}{R}$. This is often confirmed by the higher dimensional practice. Reducing the statistical error is thus expected to better pay off.

5.1.5.2   Statistical Error and Variance Reduction

In a Monte-Carlo method, the *noise* is both your best ally and your own worst enemy.

It is an *ally* because the consideration of many random realizations yields the efficiency of the approach. It usually allows one to reach quantities (averages) that otherwise would remain out of reach of deterministic approaches. The prototypical example is the computation of high-dimensional integrals, which may be obtained upon considering long-time random trajectories that, by ergodicity, explore the ambient space. In our specific context, the corrector problem, originally posed on the entire space $\mathbb{R}^d$ is truncated on a bounded domain, say the cube $[-R/2, R/2]^d$. Generating a large number of realizations ω_m of the random environment, that is of the coefficient $a(.\omega)$, we can compensate for the finite size R and improve the accuracy of the result, which would have been poor under the consideration of only one realization ω. Of course, in the limit $R \to \infty$, one such realization would be sufficient. Whatever this "one" means, since in essence in the random setting, things make only sense almost surely at best. But for $R < +\infty$, and a fortiori for a not too large value R, several realizations will help. Assuming that, all things being equal, the systematic error is small, the result will be all the more accurate.

The noise is also an *enemy*. As we have seen, the rate of convergence of statistical approximation using a Monte-Carlo method is assessed by the Central Limit Theorem and is invariably in $\dfrac{1}{\sqrt{M}}$. Trying to improve this rate in the context of *independent* draws is hopeless.

On the other hand, if the rate is immutable, the prefactor itself depends on the variance of the random variable simulated. When that variance is large, the efficiency of the Monte-Carlo method can be disappointing. The key advantage of such an approach is that the rate of convergence, arguably slow, is at least independent of the ambient dimension. This comes in sharp contrast with the competing deterministic approaches which all get less efficient, and sometimes drastically so, as the dimension grows. The universality of the rate explains the success of the Monte-Carlo method, given the numerous application contexts in the engineering sciences or life sciences where the ambient dimension is large, if not prohibitively large. Thus the many *afficionados* the method attracts. It remains that a closer look at a given problem might reveal a large variance, in which case the method might simply be impractical.

It turns out, however, that the prefactor may be improved by suitable techniques. In short, those techniques consist in putting in action the Monte-Carlo method not on the original random variable but on a variable that has a lower variance. The latter variable may in particular be simulated using draws that are not independent of one another. A final stage of the computation allows to finally return to the result originally targeted. Such approaches are called *variance reduction methods*.

We have been unable to locate in the literature any application of variance reduction methods to the context of homogenization predating the recent series of works [CLBL10b, BLBL16, LBLM16, LM15]. Such methods, on the other hand, have long been exploited in various different fields, first and foremost mathematical finance. Recent applications have also appeared in computational statistical physics.

We present in the sequel of this section a short recollection of variance reduction methods applied to homogenization. We refer to the review article [BLBL16] for more details and more references.

Let us first recall some basic notions from Sect. 1.6.1 of Chap. 1. By virtue of the strong law of large numbers (1.70), we know that, if X_i, $i \in \mathbb{N}$, is a sequence of i.i.d. random variables of finite expectation value $\mathbb{E}(|X|) < +\infty$, we almost surely have

$$\mathbb{E}(X) = \lim_{M \to +\infty} \frac{X_1(\omega) + \ldots + X_M(\omega)}{M}.$$

When in addition $\mathbb{E}(X^2) < +\infty$, the rate of convergence is given by the Central Limit Theorem (1.71): the random variable

$$\frac{\sqrt{M}}{\sigma} \left(\frac{X_1(\omega) + \ldots + X_M(\omega)}{M} - \mathbb{E}(X) \right)$$

converges in law to a random variable G (often rather denoted by $\mathcal{N}(0, 1)$, as in "normal") that follows a reduced centered gaussian law, where the variance

$$\sigma^2 = \mathbb{E}\left((X - \mathbb{E}(X))^2\right) = \mathbb{E}(X^2) - (\mathbb{E}(X))^2 \tag{5.63}$$

yields the prefactor for $\dfrac{1}{\sqrt{M}}$ we have been discussing above.

In the practical context of a Monte-Carlo method, this theoretical result is applied as follows. For $M \in \mathbb{N}$ fixed, the expectation value $\mathbb{E}(X)$ is approximated by the *empirical mean*

$$\mu_M(X) = \frac{1}{M} \sum_{m=1}^{M} X_m.$$

Similarly, the variance is approximated by the empirical variance

$$\sigma_M(X)^2 = \frac{1}{M - 1} \sum_{m=1}^{M} (X_m - \mu_M(X))^2. \tag{5.64}$$

The strong law of large numbers (1.70) tells us that $\mu_M(X) \to \mathbb{E}(X)$ almost surely as $M \to +\infty$. In addition, the Central Limit Theorem (1.71) implies the convergence $\sqrt{M} (\mu_M(X) - \mathbb{E}(X)) \to \sigma \, \mathcal{N}(0, 1)$ in law.

It is a classical technique to consider the *95th quantile* of $\mathbb{E}(X)$, that is the range of values such that, with a probability equal to or higher than 95%, the empirical mean $\mu_M(X)$ belongs to that range. The Central Limit Theorem quantifies such a range as follows:

$$|\mu_M(X) - \mathbb{E}(X)| \leq 1,96 \, \frac{\sigma_M(X)}{\sqrt{M}}. \tag{5.65}$$

This determination is actually a bit formal, since it relies upon a *double* approximation

- the approximation of the *convergence in law* (1.71) using an *equality in law*, the explicit value 1,96 of the prefactor arising in the calculation on the law $\mathcal{N}(0, 1)$;
- the approximation of the exact variance σ^2 by the empirical variance $\sigma_M(X)^2$.

The second approximation may itself be studied similarly to the first one, exactly as we are addressing the approximation of $\mathbb{E}(X)$ using $\mu_M(X)$.

In practice, things are presented and employed *the other way around:* the point in (5.65) is not that $\mu_M(X)$ should be sufficiently close to $\mathbb{E}(X)$, but rather that $\mathbb{E}(X)$, which is unknown beforehand and which we wish to approach, is sufficiently close to $\mu_M(X)$, which the computation provides. The estimation is thus rephrased as

$$\mathbb{E}(X) \in \left[\mu_M(X) - 1,96 \frac{\sigma_M(X)}{\sqrt{M}}, \ \mu_M(X) + 1,96 \frac{\sigma_M(X)}{\sqrt{M}} \right] \qquad (5.66)$$

with the probability 95%. The range (5.66) is called the *interval of confidence*.

If the expectation value is computed in a slightly different manner, so that the variance of the random variable actually simulated (often a modification of the original variable X) is smaller than that of X itself, but the two variables share the same expectation, then the accuracy is improved since the width of the interval of confidence is reduced. The *variance reduction method* is then successful.

The Example of Antithetic Variates

The *antithetic variates method* is presumably the most elementary variance reduction method. It is often taught in class in the particular context of numerical integration of a real valued function on a finite interval of the real line $\mathbb{R}$. As we briefly mentioned above, numerical integration is indeed a field of choice for Monte-Carlo methods, notably because the convergence of the method is essentially insensitive to the dimension. In sharp contrast, the computational complexity of the numerical approaches based on the integration over grids grows exponentially in the dimension. Their accuracy degrades inversely proportionally. In order to better fit our context, we directly expose the antithetic variates method on a one-dimensional case of random homogenization. We follow the presentation of [CLBL10b].

We consider the one-dimensional random corrector problem (2.59) from Sect. 2.4, in the particular case of the stationary coefficient

$$a(x, \omega) = \sum_{k \in \mathbb{Z}} a_k(\omega) \mathbf{1}_{[k-1/2, k+1/2[}(x), \qquad (5.67)$$

for i.i.d. random variables a_k, all bounded and strictly positive. We already came across this coefficient in (1.85). For the sake of simplicity, we have just shifted the periodic cell.

A straightforward calculation, repeatedly performed in this textbook and most lately in (5.53), shows that the approximation of the coefficient a^* that is obtained upon truncating the corrector problem on the interval $[-R/2, R/2]$ reads as

$$\left(a_R^*(\omega) \right)^{-1} = R^{-1} \int_{-R/2}^{R/2} (a(x, \omega))^{-1} \, dx.$$

Again for simplicity, we take $R = 2N + 1$ for $N \in \mathbb{N}$, so that

$$\left(a_{2N+1}^*(\omega) \right)^{-1} = \frac{1}{2N+1} \sum_{k=-N}^{N} \frac{1}{a_k(\omega)}.$$

The rate of convergence of the approximated coefficient $\left(a_{2N+1}^*(\omega)\right)^{-1}$ to its limit $(a^*)^{-1}$ depends on the variance of $\dfrac{1}{a_0}$. Assume now that the coefficients a_k alternately take the values α and β with probability $1/2$:

$$\mathbb{P}(a_0 = \alpha) = \frac{1}{2} \quad \text{and} \quad \mathbb{P}(a_0 = \beta) = \frac{1}{2},$$

say for $0 < \alpha < \beta$. We have

$$\mathbb{E}\left(\frac{1}{a_0}\right) = \frac{1}{2}\frac{1}{\alpha} + \frac{1}{2}\frac{1}{\beta} \quad \text{et} \quad \mathbb{V}\mathrm{ar}\left(\frac{1}{a_0}\right) = \frac{1}{4}\left(\frac{1}{\alpha} - \frac{1}{\beta}\right)^2.$$

The classical Monte-Carlo method then consists in first simulating M sets of coefficients $(a_{-N}(\omega_m), \ldots, a_N(\omega_m))$, for $1 \le m \le M$ and next calculating the average of the coefficients $\left(a_{2N+1}^*(\omega_m)\right)^{-1}$ obtained using these M realizations. The convergence analysis performed above still applies. The prefactor is proportional to the variance $\mathbb{V}\mathrm{ar}\left(\dfrac{1}{a_0}\right)$.

Instead of following the classical path, we now choose an even number of realizations M. Only half of them are drawn at random. For each of these $\dfrac{M}{2}$ realizations ω, we consider the coefficient

$$b(x, \omega) = \sum_{k \in \mathbb{Z}} b_k(\omega) \mathbf{1}_{[k-1/2, k+1/2]}(x) \tag{5.68}$$

where

$$b_k(\omega) = \alpha + \beta - a_k(\omega). \tag{5.69}$$

Put differently, we swap α and β: $a_k = \alpha \iff b_k = \beta$, and conversely. We then add up to the coefficient $\left(a_{2N+1}^*(\omega_j)\right)^{-1} = \dfrac{1}{2N+1} \displaystyle\sum_{k=-N}^{N} \dfrac{1}{a_k(\omega_j)}$ obtained for ω_j, $j = 1, \ldots, M/2$, the coefficient

$$\left(b_{2N+1}^*(\omega_{m/2+j})\right)^{-1} = \frac{1}{2N+1} \sum_{k=-N}^{N} \frac{1}{b_k(\omega_j)}$$

$$= \frac{1}{2N+1} \sum_{k=-N}^{N} \frac{1}{\alpha + \beta - a_k(\omega_j)}.$$

which is adjusted *deterministically* once the *random* realization ω_j has been drawn. The full average consisting of the coefficients $\left(a^*_{2N+1}(\omega_j)\right)^{-1}$ over the $M/2$ genuinely random realizations $(\omega_1, \ldots, \omega_{M/2})$ considered *plus* the $M/2$ realizations $(\omega_{M/2+1}, \ldots, \omega_M)$ of the coefficients $\left(b^*_{2N+1}(\omega_j)\right)^{-1}$ reads as

$$
\begin{aligned}
&\frac{1}{2}\left(\frac{1}{M/2}\sum_{m=1}^{M/2}\left(a^*_{2N+1}(\omega_m)\right)^{-1} + \frac{1}{M/2}\sum_{m=1}^{M/2}\left(b^*_{2N+1}(\omega_{M/2+j})\right)^{-1}\right) \\
&= \frac{1}{M}\sum_{m=1}^{M/2}\left(\frac{1}{2N+1}\sum_{k=-N}^{N}\frac{1}{a_k(\omega_j)} + \frac{1}{2N+1}\sum_{k=-N}^{N}\frac{1}{\alpha+\beta-a_k(\omega_j)}\right) \\
&= \frac{1}{M}\sum_{m=1}^{M/2}\frac{1}{2N+1}\sum_{k=-N}^{N}\left(\frac{1}{a_k(\omega_j)} + \frac{1}{\alpha+\beta-a_k(\omega_j)}\right) \\
&= \frac{1}{M}\sum_{m=1}^{M/2}\frac{1}{2N+1}\sum_{k=-N}^{N}\frac{1}{\alpha}+\frac{1}{\beta} \\
&= \frac{1}{2}\left(\frac{1}{\alpha}+\frac{1}{\beta}\right),
\end{aligned}
\tag{5.70}
$$

since, when $a_k = \alpha$, $b_k = \beta$ and, conversely, when $a_k = \beta$, $b_k = \alpha$. Put differently, the empirical average obtained is *exactly* equal to the ideal, asymptotic value $\mathbb{E}\left(\frac{1}{a_0}\right)$. The coefficient $(a^*)^{-1}$ is thus exactly identified. The variance of the Monte-Carlo method has been so much reduced that it has been put to zero!!

Where is the miracle? The deterministic realization considered in (5.68)–(5.69) is called the antithetic realization defined from a_k. The variable b_k defined in (5.69) is the *antithetic variate* of a_k. The sample does not any longer consist of independent realizations, so that the Central Limit Theorem is (of course!) not violated. We have of course cheated a bit, since have extensively exploited the specifics of the simple situation chosen. Let us now try and be more honest.

We revisit the above calculation for i.i.d. variables a_k now uniformly distributed on the range $[\alpha, \beta]$. Only the last two lines of (5.70) have to be modified. We obtain the average

$$
\frac{1}{M}\sum_{m=1}^{M/2}\frac{1}{2N+1}\sum_{k=-N}^{N}\left(\frac{1}{a_k(\omega_j)} + \frac{1}{\alpha+\beta-a_k(\omega_j)}\right)
\tag{5.71}
$$

which, this time, does not exactly match the desired asymptotic value $\mathbb{E}\left(\frac{1}{a_0}\right)$. That said, we observe that, since the function $x \mapsto 1/x$ is non-increasing, we have

$$\mathrm{Cov}\left(\frac{1}{a_0}, \frac{1}{b_0}\right) \leq 0, \tag{5.72}$$

where the covariance $\mathrm{Cov}(X, Y)$ of two random variables X and Y is of course defined by

$$\mathrm{Cov}(X, Y) = \mathbb{E}\left[(X - \mathbb{E}(X))\,(Y - \mathbb{E}(Y))\right] = \mathbb{E}(XY) - \mathbb{E}(X)\mathbb{E}(Y).$$

Let us indeed consider an arbitrary non-increasing function f together with two independent random variables X and Y, uniformly distributed on $[\alpha, \beta]$. Since $x \mapsto f(\alpha + \beta - x)$ is non-increasing we have

$$(f(X) - f(Y))\,(f(\alpha + \beta - X) - f(\alpha + \beta - Y)) \leq 0.$$

Taking the expectation value, we readily obtain

$$\mathbb{E}[f(X)\,f(\alpha + \beta - X)] \leq \mathbb{E}[f(X)]\,\mathbb{E}[f(\alpha + \beta - X)],$$

that is $\mathrm{Cov}[f(X), f(\alpha+\beta-X)] \leq 0$. The specific choice $f(x) = 1/x$ yields (5.72). Since

$$\mathrm{Var}\left(\frac{1}{2}\left(\frac{1}{a_{2N+1}^*} + \frac{1}{b_{2N+1}^*}\right)\right) = \frac{1}{2(2N+1)}\mathrm{Var}\left(\frac{1}{a_0}\right) + \frac{1}{2(2N+1)}\mathrm{Cov}\left(\frac{1}{a_0}, \frac{1}{b_0}\right),$$

we conclude that

$$\mathrm{Var}\left(\frac{1}{2}\left(\frac{1}{a_{2N+1}^*} + \frac{1}{b_{2N+1}^*}\right)\right) \leq \frac{1}{2}\mathrm{Var}\left(\frac{1}{a_{2N+1}^*}\right). \tag{5.73}$$

The estimator (5.71) contains $M/2$ independent realizations of $\frac{1}{2}\left(\frac{1}{a_{2N+1}^*} + \frac{1}{b_{2N+1}^*}\right)$, with an interval of confidence that has width $\sqrt{\dfrac{\mathrm{Var}\left(\frac{1}{2}\left(\frac{1}{a_{2N+1}^*}+\frac{1}{b_{2N+1}^*}\right)\right)}{M/2}}$. This is to be compared with the classical approach for which we have M realizations of $\frac{1}{a_{2N+1}^*}$ and an interval of confidence of width $\sqrt{\dfrac{\mathrm{Var}\left(\frac{1}{a_{2N+1}^*}\right)}{M}}$. Inequality (5.73) thus allows to conclude that, at equal computational cost, the antithetic variate approach is certified to improve the accuracy.

The following remarks are in order.

First and foremost, even in the one-dimensional setting and in the simplest cases, where we may indeed make a *proof*, we see that we are unable to quantify the gain. We do now know by which factor the variance is actually reduced. We may guarantee, under suitable circumstances, that the method at least does not increase the variance, but we do not know the efficiency. As surprising as it may seem to the outsider, even this minimal theoretical information is already saying much on the approach. For many numerical approaches used in practice we do not have such a certainty. At least here we are on the safe side.

Second, even if, in higher dimensions, no proof is available to date, the numerical experiments we have conducted show that, *in practice*, the variance is indeed reduced. The amplitude of the reduction is not spectacular (for this to be observed, we will have to switch to more effective approaches that we will describe in the sequel) but it is real. The moderate success of the approach is intuitively understandable. The approach does not make use of any specific "insider's" information about the problem considered. It is completely generic.

Third and last, the method is completely *for free*. Rather than randomly drawing M values of the random field a, we only draw half of them and deterministically cook up the second half from the first half. The computational cost, dominated by that of the resolution of the corrector problems, remains identical.

For the above reasons, the antithetic variate method is useful in homogenization. Figure 5.3 shows the generation of the antithetic structure for a random checkerboard. The solution procedure for the (suitably truncated, filtered, etc.) corrector problem is performed in parallel for the structure on the left and that on the right. The task is repeated for a sample of realizations. This yields an approximation of a^*. In such a case, the typical gain is a reduction of the width of the interval of confidence by a factor 2. Figure 5.4 shows a typical numerical result. The

Fig. 5.3 (reproduced from [CLBL10b, Figure 1]) Two antithetic realizations of the random checkerboard with probability 1/2. The checkerboard on the right is the "mirror image" of that on the left

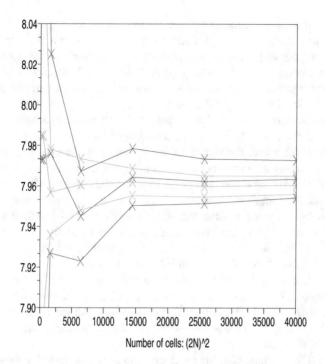

Fig. 5.4 (reproduced from [CLBL10b, Figure 2]) Approximation of the entry $[a^*]_{11}$ of the homogenized matrix for a two-dimensional situation: the empirical mean and the interval of confidence (in the sense of (5.66)) are displayed in function of the increasing size $(2N)^2$ of the truncated domain $[-N, N]^2$. Red lines: classical Monte-Carlo. Green lines: antithetic variate method

reader may notice on the figure how results are typically shown. The value of the homogenized coefficient, which is a random variable for any finite size of the truncated computational domain, is approximated by an empirical mean. An interval of confidence, estimated using a formula of the type (5.66), assesses the quality of the approximation.

For more details on the approach, we refer to the articles [CLBL10b, BLBL16] and their respective bibliography.

The Selection Approach
Although they are encouraging and they demonstrate that variance reduction methods deserve to be considered, the results of the previous section are not groundbreaking, to say the least. They suggest that we have to direct our efforts to methods that are more elaborate and less generic. The *selection approach* is one such method. It has been originally introduced in the context of materials science where it carries the name of *Special Quasi random Structures* and abbreviated in *SQS*. The original bibliographical references may be found in [LBLM16].

The approach is a representative of a larger class of approaches known in mathematical statistics as *stratified sampling*. The reader has presumably heard from such methods which are popular in the field of political science. Election polls use such an approach, namely *quota sampling*. In short, the statistical sample used to get a reliable picture of the general opinion is not entirely chosen at random. The very limited size of the sample that may actually be polled would lead to a result that is not sufficiently accurate. Put in mathematical terms, the sample is too small, the variance is too large, the interval of confidence is too wide, the empirical mean is irrelevant. Therefore, the choice of the sample is *biased*, so that the various categories of voters are represented as truthfully as possible. For instance, if a quarter of the voters are under the age of 25 years, the polling institute makes sure that the finite size sample exhibits the same proportion. And so on and so forth for all the other well known characteristic features of the general population of voters. Such features may for instance have been determined on the basis of the most recent census. We of course simplify a bit the technique for the sake of simplicity of exposition, but the spirit of the method is the correct one.

We now describe the specific method we will employ. We first do this in the specific context of materials science, for an elementary one-dimensional particle system we wish to study the statistical properties of. We have already explained in Sect. 1.3 of Chap. 1 the close connection that such interacting particle problems enjoy with homogenization problems.

Consider a linear chain of particles. Each particle, which occupies a given atomic site along the real line, may belong to one of two different atomistic species, denoted by A and B. Such particles interact through 3 different pair potentials, depending upon the combination of species involved. The three potentials are denoted by V_{AA}, V_{AB} and V_{BB}. The notation is self-explanatory.

The occupation of the atomic sites A and B are randomly chosen. A typical realization of the "material" is thus an infinite sequence $\cdots ABBAAABBAAAA \cdots$. To further simplify the problem, we assume that a particle on a given site only interacts with its two nearest neighbours, that is the two particles occupying the two nearest sites. Computing the energy of the system means determining the limit

$$\lim_{N \to \infty} \frac{1}{2N+1} \sum_{i=-N}^{N} V_{X_{i+1}(\omega)X_i(\omega)}, \tag{5.74}$$

where X_i denotes the species occupying site i.

Practically, the Monte-Carlo method consists in randomly generating a sample of chains of $2N+1$ sites (*truncated* chains in the terminology of the previous sections) and computing

$$\frac{1}{M} \sum_{m=1}^{N} \frac{1}{2N+1} \sum_{i=-N}^{N} V_{X_{i+1}(\omega_m)X_i(\omega_m)}$$

for a, possibly increasing, size M of the sample.

Can we outperform the classical approach?

To this end, we are first going to choose the sample so that it has the correct *volume fraction*, that is the correct proportion of species A and B. If, again for simplicity, the two species are equally likely to occupy a site, and that they do so independently of all the other sites, then the volume fraction is 1/2. If the m-th random draw ω_m does not show the same proportion (possibly within a certain tolerance that may be fixed beforehand), it is immediately discarded and replaced by another tentative draw. We proceed so until we actually get a satisfactory draw.

Assume next that, beyond the simple volume fraction of the system, we also know beforehand another property of the system that can be expressed using an expectation value, say

$$\mathbb{E}(f) = \lim_{N \to \infty} \frac{1}{2N+1} \sum_{i=-N}^{N} f(X_i),$$

for some function f. We may consider further biasing the statistical sample so as to reflect this new property truthfully. In our specific setting, this is indeed the case. The species A and B being equiprobable and all independent, we know that the energy reads as:

$$\lim_{N \to \infty} \frac{1}{2N+1} \sum_{i=-N}^{N} V_{X_{i+1}(\omega)X_i(\omega)} = \frac{1}{4}[V_{AA} + 2V_{AB} + V_{BB}]$$

We leave to the reader the elementary exercise to check this formula upon enumerating the various combinations possible for the "neighbourhood" of a given site and their respective probabilities. We may therefore impose that each of the draws we select to enter the statistical sample correctly reproduces not only the volume fraction, but also this particular energy, both up to a certain tolerance. Once the list of such quantities to correctly reproduce is complete, we have our biased statistical sample. We may finally proceed with the actual evaluation of our quantity of interest, exactly as we would have done using a canonical statistical sample in the classical Monte-Carlo method.

The above simple example shows everything we need to know to apply the approach to the general setting of random homogenization practically. The main computational workload there concerns the solution procedure for the corrector problem. The preliminary step necessary for this resolution is the generation of the various realizations of the random field $a(x, \omega)$ throughout the truncated domain where the problem is approximated. We are going to impose some restrictions on such realizations.

The first idea employed above in the atomistic example, namely the imposition of the correct volume fraction, is perfectly relevant in the homogenization context, say for composite materials. Except for the serendipitous one-dimensional setting where geometry does not play any role and volume fraction is "everything", we have

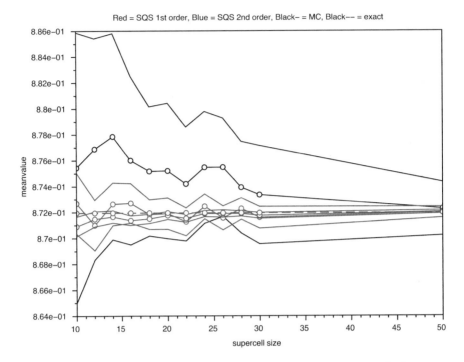

Fig. 5.5 (reproduced from [LBLM16, Figure 7]) Approximation of the entry $[a^*]_{11}$ of the homogenized matrix in a two-dimensional case: the empirical mean and the interval of confidence (in the sense of (5.66)) are displayed in function of the increasing size N of the truncated domain $[-N, N]^2$ on which the corrector problem is solved. Black lines: classical Monte Carlo method. Red lines: selection method with 1 selection criterion. Blue lines: same method with 2 criteria.

emphasized at the beginning of Chap. 3 that the homogenized matrix is not entirely fixed by the volume fraction. It is thus possible to bias the statistical sample using the volume fraction.

Other criteria may successively be imposed. We refer to [LBLM16] for more details. As shown on Fig. 5.5, the variance may typically be reduced by a factor 10, while a reduction by an even larger factor is not unseen. This evidently outperforms the results obtained by the antithetic variates method.

The obvious reason for this predominance is that the efficiency of a computational approach owes to the relevance of the ingredients employed. Involving in the selection of the realizations considered in the sample the volume fraction already implies a certain a priori understanding of the problem studied. Adding up a second criterion, as in the atomistic example addressed above, implies an even deeper knowledge of the problem. The more information about the actual system is involved, the better the result of the variance reduction approach. A generic approach such as the antithetic variates method cannot compete.

Let us also emphasize that, from a practical viewpoint, the selection approach, as the antithetic variates method, hardly increases the computational workload as compared with the classical Monte-Carlo method. Generating additional realizations of $a(x, \omega)$ is inexpensive. Possibly rejecting many of those is therefore equally inexpensive. As a matter of fact this observation may come as a surprise to the reader familiar with the application of the Monte-Carlo method to other contexts. In their full generality, Monte-Carlo methods are customized so as to generate as few realizations as possible and correspondingly to reject them on an exceptional basis. This is the typical concern in a Metropolis type algorithm, for instance. The reason is that the generation of the random variable is the most computationally expensive step within the computation. This is *not* the case in homogenization, where the computational workload is outrageously dominated by the cost of the solution procedure for the corrector problem (which is, we recall, the numerical approximation of the solution to a PDE on a d-dimensional domain). Carefully selecting the realizations of the random field, and possibly rejecting many tentative ones, is a negligible additional task.

To conclude this paragraph, we mention that the mathematical formalization of the approach may be completed using the notion of *conditional expectation*. The proof of the actual interest of the approach, under appropriate assumptions and in the specific context of homogenization, has recently been performed in [Fis19].

More Sophisticated Approaches, Such as the Control Variates Method
The *control variates method* is a general technique used for the reduction of the variance in Monte-Carlo methods. It does not directly approximate the expectation value

$$\mathbb{E}(X)$$

by the empirical mean

$$\mu_M(X) = \frac{1}{M} \sum_{m=1}^{M} X(\omega_m)$$

when this approximation is expensive, or poorly accurate because the random variable X has a large variance. The expectation is rather decomposed as

$$\mathbb{E}(X) = \mathbb{E}(Y) + \mathbb{E}(X - Y) \tag{5.75}$$

where the random variable Y, called the *control variate*, is adjusted so that, both

(i) its variance is significantly lower than that of X, so that the expectation value $\mathbb{E}(Y)$ can efficiently and accurately be computed using the empirical mean

$$\mu_M(Y) = \frac{1}{M} \sum_{m=1}^{M} Y(\omega_m)$$

(ii) the expectation value $\mathbb{E}(X - Y)$ may be, on the other hand, computed efficiently and accurately, using a different approximation that the mere approximation by its empirical mean.

The particular choice $Y = X$ of course satisfies **(ii)** but it does not satisfy **(i)**. On the other hand, $Y = 0$ ensures **(i)** but not **(ii)**. One has to find the best possible compromise...

A useful pedagogic analogy may help the reader, or at least the reader familiar with linear algebra. We like to see the control variates method as a *preconditioning* approach. Some linear algebraic systems $AX = B$ are known to be particularly difficult to solve because the matrix A is *ill-conditioned*. In short, if A is symmetric, this amounts to saying that its eigenvalues are of disparate orders of magnitude. Then, instead of solving $AX = B$ directly, the common practice is to introduce an auxiliary matrix C, called the *preconditioning matrix* or the *preconditioner* that is both better conditioned than the matrix A and as close as possible to A. The latter condition is formalized upon saying that $C^{-1} A$ is close to the Identity matrix. Somewhat in the spirit as what we mentioned above regarding the choices $Y = X$ and $Y = 0$ as control variate, the specific choice $C = A$ makes $C^{-1} A$ the identity matrix but is ill-conditioned while taking C as the identity makes $C^{-1} A$ remote from identity. The best compromise is here again to be found. Then the approach consists in successively solving

$$C Y = B \quad \text{and next} \quad (C^{-1} A) X = Y.$$

Note that we more precisely deal here with a so-called *left* preconditioning technique. Since C is better conditioned than A, the first linear system, with matrix C, may be solved more easily and accurately. Likewise, X is simpler to determine in the second linear system $(C^{-1} A) X = Y$ since the system is close to $X = Y$ and its resolution using an iterative method repeatedly involves the matrix $(C^{-1} A)$, that is in effect the resolution of a sequence of linear algebraic systems with matrix C.

The analogy goes beyond the above observation. The control variate is, as a preconditioner for an algebraic system, not determined beforehand nor explicitly given. One needs to be creative and invent one. Usually, this requires an in-depth understanding of the problem.

In the context of random homogenization, where the variable above denoted by X in (5.75) typically reads as

$$R^{-d} \int_{[-R/2,R/2]^d} a(.\,, \omega) \left(e_j + \nabla w_{e_j,R}(.\,, \omega)\right) . e_i$$

(see for instance (5.43)), the control variate Y may be the value of the coefficient $\left[a^*\right]_{ij}$ for a slightly simpler random field than the original field $a(.\,, \omega)$, for which an alternative computing strategy would be amenable. This is the case, for instance, if the geometry of the field is simple, or the randomness is small, etc.

We refer to [BLBL16, LM15] for more details on the method applied to the specific context of homogenization. We similarly refer to the bibliography for other existing variance reduction approaches that could possibly be, but have not yet been applied to this context.

What Is the Point?
We would like to conclude this somewhat lengthy Sect. 5.1 by emphasizing, using a practical example, that all the tasks we have performed are not pointless. The computation of a homogenized matrix is a precious information. It cannot be obtained on the back on an envelope. It is thus tempting to use *recipes* that bypass the significant computational cost of the approaches we have described above. Such recipes might occasionally give interesting indications under appropriate, often oversimplifying assumptions. They might provide global trends. They are however to be considered cautiously. We illustrate this on Fig. 5.6. We show there the poor accuracy that simplified recipes can provide on relevant cases. Our question to the reader is the following. Is the reader prepared to board an aircraft knowing that the composite materials its wings are built from were numerically simulated and designed using the approximate methods of Fig. 5.6? For a contrast of the order of 100 between the two coefficients characterizing the two phases of the composite material, the physical property computed varies between 2 and 60, depending upon the simplified approach selected, while, in reality, the exact result is 10.

5.2   Multiscale Computational Approaches

We now adopt, throughout Sect. 5.2, a drastically distinct perspective from that of Sect. 5.1. This perspective is actually more recent and has a slightly different purpose.

The approach we have presented in Sect. 5.1

(i) is first and foremost based upon the mathematical analysis previously performed on the problem, which is embedded in a *family* of analogous problems parameterized by the sequence $\varepsilon \to 0$ so that, under suitable assumptions,

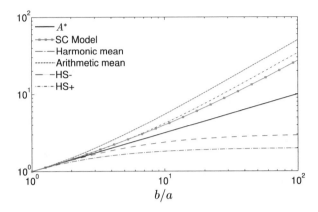

Fig. 5.6 (reproduced from [Tho12, Figure 2.1]) Comparison of some simplified "coarse" approaches for the evaluation of the homogenized coefficient. The simplified approaches (here the so-called Self-Consistent (SC) and Hashin–Shtrikman (HS+ and HS−) approaches) at best use approximations of the random microstructure of a material (here a random checkerboard) by a *periodic* microstructure. The results are shown in function of the contrast $\frac{b}{a}$ between the two coefficients in the checkerboard. The vertical axis is in log-scale. The discrepancy between the approaches and the exact result may be considerable as the contrast grows

   it admits a sufficiently explicit homogenized limit amenable to numerical
   computations,
(ii) is aimed at determining the homogenized properties, that is the homogenized
   coefficient a^* and the solution u^* in this order,
(iii) and allows for approximating the oscillatory solution u^ε, for a parameter ε
   asymptotically small, up to the first order of accuracy in ε (or up to higher
   order provided the successive terms u_k of the two-scale expansion (3.39) are
   evaluated using computational strategies similar to that for the first order).

   Many questions regarding multiscale problems in the engineering sciences and
the life sciences may be addressed effectively only knowing the homogenized properties of the material, or more generally the medium considered. The homogenized
characteristics along with a decent approximation of the oscillatory solution are
sufficient to proceed. There are however also numerous questions that require a
better approximation of the oscillatory solution to the multiscale problem itself.
   In such situations, it is not always the case that the mathematical equipment of
the problem and all the corresponding mathematical assumptions are such that there
exists a homogenized limit, that this limit may be entirely identified theoretically
and that a suitable computational strategy to evaluate it can be designed. For
instance, it can be known that there exists a small scale ε and that an explicit,
known value can be given to that scale, but the specifics of the microstructure can
be poorly known. Whether it is periodic (not even bringing up the issue of the
value of the period), quasiperiodic, randoms stationary etc. is a pure abstraction
of the mind, far too remote to admit a definite answer. A good context for this

rather poor knowledge of the microstructure is that of geoscience. Even if the soil is undoubtedly often multiscale in nature, accurately describing its microstructure and encoding it in definite mathematical properties of a coefficient a_ε is beyond reach. In some practical situations it is even an illusion to assume that the coefficient is *known* throughout the computational domain.

In the situations where

- the homogenized limit is not known, or not *yet* known, or even is not proven to exist,
- where the knowledge of only u^* would not be informative enough to solve the practical problem,
- where ε is sufficiently small so that the generic numerical approaches fail to satisfactorily capture the oscillatory solution u^ε, but on the other hand is not sufficiently small so that it might be considered asymptotically small and for the theoretical developments of homogenization theory to provide an accurate description of its behaviour,

it is practically important and relevant to come up with a computational strategy that allows to efficiently and accurately approximate u^ε, for the actual, practically relevant value of ε. In such cases, homogenization is a *theoretical guideline*.

Novel computational approaches have been developed to this very end in the past two decades. We give an overview of such approaches in the present Sect. 5.2. Since such approaches are still in their infancy, there is consequently no general consensus on which approach is best for which application. This is science in progress. So there is ample room for subjectivity and we do confess that our own choice of approaches and our presentation are entirely subjective. We wish to emphasize that the main objective of the approaches described here is paradoxically not to capture the fine scales as accurately as possible but rather to capture the large scales of the solution. As shown in Sect. 2.1.2 of Chap. 2, and in particular in Fig. 2.2, a naive approach does not even capture those large scales. The difficulty is that the small and large scales are intimately entangled. One does not know beforehand how to capture the latter without capturing as much as possible the former. But the ultimate purpose of the multiscale approaches described here is, as a matter of fact, the large scales! Note, of course, that, by this, we mean capturing the large scales typically in H^1 norm, that is the large scale structure of the solution *and* of its gradient. This task is far more demanding than capturing "only" (a) the homogenized solution and (b) in the asymptotic limit where ε vanishes.

We present in Sect. 5.2.1 the one-dimensional version and the underlying intuitive idea behind a large family of approaches, known as *Multi-Scale Finite Element Methods*, abbreviated in $MsFEM$. We consider the higher dimensional settings in Sect. 5.2.2. In each case, we provide a description of the implementation and a sketch of the numerical analysis available in a prototypical setting. As announced above, we next present in Sect. 5.2.3 the *Heterogeneous Multiscale Method* (HMM), while Sect. 5.2.4 contains one example of another approach following a slightly different line of thought, and Sect. 5.2.5 discusses some general questions.

The first textbook published on *MsFEM* (and, to the best of our knowledge, the only one entirely devoted to the method) is [EH09]. It is authored by two of the main contributors to the method, one of which is Thomas Hou, one of the founders of the domain. A second example is the monograph [MP21], focused on the approach we will be presenting in Sect. 5.2.4. Other textbooks address *MsFEM*, but in a more scattered or marginal manner. Given the novelty of the domain, this should be no surprise. A volley of research articles, on the other hand, will provide the reader with the many details we omit in our brief exposition.

Remark 5.1 The multiscale approaches we discuss throughout the present Sect. 5.2 enjoy several similarities with both *domain decomposition* approaches and *multigrid* methods. In particular, all these approaches seek to construct suitable approximation spaces to *locally* approximate the solution. The major difference, though, is that the multiscale approaches we describe here are "one-shot" methods. In sharp contrast, domain decomposition and multigrid methods are based upon the approximation of the solution using successive *iterations*. We refer, for example, to [QV99] for domain decomposition, to [Bri87] for an initiation to multigrid and to [Owh17, KPY17, KPY18] for various discussions on the similarity we mention here.   □

5.2.1   One-Dimensional *MsFEM*

In order to illustrate the new line of thought we follow throughout Sect. 5.2, we temporarily return to the one-dimensional setting of Sect. 2.1.2 in Chap. 2, that is Eq. (2.1)

$$
\begin{cases}
-\dfrac{d}{dx}\left(a_\varepsilon(x)\,\dfrac{d}{dx}u^\varepsilon(x)\right) = f(x) \text{ in } [0, 1], \\[2mm]
u^\varepsilon(0) = u^\varepsilon(1) = 0,
\end{cases}
$$

and more precisely to the variational formulation (2.10) of that equation: *Find* $u^\varepsilon \in H_0^1([0, 1])$ *such that, for any function* $v \in H_0^1([0, 1])$,

$$
\int_0^1 a_\varepsilon(x)\,(u^\varepsilon)'(x)\,v'(x)\,dx = \int_0^1 f(x)\,v(x)\,dx.
$$

We have observed in Sect. 2.1.2, using both an explicit calculation of the solution and the numerical analysis of the approach, that the classical $\mathbb{P}_1$ finite element method (where the basis functions are defined in (2.11)) performed on a mesh of size H (that is, in this one-dimensional setting, the intervals $\left[\dfrac{i}{N}, \dfrac{i+1}{N}\right]$ for $i =$ $0, \ldots, N-1$ and $H = \dfrac{1}{N}$) yields an incorrect approximation of the exact solution, unless the meshsize H is chosen such that $H \ll \varepsilon$. The latter choice is usually impractical. So a change of approach is in order.

5.2.1.1 Construction of the Multiscale Approximation

We decide, for practical purposes, to keep the meshsize H large with respect to the fine scale ε, or comparable in size, but certainly not smaller than ε. We also keep the variational formulation (2.12). We are however going to change the finite-dimensional approximation space V_H into some other space $V_{\mathrm{MsFEM},H}$, called the *multiscale approximation space*, which we now define. For each $i = 0, \ldots, N-2$, we consider the solution φ_i to the equation

$$
\begin{cases}
-\dfrac{d}{dx}\left(a_\varepsilon(x)\dfrac{d}{dx}\varphi_i(x)\right) = 0 & \text{in } \left[\dfrac{i}{N}, \dfrac{i+1}{N}\right], \\[4mm]
\varphi_i\left(\dfrac{i}{N}\right) = 0, \quad \varphi_i\left(\dfrac{i+1}{N}\right) = 1.
\end{cases}
\tag{5.76}
$$

The differential operator on the left-hand side of the equation is identical to the oscillatory operator present in (2.1). This new equation, however, has a zero right-hand side and is posed only on the interval $\left[\dfrac{i}{N}, \dfrac{i+1}{N}\right]$. It is supplied with particular Dirichlet boundary conditions. The formal similarity with the corrector problem, for instance the periodic corrector problem (2.31), associated to (2.1), is obvious. But it is to be born in mind that nothing guarantees the existence of such a corrector function in the context of a general coefficient a_ε where we are working. Homogenization theory only provides us with the intuition that, at the small scale, the right equation to introduce is of the form (5.76). Since the corrector equation is key for the theory, it is fair to think that Eq. (5.76), which looks like it, should help us in the computation.

We similarly consider

$$
\begin{cases}
-\dfrac{d}{dx}\left(a_\varepsilon(x)\dfrac{d}{dx}\psi_i(x)\right) = 0 & \text{in } \left[\dfrac{i}{N}, \dfrac{i+1}{N}\right], \\[4mm]
\psi_i\left(\dfrac{i}{N}\right) = 1, \quad \psi_i\left(\dfrac{i+1}{N}\right) = 0.
\end{cases}
\tag{5.77}
$$

Glueing the functions φ_{i-1} and ψ_i at the node $\dfrac{i}{N}$, and extending by zero outside the interval $\left[\dfrac{i-1}{N}, \dfrac{i+1}{N}\right]$, we define, for $i = 1, \ldots, N-1$, the function

$$\chi_{\text{MsFEM},i}(x) = \begin{cases} \varphi_{i-1}(x) & \text{si} \quad x \in \left[\dfrac{i-1}{N}, \dfrac{i}{N}\right] \\[2ex] \psi_i(x) & \text{si} \quad x \in \left[\dfrac{i}{N}, \dfrac{i+1}{N}\right] \\[2ex] 0 & \text{si} \quad x \in \left[0, \dfrac{i-1}{N}\right] \cup \left[\dfrac{i+1}{N}, 1\right] \end{cases} \tag{5.78}$$

We then denote by $V_{\text{MsFEM},H}$ the $N-1$-dimensional space spanned by the functions $\chi_{\text{MsFEM},i}$, $i = 1, \ldots, N-1$. This space

– has same dimension as the classical FEM space V_H; the former reads as

$$V_{\text{MsFEM},H} = \left\{ v \in C^0\left([0, 1]\right), \ -\left(a_\varepsilon(x)\, v'(x)\right)' = 0 \right.$$

$$\left. \text{on each segment} \left[\dfrac{i}{N}, \dfrac{i+1}{N}\right] \right\} \tag{5.79}$$

while the second is

$$V_H = \left\{ v \in C^0\left([0, 1]\right), \quad v \text{ is affine on each segment} \left[\dfrac{i}{N}, \dfrac{i+1}{N}\right] \right\} \tag{5.80}$$

– each element $\chi_{\text{MsFEM},i}$ of the basis has the same value at all nodes and the same support as the classical $\mathbb{P}_1$ finite element χ_i defined in (2.11) (which preserves the structure of the rigidity matrix),
– but each such element is, by definition, dependent upon the differential operator,
– thus is not generic any longer and depends on the problem (not upon its right-hand side, though)
– and finally, in the particular case $a_\varepsilon \equiv 1$, $\chi_{\text{MsFEM},i}$ turns out to coincide with the classical element χ_i.

The specific choice of the basis functions $\chi_{\text{MsFEM},i}$ to span the variational space $V_{\text{MsFEM},H}$ is motivated by the fact that, since the basis functions solve (5.76) or (5.77), they are likely to genuinely encode the fine-scale oscillations of the original problem. To a certain extent, and our analysis below will confirm this fact, they allow to essentially solve the problem at the fine scale which is not explicitly captured by the mesh of size H. There is no free lunch. The price

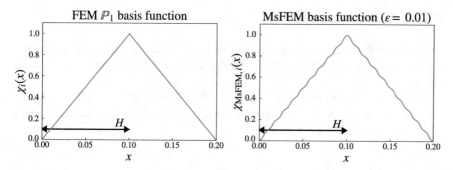

Fig. 5.7 Comparison of the profiles of the finite element basis functions over the interval $[0, 2H]$. The classical function $\mathbb{P}_1$ is piecewise affine and shown on the left. The $MsFEM$ basis function mimics the fine scale oscillations of the problem and is shown on the right. The specific discretization parameters are: $H = 10^{-1}$, $\varepsilon = 10^{-2}$. Figure courtesy of Rutger Biezemans

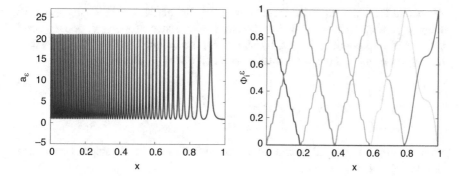

Fig. 5.8 Example of a nonperiodic one-dimensional case: on the left, the coefficient a_ε, the oscillations of which shrink from the right-hand side to the left-hand side of the unit interval; on the right, the corresponding $MsFEM$ basis functions, which exhibit a similar local behaviour

to pay is the loss of the genericity of the basis. Figure 5.7 shows the typical shape of such basis functions. Another example is given on Fig. 5.8, where the coefficient a_ε is not periodic and exhibits a varying behaviour at the small scale. The basis functions generated by the approach described above mimic the local behaviour of the coefficient a_ε. They follow it in its oscillations, so to speak. All the functions φ_i, ψ_i, $\chi_{MsFEM,i}$, which we have constructed so far in this Sect. 5.2.1.1 parametrically depend upon the meshsize H and the small lengthscale ε within the original Eq. (2.1). To simplify notation, we however do not indicate it. The $MsFEM$ approach, which we have just described in a simple setting, has been introduced by Thomas Hou and his collaborators. It first appeared in [HW97]. Many improvements, extensions and variants have been developed since, in many different contexts and application areas.

5.2.1.2 Numerical Analysis

To demonstrate the gain in efficiency brought by the approximation space $V_{\text{MsFEM},H}$, we revisit the numerical analysis performed in Sect. 2.1.2 for the approximation space V_H.

Using linearity, we may again proceed using Céa's Lemma which we express this time not in the classical H^1 norm as in (2.21), but in the so-called *energy norm*. The exact same proof indeed yields:

$$\int_0^1 a_\varepsilon(x) \left|(u^\varepsilon - u^\varepsilon_{\text{MsFEM},H})'\right|^2$$

$$\leq \inf_{v_{\text{MsFEM},H} \in V_{\text{MsFEM},H}} \int_0^1 a_\varepsilon(x) \left|(u^\varepsilon - v_{\text{MsFEM},H})'\right|^2. \qquad (5.81)$$

We bound the right-hand side from above using the particular choice $v_{\text{MsFEM},H} = R_{\text{MsFEM},H}(u^\varepsilon)$, which is the interpolant in the space $V_{\text{MsFEM},H}$ of the exact solution u^ε. This interpolant is, by definition, the unique function $R_{\text{MsFEM},H}(u^\varepsilon)$ in $V_{\text{MsFEM},H}$ that has the same value as u^ε at all nodes $\frac{i}{N}$, that is

$$R_{\text{MsFEM},H}(u^\varepsilon)(x) = \sum_{i=1}^{N-1} u^\varepsilon\left(\frac{i}{N}\right) \chi_{\text{MsFEM},i}(x).$$

We next split the integral on the right-hand side into a sum of integrals over each interval. We obtain

$$\int_0^1 a_\varepsilon(x) \left|(u^\varepsilon - R_{\text{MsFEM},H}(u^\varepsilon))'\right|^2$$

$$= \sum_{i=0}^{N-1} \int_{\frac{i}{N}}^{\frac{i+1}{N}} a_\varepsilon(x) \left|(u^\varepsilon - R_{\text{MsFEM},H}(u^\varepsilon))'\right|^2$$

$$= -\sum_{i=0}^{N-1} \int_{\frac{i}{N}}^{\frac{i+1}{N}} \left(a_\varepsilon(x)(u^\varepsilon - R_{\text{MsFEM},H}(u^\varepsilon)')\right)' (u^\varepsilon - R_{\text{MsFEM},H}(u^\varepsilon))$$

$$+ \sum_{i=0}^{N-1} \left[a_\varepsilon(x)(u^\varepsilon - R_{\text{MsFEM},H}(u^\varepsilon))'(u^\varepsilon - R_{\text{MsFEM},H}(u^\varepsilon))\right]_{\frac{i}{N}}^{\frac{i+1}{N}}$$

$$(5.82)$$

integrating by parts over each interval. In the first term on the right-hand side, we note that

$$-\left(a_\varepsilon(x)(u^\varepsilon)'\right)' = f,$$

because of the original equation restricted to the interval, while by construction of the space $V_{\text{MsFEM},H}$ using the solutions to (5.76)–(5.77),

$$- \left(a_\varepsilon(x) \, (R_{\text{MsFEM},H}(u^\varepsilon)')\right)' = 0.$$

We thus obtain

$$- \int_{\frac{i}{N}}^{\frac{i+1}{N}} \left(a_\varepsilon(x) \, (u^\varepsilon - R_{\text{MsFEM},H}(u^\varepsilon)')\right)' \, (u^\varepsilon - R_{\text{MsFEM},H}(u^\varepsilon))$$

$$= \int_{\frac{i}{N}}^{\frac{i+1}{N}} f \, (u^\varepsilon - R_{\text{MsFEM},H}(u^\varepsilon)). \qquad (5.83)$$

In the above argument, we could have equally well used directly, extending by zero the functions manipulated outside the interval, the variational formulations (2.10) of the original problem (2.1) and of problems (5.76)–(5.77) to obtain (5.83). The second term on the right-hand side of (5.82) vanishes, since the function $u^\varepsilon - R_{\text{MsFEM},H}(u^\varepsilon)$ itself vanishes at all the nodes $\dfrac{i}{N}$ by definition of the interpolant, so that all the boundary terms are zero. Consequently,

$$\int_0^1 a_\varepsilon(x) \, \left| (u^\varepsilon - R_{\text{MsFEM},H}(u^\varepsilon))' \right|^2 = \sum_{i=0}^{N-1} \int_{\frac{i}{N}}^{\frac{i+1}{N}} f \, (u^\varepsilon - R_{\text{MsFEM},H}(u^\varepsilon)).$$

$$(5.84)$$

Again since each function $u^\varepsilon - R_{\text{MsFEM},H}(u^\varepsilon)$ vanishes at both ends of the interval $\left[\dfrac{i}{N}, \dfrac{i+1}{N} \right]$, we may bound the right-hand side from above using the Poincaré inequality. We indeed have, for each $0 \leq i \leq N - 1$,

$$\int_{\frac{i}{N}}^{\frac{i+1}{N}} \left| u^\varepsilon - R_{\text{MsFEM},H}(u^\varepsilon) \right|^2 \leq (C_H)^2 \int_{\frac{i}{N}}^{\frac{i+1}{N}} \left| (u^\varepsilon - R_{\text{MsFEM},H}(u^\varepsilon))' \right|^2$$

where C_H denotes the Poincaré constant of the domain $\left[\dfrac{i}{N}, \dfrac{i+1}{N} \right]$. It is easy to see that $C_H = H \, C_1$ where C_1 is the Poincaré constant of the domain $[0, 1]$. Indeed, one considers the inequality $\int_0^1 |u|^2 \leq (C_1)^2 \int_0^1 |u'|^2$, for an arbitrary function $u \in H_0^1([0, 1])$, next define $\tilde{u}(x) = u(H^{-1} x)$ and obtain $\int_0^H |\tilde{u}|^2 \leq (C_1)^2 H^2 \int_0^H |\tilde{u}'|^2$, for an arbitrary function $\tilde{u} \in H_0^1([0, H])$. The result is then translated to the interval $\left[\dfrac{i}{N}, \dfrac{i+1}{N}\right]$. Inserting this into (5.84), using the Cauchy-Schwarz inequality on each interval, and next its discrete variant to sum over the intervals, we find:

$$
\int_0^1 a_\varepsilon(x)\left|(u^\varepsilon - R_{\text{MsFEM},H}(u^\varepsilon))'\right|^2
$$

$$
\leq C_1 H \sum_{i=0}^{N-1} \left(\int_{\frac{i}{N}}^{\frac{i+1}{N}} |f|^2\right)^{1/2} \left(\int_{\frac{i}{N}}^{\frac{i+1}{N}} \left|(u^\varepsilon - R_{\text{MsFEM},H}(u^\varepsilon))'\right|^2\right)^{1/2},
$$

$$
\leq C_1 H \left(\int_0^1 |f|^2\right)^{1/2} \left(\int_0^1 \left|(u^\varepsilon - R_{\text{MsFEM},H}(u^\varepsilon))'\right|^2\right)^{1/2}.
$$

Using that the coerciveness of a_ε is uniform with respect to ε, we obtain

$$
\int_0^1 a_\varepsilon(x)\left|(u^\varepsilon - R_{\text{MsFEM},H}(u^\varepsilon))'\right|^2 \leq \frac{(C_1)^2}{\mu} H^2 \|f\|_{L^2([0,1])}^2.
$$

We next insert this information into the right-hand side of (5.81), so that

$$
\int_0^1 a_\varepsilon(x)\left|(u^\varepsilon - u^\varepsilon_{\text{MsFEM},H})'\right|^2 \leq \frac{(C_1)^2}{\mu} H^2 \|f\|_{L^2([0,1])}^2,
$$

and thus

$$
\left\|(u^\varepsilon - u^\varepsilon_{\text{MsFEM},H})'\right\|_{L^2([0,1])} \leq \frac{C_1}{\mu} H \, \|f\|_{L^2([0,1])}. \tag{5.85}
$$

The key point is: compare (5.85) with the a priori error estimate (2.23), that is,

$$
\left\|(u^\varepsilon - u^\varepsilon_H)'\right\|_{L^2([0,1])} \leq C H \left\|(u^\varepsilon)''\right\|_{L^2([0,1])} \tag{5.86}
$$

which has been obtained for the $\mathbb{P}_1$ finite element method. From a theoretical standpoint, classical elliptic regularity results (here particularly simple since the setting is one-dimensional) on the original problem (2.1) show that the norm $\left\|(u^\varepsilon)''\right\|_{L^2([0,1])}$ is bounded by the norm$\|f\|_{L^2([0,1])}$ of the right-hand side. The two error estimates thus *look* similar. The difference is however considerable. As mentioned in Chap. 2, the second derivative $(u^\varepsilon)''$ is of the order $\dfrac{1}{\varepsilon}$, whereas the norm $\|f\|_{L^2([0,1])}$ is *independent of ε.*

Therefore, the $MsFEM$ approach is guaranteed to yield accurate results for H small and not only for $\dfrac{H}{\varepsilon}$ small! The theory is illustrated by the experimental

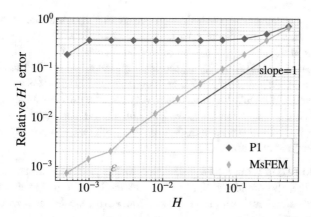

Fig. 5.9 Error estimates (5.85) and (5.86) for $\varepsilon = 2 \times 10^{-2}$. We observe that, for $H \ll \varepsilon$, both errors scale as H as it decays. Similarly, for H large, the two approaches yield a large error. In the intermediate regime of values of H, $MsFEM$ is significantly more accurate, since the asymptotic regime in H small is reached sooner than for classical finite elements. Numerical results courtesy of Rutger Biezemans

results. We see on Fig. 5.9 that for moderately small values of H (typically $H = 0.1$, that is 5 times as large as $\varepsilon = 2 \times 10^{-2}$), $MsFEM$ yields a smaller error than the classical approach. In addition, $MsFEM$ has already reached there the asymptotic regime where the error scales as H, while this asymptotic regime is only reached for much smaller values $H \ll \varepsilon$ in the classical approach. This clearly emphasizes the instrumental role of the prefactor in the estimations (5.85) and (5.86).

Interestingly, the above error analysis has been performed without any specific structure assumptions on the coefficient a_ε in (2.1), which might thus be completely *general*, provided it is bounded and coercive. In particular, we have *not* assumed $a_\varepsilon = a\left(\dfrac{\cdot}{\varepsilon}\right)$, let alone that a is periodic or stationary.

The a priori error estimate justifies, at least in dimension $d = 1$, the introduction of the multiscale computational approach. The computational cost is similar to that of a classical $\mathbb{P}_1$ finite element method, since the dimension of the approximation space is identical. Note that we neglect the cost of the integrals $\displaystyle\int f \chi_{MsFEM,i,h}$ on the right-hand sides of the discrete variational formulation. For the same cost, the approach provides much more accurate results. Nevertheless, the basis functions $\chi_{MsFEM,i}$ have to be pre-computed.

Let us also notice that, assuming that the basis functions $\chi_{MsFEM,i}$ are explicitly known and the integrals $\displaystyle\int f \chi_{MsFEM,i}$ are calculated exactly, then the numerical solution u_H^ε provided by $MsFEM$ actually satisfies $u_H^\varepsilon(x_i) = u^\varepsilon(x_i)$ at all nodes x_i of the mesh. Put differently, the method is said to be "exact at the nodes". This property can easily be established using that each $MsFEM$ basis function makes $a_\varepsilon \chi'_{MsFEM,i}$ constant between two consecutive nodes of the mesh. It follows that the vector composed of the values of the exact solution at the nodes $u^\varepsilon(x_i)$ is actually a solution to the discrete variational formulation, and this is equal to the unique solution $u^\varepsilon_{H,i} = u^\varepsilon_H(x_i)$. The reader familiar with numerical methods will recognize here a typical property of the $\mathbb{P}_1$ finite element method *in dimension 1 for the equation* $-u'' = f$ *with constant coefficients.* The proof is actually exactly identical. Even if this only happens under very specific circumstances, the fact that the approach is exact at the nodes gives a definite confidence. Of course, one should however not infer from our remark that an approach that is not exact at the nodes cannot be accurate and efficient!

We would like to remark as a conclusion to this Sect. 5.2.1, that, to the best of our knowledge, the numerical analysis conducted here, although elementary, has not appeared anywhere. As far as we are concerned, we learnt it from our close colleague Alexei Lozinski.

The next section is devoted to the *MsFEM* approach and its numerical analysis in dimension $d = 2$.

5.2.2   The *MsFEM* Approach in Dimension $d = 2$

The extension of the approach to dimensions higher than or equal to 2 requires first to understand the construction of the basis functions $\chi_{\mathrm{MsFEM},i}$ earlier defined as the solutions to (5.76)–(5.77). We proceed here in the two-dimensional setting. Higher dimensional settings would require further, however analogous developments.

It turns out that the dimension $d = 1$ possesses a miraculous property that none other dimension $d \geq 2$ possesses. The boundary of a one-dimensional mesh element consists of a pair of points and thus is of dimension zero. Any functional space of functions defined on this boundary is therefore isomorphic to $\mathbb{R}^2$. More simply stated, in the example of (5.76), one may uniquely solve the equation $-\dfrac{d}{dx}\left(a_\varepsilon(x)\dfrac{d}{dx}\varphi_i(x)\right) = 0$ on the interval $\left[\dfrac{i}{N}, \dfrac{i+1}{N}\right]$ as soon as the two scalar conditions $\varphi_i\left(\dfrac{i}{N}\right) = 0$ and $\varphi_i\left(\dfrac{i+1}{N}\right) = 1$ are given.

From dimension $d = 2$, we may try and follow the same strategy. We consider solving

$$- \operatorname{div}(a_\varepsilon(x)\,\nabla\varphi_i(x)) = 0, \quad x \in \tau_k, \tag{5.87}$$

on a generic element, say a triangle τ_k, of the mesh $\mathcal{T}_H$. But what boundary condition should we impose on the edges of that triangle? All is open....

The history of development of multiscale approaches such as *MsFEM* can be revisited and interpreted as the quest for the Holy Grail, here the best possible boundary conditions to supply the local problems (5.87) with, on each element of the mesh. The difficulty did not exist originally. It is an artefact of the

discretization technique, which has created virtual boundaries (the boundaries of
the mesh elements). This is a substantial practical difficulty.

We make precise in the sequel the simplest idea, already present in the vanilla
version of the approach, that of *linear* boundary conditions. It gives rise to the so-
called $MsFEM - lin$ version of the approach. Many alternative ideas have been
developed since. We will mention some of them later on.

5.2.2.1 Description of the Approach

We make precise here the simplest possible construction of the multiscale basis
functions $\chi_{\text{MsFEM},i}$ from local problems that are a multidimensional extension (here
a two-dimensional extension) of the one-dimensional local problems (5.76)–(5.77).
On each triangle τ_k of the mesh $\mathcal{T}_H$ (see Fig. 5.10), of which the node i is a vertex,
we solve the following equation of the type (5.87),

$$- \operatorname{div}(a_\varepsilon(x) \nabla \varphi_{\text{MsFEM},i,k}(x)) = 0, \quad x \in \tau_k, \tag{5.88}$$

which we supply with the boundary conditions

$$\varphi_{\text{MsFEM},i,k}(x) = \begin{cases} 1 & \text{at the vertex } i \text{ of the triangle } \tau_k, \\ 0 & \text{on both other vertices of the triangle } \tau_k, \\ \text{affine along the edges.} \end{cases} \tag{5.89}$$

Put differently, $\varphi_{\text{MsFEM},i,k}$ is the multiscale solution to (5.88) that agrees along the
edges of the triangle τ_k with the $\mathbb{P}_1$ finite element function classically associated
to the vertex i (see Fig. 5.11). Note that, when $a_\varepsilon \equiv 1$ in (5.88), the multiscale
function and the $\mathbb{P}_1$ finite element function exactly coincide throughout τ_k. It

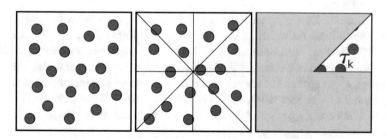

Fig. 5.10 (reproduced from [Tho12]) Schematic construction of a coarse mesh of size H on a
computational domain with small heterogeneities (the red disks, intuitively of size $\varepsilon \ll H$).
We show, from left to right, the computational domain, the coarse mesh composed of triangular
elements and such a coarse element τ_k on which the local problem of the type (5.88)–(5.89) is
solved (see Fig. 5.12 below)

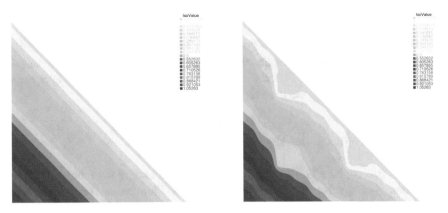

Fig. 5.11 Comparison, in dimension $d = 2$, of the profile of a classical finite element (on the left) and a $MsFEM$ basis function (on the right) defined by (5.88)–(5.89). The isovalues of the functions are shown for $H = \sqrt{2}$ and $\varepsilon = 0.4$. Figure courtesy of Rutger Biezemans

is straightforward to check that this definition extends that of dimension $d = 1$ in (5.76)–(5.77).

Similarly to what we did in (5.78) for dimension $d = 1$, we next construct $\chi_{\text{MsFEM},i}$ attaching the various functions $\varphi_{\text{MsFEM},i,k}$ for the triangles τ_k sharing the node i as a vertex. The function thus constructed is continuous across any edge $\Gamma_{k,k'}$ possibly shared by two such triangles τ_k and $\tau_{k'}$, since the two functions $\varphi_{\text{MsFEM},i,k}$ and $\varphi_{\text{MsFEM},i,k'}$ both coincide with the same affine function along that edge.

Note that, as in Sect. 5.2.1.1, we again do not make explicit the dependency of the functions $\varphi_{\text{MsFEM},i,k}$ and $\chi_{\text{MsFEM},i}$ upon the meshsize H and the small parameter ε.

The main difference with the one-dimensional construction is of course that, in any dimension $d \geq 2$, solving analytically the local problems and thus exactly identifying the basis functions $\chi_{\text{MsFEM},i}$ is hopeless. In practice, the functions $\chi_{\text{MsFEM},i}$ are numerically approximated in a pre-computing stage, called the *off-line* stage.

The discretization of each local problem (5.88)–(5.89) is performed using a discrete variational formulation, on a fine mesh $\mathcal{T}_h$ within the coarse element (see Fig. 5.12). The fine mesh elements have size $h \ll H$ (and more importantly $h \ll \varepsilon$). The approximation is a classical $\mathbb{P}_1$ finite element method, for example. Precisely since $h \ll \varepsilon$, the approach does not suffer from the drawback mentioned already in dimension $d = 1$ in Sect. 2.1.2. It provides an accurate approximation $\varphi_{\text{MsFEM},i,k,h}$ of each solution $\varphi_{\text{MsFEM},i,k}$ and thus an accurate approximation $\chi_{\text{MsFEM},i,h}$ of each exact basis function $\chi_{\text{MsFEM},i}$.

Each of the computations of the $\varphi_{\text{MsFEM},i,k,h}$ is potentially expensive. The triangles τ_k are however small as compared to the domain $\mathcal{D}$, since $H \ll |\mathcal{D}|$. The cost of each such computation is thus acceptable. More importantly, all such computations are independent of one another. They can be performed in parallel.

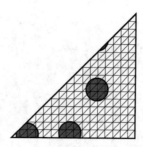

Fig. 5.12 [Tho12] Fine mesh, of size $h \ll \varepsilon$, over the coarse mesh element of size $H \gg \varepsilon$ isolated on the right of Fig. 5.10. The $MsFEM$ basis functions are approximated numerically solving problem (5.88)–(5.89). The computations are performed in parallel, for all coarse mesh elements

Neither their intrinsic complexity nor their number, which depends upon the number of nodes and triangles in the coarse mesh $\mathcal{T}_H$, is therefore an actual unsurmountable computational difficulty. In sharp contrast, a full, brute force computation over the whole domain $\mathcal{D}$, with elements of meshsize $h \ll \varepsilon \ll |\mathcal{D}|$, would be prohibitively expensive. Thus the necessity of a multiscale approach.

Once the approximations $\chi_{\mathrm{MsFEM},i,h}$ of the functions $\chi_{\mathrm{MsFEM},i}$ are known, the variational formulation of the original problem can be put in action, on the approximation space $V_{\mathrm{MsFEM},H,h}$. One proceeds exactly similarly to the one-dimensional setting, or similarly to a classical finite element approach over the mesh $\mathcal{T}_H$. The algebraic system, formally similar to (5.6) and of identical a size, is solved numerically. In sharp contrast to the first stage, the second stage of the computation we have just described depends on the right-hand side f of Eq. (1), and possibly on the specific boundary conditions the equation is supplied with if the latter are not simply (2). For this reason, the second stage is called the *online* stage. The overall algorithmic complexity of the second stage of the approach is comparable to that of a classical finite element applied on a mesh $\mathcal{T}_H$. It is observed in practice that the numerical results provided are however much more accurate, as shown for example on Fig. 5.13. As in the one-dimensional setting, we are even able to *establish* this better accuracy after a numerical analysis, under appropriate assumptions. We will proceed with this analysis below.

In spite of the efficiency and accuracy of the approach, the results obtained are far from perfect. One must always bear in mind that solving a multiscale problem is an intrinsically difficult task. One must not be too demanding. In the $MsFEM - lin$ approach described, the fact of the matter is that the linear boundary conditions imposed on the edges for the basis functions *cannot* be accurate. The exact solution u^ε is in nature oscillatory throughout the domain, and thus in particular along these edges. There is no way it is linear there. So the numerical solution procedure, potentially clever in the bulk of the triangles, is actually polluted by the linear approximation along the edges. Although it is not evident "in the norm of the eye" on Fig. 5.13, it is observed in practice that the numerical error is indeed larger on those edges.

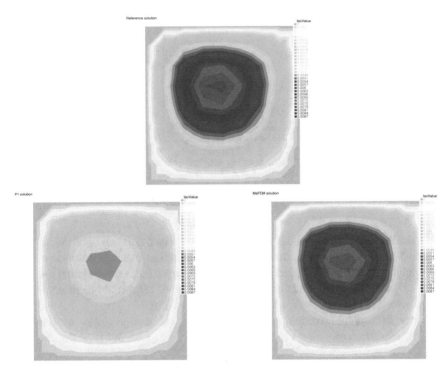

Fig. 5.13 Comparison of a $\mathbb{P}_1$ finite element method (bottom left) with $MsFEM$ (bottom right) for the approximation of the exact solution, shown on top using an accurate computation performed with FEM on a fine mesh. The right-hand side Eq. (1) is supplied with reads as $f(x) = \cos x \sin y$ on the unit square. The isovalues of each function, projected on the same coarse mesh of meshsize H, are shown for the value $H = 8\varepsilon$. The colour code is identical for the three functions. The $MsFEM$ solution is undoubtedly the best approximation. It satisfactorily captures the oscillations of the exact solution. On the other hand, classical FEM is not only unable to capture the fine oscillations, but it also provides an incorrect numerical result regarding the large scale structure of the exact solution. Numerical results courtesy of Frédéric Legoll

As we have already mentioned when we were discussing the truncation of the corrector problem on a bounded domain and were looking for a suitable boundary condition for that purpose, the ideal condition to impose would be a condition such that the exact solution u^ε indeed belongs to the approximation space thus constructed. But unfortunately, and of course, we do not know u^ε. We now present *three different strategies* to circumvent the difficulty and improve the accuracy of the numerical approach.

The most popular and efficient technique known to date is the *oversampling* technique. In the specific context of $MsFEM$, the technique proceeds from the following observation. Since the difficulty arises from the introduction of artificial boundaries (namely the edges of the triangles), why not pushing these artificial boundary further away? The oversampling technique consists in solving Eq. (5.87) not only on the triangle τ_k but on a triangle $\tilde{\tau}_k$ larger than the triangle τ_k (say 2–

Fig. 5.14 Instead of solving
the local
problem (5.88)–(5.89) posed
on the triangle τ_k considered
(see Fig. 5.12), the
oversampling method
considers the problem
analogous to (5.88)–(5.89) on
a triangle S larger than τ_k.
The basis function is then
obtained by restriction over
the original triangle. Figure
courtesy of Alexei Lozinski

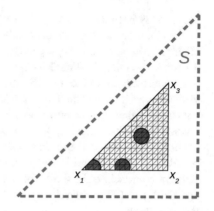

3 times as large as τ_k), homothetic to τ_k and containing τ_k (see Fig. 5.14). Some
rudimentary boundary conditions, for example linear, are then imposed on the edges
of the larger triangle $\widetilde{\tau}_k$. The solution to the boundary value problem so posed on $\widetilde{\tau}_k$
is finally *restricted* to the original triangle τ_k.

Since (5.87) is a diffusion equation, there is reasonable hope of having a trace
along the edges of τ_k that is relatively insensitive to the arbitrary choice of boundary
conditions imposed on the edges of $\partial\,\widetilde{\tau}_k$. A typical oscillatory solution is thus more
likely to be better captured. Both the numerical experiments and numerical analysis
show this is indeed the case. The rate of convergence is improved. In the error
estimate (5.108) that will be established in Sect. 5.2.2.2 below, the term $O\left(\sqrt{\dfrac{\varepsilon}{H}}\right)$
is replaced by the term $O\left(\dfrac{\varepsilon}{H}\right)$. We refer to [EH09] for further details on the
oversampling technique.

There is a price attached to the improvement. It is twofold. First, the size of
the local problems to solve is larger, so the off-line stage is more expensive. Since
the off-line stage is …off-line, and moreover parallel, this is not a major issue.
Second, and this is more concerning, the basis functions $\chi_{\mathrm{MsFEM},i}$ constructed
by the oversampling technique are not necessarily continuous across the edges.
Indeed, if two triangles τ_k and $\tau_{k'}$ share an edge $\Gamma_{kk'}$, the two local problems
solved on the corresponding enlarged triangles $\widetilde{\tau}_k$ and $\widetilde{\tau}_{k'}$ have not reason to
provide solutions $\chi_{\mathrm{MsFEM},k}$ and $\chi_{\mathrm{MsFEM},k'}$ that agree along $\Gamma_{kk'}$. Put differently,
the approximation space constructed *is not conforming*. Its elements are not
globally H^1 function throughout the domain $\mathcal{D}$. They are only piecewise H^1.
The approximation is not an *interior approximation*. The issue is, manipulating
"discontinuous" functions (or, more precisely, piecewise H^1 functions), particularly
for a diffusion problem, is neither a natural fact nor an easy matter. There are
technicalities. We again refer the reader who wishes to know more on such
approximations in general to the bibliography.

A variant of the above oversampling idea, which also aims at relaxing the
sensitivity of the problem with respect to the imposed artificial boundary conditions

has recently been developed in [LBLL13]. The case originally considered was that of Eq. (1). The method has been subsequently extended and improved to address other situations, such as an equation posed on a perforated domain [LBLL14], or the Stokes problem in Fluid Mechanics in [MNLD15, JL18], possibly with higher-order approximations like in [CEL19] or [FAO22]. In contrast to the oversampling idea, this variant keeps the local problem posed on the original triangle τ_k and does not augment that triangle into a larger one $\widetilde{\tau}_k$. It supplies the local problem in τ_k with boundary conditions on $\partial \tau_k$ that encode the *weak continuity* of the solution. Such conditions are called *Crouzeix-Raviart* type boundary conditions after the names of Michel Crouzeix and Pierre-Arnaud Raviart, their inventors for classical finite element approaches. In short, for each edge $\Gamma_{kk'}$ shared between the triangles τ_k and $\tau_{k'}$, Eq. (5.87) is solved in τ_k on the one hand and in $\tau_{k'}$ on the other hand with the conditions

$$
\begin{cases}
\fint_{\Gamma_{kk'}} \chi_{\text{MsFEM},i} = 0 \text{ or } 1 \\[2mm]
(a_\varepsilon \nabla \chi_{\text{MsFEM},i}) \cdot \mathbf{n}_{\Gamma_{kk'}} = \text{constant along the edge } \Gamma_{kk'}.
\end{cases}
\tag{5.90}
$$

In (5.90), the value 0 or 1 imposed to the average of $\chi_{\text{MsFEM},i}$ depends upon the three indices i, k, k' (exactly as in $MsFEM - lin$ the value 0 or 1 is assigned to each vertex of the triangle considered). The value of the normal flux of $a_\varepsilon \nabla \chi_{\text{MsFEM},i}$ is left free but forced to stay constant along each edge. The constant values may be different from one edge to the other. As in the oversampling technique, the approximation is not conforming. Each local problem has, however, kept its original size. Therefore, the enlargement factor, that is the homothetic ratio to transform τ_k into $\widetilde{\tau}_k$, which is a delicate discretization parameter to adjust, is not relevant any longer. With this approach, the hope is that, upon relaxing the strong imposition of assigning a definite arbitrary value to the numerical solution along the edges, more flexibility is left and an oscillation of the numerical solution is allowed also along these edges. The numerical experiments again confirm this hope. This is particularly the case for instance for problems posed on *perforated* domains, a category of problems that are practically relevant in many contexts. The numerical analysis that is available to date on the approach has however not allowed so far to theoretically confirm a better rate of convergence than with $MsFEM - lin$. More details are provided in [LBLL13].

To conclude, we mention a third and last variant. This extremely elegant variant has been originally introduced in a joint work [AB05] by Grégoire Allaire and Robert Brizzi. It also aims at improving the accuracy of the approximation specifically along the edges thus eventually everywhere throughout the computational domain. It however proceeds differently from the two variants we have just described.

In a classical finite element approach for a boundary value problem similar to (1)–(2) but with only *one* scale (that is, $\varepsilon \approx 1$), two options are mainly considered if the numerical accuracy is deemed unsatisfactory. The first option is to reduce the meshsize H. The second option is to replace the affine approximation of the $\mathbb{P}_1$ elements by a higher order approximation $\mathbb{P}_k$, of order $k \geq 2$. The idea introduced in [AB05] follows the second option.

For each $i = 1, \ldots, d$, we define, on the triangle τ_k, a function $\theta_{\text{MsFEM},i,k}$, which as $\varphi_{\text{MsFEM},i,k}$ satisfies Eq. (5.88) in τ_k,

$$- \operatorname{div}(a_\varepsilon(x) \, \nabla\theta_{\text{MsFEM},i,k}(x)) = 0, \quad x \in \tau_k \tag{5.91}$$

but we in addition impose that it satisfies

$$\theta_{\text{MsFEM},i,k} = x_i \quad \text{on } \partial \tau_k, \text{ for } 1 \leq i \leq d. \tag{5.92}$$

We next compose the solution to (5.91)–(5.92) with the polynomials $\chi_j^{\text{degree } K}$ of degree K that generate the approximation $\mathbb{P}_K$ on this triangle. We denote by $\Theta_{\text{MsFEM},k}$ the vector valued map with i-th component $\theta_{\text{MsFEM},i,k}$. Put differently,

$$\forall x \in \tau_k, \quad \Theta_{\text{MsFEM},k}(x) = \begin{pmatrix} \theta_{\text{MsFEM},1,k} \\ \vdots \\ \theta_{\text{MsFEM},d,k} \end{pmatrix}.$$

We eventually define the multiscale basis function as the composition

$$\chi_j^{\text{degré } K} \circ \Theta_{\text{MsFEM},k}. \tag{5.93}$$

When $a_\varepsilon \equiv 1$ in (5.91), $\theta_{\text{MsFEM},i,k} = x_i$ throughout τ_k, for all k and all i. The approach then coïncides with the classical $\mathbb{P}_K$ finite elements, just as $MsFEM - lin$ then coincides with $\mathbb{P}_1$. In addition, since $\theta_{\text{MsFEM},i,k}$ satisfies (5.92), the composed function (5.93) agrees with $\chi_j^{\text{degree } K}(x)$ on the boundary of τ_k. The basis thus constructed consists of functions that are continuous across the edges. In addition, the approximation along each edge is now a polynomial of degree K. It is likely to be more accurate than the affine approximation provided by $MsFEM - lin$. The specific choice $K = 1$ yields a composition with the finite element $\chi_j^{\text{degree } K=1} \in \mathbb{P}_1$ and the basis function coincides with the basis function $\varphi_{\text{MsFEM},j,k}$ of the $MsFEM - lin$ approach. Indeed, $\chi_j^{\text{degree } 1} \circ \Theta_{\text{MsFEM},k}$ solves (5.88) and agrees with $\chi_j^{\text{degree } 1}(x) = \chi_j(x)$ on the boundary τ_k, thus satisfies (5.89). Degrees $K \geq 2$ are likely to improve the approach.

Several other ideas besides the above mentioned three variants may be developed. This is still research in progress. The reader may wish to have a look at recently published research articles on the topic to realize this. We show on Fig. 5.15 an example of numerical results obtained with an advanced variant of $MsFEM$ in a somewhat complex situation.

We now turn to the analysis of $MsFEM - lin$.

Fig. 5.15 Example of an advanced application of $MsFEM$. A fluid flows in a channel obstructed by obstacles. The flow is modelled by the Stokes equation. We show on the left the geometry of the domain with a periodic array of obstacles (in red), in the center the reference solution and on the right the solution obtained by a certain variant of $MsFEM$ specifically adjusted to the case at hand. Figures courtesy of Alexei Lozinski and Gaspard Jankowiak, reproduced from [JL18]

5.2.2.2 Numerical Analysis of $MsFEM - lin$

In contrast to the one-dimensional setting studied in Sect. 5.2.1 and for which we were able to prove convergence and assess the rate of convergence without any structure assumption of the coefficient a_ε, we only know how to complete the same analysis in dimension $d = 2$ when we use homogenization theory and when we assume some specific form of the coefficient. We indeed need to assume that $a_\varepsilon = a_{per}\left(\dfrac{\cdot}{\varepsilon}\right)$ for some *periodic* coefficient a_{per}, satisfying the usual boundedness and coerciveness conditions. Such an assumption is frustrating, since we have emphasized that $MsFEM$ type methods have precisely been developed as alternatives to the classical homogenization approach. Such methods may be practically implemented and are successful far beyond the restricted setting and regime of homogenization. But, to the best of our knowledge, the only proofs of convergence available to date are valid under restrictive assumptions and exploit homogenization theory. Let us mention that, on the other hand, a "general" proof of convergence is however known for a variant of $MsFEM$, namely LOD, which we will briefly present in Sect. 5.2.4.

In addition to the assumption of periodicity, we also assume that the coefficient a_{per} is sufficiently regular (typically that it is Hölder continuous) so that the corrector function $w_{p,per}$ belongs to $W^{1,\infty}$. We acknowledge this is a technicality, indeed restrictive for the practice (composite materials typically correspond to piecewise constant coefficients that do not enjoy this regularity) but far less restrictive than the periodicity assumption indeed. In any event, it is unclear, in our present understanding, how to perform the proof in the absence of this additional assumption.

Another couple of assumptions is necessary. Those are "ad hoc" properties that we assume for simplicity of exposition.

We assume that the right-side f of Eq. (1) is $C^{0,\alpha}$—Hölder continuous, so that the homogenized solution u^* is, by elliptic regularity, $W^{2,\infty}$—Sobolev regular. The proof can be performed for $f \in L^2$ but it is slightly more technical. We also finally assume that the meshsize h of the fine mesh is zero. Put differently, the multiscale basis functions are assumed to be known exactly. This amounts to omitting in the global error estimate an h-dependent term. This term scales linearly in h when the finite elements used in the off-line stage for the approximation of the basis functions are $\mathbb{P}_1$ finite elements. More precisely, it is actually proportional to the ratio $\frac{h}{\varepsilon}$. If $h \neq 0$ and $\frac{h}{\varepsilon}$ is of comparable a size with the right-hand side of our error estimation (5.108) below, then this error term must be accounted for. It may be estimated using standard arguments of error estimation for classical finite element approximation. In our proof, we anyhow take $h = 0$ and, in our previous notation, our approximation space reads as $V_{\text{MsFEM},H,h=0}$, which we henceforth in this section denote by $V_{\text{MsFEM},H}$ for simplicity.

To start with, we apply Céa's Lemma in our specific context to assess the quality of the best approximation in $V_{\text{MsFEM},H}$. Since u^* is assumed H^2 (we actually have here $u^* \in W^{2,\infty}$), it is continuous in dimension $d = 2$. We may therefore consider its point values u_i^* at the nodes i. It is standard that

$$\left\| u^* - \sum_{i=1}^{N} u_i^* \chi_i \right\|_{H^1(\mathcal{D})} \leq CH \left\| u^* \right\|_{H^2(\mathcal{D})} \leq CH \left\| f \right\|_{L^2(\mathcal{D})}, \tag{5.94}$$

where, we recall, χ_i is the classical $\mathbb{P}_1$ finite element defined in (5.3). We now choose the interpolant of u^* in $V_{\text{MsFEM},H}$, that is $\bar{u} = \sum_{i=1}^{N} u_i^* \chi_{\text{MsFEM},i}$, as test function to insert on the right-hand side of Céa's Lemma. Using the triangle inequality we obtain

$$\left\| u^\varepsilon - u^\varepsilon_{\text{MsFEM},H} \right\|_{H^1(\mathcal{D})} \leq C \left(\left\| u^\varepsilon - u^{\varepsilon,1} \right\|_{H^1(\mathcal{D})} + \left\| u^{\varepsilon,1} - \bar{u} \right\|_{H^1(\mathcal{D})} \right). \tag{5.95}$$

Our theoretical results of Sect. 3.4.3.4, with (3.131), yield

$$\left\| u^\varepsilon - u^{\varepsilon,1} \right\|_{H^1(\mathcal{D})} \leq C\sqrt{\varepsilon} \left\| \nabla u^* \right\|_{W^{1,\infty}(\mathcal{D})}. \tag{5.96}$$

This result simply comes from periodic homogenization theory, and is not related to the numerical approximation. The sequel of the proof now consists in focusing on the numerical approximation itself, which is estimated by the term $\left\| u^{\varepsilon,1} - \bar{u} \right\|_{H^1(\mathcal{D})}$ in (5.95).

We note that, in order to repeatedly and legitimately apply above the elliptic regularity results of Chap. 3, we actually need a better regularity of the boundary $\partial \mathcal{D}$, namely a C^2 regularity typically. This regularity does *not* hold here since the domain $\mathcal{D}$ is assumed to be a polygon. Nevertheless, as mentioned previously (see the paragraph following (5.12)), it is possible to adapt the above arguments to our case. Here again, the details of such a proof go beyond the scope of the present textbook.

Again applying periodic homogenization theory, this time to the multiscale basis functions $\chi_{\text{MsFEM},i}$, we know that the functions obtained in the homogenized limit are precisely the classical $\mathbb{P}_1$ finite element denoted by χ_i. The periodic corrector problem being associated to the local problems (5.88)–(5.89) solved by the functions $\chi_{\text{MsFEM},i}$, we thus know that the two-scale expansion of the function $\chi_{\text{MsFEM},i}$ reads as

$$\chi_{\text{MsFEM},i} = \chi_i + \sum_{j=1}^{2} \varepsilon w_{e_j,per}\left(\frac{\cdot}{\varepsilon}\right) \partial_j \chi_i + \theta_i^{\varepsilon}, \quad i = 1, \ldots, N, \tag{5.97}$$

where θ_i^{ε} denotes the error term. Subtracting this expansion from the analogous expansion $u^{\varepsilon,1}$ of u^{ε}, we obtain

$$u^{\varepsilon,1} - \bar{u} = \left(u^* - \sum_{i=1}^{N} u_i^* \chi_i \right) + \sum_{j=1}^{2} \varepsilon w_{e_j,per}\left(\frac{\cdot}{\varepsilon}\right) \left[\partial_j u^* - \sum_{i=1}^{N} u_i^* \partial_j \chi_i \right]$$

$$- \sum_{i=1}^{N} u_i^* \theta_i^{\varepsilon}. \tag{5.98}$$

The first term of (5.98) is directly estimated using (5.94). As for the second term of (5.98), we combine (5.94) with the regularity $w_{e_j,per} \in W^{1,\infty}(Q)$, in order to first obtain, in the L^2 norm,

$$\left\| \sum_{j=1}^{2} \varepsilon w_{e_j,per}\left(\frac{\cdot}{\varepsilon}\right) \left[\partial_j u^* - \sum_{i=1}^{N} u_i^* \partial_j \chi_i \right] \right\|_{L^2(\mathcal{D})} \leq \varepsilon\, C\, H\, \|f\|_{L^2(\mathcal{D})}. \tag{5.99}$$

In order to similarly estimate the H^1 norm, we compute the gradient of the second term in (5.98). The first contribution is easy: as above we have

$$\left\| \sum_{j=1}^{2} \nabla w_{e_j,per}\left(\frac{\cdot}{\varepsilon}\right) \left[\partial_j u^* - \sum_{i=1}^{N} u_i^* \partial_j \chi_i \right] \right\|_{L^2(\mathcal{D})} \leq C\, H\, \|f\|_{L^2(\mathcal{D})}, \tag{5.100}$$

In the second contribution, the term

$$\left\| \sum_{j=1}^{2} \varepsilon w_{e_j,per}\left(\frac{\cdot}{\varepsilon}\right) \partial_j \nabla u^* \right\|_{L^2(\mathcal{D})} \leq C\varepsilon\, \|f\|_{L^2(\mathcal{D})}. \tag{5.101}$$

may be easily estimate proceeding as above. The second term

$$\varepsilon w_{e_j,per}\left(\frac{\cdot}{\varepsilon}\right)\nabla\left[\sum_{i=1}^{N}u_i^*\partial_j\chi_i\right],$$

however requires more care. The gradients $\nabla\chi_i$ of the finite element $\mathbb{P}_1$ are indeed not globally H^1 functions. We thus unite this term and the gradient of the third and last term of (5.98). We then have

$$\sum_{i=1}^{N}u_i^*\left(\nabla\theta_i^\varepsilon + \sum_{j=1}^{2}\varepsilon w_{e_j,per}(\cdot/\varepsilon)\partial_j\nabla\chi_i\right) \in L^2(\mathcal{D}),$$

since the left-hand side and the first term of (5.98) are $H^1(\mathcal{D})$ functions. Their gradients are thus $L^2(\mathcal{D})$. So is the term estimated in (5.100). Estimating (5.98) in H^1 norm thus amounts to estimating

$$\left\|\sum_{i=1}^{N}u_i^*\theta_i^\varepsilon\right\|_{L^2(\mathcal{D})} \tag{5.102}$$

and

$$\left\|\sum_{i=1}^{N}u_i^*(\nabla\theta_i^\varepsilon + \sum_{j=1}^{2}\varepsilon w_{e_j,per}(\cdot/\varepsilon)\partial_j\nabla\chi_i)\right\|_{L^2(\mathcal{D})}. \tag{5.103}$$

The delicate term is the second term (5.103). Since we *already* know that the function is $L^2(\mathcal{D})$, we may rightfully decompose its $L^2(\mathcal{D})$ norm on the set of all triangles τ_k of the mesh:

$$\left\|\sum_{i=1}^{N}u_i^*\left(\nabla\theta_i^\varepsilon + \sum_{j=1}^{2}\varepsilon w_{e_j,per}(\cdot/\varepsilon)\partial_j\nabla\chi_i\right)\right\|_{L^2(\mathcal{D})}^2$$

$$= \sum_{\tau_k\in\mathcal{T}_H}\left\|\sum_{i=1}^{N}u_i^*\left(\nabla\theta_i^\varepsilon + \sum_{j=1}^{2}\varepsilon w_{e_j,per}(\cdot/\varepsilon)\partial_j\nabla\chi_i\right)\right\|_{L^2(\tau_k)}^2$$

$$= \sum_{\tau_k\in\mathcal{T}_H}\left\|\sum_{i=1}^{N}u_i^*\nabla\theta_i^\varepsilon\right\|_{L^2(\tau_k)}^2, \tag{5.104}$$

since the gradients $\nabla \chi_i$ are constant on each triangle. Associating this to (5.102), we observe that, to be able to conclude, we only need to estimate

$$\left\| \sum_{i=1}^{N} u_i^* \theta_i^\varepsilon \right\|_{H^1(\tau_k)} . \tag{5.105}$$

The remainder (5.105) is a combination of all the remainders obtained in (5.97) when homogenizing each basis function $\chi_{\mathrm{MsFEM},i}$ into the corresponding function χ_i on each triangle τ_k, in the limit $\varepsilon \to 0$. Assessing the rate of this convergence requires to revisit step by step our study of Sect. 3.4.3.4 that leads to (3.131), while continually keeping track of the dependency of all constants with respect to the size of the domain, that is here H. We omit this part of the proof, which is not difficult but a bit tedious. Including it would only make our already long proof unnecessarily longer. We refer to the bibliography. The outcome of the full analysis is the following:

$$\left\| \sum_{i=1}^{N} u_i^* \theta_i^\varepsilon \right\|_{H^1(\tau_k)} \leq C \sqrt{\varepsilon} \sqrt{H} \left\| \sum_{i=1}^{N} u_i^* \nabla \chi_i \right\|_{W^{1,\infty}(\tau_k)}$$

$$= C \sqrt{\varepsilon} \sqrt{H} \left\| \sum_{i=1}^{N} u_i^* \nabla \chi_i \right\|_{L^\infty(\tau_k)} .$$

The latter equality originates from the fact that the $\nabla \chi_i$ are all constants on τ_k. Again because of this property, we have

$$\left\| \sum_{i=1}^{N} u_i^* \nabla \chi_i \right\|_{L^\infty(\tau_k)} \leq \frac{C}{H} \left\| \sum_{i=1}^{N} u_i^* \nabla \chi_i \right\|_{L^2(\tau_k)} ,$$

where the constant C is uniform in all triangles and in H, given the assumed regularity of the mesh $\mathcal{T}_H$. Summing on all the triangles, we obtain

$$\sqrt{\sum_{\tau_k \in \mathcal{T}_H} \left\| \sum_{i=1}^{N} u_i^* \theta_i^\varepsilon \right\|_{H^1(\tau_k)}^2} \leq C \sqrt{\frac{\varepsilon}{H}} \sqrt{\sum_{\tau_k \in \mathcal{T}_H} \left\| \sum_{i=1}^{N} u_i^* \nabla \chi_i \right\|_{L^2(\tau_k)}^2}$$

$$= C \sqrt{\frac{\varepsilon}{H}} \left\| \sum_{i=1}^{N} u_i^* \nabla \chi_i \right\|_{L^2(\mathcal{D})} .$$

Using first the triangle inequality and next (5.94), we bound the right-hand side from above:

$$\left\| \sum_{i=1}^{N} u_i^* \nabla \chi_i \right\|_{L^2(\mathcal{D})} \leq \left\| \nabla \left(u^* - \sum_{i=1}^{N} u_i^* \chi_i \right) \right\|_{L^2(\mathcal{D})} + \left\| \nabla u^* \right\|_{L^2(\mathcal{D})}$$

$$\leq C(H+1) \left\| f \right\|_{L^2(\mathcal{D})}.$$

We obtain

$$\sqrt{ \sum_{\tau_k \in \mathcal{T}_H} \left\| \sum_{i=1}^{N} u_i^* \theta_i^\varepsilon \right\|_{H^1(\tau_k)}^2 } \leq C \left(\sqrt{\varepsilon H} + \sqrt{\frac{\varepsilon}{H}} \right) \left\| f \right\|_{L^2(\mathcal{D})}. \tag{5.106}$$

Inserting the six estimates (5.94), (5.96), (5.99), (5.100), (5.101) and (5.106) into (5.95), we deduce

$$\left\| u^\varepsilon - u_{\mathrm{MsFEM},H}^\varepsilon \right\|_{H^1(\mathcal{D})} \leq C \sqrt{\varepsilon} \left\| \nabla u^* \right\|_{W^{1,\infty}(\mathcal{D})}$$

$$+ C(H + \varepsilon H + \varepsilon + \sqrt{\varepsilon H} + \sqrt{\varepsilon/H}) \left\| f \right\|_{L^2(\mathcal{D})}. \tag{5.107}$$

As the parameter ε is much smaller than the size of the domain $\mathcal{D}$, we have, for a constant C independent of ε, H and f:

$$\left\| u^\varepsilon - u_{\mathrm{MsFEM},H}^\varepsilon \right\|_{H^1(\mathcal{D})} \leq C \left(\sqrt{\varepsilon} + H + \sqrt{\frac{\varepsilon}{H}} \right) \left\| f \right\|_{L^2(\mathcal{D})}$$

$$+ C \sqrt{\varepsilon} \left\| \nabla u^* \right\|_{W^{1,\infty}(\mathcal{D})}. \tag{5.108}$$

The term $\sqrt{\dfrac{\varepsilon}{H}}$ deserves some comments. It is called the *resonance error* for a reason that, after the many years we have been working in the field and in spite of repeated explanations from colleagues, still eludes us. However it is called, the term is a consequence of the estimate (5.106), which in turn comes from the estimation of (5.103). It is important to note that its presence in the error estimate is not an artefact of the proof. Except in the miraculous one-dimensional setting where it disappears, such an error is indeed observed in the numerical practice. It originates from the poor approximation of the oscillatory solution by the $MsFEM-lin$ basis functions along and around the edges of the triangles. In the bulk of a triangle τ_k, the function $\chi_{\mathrm{MsFEM},i}$ accurately captures the oscillations of the exact solution, but the linear boundary conditions imposed impair the approximation locally at the vicinity of edges. As mentioned above, this particular error is improved when, say, the *oversampling* technique is employed, and this is reflected in the corresponding numerical analysis. In an ideal world, the numerical error should only be in $O(H)$, or, which in practice is the same since most often $\varepsilon \ll H$, in $O(\varepsilon + H)$. In the real world, the error observed in (5.108), and the analogous errors for other variants of $MsFEM$, are also dependent on $\dfrac{\varepsilon}{H}$.

We conclude this section upon mentioning that the convergence result (5.108) may be improved for $MsFEM - lin$ in various directions. This comes at the price of some sophisticated considerations. We will not say anymore on this.

An error estimate similar to (5.108) holds for the oversampling technique. It usually reads as the exact same estimate where $\dfrac{\varepsilon}{H}$ replaces $\sqrt{\dfrac{\varepsilon}{H}}$ (thereby confirming the above claim regarding the reduction of the resonance error). As for the approach introduced by Grégoire Allaire and Robert Brizzi and briefly presented in Sect. 5.2.2.1, the numerical analysis, again performed in the setting of periodic homogenization, also provides an error estimate similar to (5.108) where H^k replaces H, see [AB05, Theorem 4.1].

5.2.3 A Brief Description of HMM

The *Heterogeneous Multiscale Method*, abbreviated in HMM, is the principal competitor of $MsFEM$. Both methods are equally often used in practice by the experts of the topic. Most experts have a particular inclination to one method or the other, with all the reasons in the world to motivate their personal choice. In any event, the two methods are equally versatile. Both carry over to a large spectrum of contexts and equations. One should in effect speak about HMM-type methods and $MsFEM$-type methods, in the plural form, but we will keep things simple here.

As we shall see in this section, the two methods definitely have their differences, although they share some common features and success in terms of popularity and performance.

HMM has been introduced by Björn Engquist and Weinan E in [EE03], slightly after the introduction of $MsFEM$, although roughly around the same period. The method is more exhaustively reviewed in [AEEVE12]. Interestingly, in a different literature than that of applied mathematics, namely that of *Computational Engineering*, a similar, very close approach has been developed under the name of FE^2 (as in *Finite Elements squared*), by Frédéric Feyel and Jean-Louis Chaboche in [FC00].

The intuitive idea that underlies HMM is as elegant and cunning as that for $MsFEM$. Passing information from the finer scale to the larger one is absolutely necessary in a multiscale problem. Otherwise, if the finer scale is entirely discarded, the result obtained at the large scale is not even approximate, it is simply incorrect. On the other hand, it might not be necessary to pass *all* the information. Perhaps some information from the small scale could be sufficient to proceed, particularly

(i) when the structure at the fine scale has no wild variation (presumably a fair assumption in many contexts),

(ii) when the information is inserted in the numerical simulation at the large scale which in any event is only approximate
(iii) and when the purpose is only to obtain some averaged (or, more precisely stated, homogenized) information.

This is of course walking a fine line.   Practically calibrating the amount of information to pass is delicate.

In any event, the primary purpose of HMM is to *approximate the homogenized matrix A^* and the homogenized solution u^** inserting in the computation at the large scale only *some* information from the small scale. It is expected that

(i) the computation is approximately accurate (in contrast to a computation totally ignoring the small scale that would be incorrect),
(ii) but that the computational workload remains somewhat economical, both in the off-line and in the on-line stages, but particularly in the latter one.

The above claims hold for the original, vanilla version of HMM.   This allows to clearly discriminate the method from $MsFEM$, the purpose of which is to approximate the oscillatory solution u^ε itself (and not only u^*), at the price of accounting for "all" the information at the small scale (*via* the multiscale basis functions $\chi_{\text{MsFEM},i}$). In the subsequent development of HMM, however, a stage of *post-treatment* has been introduced.   This stage allows for also obtaining an approximation of u^ε, at some additional price.   We shall comment on that final stage later on.

We now describe the essence of the method.   We proceed in the specific case when the multiscale problem (1) *admits a homogenized limit*, namely (5.11), and we have at our disposal an expression, say of the type (3.54), which defines the homogenized coefficient. It is possible to adapt the approach in some other cases.

Following the pattern of the numerical resolution by a classical finite element method as recalled in Sect. 5.1.1, and in particular using the discrete variational formulation (5.5), we know that the homogenized problem (5.11)

$$\begin{cases} -\operatorname{div}(a^*(x)\,\nabla u^*(x)) = f(x) \text{ in } \mathcal{D}, \\ \qquad\quad u^*(x) \qquad\quad = 0 \qquad \text{on } \partial\mathcal{D}, \end{cases}$$

may be formulated, for say $\mathbb{P}_1$ finite elements, as

$$\sum_{j=1}^{N} \left[\int_{\mathcal{D}} a^*(x)\,\nabla\chi_i(x)\,.\,\nabla\chi_j(x)\,dx \right] (u_H^*)_j \; = \; \int_{\mathcal{D}} f(x)\,\chi_i(x)\,dx, \qquad (5.109)$$

for all $i = 1, \ldots, N$. Knowing the value of the integrand on the left-hand side at all points of the computational domain is not straightforward. This is also true generally for finite element methods. This is all the more true here as the matrix $a^*(x)$ is not "yet" known. In such a situation, it is customary to use *numerical quadrature formulae*, exactly as we did in Sect. 5.1.3 (and also in Sect. 5.1.1, page 263, when

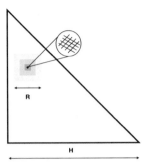

Fig. 5.16 *HMM* only approximates the value of the homogenized matrix at the Gauss nodes of the elements of the coarse mesh (here a triangle, of size H). A description of the microstructure at such points is considered, over a representative subdomain (here a square, of size R, smaller than H but larger than the characteristic size ε of the fine scale structure (pictured in the magnifying glass). On the representative subdomain is solved a problem that looks like the corrector problem, independent of the right-hand side of the equation originally considered. This yields the homogenized matrix at the Gauss point

evaluating the right-hand side). Using such a formula, each integral of the left-hand side is approximated by

$$\int_{\mathcal{D}} a^*(x)\, \nabla\chi_j(x) \cdot \nabla\chi_i(x)\, dx \;\approx\; \sum_{x_G} \omega_{x_G} a^*(x_G)\, \nabla\chi_j(x_G) \cdot \nabla\chi_i(x_G),$$

(5.110)

where the scalar parameters ω_G are called the *weights* of the quadrature formula and where the points x_G are the *nodes* of the quadrature, a.k.a. the *Gauss points*, thus the subscript G. Various such quadrature formulae exist, at different precision orders. Figure 5.16 illustrate the case of a triangle within a mesh of meshsize H that contains several quadrature points x_G. The essence of HMM is then to substitute for (5.109) the following formulation

$$\sum_{j=1}^{N} \left[\sum_{x_G} \omega_{x_G} a^*(x_G)\, \nabla\chi_i(x_G) \cdot \nabla\chi_j(x_G) \right] (u_H^{\mathrm{HMM}})_j = \int_{\mathcal{D}} f(x)\, \chi_i(x)\, dx,$$

(5.111)

and to only evaluate $a^*(x_G)$ at the quadrature nodes x_G. This is achieved using a local problem at the fine scale, literally mimicking the expression (3.54) of the homogenized matrix a^* from the periodic corrector problem (3.67). More explicitly, HMM assumes

$$a^*(x_G)_{ij} = \frac{1}{|Q_R(x_G)|} \int_{Q_R(x_G)} \big(a(y)\, (e_j + \nabla_y w_{e_j, R, x_G}(y))\big) \cdot e_i \, dy, \qquad (5.112)$$

where the integral, once suitably normalized, is taken on a small domain $Q_R(x_G)$, of characteristic size R (typically, $R \geq \varepsilon$ is a fraction of H) surrounding x_G, and the "corrector" w_{e_j,R,x_G} is defined as the solution to

$$\begin{cases} -\operatorname{div}\left(a(y)\left(e_j + \nabla w_{e_j,R,x_G}(y)\right)\right) = 0 \text{ in } Q_R(x_G), \\[2mm] w_{e_j,R,x_G} = 0 \qquad\qquad\qquad\quad \text{ on } \partial Q_R(x_G). \end{cases} \qquad (5.113)$$

The specific choice of a homogeneous Dirichlet boundary condition on $\partial Q_R(x_G)$ is of course entirely arbitrary. Other options exist.

It is immediately seen, from the above description, that *if* the microstructure were periodic, *if* $Q_R(x_G)$ were chosen as the periodic cell with $R = \varepsilon$ or one of its multiples, and *if* the boundary condition imposed were periodic, then irrespective of the choice of the points x_G, the value of $a^*(x_G)$ would be exact. All quadrature formulae being also exact for constant functions (argue triangle by triangle and use that the $\mathbb{P}_1$ finite element have constant gradient there), the variational formulation (5.110) would coincide with that of the homogenized equation (5.11).

For a more general coefficient $a(x)$ this is not the case. The approach only provides an approximation. In practice (see again Fig. 5.16), the size R of the domains $Q_R(x_G)$ where the problem at the small scale is *sampled* is taken *larger than* ε, so that the behaviour in those domains is accurately observed, but also *smaller than* H, so that there is a substantial gain in computational workload. All the computations for (5.112)–(5.113) are independent of the right-hand side f of the equation. They can be completed off-line. Each solution procedure for (5.113), say using classical finite elements on a fine mesh of meshsize h (as for *MsFEM*), is independent of all the other ones. They can all be processed in parallel.

It is interesting to note that, to some extent, *HMM* contains *MsFEM* as a particular case, choosing for $Q_R(x_G)$ the triangle x_G belongs to. Of course some details have to be adjusted, such as, besides the domains $Q_R(x_G)$, the boundary condition in (5.113) so that $x_j + w_{e_j,R,x_G}$ agrees with the multiscale basis function $\chi_{\text{MsFEM},i}$. One has to observe how much (5.113) ressembles the local problem (5.88), that is

$$-\operatorname{div}(a_\varepsilon(x) \nabla \varphi_{\text{MsFEM},i,k}(x)) = 0, \quad x \in \tau_k,$$

supplied with the suitable boundary conditions such as (5.89). Inserting this similarity into the variational formulation (5.109), the left-hand side is then recognized as that of the variational formulation of (1) on the basis functions $\chi_{\text{MsFEM},i}$. To some extent in this interpretation, the classical finite elements $\nabla \chi_i$ of the $\mathbb{P}_1$ method are only auxiliary quantities. As for the comparison between the right-hand side of (5.109) and the right-hand side $\int_{\mathcal{D}} f(x) \chi_{\text{MsFEM},i} \, dx$ of the variational formulation in *MsFEM*, we remark that, for ε small, the two-scale expansion (5.97) of the function $\chi_{\text{MsFEM},i}$ in particular shows that $\chi_{\text{MsFEM},i} \xrightarrow{\varepsilon \to 0} \chi_i$ weakly. In any event, up to such details, and with all the necessary precautions, $MsFEM$ may be indeed seen as a particular instance of HMM.

The above remark explains the existence, in HMM, of a possible *post-treatment* stage we have mentioned above. Once the approximation of u^* has been obtained by a standard HMM computation, it is a natural question to try and approximate u^ε itself. To that end, a possibility is to solve a specific local problem, again with classical finite elements on a fine scale mesh of meshsize h, around each point x_G. The domains are chosen such that, eventually collecting all of them, the entire coarse mesh is essentially covered. The local problem reads as

$$\begin{cases} -\operatorname{div}(a_\varepsilon(x)\,\nabla v^\varepsilon(x)) = f(x) & \text{around } x_G \\[2mm] v^\varepsilon = u_H^{\text{HMM}} & \text{on the boundaries,} \end{cases} \tag{5.114}$$

where the boundary value imposed is that which has just been computed using the standard HMM computation. An approximation of u^ε is then constructed from the local problems. Although described as efficient and accurate in the literature, this post-treatment is slightly counter intuitive. The major purpose, and the major asset of HMM, that is, satisfactorily approximating u^* at a limited cost and thereby striking a nice compromise between workload and accuracy, is somewhat impaired. A post-treatment stage that is eventually essentially as expensive as a full direct computation using $MsFEM$ is a nice option, but it requires a specific motivation.

There exists in the literature, as is the case for $MsFEM$, a numerical analysis of HMM in the case of the diffusion equation (in particular). Again as for $MsFEM$, we are unaware of any such analysis besides the periodic setting. The unfortunate fact is, again, that this idealized setting is certainly not the setting for which the approach is primarily meant. In addition, and specifically for HMM, the periodic setting is twice unfortunate since the sample at Gauss points is then fully representative of the microstructure (precisely by periodicity...). This clearly does not give a truthful idea of the behaviour of the approach in the generic case.

The main error estimate that plays for HMM the role that (5.108) plays for $MsFEM$ concerns this time the homogenized solution (and not the oscillatory solution) and reads as

$$\left\| u^* - u_H^{\text{HMM}} \right\|_{H^1(\mathcal{D})} \le C\left(H + \frac{\varepsilon}{R} \right) \tag{5.115}$$

where constant C depends on some norm of the right-hand side f, and where R is, we recall, chosen in $[\varepsilon, H]$ as a fraction of H. Since H itself is in practice chosen as a multiple of ε, say $H = 10\,\varepsilon$, the above error estimate scales as $O\left(H + \dfrac{\varepsilon}{H} \right)$. For instance choosing $H \propto \sqrt{\varepsilon}$, this estimate implies in particular the convergence of u_H^{HMM} to the homogenized solution u^*. Capturing u^ε requires to resort to a post-

treatment, as that mentioned above. We will not proceed with the numerical analysis
of that additional portion of the computation.

5.2.4 A Variant Slightly Different in Nature

We overview here an appealing approach that proceeds from a slightly different
perspective from $MsFEM - lin$. It shares, however, a number of common
features with $MsFEM - lin$. The approach carries the name of *Localized
Orthogonal Decomposition*, abbreviated in LOD. It has been originally introduced
by Axel Målqvist and Daniel Peterseim in [MP14]. Like the so-called Variational
Multiscale Method [HFMQ98], the LOD approach is based upon a decomposition
of the original problem into a coarse and (possibly several) fine scales problems.
Similarly to $MsFEM - lin$, it has been since adapted to various settings. We
only present here the original version of the approach. We refer the reader to the
recent monograph [MP21] and to the survey article [AHP21] for a comprehensive
presentation of the approach together with its multiple, recent extensions. We also
refer to [HP13] for an attempt to make connections between the LOD approach and
$MsFEM$.

We start from the variational formulation (3.2) of the original problem (2.1),
which we write here as

$$\mathcal{A}_\varepsilon (u, v) = \mathcal{L}(v), \quad \forall v \in H_0^1 (\mathcal{D}), \tag{5.116}$$

with

$$\mathcal{A}_\varepsilon (u, v) = \int_\mathcal{D} a_\varepsilon(x) \, \nabla u^\varepsilon (x) . \nabla v(x) \, dx, \tag{5.117}$$

and

$$\mathcal{L}(v) = \int_\mathcal{D} f v.$$

The right-hand side f has been taken in $L^2(\Omega)$, for simplicity.

We next introduce an *interpolation operator* $\mathcal{I}_H$, which is similar to the so-
called *Clément interpolation operator* at the nodes of the mesh $\mathcal{T}_H$ for the $\mathbb{P}_1$ finite
elements. There are several variants of such an interpolation operator that can be
used with the LOD approach, as for instance the operators defined in [MP14,
Section 2.2], [MP21, Section 3.3, Example 3.1] and also the *classical* Clément
interpolation operator as rigorously defined for instance in [Qua18, p 105]. We refer
to the bibliography for more details. In any event, this operator allows to assign to
each function $v \in H_0^1 (\mathcal{D})$ a "natural" value at each node. Strictly speaking, such
values do not necessarily exist, except in the particular case of dimension $d = 1$.

The underlying idea to define the operator $\mathcal{I}_H$ is both elementary and ingenious. The function is averaged over the mesh elements that share this node as a vertex. The value of the average is then taken as value at the given node. More precisely, for each node i in the mesh, $\mathcal{I}_H(v)(i)$ is defined by

$$\mathcal{I}_H(v)(i) = \left(\int_{\mathcal{D}} \chi_i(x)dx \right)^{-1} \int_{\mathcal{D}} v(x)\,\chi_i(x)\,dx, \qquad (5.118)$$

where χ_i is the nodal basis function associated to the node i, which we assume interior to the domain $\mathcal{D}$, in the sense of definition (5.3). As for the nodes on the boundary of the domain, we simply set $\mathcal{I}_H(v)(i) = 0$. For the approach we consider here, we refer to [MP14] for more details on the construction of this operator, defined from $H_0^1(\mathcal{D})$ to the classical $\mathbb{P}_1$ finite element space V_H on the mesh $\mathcal{T}_H$.

We next define the "fine-scale space":

$$V_f = \{ v \in H_0^1(\mathcal{D}),\, \mathcal{I}_H(v) = 0, \text{that is,}$$

$$v \text{ "vanishes" at the nodes of the mesh} \mathcal{T}_H \}, \qquad (5.119)$$

which is in fact the kernel

$$V_f = \text{Ker}\,\mathcal{I}_H \qquad (5.120)$$

of the operator $\mathcal{I}_H$. We then supplement this space V_f so that

$$H_0^1(\mathcal{D}) = V_f \oplus^{\perp^{\mathcal{A}_\varepsilon}} V_{\text{Ms}}, \qquad (5.121)$$

where the operator $\oplus^{\perp^{\mathcal{A}_\varepsilon}}$ denotes the orthogonal direct sum *for the scalar product defined by the symmetric bilinear form* (5.117). The relation (5.121) defines the space V_{Ms}, which will be the multiscale approximation space in the method.

We note in passing that V_f is an infinite dimensional space that has *finite codimension*. The dimension of the space V_{Ms} is indeed equal to the number of nodes in the mesh (excluding the nodes on the boundary $\partial\mathcal{D}$ since we approximate here the space $H_0^1(\mathcal{D})$).

In short, the functions in V_f are arbitrary but "vanish" at all nodes. The functions in the orthogonal space V_{Ms}, on the other hand, have arbitrary values at the nodes but, in spirit at least and in some particular settings exactly, solve equation $-\text{div}(a_\varepsilon(x)\nabla.) = 0$ outside the nodes. Put differently, the space V_{Ms} is the natural exact extension to the multidimensional setting of the space (5.79), namely

$$V_{\text{MsFEM},H} = \{ v \in C^0([0, 1]),$$

$$-\left(a_\varepsilon(x)\,v'(x) \right)' = 0 \quad \text{on each segment} \left[\frac{i}{N}, \frac{i+1}{N} \right] \}$$

in the one-dimensional version of $MsFEM$. Or, alternatively, the space V_{Ms} has the same dimension as the classical $\mathbb{P}_1$ finite element space V_H associated to the mesh $\mathcal{T}_H$, but is a refinement of that space since between the nodes it contains much finer and better adjusted approximations of tentative solutions of the oscillatory problem considered.

We now introduce P_H, the orthogonal projector (again in the sense of the scalar product defined by (5.117)) on the space V_f. This operator is defined by

$$\mathcal{A}_\varepsilon (v, P_H w) = \mathcal{A}_\varepsilon (v, w) \quad \forall v \in V_f, \quad \forall w \in H_0^1(\mathcal{D}).$$

We notice that the space V_{Ms} also reads as

$$V_{Ms} = (\mathrm{Id} - P_H) V_H. \tag{5.122}$$

Indeed, the inclusion $(\mathrm{Id} - P_H) V_H \subset V_{Ms}$ is clear, by definition of the projection operator P_H. Since V_H and V_{Ms} have identical dimension, (5.122) will be established as soon as we show that the operator $(\mathrm{Id} - P_H)$ is one-to-one on V_H. If $v_H \in V_H$ satisfies $(\mathrm{Id} - P_H) v_H = 0$, then $v_H = P_H v_H$, thus $v_H \in V_f$. It follows that $\mathcal{I}_H v_H = 0$ thus, using (5.118), that the scalar products $\int_{\mathcal{D}} v_H \chi_i$ vanish for all finite elements χ_i. This yields $v_H = 0$ and concludes the proof of (5.122).

The variational formulation chosen for the approximation of (5.116) now reads as: *Find $u_{Ms} \in V_{Ms}$ such that, for all $v_{Ms} \in V_{Ms}$,*

$$\mathcal{A}_\varepsilon (u_{Ms}, v_{Ms}) = \mathcal{L}(v_{Ms}). \tag{5.123}$$

This formulation is implemented using a numerical approximation of a basis of V_{Ms}. Given (5.122), the natural strategy is to consider a basis of V_H, that is the nodal basis of finite elements χ_i defined in (5.3), and to choose as basis of V_{Ms} the functions $w_i = \chi_i - P_H \chi_i$, where the functions $P_H \chi_i$ are defined by

$$\mathcal{A}_\varepsilon (v, P_H \chi_i) = \mathcal{A}_\varepsilon (v, \chi_i) \quad \forall v \in V_f. \tag{5.124}$$

Formally (that is to say, temporarily assuming that the test functions in (5.124) are "all" functions and not only the elements of V_f), such functions satisfy, for each i,

$$- \mathrm{div} (a_\varepsilon(x) \nabla (P_H \chi_i)) = - \mathrm{div}(a_\varepsilon(x) \nabla \chi_i) \quad \text{in} \quad \mathcal{D}, \tag{5.125}$$

where, since χ_i is a $\mathbb{P}_1$ finite element, the gradient $\nabla \chi_i$ is a constant vector throughout the mesh element. We have to point out, though, that the functions w_i have no reason to be compactly supported (actually they are not, unless we work in dimension $d = 1$ where the two approaches $MsFEM$ and LOD in fact coincide), let alone have support in the mesh elements adjacent to the node i. In practice, it is customary to address this difficulty using a *truncation*. A theoretical analysis, which we do not present here, helps calibrating the truncation and assessing how it affects the accuracy. In our simplified presentation, we deliberately ignore this aspect and directly proceed with a discretization space spanned by the actual functions w_i themselves. We similarly ignore the fact, as we previously did, that, irrespective of the possible truncation, the functions w_i have to be *approximated* numerically. This is typically performed with a fine discretization of meshsize h. An additional numerical error arises from this particular discretization.

The numerical analysis of LOD is both elementary and illuminating. This is one of the reasons why we are so keen on presenting this approach in this introductory textbook. Although LOD is, for the time being at least, significantly less popular than, say, $MsFEM$, this Sect. 5.2.4 is fully justified in view of the peculiar theoretical properties of the approach.

We first decompose the exact solution u^ε in

$$u^\varepsilon = u_f + u_{\mathrm{Ms}}$$

along the orthogonal direct sum (5.121). We immediately remark that (and this motivates our particular notation) u_{Ms} is indeed the numerical solution defined in (5.123). Indeed, for all $v_{\mathrm{Ms}} \in V_{\mathrm{Ms}}$,

$$\mathcal{A}_\varepsilon (u_{\mathrm{Ms}}, v_{\mathrm{Ms}}) = \mathcal{A}_\varepsilon (u^\varepsilon - u_f, v_{\mathrm{Ms}}),$$
$$= \mathcal{A}_\varepsilon (u^\varepsilon, v_{\mathrm{Ms}}) - 0,$$
$$= \mathcal{L}(v_{\mathrm{Ms}}),$$

since, successively, $\mathcal{A}_\varepsilon (u_f, v_{\mathrm{Ms}}) = 0$ by orthogonality, and u^ε is the exact solution.

It follows that the numerical error, estimated in energy norm $\mathcal{A}_\varepsilon (., .)$, also reads as

$$\mathcal{A}_\varepsilon (u^\varepsilon - u_{\mathrm{Ms}}, u^\varepsilon - u_{\mathrm{Ms}}) = \mathcal{A}_\varepsilon (u_f, u_f),$$
$$= \mathcal{A}_\varepsilon (u_f + u_{\mathrm{Ms}}, u_f),$$
$$(\text{since } \mathcal{A}_\varepsilon (u_{\mathrm{Ms}}, u_f) = 0),$$
$$= \mathcal{L}(u_f),$$
$$(\text{since } u^\varepsilon = u_f + u_{\mathrm{Ms}} \text{ is the exact solution}),$$
$$= \mathcal{L}((\mathrm{Id} - \mathcal{I}_H)\, u_f),$$
$$(\text{since } \mathcal{I}_H u_f = 0),$$
$$= \mathcal{L}((\mathrm{Id} - \mathcal{I}_H)\, (u^\varepsilon - u_{\mathrm{Ms}})).$$

We obtain

$$\mathcal{A}_\varepsilon \left(u^\varepsilon - u_{\mathrm{Ms}}, \, u^\varepsilon - u_{\mathrm{Ms}}\right) \leq C \, \|f\|_{L^2(\mathcal{D})} \, \left\|(\mathrm{Id} - \mathcal{I}_H)\left(u^\varepsilon - u_{\mathrm{Ms}}\right)\right\|_{L^2(\mathcal{D})},$$

$$\leq C H \, \|f\|_{L^2(\mathcal{D})} \, \left\|u^\varepsilon - u_{\mathrm{Ms}}\right\|_{H^1(\mathcal{D})},$$

using a classical property of the interpolation operator for which we refer to the bibliography. The coerciveness of the bilinear form being uniform in ε, we have:

$$\left\|u^\varepsilon - u_{\mathrm{Ms}}\right\|_{H^1(\mathcal{D})} \leq C H \, \|f\|_{L^2(\mathcal{D})}, \tag{5.126}$$

for some constant C independent of ε. As for $MsFEM-lin$, we find, with (5.126), the convergence with respect to the coarse meshsize H, and this *uniformly in ε*. There are, however, two main differences with the analysis performed in Sect. 5.2.2.2 for $MsFEM-lin$. First, in contrast to (5.108), the estimation (5.126) does not involve any term in $\dfrac{\varepsilon}{H}$. It comes at a price. The local problems solved in LOD are, no matter how they are truncated, not as local as those in $MsFEM$, which implies an additional computational cost. Second, no homogenized limit has been used within the proof, and no particular assumption neither on the structure nor on the form of the coefficient a_ε has been necessary. We indeed do not need to assume that a_ε is a rescaled function, let alone that of a periodic function. Methodologically, the numerical error has not been bounded from above using a triangle inequality of the type

$$\left\|u^\varepsilon - u_H\right\| \leq \left\|u^\varepsilon - u^{\varepsilon,1}\right\| + \left\|u^{\varepsilon,1} - u_H\right\|.$$

The *key* ingredient in our above proofs is, on the other hand, the symmetry of the form $\mathcal{A}_\varepsilon(.,.)$. It allows to define a scalar product which we extensively use in the course of the proof. To a certain extent, this is also a limitation of the original version [MP14] of the LOD approach itself. Some recent extensions, for which we refer to [AHP21] and the references therein, carry over to the non-symmetric setting.

5.2.5 Some General Comments on Multiscale Approaches

The three examples of modern numerical approaches we have presented above, namely $MsFEM$ approaches, which we discussed at length, and HMM and LOD approaches which we only gave a flavour of, clearly show that the gain in terms of computational time is obtained upon a possibly intrusive implementation. This implementation is typically heavier than that for a classical approach put in action on a problem that involves only one scale. Its ingredients are problem-dependent. Two of the major assets of the original, classical finite element method therefore essentially vanish: the versatile character of the approach and its relative simplicity of implementation.

Another asset of the finite element approach is also considerably weakened, namely accuracy.

We have indeed noted (but presumably not sufficiently emphasized) that $MsFEM$ approaches typically provide results that are only a few percents accurate. It may even happen, sometimes, that the relative accuracy only reaches a few tens of percents. It is unclear yet whether we should be disappointed with such results, or, on the other hand, we should appreciate them favorably.

We devote this short section to the above issues. To fix the ideas, we will again discuss our favorite approach, namely $MsFEM$, which to some extent concentrates the issues, and even amplifies some of them. It should be borne in mind, however, that our discussion below also applies to the other approaches, either identically, or more critically, or to a slightly lesser extent.

It is rarely the case in scientific computing that one starts from scratch when developing a numerical simulation software. Every single scientist in this discipline, even every single graduate student, has accumulated on their laptop at least one piece, not to say a portfolio of pieces of simulation software. In the engineering sciences, that software is often a finite element code. Modifying this code, and a fortiori modifying it in depth, is not a decision taken lightly, because it may endanger thousands of hours of meticulous work.

Outside the academia, an in particular in the industrial sector, this is even more so. A piece of software usually capitalizes hundreds of years of engineering work. No one would dare to jeopardize this legacy code by allowing for unintentional mistakes. In addition, even if, for some good reasons, one were willing to risk this and introduce some modifications in the implementation or some new functionalities, the development procedures in place may forbid it. In a large variety of industrial activities indeed, all software issues are extremely sensitive. In nuclear engineering, in military applications, in aerospace engineering, software is subject to certification, which substantially complicates the landscape and drastically regulates all implementation efforts.

Modifying an existing finite element software in depth is however what is a priori required when implementing $MsFEM$ approaches. The basis functions are not any longer piecewise polynomials. They are the numerical solution to the local fine scale problems specifically defined in the approach. Each of these local problems should be solved separately, so that the numerical basis functions obtained are in turn inserted in the computation at the coarse scale. Intrusiveness is evident. This has motivated some recent works such as [BLLL22] where research efforts are directed toward implementing the existing $MsFEM$ approaches in a manner that is as little intrusive as possible, or specifically adjusting $MsFEM$ approaches in this spirit. We mention in passing that HMM approaches are naturally lighter and less intrusive to implement. On the other hand, LOD approaches are likely to be more delicate in this respect.

We now address the accuracy issue. In multiscale computational science, relative accuracies of a few, not to say a few tens, of percents are commonly observed. The outsiders, in particular if they are familiar with other categories of problems in scientific computing for which typical accuracies are a thousand times better, may judge such an accuracy disappointing and unsatisfactory. It is not reason enough, however, to enter an endless quest for a better accuracy that would unnecessarily complicates the approaches taken and would place a rising burden on implementation issues.

The reason is twofold. Multiscale problems are intrinsically difficult problems, and the practical concern is not necessarily to solve them with considerable accuracy.

In the engineering sciences, say for instance in mechanical engineering or in materials science, it is often the case that calculations scribbled on the back of an envelope, hasty empirical observations, numerical simulations conducted sloppily are all off by a factor 10, a factor 100, or even a factor one thousand. Gross errors in modelling, in the string of theoretical arguments, in the measurements, in the implementation are responsible for this. Eliminating these sources of errors using a rough but overall correct numerical simulation only 50% accurate is already a major contribution to the research effort. Similarly, computing the order of magnitude of the result expected when the equations or the parameters therein are only known in terms of orders of magnitude is already quite an achievement. In such circumstances, searching for an accurate, almost perfect result is an illusion, which, in the worst case scenario, may even have a counterproductive effect.

The above considerations are even more relevant in the multiscale context. Results with a 30% relative accuracy must not be swept aside and must, in sharp contrast, be considered seriously. The outsider should realize that, say, for a fluid mechanics problem where the Reynolds number is large and a boundary layer is observed at the vicinity of the walls, even the most sophisticated multiscale approach cannot capture the velocity profile within the boundary layer better than with a 50% accuracy.

The situation is all the more critical when implementation issues are accounted for. Using a multiscale approach (a) which is versatile, (b) the implementation of which is easy and as little as possible intrusive, and (c) which provides decently accurate results, is a perfectly honorable strategy. It is especially so if the numerical results obtained, however coarse, are certified using a numerical analysis and if error bounds are provided for. All is then in place for a concrete, reliable, even industrial exploitation of these results.

The strategy above may then be, in a second stage, usefully complemented using a more accurate, possibly sophisticated, intrusive, memory and time-consuming, numerical multiscale approach.

We are convinced that the future of multiscale approaches resides somewhere in a variety of approaches, ranging from the coarsest and the lightest to implement approaches to the most accurate and possibly heaviest to implement approaches.

In addition to the above remarks, we wish to conclude this section emphasizing that in this relatively recent, increasingly practically relevant and constantly evolving field of multiscale scientific computing, nothing is carved in stone yet. A lot of theoretical issues remain to be understood. A lot of practical problems remain to be suitably addressed. A perfect example is provided by random multiscale problems. All the multiscale approaches presented in this Sect. 5.2 a priori carry over to the random setting. Their efficiency originating from the adjustment of the basis functions employed to the specific problem solved, a random problem would in principle require adjusting the basis functions to each particular realization. It is unclear how to achieve this at reasonable implementation and computational costs. Despite some attempts in this direction in the literature, the random setting essentially remains, from that perspective, an open, challenging and practically relevant question.

5.3 Weakly Stochastic Problems

In this final section of this Chap. 5, we return to the classical strategy described in Sect. 5.1, consisting in computing the homogenized coefficient and the homogenized solution. We investigate some new settings

- that are random perturbations of the periodic setting and
- that may be addressed computationally using very efficient techniques.

The developments we present have been originally introduced in a series of relatively recent works [BLL07a, CLBL10a, ALB11]. Such works all proceed from the following observation: several practically relevant random situations are close to the periodic framework and may be efficiently addressed without applying the

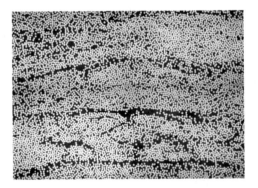

Fig. 5.17 (reproduced from [Tho08]) Real-world random structure: two-dimensional cut of a fiber-reinforced composite material used in aeronautics. The fibers are essentially perpendicular to the plane of the figure. Their cross-sections form a two-dimensional array, close but not identical to a periodic array, with various regions of defects where the distance between fibers varies. Fibers may be denser than in a periodic structure, or, on the other hand, space out

brute force general approach of the random case (random realizations, variance issues, etc.). Some methodological shortcuts may greatly reduce the computational workload, provided only approximative answers are sought. A prototypical example in materials science is shown on Fig. 5.17.

We successively present two specific modelling strategies: in Sect. 5.3.1, for periodic microstructures that are *randomly deformed* and in Sect. 5.3.2, for periodic microstructures that are *randomly perturbed*.

5.3.1 Randomly Deformed Periodic Structures

The first setting we consider makes use of what we call *random diffeomorphisms*. We will give the detailed mathematical setting in the next paragraph. Let us at once explain the *motivation*.

Consider a microstructure, that is, a particular coefficient a, that although somewhat messy, is, in a certain sense that will be made precise below, the presumably small *deformation* of a periodic microstructure, that is a periodic coefficient a_{per}. If the deformation that transforms the periodic microstructure into the actual one were entirely known, then the latter would essentially be identical to the former, once read in a certain new system of geometric coordinates (in the "right map", as the geometers say). The setting could be addressed simply using a straightforward adaptation of periodic homogenization. This does not require any particular new theoretical development.

On the other hand, assume that the deformation is not entirely known. We do know it enjoys nice regularity properties, and that it is likely to be only a small deformation. But that is essentially all we know. Then we may encode our lack of complete information on that deformation upon assuming it is *random* in nature. To some extent, shifting the viewpoint, it is *as if* we were looking at a periodic

Fig. 5.18 (reproduced from [CLBL10a, Figure 1]) Deformation by a random diffeomorphism of a periodic structure, composed of an array of disks centered in squares. The deformed disks are all different in shape, with a slightly diffuse boundary. They are however equal "in law". In some average sense, the deformed structure should intuitively inherit most homogenization properties of the original periodic structure. The parameters are $N = 5$ and $\eta = 0.05$

microstructure using a distorting glass. The problem is then both more delicate and interesting. An example is shown on Fig. 5.18, which looks like an eye examination in the practice of an ophtalmologist . . .

Of course, such a model is per se an *idealization*, which allows to mimic some of the features of real microstructures. In our above example of fiber reinforced material of Fig. 5.17, some of the areas, which may be easily identified, look like such a deformation of a periodic array of fibers. Some other areas, where the fibers have practically locally disappeared, will be addressed with a different modelling strategy, which we will describe in Sect. 5.3.2.

5.3.1.1 Mathematical Setting

Consider a periodic coefficient a_{per}. Assume, as usual in this textbook, that it has periodic cell $Q = [0, 1]^d$ and that it satisfies the boundedness and the coerciveness (3.1). Pick then, as coefficient to insert in the random multiscale equation (4.115), the random coefficient

$$a_\varepsilon(x, \omega) = a\left(\frac{x}{\varepsilon}, \omega\right) \qquad \text{for} \quad a(x, \omega) = a_{per}\left(\Phi^{-1}(x, \omega)\right), \qquad (5.127)$$

where the function $\Phi(\cdot, \omega)$ is assumed to be a *random diffeomorphism* from $\mathbb{R}^d$ to $\mathbb{R}^d$, that is, the application $\Phi(.\,\omega)$ is, almost surely, a diffeomorphism from $\mathbb{R}^d$ to $\mathbb{R}^d$. The notation Φ^{-1} designates the *reciprocal* function of the function Φ (and not the inverse function). The diffeomorphism $\Phi(\cdot, \omega)$ is, in addition, assumed to satisfy the following conditions

$$\inf_{\omega \in \Omega,\, x \in \mathbb{R}^d} \text{ess} \;\; [\det(\nabla\Phi(x, \omega))] = \nu > 0, \qquad (5.128)$$

$$\sup_{\omega \in \Omega,\, x \in \mathbb{R}^d} \text{ess} \;\; (|\nabla\Phi(x, \omega)|) = M < \infty, \qquad (5.129)$$

$$\nabla\Phi(x, \omega) \quad \text{is stationary (in the sense of (1.74)).} \qquad (5.130)$$

We call Φ a *stationary* random diffeomorphism. By calling it "stationary", we actually mean that the *gradient* $\nabla\Phi$ is stationary, not Φ itself. We will return to this. We should in fact more precisely speak of a *random diffeomorphism with stationary gradient*.

The two properties (5.128)–(5.129) respectively guarantee that the periodic environment deformed by Φ is nowhere arbitrarily compressed nor expanded. These assumptions are in the vein of our assumptions (H1)–(H2) in Sect. 1.3.3 regarding systems of interacting particles. Property (5.130), on the other hand, allows to carry over the homogenization properties of the periodic setting to the setting defined in (5.127). It is important to note that, even under the assumptions (5.128)–(5.129)–(5.130), a function such as (5.127) *is not necessarily stationary*. The theory we present here is therefore a *variant* of the random stationary theory. It is neither a particular case nor an extension thereof.

In order to intuitively understand the role of the stationarity property (5.130) and why it concerns $\nabla\Phi$ and not Φ, it is interesting to temporarily return to the setting of Chap. 1 and the issues we examined there regarding the calculation of *averages*. For a function of the type (5.127) defined on the real line $\mathbb{R}$, consider its average

$$\frac{1}{R}\int_0^R a(x,\omega)\,dx = \frac{1}{R}\int_0^R a_{per}\left(\Phi^{-1}(x,\omega)\right)\,dx.$$

Using the change of variable $y = \Phi^{-1}(x,\omega)$ in the integral, we obtain

$$\frac{\displaystyle\int_0^R a(x,\omega)\,dx}{R} = \frac{\dfrac{1}{\Phi^{-1}(R,\omega)-\Phi^{-1}(0,\omega)}\displaystyle\int_{\Phi^{-1}(0,\omega)}^{\Phi^{-1}(R,\omega)} a_{per}(y)\,\Phi'(y,\omega)\,dy}{\dfrac{1}{\Phi^{-1}(R,\omega)-\Phi^{-1}(0,\omega)}\displaystyle\int_{\Phi^{-1}(0,\omega)}^{\Phi^{-1}(R,\omega)}\Phi'(y,\omega)\,dy}.$$

Our assumption (5.129) on Φ now implies that, almost surely, the reciprocal function Φ^{-1} satisfies $\Phi^{-1}(R,\omega) - \Phi^{-1}(0,\omega) \geq M^{-1}R$. The function $\Phi'(y,\omega)$ *being stationary*, that is (5.130) being satisfied, the denominator of the above quotient converges, as $R \to +\infty$, to the expectation value $\mathbb{E}\displaystyle\int_0^1 \Phi'(y,\omega)\,dy$. On the other hand, since a_{per} is a deterministic periodic coefficient and $\Phi'(.,\omega)$ is stationary, the product $a_{per}\,\Phi'(.,\omega)$ is also stationary. The numerator thus converges to $\mathbb{E}\displaystyle\int_0^1 a_{per}(y)\,\Phi'(y,\omega)\,dy$. This yields:

$$\langle a\rangle = \left[\mathbb{E}\int_0^1 \Phi'(y,\omega)\,dy\right]^{-1}\mathbb{E}\int_0^1 a_{per}(y)\,\Phi'(y,\omega)\,dy.$$

We therefore observe that the key assumption that guarantees the existence of suitable averages in the setting (5.127) is (5.130) and indeed concerns the gradient of Φ and not Φ itself. We could have equally well introduced this setting as early as in Chap. 1 among the other settings that allow for the existence of averages. We prefer to only do this now. We also wish to notice that, actually, assuming that a_{per} is stationary rather than periodic is also a sufficient assumption to argue as above, since, then, the function $a\,\Phi'$ is stationary.

The argument conducted above on the coefficient a evidently applies to the *inverse* coefficient a^{-1}. We have therefore actually *proven* that homogenization theory is valid, at least in the one-dimensional setting, for (5.127)–(5.128)–(5.129)–(5.130), and that it yields the homogenized coefficient

$$\left(a^*\right)^{-1} = \left[\mathbb{E}\int_0^1 \Phi'(y,\omega)\,dy\right]^{-1}\mathbb{E}\int_0^1 \left(a_{per}\right)^{-1}(y)\,\Phi'(y,\omega)\,dy.$$

The question is now to understand whether a similar study may be completed for dimensions $d \geq 2$. The answer is yes. We refer the reader to [BLL07a] for such a study. Suffice it to say, in this introductory textbook, that the problem obtained in the homogenized limit reads as (3.8) where the homogenized coefficient is defined by

$$[a^*]_{ij} = \det \left(\mathbb{E} \left(\int_Q \nabla \Phi(z, \cdot) dz \right) \right)^{-1}$$

(5.131)

$$\times \mathbb{E} \left(\int_{\Phi(Q, \cdot)} e_i^T a_{per} \left(\Phi^{-1}(y, \cdot) \right) \left(e_j + \nabla w_{e_j}(y, \cdot) \right) dy \right),$$

with, for $p \in \mathbb{R}^d$,

$$\begin{cases} -\text{div} \left[a_{per} \left(\Phi^{-1}(y, \omega) \right) \left(p + \nabla w_p \right) \right] = 0, \\[2mm] w_p(y, \omega) = \widetilde{w}_p \left(\Phi^{-1}(y, \omega), \omega \right), \\ \qquad \nabla \widetilde{w}_p \text{ is stationary in the sense of (1.74),} \\[2mm] \mathbb{E} \left(\int_{\Phi(Q, \cdot)} \nabla w_p(y, \cdot) dy \right) = 0. \end{cases}$$

(5.132)

Observing the quite intricate form of Eqs. (5.131)–(5.132) makes one wonder whether the formalism introduced is not an unnecessary complication that we could have spared the reader. The fact of the matter is that the formalism is essentially useful because it involves a new mathematical object, namely the random diffeomorphism Φ, which, in a second stage which we now approach, may be conveniently adjusted. Indeed, we are now going to assume that Φ is close to the Identity map, and see how this allows us to significantly simplify our problem.

5.3.1.2 Small Deformation Theory

We now make the additional assumption that the random deformation introduced in the previous section is *small*. This encodes mathematically that, as mentioned above, the microstructure considered is close to a periodic microstructure. Assume

$$\Phi(x, \omega) = x + \eta \, \Psi(x, \omega).$$

(5.133)

In (5.133), the function $\Psi(., \omega)$ is a random function the gradient of which satisfies (5.130) and is supposed bounded. The coefficient η is a deterministic coefficient which is chosen small. Then, Φ satisfies (5.128)–(5.129). In the extreme situation where $\eta = 0$, the random diffeomorphism Φ is the identity map and the coefficient (5.127) is periodic.

It is then possible to expand the solution to the corrector problem (5.132) in powers of the small parameter η. We formally postulate that

$$\widetilde{w}_p(x, \omega) = w_{p,per}(x) + \eta\, w_p^1(x, \omega) + O(\eta^2),$$

where $w_{p,per}$ is the (deterministic) periodic corrector and w_p^1 is a certain first order term modification thereof. We next insert this expansion (called an "Ansatz" in Chap. 3) into Eq. (5.132). The equation necessarily satisfied by w_p^1 reads as:

$$\begin{cases} -\mathrm{div}\left[a_{per}\, \nabla w_p^1\right] = \mathrm{div}\left[-a_{per}\, \nabla\Psi\, \nabla w_{p,per}\right. \\ \qquad\qquad\qquad\qquad \left. -(\nabla\Psi^T - (\mathrm{div}\,\Psi)\mathrm{Id})\, a_{per}\, (p + \nabla w_{p,per})\right], \\ \nabla w_p^1 \text{ is stationary and } \mathbb{E}\left(\displaystyle\int_Q \nabla w_p^1\right) = 0. \end{cases}$$

$$(5.134)$$

Problem (5.134) defining w_p^1 (or, more precisely, its gradient) is a random problem. It enjoys a noticeable structure that greatly simplifies our task when solving it. It is indeed easy to check that the expectation $\overline{w}_p^1 = \mathbb{E}(w_p^1)$ is, by construction, a periodic function (since w_p^1 is stationary) that solves the *deterministic* problem

$$-\mathrm{div}\left[a_{per}\, \nabla\overline{w}_p^1\right] = \mathrm{div}\left[-a_{per}\, \mathbb{E}(\nabla\Psi)\, \nabla w_{p,per}\right.$$

$$\left. -(\mathbb{E}(\nabla\Psi^T) - \mathbb{E}(\mathrm{div}\,\Psi)\mathrm{Id})\, a_{per}\, (p + \nabla w_{p,per})\right]. \qquad (5.135)$$

obtained upon taking the expectation value in (5.134). This observation is key. The only knowledge of (the gradients of) $w_{p,per}$ and $\overline{w}_p^1$ is indeed sufficient to obtain the expansion at first order in η of the homogenized coefficient a^* defined in (5.131). The precise results established in [BLL07a] confirm that

$$a^* = a_{per}^* + \eta\, a^1 + O(\eta^2), \qquad (5.136)$$

for

$$a_{ij}^1 = -\int_Q \mathbb{E}(\mathrm{div}\,\Psi)\, [a_{per}^*]_{ij} + \int_Q (e_i + \nabla w_{e_i,per})^T a_{per}\, e_j\, \mathbb{E}(\mathrm{div}\,\Psi)$$

$$+ \int_Q \left(\nabla\overline{w}_{e_i}^1 - \mathbb{E}(\nabla\Psi)\nabla w_{e_i,per}\right)^T a_{per}\, e_j. \qquad (5.137)$$

We leave to the reader the easy exercise to check those formulae in dimension $d = 1$ so that it is conceivable they hold in any dimension. We emphasize that the above argument relies on the fact that we have started from a *bilinear* problem. In (4.115) and all the equations that follow, the only genuinely nonlinear term is of the form $a\, \nabla u$ where the coefficient multiplies the solution. The first order expansion of each of the two factors consequently involves equations that have only one random factor and are thus linear in the randomness. Taking the expectation value is therefore possible and effective. From the second order term on, new substantial difficulties would arise, where delicate *correlation* terms have to be addressed.

The remarkable consequence of the above development is a drastic simplification of the computational workload. In order to approximate a "slightly" deformed periodic microstructure, that is (5.127) and (5.133) for a small value of the parameter η, it is enough (provided an approximate result accurate at the second order η^2 is deemed acceptable) to only successively solve the periodic corrector problem, next a similar periodic problem (5.135) and finally to conclude with the elementary computations (5.137) and (5.136).

In view of the complexity and the computational cost of a direct resolution of (5.132) which are evidently comparable to that of the random stationary corrector problem, the gain is obvious. Solving *two deterministic periodic* problems is orders of magnitude less expensive than solving *one random stationary* problem. This comes at the price of only getting an *approximation* in η^2 of the result. The choice between the two options is the practitioner's to make.

We now turn to a second modelling strategy.

5.3.2   Randomly Perturbed Periodic Structures

We introduce here another possible mathematical formalization of the concept of "random perturbation" of a periodic microstructure, different from that of Sect. 5.3.1. A limitation of the previous approach is that the distance between the modified microstructure and the original one, measured say by the L^p norm of the difference $a(., \omega) - a_{per}$, should be small. Thus the deterministic coefficient η in (5.133). In the present section, the difference $a(., \omega) - a_{per}$ is kept of order 1 in all L^p norms, and in particular in L^∞ norm. It is only small in another type of distance, intrinsically random in nature. The setting has been introduced for random homogenization in [ALB11]. As later noticed in [DG16], the approach may be seen as an extension to the random perturbation of a periodic medium, both of the famous Rayleigh-Maxwell formula (see [ZKO94, page 45]), which addresses small deterministic perturbations in a specific deterministic setting, and of the no less famous Clausius-Mossotti formula, which deals with the random perturbation of a homogeneous medium. Consider

$$a_\eta(x, \omega) = a_{per}(x) + b_\eta(x, \omega) \left(c_{per}(x) - a_{per}(x) \right), \qquad (5.138)$$

instead of the coefficient (5.127) where the diffeomorphism is expanded as in (5.133). The coefficient c_{per} is typically another periodic structure, which is

Fig. 5.19 (reproduced from [ALB11, Figure 4.2]) Random perturbation of a periodic array of inclusions. Some inclusions were eliminated. Some others were left in place. Copyright ©2011 Society for Industrial and Applied Mathematics. Reprinted with permission. All rights reserved

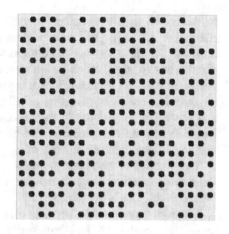

randomly mixed with a_{per}, using the random field

$$b_\eta(x, \omega) = \sum_{k \in \mathbb{Z}^d} \mathbf{1}_{Q+k}(x) B_\eta^k(\omega), \tag{5.139}$$

where the variables B_η^k are i.i.d. variables that take value 0 or 1. More precisely, using such a field, we locally and randomly replace the structure a_{per} with the structure c_{per}. The practically relevant case we study in depth is that where the B_η^k are Bernoulli variables of small parameter $\eta > 0$, that is

$$\mathbb{P}(B_\eta^k = 1) = \eta, \quad \mathbb{P}(B_\eta^k = 0) = 1 - \eta.$$

Figure 5.19 shows the structure we consider. It displays some areas where the inclusions are missing, exactly in the spirit of the regions where actual fibers are missing on Fig. 5.17.

Figure 5.19 is generated flipping a coin at each periodic cell. The inclusion within the cell is either kept or removed, depending on the outcome of the draw (which is a realization of that Bernoulli variable). Since the coin is unfair, because the parameter η is small, we keep most of the inclusions. But even if we were removing only one of them, the L^∞ distance with the original structure would be 1. On the other hand, it is still a small perturbation.

We now formalize this.

5.3.2.1 Formal Description of the Approach

The corrector problem obtained for the choice (5.138) of the random coefficient a_η reads as

$$-\operatorname{div}\left[a_\eta(y, \omega)\left(p + \nabla w_p(y, \omega)\right)\right] = 0. \tag{5.140}$$

In the above equation, randomness only appears through the presence of the coefficient $a_\eta(.\,,\omega)$. For each random realization of that coefficient, the solution $w_p(.\,,\omega)$ is otherwise entirely known deterministically (of course up to the addition of a constant, which we ignore in our discussion). It follows that, as soon as the law of $a_\eta(.\,,\omega)$ is known, that of the solution $w_p(.\,,\omega)$ is also known. We may next consider computing the corresponding homogenized coefficient a_η^*. The law of that coefficient only depends on the joint law of the pair $(a_\eta(.\,,\omega)\,,\ w_p(.\,,\omega))$, which in turn only depends on that of $a_\eta(.\,,\omega)$ since we have just established that w_p is a deterministic function of a_η.

The problem is that, apart from the law of a_η, nothing is explicitly known in the above discussion. Nevertheless, *for η small*, we may efficiently approximate all these objects. This is our purpose now.

Consider a presumably large cube $Q_N = [0, N]^d$, consisting of N^d periodic cells. The probability to find the periodic structure a_{per} throughout this cube is $(1 - \eta)^{N^d} \approx 1 - N^d \eta + O(\eta^2)$. Similarly, that of finding that periodic structure only modified in exactly one cell, where the structure a_{per} has been locally replaced by the structure c_{per}, reads as $N^d (1 - \eta)^{N^d - 1} \eta \approx N^d \eta + O(\eta^2)$. All other realizations contain at least two modified periodic cells and thus contribute in probability at most at a formal order η^2 (see Fig. 5.20). This estimation done on the back on an envelope gives the intuition that the homogenized coefficient might be expanded as

$$a_\eta^* = a_{per}^* + \eta\, a_{1,*} + o(\eta), \tag{5.141}$$

where a_{per}^* of course denotes the periodic homogenized coefficient. The first order term in η is obtained subtracting the periodic coefficient from the coefficient for

Fig. 5.20 (reproduced from [ALB11, Figure 3.1]) At the first order in the small parameter η, the only realizations to consider are those modified at only *one* place (on the left). At the second order, *couples* of modifications have to be considered (on the right). Copyright ©2011 Society for Industrial and Applied Mathematics. Reprinted with permission. All rights reserved

exactly one modified cell, that is

$$[a_{1,*}]_{ji} = \lim_{N \to +\infty} \int_{Q_N} \left[(a_{per} + \mathbf{1}_Q(c_{per} - a_{per}))(e_i + \nabla w_{e_i}^N) \right.$$

$$\left. - a_{per}(e_i + \nabla w_{e_i, per}) \right] . e_j, \qquad (5.142)$$

where $w_{e_i}^N$ solves

$$\begin{cases} -\text{div}\left((a_{per} + \mathbf{1}_Q(c_{per} - a_{per}))(e_i + \nabla w_{e_i}^N)\right) = 0 \quad \text{in} \quad Q_N, \\ \\ w_{e_i}^N \text{ is } Q_N - \text{periodic.} \end{cases} \qquad (5.143)$$

The fact that we locate the unique modified cell at the origin to formulate (5.142) is irrelevant, because of the periodic conditions imposed in (5.143). We emphasize that the integral on the right-hand side of (5.142) *is not normalized* by the volume of Q_N. The two contributions to its integrand compensate for one another, although they would separately yield a divergent integral that scales as the volume N^d of Q_N.

The expressions derived above on a sheer intuitive basis are confirmed by both the numerical practice and the theoretical analysis.

Practically, the two deterministic periodic problems, respectively the corrector problem posed on the periodic cell Q and the problem (5.143) posed on the domain Q_N, are approximated numerically. The coefficients a_{per} and (5.142) are next computed, and inserted in the expansion (5.141). Should need be, if this first order approximation is deemed insufficiently accurate, the second order coefficient $a_{2,*}$, which would appear next in (5.141), is similarly computed, considering twice modified realizations as on the right of Fig. 5.20 and the corresponding corrector problems similar to (5.143). All such problems form a *family* of partial differential equations of similar type, *parameterized* by the number and particular location of the modified cells. The numerical resolution can thus be accelerated using typical techniques for this category of problems, such as *reduced basis techniques*. We refer to [Qua18, Chapter 19] for a general pedagogic introduction to these techniques and to [LBT12] for the specific application to the context at hand. From a more general perspective, seeing the local problems as a *family* of parameterized problems related to one another is useful in several contexts. For example in the setting of the multiscale approaches described in Sect. 5.2 above, recent works proceed in this direction, such as [HKM20, HM19, MV22] in the context of the LOD method introduced in Sect. 5.2.4.

The size of the domain Q_N has to be chosen sufficiently large, in order to get an accurate result. Even for such a large value of N, the approach is efficient, since, as that of the previous section, it uses *deterministic* problems to bypass the extremely expensive computation of the original random problem. Each deterministic problem actually corresponds to a configuration locally modified and otherwise periodic, exactly as the problems with localized defects which were considered in Sect. 5.1.4.1. This establishes a nice and natural connection between these problems.

Also, and interestingly, we may use such expansions in powers of the small parameter η as a practical mean to *construct* a control variate. Then, as explained in Sect. 5.1.5.2, we may use that control variate to reduce the variance of a genuinely random problem with η *not* small, which is itself inaccessible using an expansion. We refer to [BLBL16, LM15] for more details on this approach.

5.3.2.2 Proof of the Expansion at First Order

We *prove* here that the first order expansion (5.141) formally derived above with the expression (5.142) of the homogenized coefficient at first order is indeed valid. The strategy of proof we present may actually be extended to an expansion at all orders in η and in more general settings (such as for a non-symmetric matrix valued coefficient a_η). It is borrowed from the work [DG16]. We only present here a sketch of the proof, in the simplified setting of a symmetric a_η.

Knowing the periodic corrector $w_{p,per}$ solution to (3.67), that is

$$-\operatorname{div}\left(a_{per}(y)\left(p+\nabla w_{p,per}(y)\right)\right)=0,$$

and the random corrector $w_{p,\eta}(.,\omega)$ solution to (4.116) for the coefficient $a_\eta(.,\omega)$, that is

$$\begin{cases} -\operatorname{div}(a_\eta(x,\omega)\left(p+\nabla w_{p,\eta}(x,\omega)\right))=0 \text{ in } \mathbb{R}^d, \\[2mm] \nabla w_{p,\eta}(x,\omega) \text{ stationary}, \\[2mm] \mathbb{E}\int_Q \nabla w_{p,\eta}(x,\omega)\,dx=0, \end{cases}$$

we may express, for $i,j \in \{1,\ldots,d\}$, the difference between the periodic homogenized coefficient a^*_{per} defined in (3.54)

$$\left[a^*_{per}\right]_{ij}=\int_Q a_{per}\left(e_j+\nabla w_{e_j,per}\right)\cdot e_i,$$

and the random homogenized coefficient a^*_η defined in (4.134)

$$\left[a^*_\eta\right]_{ij}=\mathbb{E}\int_Q a_\eta\left(e_j+\nabla w_{e_j,\eta}\right)\cdot e_i,$$

as

$$\left[a_\eta^* - a_{per}^* \right]_{ij}$$

$$= \mathbb{E} \int_Q \left(e_i + \nabla w_{e_i,per} \right) \cdot \left((a_\eta(.,\omega) - a_{per}) \left(e_j + \nabla w_{e_j,\eta}(.,\omega) \right) \right). \tag{5.144}$$

Checking the above formula is only a matter of expanding the product in the integrand, using the symmetry of a_{per} and the two Eqs. (3.67) and (4.116) and a sequence of integrations by parts.

The explicit expressions (5.138)–(5.139) of the coefficient a_η then yield

$$\left[a_\eta^* - a_{per}^* \right]_{ij} = \mathbb{E} \int_Q \left(e_i + \nabla w_{e_i,per} \right)$$

$$\cdot \left(\left(\sum_{k\in\mathbb{Z}^d} \mathbf{1}_{Q+k}(x) B_\eta^k(\omega)(c_{per} - a_{per}) \right) \left(e_j + \nabla w_{e_j,\eta}(.,\omega) \right) \right). \tag{5.145}$$

The key ingredient of the proof is the following shrewd observation by Mitia Duerinckx and Antoine Gloria in [DG16]. For $k \in \mathbb{Z}^d$ fixed, it is possible, in the product

$$B_\eta^k(\omega) \left(c_{per} - a_{per} \right) \left(e_j + \nabla w_{e_j,\eta}(.,\omega) \right), \tag{5.146}$$

to substitute for the corrector $w_{e_j,\eta}$ the modified corrector $w_{e_j,\eta}^{\cup\{k\}}$, solution to the same problem as (4.116) where the coefficient (5.138)–(5.139) is modified in

$$a_\eta^{\cup\{k\}}(.,\omega) = a_{per}(x) + \mathbf{1}_{Q+k}(x) \left(c_{per}(x) - a_{per}(x) \right)$$

$$+ \left(\sum_{k'\neq k\in\mathbb{Z}^d} \mathbf{1}_{Q+k'}(x) B_\eta^{k'}(\omega) \right) \left(c_{per}(x) - a_{per}(x) \right)$$

$$= a_\eta(x) + \mathbf{1}_{Q+k}(x) \left(1 - B_\eta^k(\omega) \right) \left(c_{per}(x) - a_{per}(x) \right). \tag{5.147}$$

The difference between the two coefficients $a_\eta(.,\omega)$ in (5.138)–(5.139) and $a_\eta^{\cup\{k\}}(.,\omega)$ in (5.147) is that, in the latter, the structure a_η of the k-th cell is almost surely replaced by the structure c_{per}. The product (5.146) is insensitive to the substitution, which would only modify $w_{e_j,\eta}$ if $B_\eta^k(\omega) = 0$, in which case the product vanishes no matter what. The simplification, however, is that the two factors $B_\eta^k(\omega) \left(c_{per} - a_{per} \right)$ and $e_j + \nabla w_{e_j,\eta}^{\cup\{k\}}(.,\omega)$ within the *new* product (5.146) are now *independent* random variables. This will be most useful.

We insert this modification (5.147) in (5.145) and obtain:

$$\left[a^*_\eta - a^*_{per} \right]_{ij}$$

$$= \mathbb{E} \int_Q \left(e_i + \nabla w_{e_i,per} \right)$$

$$\cdot \left(\sum_{k \in \mathbb{Z}^d} \mathbf{1}_{Q+k}(x) B^k_\eta(\omega)(c_{per} - a_{per}) \left(e_j + \nabla w^{\cup\{k\}}_{e_j,\eta}(.,\omega) \right) \right)$$

$$= \mathbb{E} \int_Q \left(e_i + \nabla w_{e_i,per} \right)$$

$$\cdot \left(\sum_{k \in \mathbb{Z}^d} \mathbf{1}_{Q+k}(x) B^k_\eta(\omega)(c_{per} - a_{per}) \left(e_j + \nabla w^{\{k\}}_{e_j}(.,\omega) \right) \right) \qquad (5.148)$$

$$+ \mathbb{E} \int_Q \left(e_i + \nabla w_{e_i,per} \right)$$

$$\cdot \left(\sum_{k \in \mathbb{Z}^d} \mathbf{1}_{Q+k}(x) B^k_\eta(\omega)(c_{per} - a_{per}) \nabla \left(w^{\cup\{k\}}_{e_j,\eta} - w^{\{k\}}_{e_j} \right)(.,\omega) \right).$$

$$(5.149)$$

We have denoted by $w^{\{k\}}_{e_j} = w^{\{0\}}_{e_j}(. - k)$ where $w^{\{0\}}_{e_j,\eta}$ solves the deterministic problem

$$- \operatorname{div}\left(\left(a_{per}(x) + \mathbf{1}_Q(x) \left(c_{per}(x) - a_{per}(x) \right) \right) \left(e_j + \nabla w^{\{0\}}_{e_j}(x) \right) \right) = 0 \qquad (5.150)$$

in $\mathbb{R}^d$, that is exactly the corrector problem for the microstructure on the left of Fig. 5.20. The function $w^{\{k\}}_{e_j}$, which is the deterministic corrector for one defect localized in the k-th cell, must not be mistaken for $w^{\cup\{k\}}_{e_j,\eta}$, which is the random corrector for a microstructure that has *at least* a defect in the k-th cell. Clearly enough, (5.143) is the truncated approximation on the cube Q_N of Eq. (5.150). So that $w^N_{e_i}$ is an approximation of $w^{\{0\}}_{e_i}$ on that domain.

Denote by Z_{ij} the integrand of (5.148), that is

$$Z_{ij}(x,\omega) =$$

$$\left(e_i + \nabla w_{e_i,per} \right) \cdot \left(\sum_{k \in \mathbb{Z}^d} \mathbf{1}_{Q+k}(x) B^k_\eta(\omega)(c_{per} - a_{per}) \left(e_j + \nabla w^{\{0\}}_{e_j}(. - k) \right) \right).$$

The integration of Z_{ij} in x over the cell Q only yields one contribution, that for the term $k = 0$. Therefore, (5.148) reads as

$$\mathbb{E} \int_Q Z_{ij}(x, \omega) dx$$

$$= \mathbb{E}(B_0(\omega)) \int_Q (e_i + \nabla w_{e_i, per}) \cdot \left((c_{per} - a_{per}) \left(e_j + \nabla w_{e_j}^{\{0\}} \right) \right)$$

$$= \eta \int_Q (e_i + \nabla w_{e_i, per}) \cdot \left((c_{per} - a_{per}) \left(e_j + \nabla w_{e_j}^{\{0\}} \right) \right). \qquad (5.151)$$

We now denote by

$$[a_{1,*}]_{ij} = \int_Q (e_i + \nabla w_{e_i, per}) \cdot \left((c_{per} - a_{per}) \left(e_j + \nabla w_{e_j}^{\{0\}} \right) \right). \qquad (5.152)$$

If we establish that the term (5.149) is of order $o(\eta)$, the expansion (5.141) will indeed be proven at the right order. Let us note that the coefficient (5.152) actually identically agrees with that formally defined in (5.142). Indeed, in (5.142),

$$\int_{Q_N} \left[(a_{per} + \mathbf{1}_Q(c_{per} - a_{per}))(e_i + \nabla w_{e_i}^N) - a_{per}(e_i + \nabla w_{e_i, per}) \right] \cdot e_j$$

$$= \int_{Q_N} \left[(a_{per} + \mathbf{1}_Q(c_{per} - a_{per}))(e_i + \nabla w_{e_i}^N) \right.$$

$$\left. - a_{per}(e_i + \nabla w_{e_i, per}) \right] \cdot (e_j + \nabla w_{e_j, per}) \qquad (5.153)$$

in view of Eq. (5.143), of the periodicity of $w_{e_i, per}$ on Q thus also on Q_N, and of the equation satisfied by the periodic corrector $w_{e_i, per}$. The right-hand side of (5.153) may therefore read as

$$\int_Q \left[(c_{per} - a_{per})(e_i + \nabla w_{e_i}^N) \right] \cdot (e_j + \nabla w_{e_j, per})$$

$$+ \int_{Q_N} \left[a_{per} \nabla(w_{e_i}^N - w_{e_i, per}) \right] \cdot (e_j + \nabla w_{e_j, per}), \qquad (5.154)$$

where, integrating by parts on Q_N, we observe that because of the periodic corrector equation (3.67), the second term is

$$\int_{\partial Q_N} (w_{e_i}^N - w_{e_i, per}) \cdot \left[a_{per} (e_j + \nabla w_{e_j, per}) \right] \cdot \mathbf{n}.$$

This second term vanishes, at least formally, in the limit $N \to +\infty$, since the solution $w_{e_i}^N$ to (5.143) converges to the solution $w_{e_i}^{\{0\}}(x)$ to (5.150), which in turns converges to the periodic corrector $w_{e_i,per}(x)$ as $|x| \to +\infty$.

In the limit $N \to +\infty$, only the first term of (5.154) survives, that is

$$\int_Q \left[(c_{per} - a_{per})(e_i + \nabla w_{e_i}^N) \right] \cdot (e_j + \nabla w_{e_j,per}), \tag{5.155}$$

rewriting (5.142). Again using that, in the limit $N \to +\infty$, $w_{e_i}^N$ converges to $w_{e_i}^{\{0\}}$, we obtain (5.152).

The above string of arguments is formal. It may be established rigorously, upon more precisely studying the behaviour at infinity of the solutions to the various partial differential equations involved. The sketch of proof we have just made bypasses this step upon using the definition (5.152) of the first order coefficient, rather than its "intuitive and algorithmic" expression (5.142). Collecting our arguments from (5.152) to (5.155) and justifying them step by step mathematically, the equivalence of the two expressions may be established.

Concluding our sketch of proof amounts to studying the term (5.149). We are going to see it is not only $o(\eta)$ but $O(\eta^2)$. To this end, we argue similarly as above. Subtracting the equation satisfied by $w_{e_j,\eta}^{\cup\{k\}}$ to that satisfied by $w_{e_j}^{\{k\}}$ yields

$$- \operatorname{div} \left((a_{per} + \mathbf{1}_{Q+k}(c_{per} - a_{per})) \nabla \left(w_{e_j,\eta}^{\cup\{k\}} - w_{e_j}^{\{k\}} \right) \right)$$

$$= \operatorname{div} \left(\sum_{k' \neq k \in \mathbb{Z}^d} B_\eta^{k'}(\omega) \mathbf{1}_{Q+k'}(c_{per} - a_{per}) \left(e_j + \nabla w_{e_j,\eta}^{\cup\{k\}} \right) \right). \tag{5.156}$$

Likewise, subtracting the equation satisfied by $w_{e_i}^{\{k\}}$ to that satisfied by $w_{e_i,per}$, we have

$$- \operatorname{div} \left((a_{per} + \mathbf{1}_{Q+k}(c_{per} - a_{per})) \nabla \left(w_{e_i,\eta}^{\{k\}} - w_{e_i,per} \right) \right)$$

$$= \operatorname{div} \left(\mathbf{1}_{Q+k}(c_{per} - a_{per}) \left(e_i + \nabla w_{e_i,per} \right) \right). \tag{5.157}$$

We next multiply (5.156) par $w_{e_i}^{\{k\}} - w_{e_i,per}$ and (5.157) by $w_{e_j,\eta}^{\cup\{k\}} - w_{e_j}^{\{k\}}$, again subtract the latter result from the former one, and integrate on the unit cube. An integration by parts yields

$$\int_Q \left(e_i + \nabla w_{e_i,per}\right) \cdot \mathbf{1}_{Q+k}(x)(c_{per} - a_{per})\, \nabla(w_{e_j,\eta}^{\cup\{k\}}(.,\omega) - w_{e_j}^{\{k\}}) =$$

$$\int_Q \left(\nabla(w_{e_i}^{\{k\}} - w_{e_i,per})\right)$$

$$\cdot \left(\sum_{k' \neq k \in \mathbb{Z}^d} \mathbf{1}_{Q+k'}(x) B_\eta^{k'}(\omega)(c_{per} - a_{per})\right) \left(e_j + \nabla w_{e_j,\eta}^{\cup\{k\}}(.,\omega)\right).$$

The boundary terms in the integration by parts do not vanish. Using the equations satisfied by $w_{e_i,per}$, $w_{e_i}^{\{k\}}$ and $w_{e_i,\eta}^{\cup\{k\}}$, we may however show they overall cancel out. Note that we have also used the symmetry of the matrices a_{per} and c_{per} to collect all the terms in volume. We now successively multiply this identity by $B_\eta^k(\omega)$, sum over $k \in \mathbb{Z}^d$ and take the expectation value. We obtain the following new expression for the term (5.149)

(5.149)

$$= \mathbb{E}\int_Q \sum_{k \in \mathbb{Z}^d} \sum_{k' \neq k \in \mathbb{Z}^d} \left(\nabla(w_{e_i}^{\{k\}} - w_{e_i,per})\right)$$

$$\left(\mathbf{1}_{Q+k'}(x) B_\eta^k(\omega) B_\eta^{k'}(\omega)(c_{per} - a_{per})\right) \left(e_j + \nabla w_{e_j,\eta}^{\cup\{k\}}(.,\omega)\right).$$

Using exactly the same observation as above, we may substitute for the corrector $w_{e_j,\eta}^{\cup\{k\}}$ the corrector $w_{e_j,\eta}^{\cup\{k\}\cup\{k'\}}$ solution to (4.116) with at least *two* defects for the coefficient

$$a_\eta^{\cup\{k\}\cup\{k'\}}(.,\omega) = a_{per}(x) + \left(\mathbf{1}_{Q+k}(x) + \mathbf{1}_{Q+k'}(x)\right)\left(c_{per}(x) - a_{per}(x)\right)$$

$$+ \left(\sum_{k'' \neq k, k' \in \mathbb{Z}^d} \mathbf{1}_{Q+k''}(x) B_\eta^{k''}(\omega)\right)\left(c_{per}(x) - a_{per}(x)\right), \qquad (5.158)$$

since this substitution only matters when $B_\eta^k(\omega) B_\eta^{k'}(\omega) = 0$ and nowhere affects the sum. We obtain

(5.149)

$$= \mathbb{E}\int_Q \sum_{k \in \mathbb{Z}^d} \sum_{k' \neq k \in \mathbb{Z}^d} \left(\nabla(w_{e_i}^{\{k\}} - w_{e_i,per})\right)$$

$$\left(\mathbf{1}_{Q+k'}(x) B_\eta^k(\omega) B_\eta^{k'}(\omega)(c_{per} - a_{per})\right) \left(e_j + \nabla w_{e_j,\eta}^{\cup\{k\}\cup\{k'\}}(.,\omega)\right).$$

The variables $B_\eta^k(\omega)B_\eta^{k'}(\omega)$ and $\nabla w_{e_j,\eta}^{\cup\{k\}\cup\{k'\}}(.\,,\omega)$ are *independent* (the latter only depends on the variable $B_\eta^{k''}(\omega)$ for $k'' \neq k, k'$, the two defects at the locations k, k' being present in any event). The other factors within the product are, on the other hand, deterministic. We may therefore "distribute" the expectation value. We obtain, for all $k, k' \in \mathbb{Z}^d$,

$$\mathbb{E}\int_Q \left(\nabla(w_{e_i}^{\{k\}} - w_{e_i,per})\right)$$

$$\cdot \left(\mathbf{1}_{Q+k'}(x)B_\eta^k(\omega)B_\eta^{k'}(\omega)(c_{per} - a_{per})\right)\left(e_j + \nabla w_{e_j,\eta}^{\cup\{k\}\cup\{k'\}}(.\,,\omega)\right)$$

$$= \eta^2\,\mathbb{E}\left(\int_Q \left(\nabla(w_{e_i}^{\{k\}} - w_{e_i,per})\right)\right.$$

$$\cdot \left(\mathbf{1}_{Q+k'}(x)(c_{per} - a_{per})\right)\left(e_j + \nabla w_{e_j,\eta}^{\cup\{k\}\cup\{k'\}}(.\,,\omega)\right)\bigg)$$

$$= \eta^2\,\delta_{0k'}\,\mathbb{E}\left(\int_Q \left(\nabla(w_{e_i}^{\{k\}} - w_{e_i,per})\right)\right.$$

$$\cdot \left((c_{per} - a_{per})\right)\left(e_j + \nabla w_{e_j,\eta}^{\cup\{k\}\cup\{0\}}(.\,,\omega)\right)\bigg),$$

(where δ is the Kronecker symbol). This holds since, on the one hand,

$$\mathbb{E}\left(B_\eta^k(\omega)B_\eta^{k'}(\omega)\right) = \mathbb{E}\left(B_\eta^k(\omega)\right)\mathbb{E}\left(B_\eta^{k'}(\omega)\right) = \eta^2,$$

while, on the other hand, $\mathbf{1}_{Q+k'}(x)$ is only non-zero on Q when $k' = 0$. Summing on k and k', we find

$$(5.149)$$

$$= \eta^2\,\mathbb{E}\left(\sum_{k\in\mathbb{Z}^d}\int_Q \left(\nabla(w_{e_i}^{\{k\}} - w_{e_i,per})\right)\right.$$

$$\cdot \left((c_{per} - a_{per})\right)\left(e_j + \nabla w_{e_j,\eta}^{\cup\{k\}\cup\{0\}}(.\,,\omega)\right)\bigg).$$

The series converges. It can indeed be proven arguing as we mentioned above. For large $|k|$, the corrector $w_{e_i}^{\{k\}}$ converges to $w_{e_i,per}$ on the cell Q, while the other factors stay uniformly bounded there. The convergence is fast because of the ellipticity of the equation. These facts may again be rigorously established. We finally obtain

$$\left[a_\eta^* - a_{per}^*\right]_{ij} - \eta\,[a_{1,*}]_{ij} = (5.149) = O\left(\eta^2\right),$$

which concludes our sketch of proof of the first order expansion (5.141).

As a matter of fact, this also concludes Chap. 5. All the chapters so far, from Chaps. 2 to 4 and the current chapter, have entirely been devoted to the diffusion equation (1), with the rare exception of our Sect. 2.5.

The final chapter of this textbook addresses other equations.

Chapter 6
Beyond the Diffusion Equation and Miscellaneous Topics

Abstract We present in this final chapter a selection of topics in homogenization theory. We consider other equations than the diffusion equation (1), or, for this very equation, techniques different in nature.

Keywords Advection-diffusion equation · Invariant measure · Γ-convergence theory · Stochastic process · Stochastic differential equation · Hamilton-Jacobi equation

We present in this final chapter a selection of topics in homogenization theory. We consider other equations than the diffusion equation (1), or, for this very equation, techniques different in nature.

The topics addressed are the following.

(i) We will overview in Sect. 6.1 homogenization theory for the *advection-diffusion equation*. The important mathematical tool for such linear equations, namely that of the *invariant measure*, will appear there.

(ii) We will introduce in Sect. 6.2 the *variational* techniques useful for homogenization when the equation under consideration admits an associated energy functional. For most of the chapters so far, this is the case and the energy functional is then quadratic. We will mention Γ-convergence theory, which is particularly well suited for a large class of well more general problems.

(iii) Section 6.3 will be devoted to homogenization using the interpretation of the diffusion equation in terms of *stochastic processes*. We will explain how the homogenized limit may be established, for a given deterministic equation, in the stochastic language as well as in the deterministic language. In passing, we will recall in Sect. 6.3.2 the intimate links between (possibly stochastic) differential equations and partial differential equations.

(iv) The homogenization of some "difficult", called *fully nonlinear* equations, namely *Hamilton-Jacobi* type equations will be briefly approached in Sect. 6.4, using the illustrative example of the one-dimensional Hamilton-Jacobi equation for a quadratic periodic potential.

X. Blanc, C. Le Bris, *Homogenization Theory for Multiscale Problems*, MS&A 21, https://doi.org/10.1007/978-3-031-21833-0_6

As a rule, which will be repeatedly emphasized in the chapter, we only provide the reader with some *initiation* to each topic. Our presentation is deliberately brief and simplified. Each topic is a world on its own. We only wish to mention some key ideas and phenomena that have not appeared so far in this textbook.

The very final section of the chapter, and thus that of the textbook, is Sect. 6.4.2. We express there some general considerations regarding the current status of homogenization theory and some feelings regarding its possible future. We are grateful to Pierre-Louis Lions for sharing with us some of his thoughts on these issues during some enlightening discussions. Our own views on the subject, which we express in this final section, in particular originate from these discussions.

6.1 The Advection-Diffusion Equation

Since we have now covered at length homogenization theory for Eq. (1), which, in an expanded form, reads as

$$- (a_\varepsilon)_{ij} (x) \, \partial^2_{ij} u^\varepsilon(x) - (\partial_i (a_\varepsilon)_{ij}) (x) \, \partial_j u^\varepsilon(x) = f(x),$$

where we use the convention of summation over repeated indices (see (4.43)), it is somewhat natural to wonder whether we may similarly proceed for equations of the type

$$- (a_\varepsilon)_{ij} (x) \, \partial^2_{ij} u^\varepsilon(x) + (b_\varepsilon)_j (x) \, \partial_j u^\varepsilon(x) = f(x), \qquad (6.1)$$

where $(b_\varepsilon)_j = - \partial_i (a_\varepsilon)_{ij}$ does not necessarily hold. Equation (6.1) is called an *advection-diffusion* equation. Its first term, that is $-a_{ij}(x) \, \partial^2_{ij} u$, is the *diffusion* term. The second one, $b_j \, \partial_j u$, is the *advection* (or *convection*) term. The particular case when $(a_\varepsilon)_{ij} (x) = a_{ij} \left(\dfrac{x}{\varepsilon} \right)$, for which

$$(\partial_i (a_\varepsilon)_{ij}) (x) = \frac{1}{\varepsilon} (\partial_i a_{ij}) \left(\frac{x}{\varepsilon} \right),$$

suggests, when the setting is oscillatory, to more precisely consider

$$- a_{ij} \left(\frac{x}{\varepsilon} \right) \partial^2_{ij} u^\varepsilon(x) + \frac{1}{\varepsilon} b_j \left(\frac{x}{\varepsilon} \right) \partial_j u^\varepsilon(x) = f(x) \qquad (6.2)$$

instead of (6.1) taken at $\dfrac{x}{\varepsilon}$.

Furthermore, the even more particular periodic setting, for which

$$\left\langle \partial_i \left(a_{per,\varepsilon} \right)_{ij} \right\rangle = 0,$$

also suggests to consider adding to Eq. (6.2) a condition on the average that somehow reads as

$$\langle b_j \rangle = 0, \tag{6.3}$$

or a condition in this vein. The present Sect. 6.1 is devoted to homogenization theory for Eq. (6.2). Most of our developments assume that all coefficients a_{ij} and b_j are periodic. We only temporarily leave this setting when considering some, now well known, perturbations of the periodic structure in Sect. 6.1.3.

The reader is, at this stage, familiar with our good habit to first examine a given problem in the one-dimensional setting before addressing the more general higher dimensional setting. We will not deviate here from this good habit. Sect. 6.1.1 deals with the one-dimensional setting. Section 6.1.2 considers dimensions $d \geq 2$.

6.1.1   The One-Dimensional Setting

Most of the arguments of this section have been already presented in Sect. 2.5.2, where we have first briefly introduced the advection-diffusion equation. We however give some more details here.

Expanding (1) has led us above to introduce Eq. (6.2), that is, for periodic coefficients,

$$- \left(a_{per} \right)_{ij} \left(\frac{x}{\varepsilon} \right) \partial_{ij}^2 u^{\varepsilon}(x) + \frac{1}{\varepsilon} \left(b_{per} \right)_j \left(\frac{x}{\varepsilon} \right) \partial_j u^{\varepsilon}(x) = f(x). \tag{6.4}$$

We consider this equation on the interval $[0, 1]$, and supply it with the homogeneous Dirichlet boundary conditions $u^{\varepsilon}(0) = u^{\varepsilon}(1) = 0$. For simplicity, we assume henceforth that

$$a_{ij,per} = \delta_{ij},$$

so that we may concentrate our attention on the term $- \dfrac{1}{\varepsilon} \left(b_{per} \right)_j \left(\dfrac{x}{\varepsilon} \right) \partial_j u^{\varepsilon}(x)$. In dimension $d = 1$, the above equation reads as

$$- (u^{\varepsilon})''(x) + \frac{1}{\varepsilon} b_{per} \left(\frac{x}{\varepsilon} \right) (u^{\varepsilon})'(x) = f(x). \tag{6.5}$$

Our above discussion has also suggested that we may wish to complement this equation with the condition (6.3), namely

$$\langle b_{per} \rangle = 0. \tag{6.6}$$

Let us first wonder whether (6.2) and (6.3), which we have just derived on the basis of a somewhat vague argument, is the right problem to consider. It is indeed the case. Other regimes for this particular equation would be much less interesting mathematically. We have already argued in this direction in Sect. 2.5.2. Let us reproduce here our argument for the reader's convenience.

To start with, we keep the condition (6.6) but *delete* the prefactor $\dfrac{1}{\varepsilon}$ in (6.5):

$$- (u^\varepsilon)'' (x) + b_{per} \left(\frac{x}{\varepsilon}\right) (u^\varepsilon)' (x) = f(x). \tag{6.7}$$

Introduce the primitive function $B_\varepsilon (x) = \displaystyle\int_0^x b_{per} \left(\frac{t}{\varepsilon}\right) dt$ of the function $b_{per} \left(\dfrac{x}{\varepsilon}\right)$. In the notation of Chap. 2, this function reads as $B_\varepsilon (x) = \varepsilon B \left(\frac{x}{\varepsilon}\right)$, where B is the primitive of b_{per} defined in Sect. 2.5.2 and appearing in (2.80). We leave to the reader the easy exercise to check that our formulae in the present section agree with those of Sect. 2.5.2.

Variation of parameters yield

$$(u^\varepsilon)' (x) = \left(\lambda_\varepsilon - \int_0^x f(t) \exp\left(-B_\varepsilon (t)\right) dt\right) \exp B_\varepsilon (x). \tag{6.8}$$

The constant λ_ε is adjusted so that $\displaystyle\int_0^1 (u^\varepsilon)' = 0$, that is

$$\lambda_\varepsilon = \left(\int_0^1 \exp B_\varepsilon (x) \, dx\right)^{-1} \int_0^1 \left(\int_0^x f(t) \exp\left(-B_\varepsilon (t)\right) dt\right) \exp B_\varepsilon (x) \, dx.$$
$$\tag{6.9}$$

The homogeneous Dirichlet conditions $u^\varepsilon (0) = u^\varepsilon (1) = 0$ have to be satisfied after a second integration. Therefore

$$(u^\varepsilon)'' (x) = - f(x) + b_{per} \left(\frac{x}{\varepsilon}\right) \left(\lambda_\varepsilon - \int_0^x f(t) \exp\left(-B_\varepsilon (t)\right) dt\right) \exp B_\varepsilon (x). \tag{6.10}$$

As $\varepsilon \to 0$, we know, for instance using (1.12) in the proof of Proposition 1.1, that $B_\varepsilon (x) \to \langle b_{per}\rangle x$, thus to zero, uniformly in the interval $[0, 1]$. We obtain that, in the limit $\varepsilon \to 0$,

$$\left(\lambda_\varepsilon - \int_0^x f(t) \exp\left(-B_\varepsilon (t)\right) dt\right) \exp B_\varepsilon (x) \longrightarrow \int_0^1 \int_0^x f(t) \, dt \, dx - \int_0^x f$$

uniformly. Inserting this convergence into the expression of $(u^\varepsilon)'' (x)$, we obtain

$$(u^\varepsilon)'' \longrightarrow - f,$$

in $L^2([0, 1])$ for example, if we have assumed that the right-hand side is indeed in that space. Put differently, the homogenized limit of (6.7) is $- (u^*)'' = f$. It agrees with the homogenized limit in the absence of the advection term $b_{per} \left(\dfrac{x}{\varepsilon} \right) (u^\varepsilon)'(x)$. The reader may easily check that an analogous conclusion holds upon replacing in (6.7) the diffusion term $- (u^\varepsilon)''(x)$ by $- a_{per} \left(\dfrac{x}{\varepsilon} \right) (u^\varepsilon)''(x)$ or $- \left(a_{per} \left(\dfrac{x}{\varepsilon} \right) (u^\varepsilon)'(x) \right)'$.

The combination of the elimination of the prefactor $\dfrac{1}{\varepsilon}$ and the average condition (6.6) makes the advection term negligible with respect to the diffusion term in the homogenized limit. This is the very motivation of amplifying the advection using to coefficient $\dfrac{1}{\varepsilon}$ to compensate for this. A competition between the two terms is then possible. As a matter of fact, even if we eliminate condition (6.6) while again discarding the prefactor $\dfrac{1}{\varepsilon}$, nothing interesting happens. Using the same expressions (6.8)–(6.9)–(6.10), this time with $\langle b_{per} \rangle \neq 0$, we literally *read* on these expressions that the homogenized limit of each term is again identical, except for the fact that the limit of B_ε does no longer vanish but is the linear function $\langle b_{per} \rangle x$. A similar calculation yields the homogenized equation

$$- (u^*)''(x) + \langle b_{per} \rangle (u^*)'(x) = f(x),$$

Nothing fancy there.

Even in the presence of $- \left(a_{per} \left(\dfrac{x}{\varepsilon} \right) (u^\varepsilon)'(x) \right)'$, we would have found

$$- \left(\langle a_{per}^{-1} \rangle \right)^{-1} (u^*)''(x) + \langle b_{per} \rangle (u^*)'(x) = f(x).$$

The two terms homogenize *"separately"* and there is no interesting competition nor combination.

The conclusion is that the prefactor $\dfrac{1}{\varepsilon}$ must be present in the first order term so that this term becomes comparable to a second order term in the homogenized limit. The alchemy between the two different terms then deserves mathematical attention.

Somewhat symmetrically to what we just did, let us now keep the prefactor $\dfrac{1}{\varepsilon}$ but discard condition (6.6). We do so in the particular case $b_{per} \equiv 1$ and $f \equiv 1$. This is enough to understand. We have

$$- (u^\varepsilon)''(x) + \frac{1}{\varepsilon} (u^\varepsilon)'(x) = 1. \tag{6.11}$$

It is immediate to see that

$$(u^\varepsilon)'(x) = - \left(\frac{\exp \left(\frac{x}{\varepsilon} \right)}{\exp \left(\frac{1}{\varepsilon} \right) - 1} \right) + \varepsilon, \tag{6.12}$$

where the integration constant to identify the derivative $(u^\varepsilon)'$ has been adjusted so that $\int_0^1 (u^\varepsilon)' = 0$, given that $u^\varepsilon(0) = u^\varepsilon(1) = 0$. We read on that expression that $(u^\varepsilon)' \xrightarrow[\varepsilon \to 0]{} 0$, uniformly for all $x \in [0, 1]$. Likewise,

$$u^\varepsilon \xrightarrow[\varepsilon \to 0]{} 0, \quad \text{uniformly in } x \in [0, 1],$$

again because of the boundary condition. Consequently, (6.11) cannot admit a homogenized limit with a right-hand side equal to the function 1. The coefficient $\frac{1}{\varepsilon}$ blows up asymptotically and is not compensated for. The solution is correspondingly forced to identically vanish.

We may make this convergence more precise. Because of (6.12), we actually have

$$\frac{1}{\varepsilon}(u^\varepsilon)' \xrightarrow[\varepsilon \to 0]{} 1 - \delta_1, \tag{6.13}$$

in the distribution sense. Note that we have extended u^ε by the null function outside the interval $[0, 1]$. To establish this fact starting from (6.12), we only need to show that $\dfrac{1}{\varepsilon} \dfrac{\exp\left(\frac{x}{\varepsilon}\right)}{\exp\left(\frac{1}{\varepsilon}\right) - 1} \longrightarrow \delta_1$, in the distribution sense, that is

$$\int_0^1 \frac{1}{\varepsilon} \frac{\exp\left(\frac{x}{\varepsilon}\right)}{\exp\left(\frac{1}{\varepsilon}\right) - 1} \varphi(x)\,dx \xrightarrow[\varepsilon \to 0]{} \varphi(1),$$

for all smooth, compactly supported functions. Using the change of variables $x = 1 - \varepsilon y$ into the integral, we obtain

$$\int_0^1 \frac{1}{\varepsilon} \frac{\exp\left(\frac{x}{\varepsilon}\right)}{\exp\left(\frac{1}{\varepsilon}\right) - 1} \varphi(x)\,dx = \int_0^{\frac{1}{\varepsilon}} \frac{\exp(-y)}{1 - \exp\left(-\frac{1}{\varepsilon}\right)} \varphi(1 - \varepsilon y)\,dy.$$

The integrand on the right-hand side almost everywhere converges to $e^{-y}\varphi(1)$. It is bounded from above by $2\exp(-y)\|\varphi\|_{L^\infty}$, for sufficiently small ε. Lebesgue's dominated convergence Theorem thus implies the desired convergence, from where (6.13) follows. The argument easily carries over to a general, non-constant right-hand side f:

$$\frac{1}{\varepsilon}(u^\varepsilon)' \xrightarrow[\varepsilon \to 0]{} f(x) - \delta_1 \int_0^1 f.$$

This limit mathematically encodes a boundary layer effect. The function $(u^\varepsilon)'$ concentrates at the boundary (here on the left of the point 1) when $\varepsilon \to 0$. Note that the limit is not the solution of any differential equation posed on the interval $]0, 1[$. It is not even a distribution on the open interval $]0, 1[$.

It follows from the above discussion that the regime of Eqs. (6.5) and (6.6) is the only interesting setting to consider. Its study in the one-dimensional setting is easy, because we may "cheat", that is, we solve explicitly the equation and observe what happens.

This is exactly what we have accomplished in (6.8) and (6.9), except for the fact that b_{per} there should be replaced by $\frac{1}{\varepsilon} b_{per}$ and, likewise, B_ε by $\frac{1}{\varepsilon} B_\varepsilon$. We observe that the function $\frac{1}{\varepsilon} B_\varepsilon$ also reads as

$$\frac{1}{\varepsilon} B_\varepsilon(x) = \frac{1}{\varepsilon} \int_0^x b_{per}\left(\frac{t}{\varepsilon}\right) dt = B_{per}\left(\frac{x}{\varepsilon}\right),$$

where

$$B_{per}(x) = \int_0^x b_{per}(t)\, dt \tag{6.14}$$

is the primitive of b_{per}, and is indeed a periodic function given the condition (6.6) of zero mean. Formulae (6.8) and (6.9), once adapted, yield

$$(u^\varepsilon)'(x) = \left(\lambda_\varepsilon - \int_0^x f(t) \exp\left(-B_{per}\left(\frac{t}{\varepsilon}\right)\right) dt\right) \exp\left(B_{per}\left(\frac{x}{\varepsilon}\right)\right),$$
$$\tag{6.15}$$

which agrees with (2.80), and

$$\lambda_\varepsilon = \left(\int_0^1 \exp\left(B_{per}\left(\frac{x}{\varepsilon}\right)\right) dx\right)^{-1}$$
$$\int_0^1 \left(\int_0^x f(t) \exp\left(-B_{per}\left(\frac{t}{\varepsilon}\right)\right) dt\right) \left(\exp B_{per}\left(\frac{x}{\varepsilon}\right)\right) dx, \tag{6.16}$$

which is also (2.81). We may now identify the homogenized limit. Indeed, since $\exp B_{per}$ is periodic, the weak (e.g. $L^2([0, 1])$) convergence

$$\exp B_{per}\left(\frac{x}{\varepsilon}\right) \longrightarrow \langle \exp B_{per} \rangle,$$

holds as $\varepsilon \to 0$. The limit quantity should of course not be mistaken for the *different* value $\exp \langle B_{per} \rangle$. Similarly,

$$\exp\left(-B_{per}\left(\frac{x}{\varepsilon}\right)\right) \longrightarrow \langle \exp\left(-B_{per}\right) \rangle.$$

We first infer from this that

$$\lambda_\varepsilon \longrightarrow \langle \exp\left(-B_{per}\right)\rangle \int_0^1 \int_0^x f(t)\,dt\,dx,$$

and next that

$$(u^\varepsilon)'(x) \longrightarrow \langle \exp\left(-B_{per}\right)\rangle \langle \exp\left(B_{per}\right)\rangle$$

$$\times \left(\int_0^1 \int_0^x f(t)\,dt\,dx - \int_0^x f(t)\,dt \right).$$

The homogenized equation thus reads as

$$- a^* (u^*)'' = f. \tag{6.17}$$

It is supplied with the homogeneous Dirichlet boundary condition and with the value

$$a^* = \left(\langle \exp\left(-B_{per}\right)\rangle \langle \exp\left(B_{per}\right)\rangle \right)^{-1} \tag{6.18}$$

of the homogenized coefficient. We recognize (2.83). Equation (6.17) is a pure diffusion equation. The advection term has formally disappeared in the homogenized limit, *but* some information regarding the advection is hidden in the value (6.18) of the diffusion coefficient a^* of that equation. It is no longer 1, as in the original Eq. (6.5).

Let us notice, in passing, that we have defined B_{per} as the primitive of b_{per} that vanishes at 0, but that this specific choice is both arbitrary and irrelevant. Any other primitive is suitable. This does not modify the value a^* given by (6.18).

We are going to see that all our one-dimensional observations survive in higher dimensions.

6.1.2   The General Periodic Setting

The *explicit* one-dimensional calculations of the previous section in the one-dimensional setting cannot be performed in dimensions $d \geq 2$. As in, Sect. 3.3.1 on the diffusion equation, we first need to develop our intuition on what we expect to find in the homogenized limit. We therefore formally insert into (6.2), that is

$$- \left(a_{per}\right)_{ij} \left(\frac{x}{\varepsilon}\right) \partial_{ij}^2 u^\varepsilon(x) + \frac{1}{\varepsilon} \left(b_{per}\right)_j \left(\frac{x}{\varepsilon}\right) \partial_j u^\varepsilon(x) = f(x),$$

the two-scale expansion (3.39), namely

$$u^\varepsilon(x) = u_0\left(x, \frac{x}{\varepsilon}\right) + \varepsilon\, u_1\left(x, \frac{x}{\varepsilon}\right) + \varepsilon^2 u_2\left(x, \frac{x}{\varepsilon}\right) + \dots,$$

where the successive functions $u_k(x, y)$, $k = 0, 1, 2 \dots$, are periodic in their second argument y. This "mechanical" calculation is left to the reader. It is easily seen that instead of (3.43), (3.46) and (3.49), we have

$$- \left(a_{per}\right)_{ij}(y)\, \partial^2_{y_i y_j} u_0(x, y) + \left(b_{per}\right)_j(y)\, \partial_{y_j} u_0(x, y) = 0, \qquad (6.19)$$

$$- \left(a_{per}\right)_{ij}(y)\, \partial^2_{y_i y_j} u_1(x, y) + \left(b_{per}\right)_j(y)\, \partial_{y_j} u_1(x, y)$$

$$- \left(a_{per}\right)_{ij}(y)\, \partial^2_{y_i x_j} u_0(x, y) - \left(a_{per}\right)_{ij}(y)\, \partial^2_{x_i y_j} u_0(x, y)$$

$$+ \left(b_{per}\right)_j(y)\, \partial_{x_j} u_0(x, y) = 0, \qquad (6.20)$$

$$- \left(a_{per}\right)_{ij}(y)\, \partial^2_{y_i y_j} u_2(x, y) + \left(b_{per}\right)_j(y)\, \partial_{y_j} u_2(x, y)$$

$$- \left(a_{per}\right)_{ij}(y)\, \partial^2_{y_i x_j} u_1(x, y) - \left(a_{per}\right)_{ij}(y)\, \partial^2_{x_i y_j} u_1(x, y)$$

$$+ \left(b_{per}\right)_j(y)\, \partial_{x_j} u_1(x, y) - \left(a_{per}\right)_{ij}(y)\, \partial^2_{x_i x_j} u_0(x, y) = f(x). \qquad (6.21)$$

In all likelihood, Eq. (6.19), posed for a fixed x, should imply that u_0 is independent of y:

$$u_0(x, y) = u^*(x). \qquad (6.22)$$

We temporarily admit this fact, which we will establish in (6.32) below. The independence (6.22) allows us to write (6.20) as

$$- \left(a_{per}\right)_{ij}(y)\, \partial^2_{y_i y_j} u_1(x, y) + \left(b_{per}\right)_j(y)\, \partial_{y_j} u_1(x, y)$$

$$+ \left(b_{per}\right)_j(y)\, \partial_{x_j} u^*(x) = 0. \qquad (6.23)$$

Again for a fixed x, the solution reads as

$$u_1(x, y) = \sum_{j=1}^{d} \partial_{x_j} u^*(x)\, w_{e_j, per}(y), \qquad (6.24)$$

where, for $p \in \mathbb{R}^d$, $w_{p,per}$ is the periodic solution to

$$- \left(a_{per} \right)_{ij} (y) \, \partial^2_{y_i y_j} w_{p,per}(y) + b_{per}(y) . \nabla w_{p,per}(y) = - b_{per}(y) . p.$$

(6.25)

We expect to be able to infer the homogenized equation from that and (6.21).

The point is therefore to solve (6.25). As is not unexpected, this equation is the corrector equation associated to the periodic advection-diffusion equation. The classical argument to show the well-posedness of (6.25) under suitable assumptions is to apply the *Fredholm alternative*. In the specific case at hand, the condition reads as follows. Equation (6.25) is uniquely solvable if and only if

$$\int_Q m_{per} \, b_{per}(y) = 0,$$

(6.26)

where the function m_{per}, called the *invariant measure*, solves

$$\begin{cases} - \partial^2_{y_i y_j} \left(\left(a_{per} \right)_{ij} m_{per} \right) - \partial_{y_j} \left(\left(b_{per} \right)_j m_{per} \right) = 0, \quad \text{in } Q, \\ \\ m_{per} \quad \text{periodic}, \quad m_{per} \geq 0, \quad \int_Q m_{per} = 1. \end{cases}$$

(6.27)

The differential operator $- \partial^2_{y_i y_j} \left(\left(a_{per} \right)_{ij} \cdot \right) - \partial_{y_j} \left(\left(b_{per} \right)_j \cdot \right)$ on the left-hand side of (6.27) is the *adjoint* operator of the operator $- \left(a_{per} \right)_{ij} (y) \, \partial^2_{y_i y_j} \cdot + b_{per}(y) . \nabla$ of the left-hand side of (6.25). The invariant measure m_{per} is also called the *stationary measure*. We will however not use the latter terminology, in order to avoid any confusion with the notion of stationarity used in random homogenization (not even bringing up the fact that the invariant measure exists in the stationary setting and is thus a "stationary stationary" measure!). Irrespective of the specific choice "invariant" or "stationary", the terminology originates from the interpretation of this measure in terms of the underlying dynamical system and more precisely here, of the stochastic processes hidden behind the diffusion operator present in the partial differential equations we manipulate. We will return to this in Sect. 6.3 (see in particular there the last paragraph of Sect. 6.3.3 devoted to ergodicity).

It is far from obvious that there exists a solution to (6.27). We *admit* this fact and refer to the proof in [BLP11, chapter 3, section 3.3]. The only setting where it is straightforward to establish the existence is that where

$$- \left(a_{per} \right)_{ij} = \delta_{ij}, \quad b_{per} = \nabla V_{per},$$

that is the case where the diffusion operator is the Laplacian and the advection field is the gradient of a periodic function (called the potential). Equation (6.27) then also reads as

$$- \operatorname{div}\left(\nabla m_{per} + m_{per}\,\nabla V_{per}\right) = 0,$$

and has solution $m_{per} = \langle\exp\left(- V_{per}\right)\rangle^{-1}\exp\left(- V_{per}\right)$. Remarkably, condition (6.26) is then satisfied since

$$\langle\exp\left(- V_{per}\right)\nabla V_{per}\rangle = -\langle\nabla\left[\exp\left(- V_{per}\right)\right]\rangle = 0.$$

Given the existence of the invariant measure, it is then possible to multiply (6.25) by m_{per} and write the equation as

$$- \operatorname{div}\left(m_{per}\,a_{per}\,(p + \nabla w_{per})\right) + \overline{b_{per}}\cdot(p + \nabla w_{per}) = 0,$$

where

$$\overline{b_{per}} = \operatorname{div}\left(m_{per}\,a_{per}\right) + m_{per}\,b_{per}. \tag{6.28}$$

The field $\overline{b_{per}}$ is, by definition, a *periodic* field, that in addition is *divergence free* (since its divergence is precisely the left-hand side of Eq. (6.27)) and of zero mean, since

$$\langle\overline{b_{per}}\rangle = \langle\operatorname{div}\left(m_{per}\,a_{per}\right)\rangle + \langle m_{per}\,b_{per}\rangle = 0,$$

where both term vanishes, the former by periodicity and the latter because of condition (6.26). Arguing as in Sect. 3.4.3.1, we may write $\overline{b_{per}}$ as the gradient of an antisymmetric matrix (or, in simpler terms when the ambient dimension $d = 3$, as the curl of another vector field, see Remark 3.10). We write

$$\overline{b_{per}} = \operatorname{div} g_{per}, \tag{6.29}$$

where g_{per} is a periodic, zero-mean function valued in the space of antisymmetric matrices. Equation (6.25) may thus be put under the form

$$- \operatorname{div}\left(\mathcal{A}_{per}\,(p + \nabla w_{per})\right) = 0, \tag{6.30}$$

with the matrix valued coefficient

$$\mathcal{A}_{per} = m_{per}\,a_{per} - g_{per}. \tag{6.31}$$

It is then solved as such a diffusion equation.

Using a similar manipulation, we are in position to prove (6.22). Indeed, multiplying (6.19) by m_{per}, we write it as

$$- \operatorname{div}_y\left(m_{per}(y)\,a_{per}(y)\,\nabla_y u_0(x, y)\right) + \overline{b_{per}}(y)\cdot\nabla_y u_0(x, y) = 0. \tag{6.32}$$

This partial differential equation in the variable y is well posed, for each x fixed, because the first term is elliptic and $\overline{b}_{per}$ is divergence-free. Its only solution is thus constant in y, which proves (6.22).

The well-posedness of (6.25) allows us to define u^1 by (6.24), and to return to (6.21). The latter equation reads as

$$- \left(a_{per}\right)_{ij} (y)\, \partial^2_{y_i y_j} u_2(x,\, y) + \left(b_{per}\right)_j (y)\, \partial_{y_j} u_2(x,\, y) =$$

$$\left(a_{per}\right)_{ij} (y)\, \partial^2_{x_j x_k} u^*(x) \partial_{y_i} w_{ek,per}(y) + \left(a_{per}\right)_{ij} (y)\, \partial^2_{x_i x_k} u^*(x) \partial_{y_j} w_{ek,per}(y)$$

$$- \left(b_{per}\right)_j (y)\, \partial^2_{x_j x_k} u^*(x) w_{ek,per}(y) + \left(a_{per}\right)_{ij} (y)\, \partial^2_{x_i x_j} u^*(x) + f(x).$$
$$(6.33)$$

Using again the Fredholm alternative, the existence and uniqueness of u_2 is obtained under the condition (6.26) expressed for this particular equation, that is, the integral over Q of the right-hand side multiplied by m_{per} should vanish. This yields

$$- \left\langle m_{per} \left(a_{per}\right)_{ij} \partial_{y_i} w_{ek,per}\right\rangle \partial^2_{x_j x_k} u^*(x) - \left\langle m_{per} \left(a_{per}\right)_{ij} \partial_{y_j} w_{ek,per}\right\rangle \partial^2_{x_i x_k} u^*(x)$$

$$+ \left\langle m_{per} \left(b_{per}\right)_j w_{ek,per}\right\rangle \partial_{x_j x_k} u^*(x) - \left\langle m_{per} \left(a_{per}\right)_{ij}\right\rangle \partial^2_{x_i x_j} u^*(x) = \left\langle m_{per}\right\rangle f.$$

Since $\left\langle m_{per}\right\rangle = 1$, the right-hand side is equal to f. The equation thus reads as

$$- \operatorname{div}\left(\mathcal{A}^* \nabla u^*\right) = f,$$
$$(6.34)$$

where the (constant) diffusion matrix is

$$\left[\mathcal{A}^*\right]_{ij} = \left\langle m_{per} \left(a_{per}\right)_{ij}\right\rangle + \left\langle m_{per} \left(a_{per}\right)_{ik} \partial_{y_k} w_{ej,per}\right\rangle$$

$$+ \left\langle m_{per} \left(a_{per}\right)_{ki} \partial_{y_k} w_{ej,per}\right\rangle - \left\langle m_{per} \left(b_{per}\right)_i w_{ej,per}\right\rangle.$$
$$(6.35)$$

We use (6.28) and (6.29), that is

$$\partial_{y_k} \left(m_{per} \left(a_{per}\right)_{ki}\right) + m_{per} \left(b_{per}\right)_i = \partial_k \left(g_{per}\right)_{ki}.$$

to simplify (6.35). We multiply the equality by $w_{ej,per}$ and integrate over the unit cube Q. After an integration by parts, this yields

$$- \left\langle m_{per} \left(a_{per}\right)_{ki} \partial_{y_k} w_{ej,per}\right\rangle + \left\langle m_{per} \left(b_{per}\right)_i w_{ej,per}\right\rangle = - \left\langle \left(g_{per}\right)_{ki} \partial_{y_k} w_{ej,per}\right\rangle.$$

We next insert this expression into (6.35), so that, given that g_{per} is antisymmetric,

$$[\mathcal{A}^*]_{ij} = \left\langle m_{per}\,(a_{per})_{ij} \right\rangle + \left\langle m_{per}\,(a_{per})_{ik}\,\partial_{y_k} w_{e_j,per} \right\rangle - \left\langle (g_{per})_{ik}\,\partial_{y_k} w_{e_j,per} \right\rangle$$

$$= \left\langle \left(m_{per}\,(a_{per})_{ij} - (g_{per})_{ik} \right) \left(\delta_{kj} + \partial_{y_k} w_{e_j,per} \right) \right\rangle$$

$$= \left\langle \left(m_{per}a_{per} - g_{per} \right) \left(e_j + \nabla w_{e_j,per} \right) . e_i \right\rangle.$$

This also reads as

$$[\mathcal{A}^*]_{ij} = \int_Q \left(m_{per}(y)\,(a_{per})\,(y) - g_{per}(y) \right) \left(e_j + \nabla w_{e_j,per}(y) \right)$$

$$. \left(e_i + \nabla w_{e_i,per}(y) \right) dy. \qquad (6.36)$$

The homogenized limit (6.34) and (6.36) is not unexpected. We have already explained that, under suitable conditions which are all fulfilled here, the corrector equation may be written in divergence form. The homogenized equation is thus likely to satisfy the same property. And it does. In addition, the explicit expression (6.36) of the homogenized matrix is reminiscent from forms repeatedly encountered in the previous chapters.

It is actually possible, now that we have in our possession the powerful tool of the invariant measure m_{per}, to use it earlier. We may indeed multiply the original Eq. (6.2) by $m_{per}\left(\frac{x}{\varepsilon}\right)$ from the very beginning. The above manipulations are still valid and allow us to write (6.2) in the form $-\operatorname{div}\left(\mathcal{A}_{per}\left(\frac{\cdot}{\varepsilon}\right)\nabla u^\varepsilon\right) = m_{per}\left(\frac{\cdot}{\varepsilon}\right)f$, where the matrix $\mathcal{A}_{per}$ is defined in (6.31). Homogenization theory, as seen in Chap. 3, is then readily applied. The only modification one needs to be cautious about is that the right-hand side of the equation now depends upon ε. Since this right-hand side converges in $H^{-1}(\mathcal{D})$, the theory however applies, as noticed the e.g. in [All02, Proposition 1.2.19].

We next examine the result provided by these two approaches on an extremely simple example and check that they eventually (fortunately!) yield the same conclusion. We temporarily return to the one-dimensional setting, with $a = 1$ as diffusion coefficient and $b = b_{per}$, periodic and of zero mean. The equation reads as

$$-\left(u^\varepsilon\right)'' + \frac{1}{\varepsilon}b_{per}\left(\frac{x}{\varepsilon}\right)\left(u^\varepsilon\right)' = f, \qquad (6.37)$$

while system (6.27) becomes

$$-m''_{per} - \left(b_{per}m_{per}\right)' = 0, \qquad (6.38)$$

with m_{per} periodic, positive and of zero mean. It is straightforward that

$$m_{per}(x) = \frac{e^{B_{per}(x)}}{\langle e^{B_{per}} \rangle},$$

where B_{per} is a primitive of b_{per}, as in (6.14). In addition, the function g_{per} within (6.31) vanishes, so that

$$\mathcal{A}_{per}(x) = m_{per}(x) = \frac{e^{B_{per}(x)}}{\langle e^{B_{per}} \rangle}.$$

On the other hand, the corrector w_{per} solution to (6.25), that is (6.30), satisfies $1 + w'_{per} = \frac{C_0}{\mathcal{A}_{per}}$, where the integration constant is $C_0 = \left\langle \mathcal{A}_{per}^{-1} \right\rangle^{-1}$. The expression (6.36) thus reads as

$$\mathcal{A}^* = \left\langle \frac{1}{\mathcal{A}_{per}} \right\rangle^{-1} = \left\langle \frac{1}{m_{per}} \right\rangle^{-1} = \left\langle e^{B_{per}} \right\rangle^{-1} \left\langle e^{-B_{per}} \right\rangle^{-1}.$$

Directly multiplying Eq. (6.37) by $m_{per}\left(\frac{x}{\varepsilon}\right)$, as we mentioned above, we obtain

$$-\left(\mathcal{A}_{per}\left(u^{\varepsilon}\right)'\right)' = m_{per}\left(\frac{x}{\varepsilon}\right) f.$$

On the latter equation, a new two-scale expansion, as we did in Chap. 2, yields the homogenized equation (6.34), with

$$\mathcal{A}^* = \left\langle \frac{1}{\mathcal{A}_{per}} \right\rangle^{-1},$$

since

$$m_{per}\left(\frac{x}{\varepsilon}\right) \xrightarrow[\varepsilon \to 0]{} \langle m_{per} \rangle = 1.$$

The results agree with one another.

We will not *prove* that the homogenized limit (6.34)–(6.36) is indeed correct. It is intuitively clear, at least, that the same type of arguments as those of Sect. 3.4 allow to make the proof. We prefer to focus on some modifications of the periodic setting.

6.1.3 Beyond the Periodic Setting

We consider again (6.4), however in the setting where its periodic coefficients are perturbed by localized defects, in the vein of our developments of Sect. 4.2.1. Equation (6.4) thus becomes

$$- \left(a_{per} + \widetilde{a} \right)_{ij} \left(\frac{x}{\varepsilon} \right) \partial_{ij}^2 u^\varepsilon (x) + \frac{1}{\varepsilon} \left(b_{per} + \widetilde{b} \right)_j \left(\frac{x}{\varepsilon} \right) \partial_j u^\varepsilon (x) = f(x), \quad (6.39)$$

where $\widetilde{a}$ and $\widetilde{b}$ satisfy the following assumptions

$$\begin{cases} a^{per}(x) + \widetilde{a}(x) \text{ and } a^{per}(x) \text{ are both uniformly coercive,} \\ \widetilde{a} \in L^r(\mathbb{R}^d)^{d \times d}, \quad \widetilde{b} \in L^s(\mathbb{R}^d)^d, \quad \text{for some } 1 \le r, s < +\infty, \\ a^{per}, \widetilde{a} \in \left(C^{0,\alpha}_{unif}(\mathbb{R}^d) \right)^{d \times d}, \quad b^{per}, \widetilde{b} \in \left(C^{0,\alpha}_{unif}(\mathbb{R}^d) \right)^d, \end{cases} \quad (6.40)$$

for some $\alpha > 0$. As we saw in Sect. 6.1.2 for the periodic setting, the existence of an invariant measure allows one to write the problem in divergence form. Admit temporarily that there exists some measure m, solution to

$$\begin{cases} - \partial_{y_i y_j}^2 \left(\left(a_{per} + \widetilde{a} \right)_{ij} m \right) - \partial_{y_j} \left(\left(b_{per} + \widetilde{b} \right)_j m \right) = 0, \quad \text{in } \mathbb{R}^d, \\ m \ge 0, \quad \langle m \rangle = 1, \end{cases} \quad (6.41)$$

where the condition on the average will be made precise below. Then, multiplying (6.39) by $m_\varepsilon(x) = m \left(\frac{x}{\varepsilon} \right)$ and arguing as above in the periodic setting, we obtain

$$- \operatorname{div} \left(\mathcal{A}_\varepsilon \nabla u^\varepsilon \right) = m_\varepsilon f, \quad (6.42)$$

where

$$\mathcal{A}_\varepsilon(x) = \mathcal{A} \left(\frac{x}{\varepsilon} \right), \quad \text{avec} \quad \mathcal{A} = m \left(a_{per} + \widetilde{a} \right) - g. \quad (6.43)$$

The antisymmetric matrix g is assumed to solve

$$\operatorname{div}(g) = m \left(b_{per} + \widetilde{b} \right) + \operatorname{div} \left(m \left(a_{per} + \widetilde{a} \right) \right).$$

For the existence of such a matrix g, we refer for instance to [BJL20, Lemma 5.3], where it is also shown that, in fact, $g = g_{per} + \widetilde{g}$, with g_{per} solution to (6.29) and $\widetilde{g} \in L^q(\mathbb{R}^d)$ for a certain exponent $q \in]1, +\infty[$. As soon as a similar decomposition is shown for the matrix $\mathcal{A}$, that is, $\mathcal{A} = \mathcal{A}_{per} + \widetilde{\mathcal{A}}$, it will be straightforward to address Eq. (6.42) using the results of Sect. 4.1.

The key point of the study is therefore to establish that m exists and reads as

$$m = m_{per} + \widetilde{m}, \qquad (6.44)$$

where $\widetilde{m}$ has suitable properties of integrability. Inserting (6.44) into (6.41) and using that m_{per} solves (6.27), we readily see that $\widetilde{m}$ should be solution to

$$- \partial_{y_i} \left(\partial_{y_j} \left((a_{per} + \widetilde{a})_{ij} \, \widetilde{m} \right) + \left(b_{per} + \widetilde{b} \right)_i \widetilde{m} \right) = \partial_{y_i} \left(\partial_{y_j} \left(\widetilde{a}_{ij} \, m_{per} \right) + \widetilde{b}_i \, m_{per} \right). \qquad (6.45)$$

This equation has of course to be understood in the weak sense, that is,

$$\forall \varphi \in C_c^\infty \left(\mathbb{R}^d \right), \quad \int_{\mathbb{R}^d} - (a_{per} + \widetilde{a})_{ij} \, \widetilde{m} \, \partial_{ij}^2 \varphi + \left(b_{per} + \widetilde{b} \right)_i \widetilde{m} \, \partial_i \varphi$$

$$= \int_{\mathbb{R}^d} \widetilde{a}_{ij} \, m_{per} \, \partial_{ij}^2 \varphi - \widetilde{b}_i \, m_{per} \, \partial_i \varphi.$$

Given the assumptions on $\widetilde{a}$ and $\widetilde{b}$, we know that, on the right-hand side, $\widetilde{a} \, m_{per} \in L^r$ and $\widetilde{b} \, m_{per} \in L^s$. The issue is therefore to solve Eq. (6.45) in a given L^q space, for some $q \in]1, +\infty[$. To this end, we will actually solve the *dual* problem. The result reads as follows.

Proposition 6.1 *Assume that (6.40) holds true for some $1 \le r < d$ and $1 \le s < d$. Assume that the periodic invariant measure m_{per} associated to the periodic operator $(a_{per})_{ij} \, \partial_{ij}^2 + (b_{per})_j \, \partial_j$, that is the solution to (6.27), satisfies the average condition $\langle m_{per} b_{per} \rangle = 0$. Take $1 < q < d$ and define $\dfrac{1}{q^*} = \dfrac{1}{q} - \dfrac{1}{d}$.*

Then, for all $f \in \left(L^{q^*} \cap L^q \right) (\mathbb{R}^d)$, there exists $u \in L^1_{loc}(\mathbb{R}^d)$, such that $D^2 u \in L^q(\mathbb{R}^d)$, solution to

$$- (a_{per} + \widetilde{a})_{ij} \, \partial_{ij}^2 u + \left(b_{per} + \widetilde{b} \right)_j \, \partial_j u = f \quad \text{dans} \quad \mathbb{R}^d. \qquad (6.46)$$

Such a solution u is unique up to the addition of a constant (or affine) function.

In addition, there exists some constant C_q, depending only on q, d and on the coefficients a_{per}, $\widetilde{a}$, b_{per} and $\widetilde{b}$, such that u satisfies

$$\left\| D^2 u \right\|_{(L^{q^*}(\mathbb{R}^d))^{d \times d}} + \| \nabla u \|_{(L^{q^*}(\mathbb{R}^d))^d} \le C_q \, \| f \|_{(L^{q^*} \cap L^q)(\mathbb{R}^d)}. \qquad (6.47)$$

In the above result, the norm equipping the space $\left(L^{q^*} \cap L^q \right) (\mathbb{R}^d)$ is of course

$$\| f \|_{(L^{q^*} \cap L^q)(\mathbb{R}^d)} = \| f \|_{L^{q^*}(\mathbb{R}^d)} + \| f \|_{L^q(\mathbb{R}^d)}.$$

On the other hand, we have to comment upon uniqueness in the statement of Proposition 6.1. Evidently, the notion of solution is stable under addition of a constant. That said, in the particular case when $b = b_{per} + \widetilde{b}$ satisfies $\forall x \in \mathbb{R}^d$, $b(x) . v_0 = 0$ for a certain constant vector v_0, then adding to u the affine function $v_0 . x + \alpha$, for any $\alpha \in \mathbb{R}$, yields another solution. An example of such a case is: $b = 0$.

We note that Proposition 6.1 has been originally introduced and proven in the article [BLL19]. The particular case of the exponent $q = 1$ was erroneously included in the statement of the result therein. We are thankful to Sylvain Wolf for pointing out this error, which has subsequently been corrected in the erratum [BLBL20].

Remark 6.1 Estimation (6.47) is, in full generality, sharp. In the particular case when $b = 0$, it is however not. And this fact is already true in the purely periodic setting. Proposition 3.1 of [BLL18] indeed yields, in the case $b = 0$,

$$\left\| D^2 u \right\|_{\left(L^q(\mathbb{R}^d)\right)^{d \times d}} \leq C_q \, \|f\|_{L^q(\mathbb{R}^d)} ,$$

for all $f \in L^q(\mathbb{R}^d)$, where u solves (6.46) (with $b = 0$). On the other hand, Theorem B of [AL91] establishes that this estimation holds true (in the periodic case) if and only if $b = 0$. Put differently, as soon as b does not identically vanish, there is a loss in the decay properties at infinity (thus q^* replaces q). We refer to [BLL19, Remarks 4 and 5] for more details on this issue.   □

An immediate consequence of Proposition 6.1 is the existence of the invariant measure stated in the following proposition.

Proposition 6.2 *Assume* (6.40) *with* $1 \leq r, s < d$. *Assume that* $\left(m_{per} b_{per}\right) = 0$, *where* m_{per} *is the periodic invariant measure associated to the periodic operator* $- \left(a_{per}\right)_{ij} \partial_{ij}^2 + \left(b_{per}\right)_j \partial_j$, *that is the solution to* (6.27). *Then there exists a solution* m *to* (6.41). *It reads as* $m = m_{per} + \widetilde{m}$ *and* $\widetilde{m} \in \left(L^{q'} \cap L^\infty\right)(\mathbb{R}^d)$ *where* q' *satisfies*

$$\frac{1}{q'} = \min\left(\frac{1}{r} - \frac{1}{d}, \frac{1}{s} - \frac{1}{d}\right). \tag{6.48}$$

The function m *is Hölder continuous and is bounded away from zero by a strictly positive constant.*

We readily prove Proposition 6.2 from Proposition 6.1. Take $q \in]1, d[$, which is temporarily arbitrary but will be fixed below. Define the linear map

$$T : \left(L^q \cap L^{q^*}\right)(\mathbb{R}^d) \longrightarrow \left(L^{q^*}(\mathbb{R}^d)\right)^{d \times d} \times \left(L^{q^*}(\mathbb{R}^d)\right)^d$$
$$f \longmapsto \left(D^2 u, \nabla u\right),$$

where u solves (6.46). Proposition 6.1 yields the continuity of this map. The adjoint map T^* is thus linear continuous from the dual space of $\left(L^{q^*}(\mathbb{R}^d)\right)^{d\times d} \times \left(L^{q^*}(\mathbb{R}^d)\right)^d$, that is $\left(L^{(q^*)'}(\mathbb{R}^d)\right)^{d\times d} \times \left(L^{(q^*)'}(\mathbb{R}^d)\right)^d$, to the dual space of $\left(L^q \cap L^{q^*}\right)(\mathbb{R}^d)$, that is $\left(L^{q'} + L^{(q^*)'}\right)(\mathbb{R}^d)$. We recall that the latter space contains the functions of the form $\varphi = \chi + \psi$ with $\chi \in L^{q'}(\mathbb{R}^d)$ and $\psi \in L^{(q^*)'}(\mathbb{R}^d)$. The associated norm reads as

$$\|\varphi\|_{\left(L^{q'}+L^{(q^*)'}\right)(\mathbb{R}^d)} = \inf_{\varphi=\chi+\psi} \left(\|\chi\|_{L^{q'}(\mathbb{R}^d)} + \|\psi\|_{L^{(q^*)'}(\mathbb{R}^d)}\right).$$

Take now $\overline{a} \in \left(L^{(q^*)'}(\mathbb{R}^d)\right)^{d\times d}$ and $\overline{b} \in \left(L^{(q^*)'}(\mathbb{R}^d)\right)^d$. Denote by $\mu = T^*\left(\overline{a},\overline{b}\right)$. We claim that μ solves

$$-\partial_{y_i}\left(\partial_{y_j}\left((a_{per}+\widetilde{a})_{ij}\,\mu\right) + (b_{per}+\widetilde{b})_i\,\mu\right) = \partial_{y_i}\left(\partial_{y_j}\overline{a}_{ij} - \overline{b}_i\right), \qquad (6.49)$$

in the weak sense. Indeed, consider φ a smooth, compactly supported test function and define $f = -(a_{per}+\widetilde{a})_{ij}\,\partial^2_{y_iy_j}\varphi + (b_{per}+\widetilde{b})_i\,\partial_{y_i}\varphi$, which belongs to $L^q \cap L^{q^*}$ and satisfies $Tf = \left(D^2\varphi, \nabla\varphi\right)$. We have

$$\int_{\mathbb{R}^d} \mu\left[-(a_{per}+\widetilde{a})_{ij}\,\partial^2_{y_iy_j}\varphi + (b_{per}+\widetilde{b})_i\,\partial_{y_i}\varphi\right] = \int_{\mathbb{R}^d} \mu f = \langle\mu, f\rangle_{L^{q'}+L^{(q^*)'},L^q\cap L^{q^*}},$$

and

$$\begin{aligned}
\langle\mu, f\rangle_{L^{q'}+L^{(q^*)'},L^q\cap L^{q^*}} &= \left\langle T^*(\overline{a},\overline{b}), f\right\rangle_{L^{q'}+L^{(q^*)'},L^q\cap L^{q^*}} \\
&= \left\langle(\overline{a},\overline{b}), T(f)\right\rangle_{(L^{q^*})^{d\times d}\times(L^{q^*})^d,\left(L^{(q^*)'}\right)^{d\times d}\times\left(L^{(q^*)'}\right)^d} \\
&= \left\langle(\overline{a},\overline{b}), \left(D^2\varphi, \nabla\varphi\right)\right\rangle_{(L^{q^*})^{d\times d}\times(L^{q^*})^d,\left(L^{(q^*)'}\right)^{d\times d}\times\left(L^{(q^*)'}\right)^d}.
\end{aligned}$$

Expanding the latter duality backet, we obtain

$$\int_{\mathbb{R}^d} \mu\left[-(a_{per}+\widetilde{a})_{ij}\,\partial^2_{y_iy_j}\varphi + (b_{per}+\widetilde{b})_i\,\partial_{y_i}\varphi\right] = \int_{\mathbb{R}^d} \overline{a}_{ij}\partial^2_{y_iy_j}\varphi + \overline{b}_i\partial_{y_i}\varphi,$$

which is indeed the weak formulation of (6.49). Since the continuity of T implies that of T^* with the same norm, we thus have

$$\|\mu\|_{\left(L^{q'}+L^{(q^*)'}\right)(\mathbb{R}^d)} = \left\|T^*\left(\overline{a}, \overline{b}\right)\right\|_{\left(L^{q'}+L^{(q^*)'}\right)(\mathbb{R}^d)}$$

$$\leq C\left(\|\overline{a}\|_{L^{(q^*)'}(\mathbb{R}^d)} + \|\overline{b}\|_{L^{(q^*)'}(\mathbb{R}^d)}\right), \qquad (6.50)$$

where the constant C is identical to that within (6.47). In particular, it depends neither on $(\overline{a}, \overline{b})$ nor on f and μ.

We now wish to apply this result to

$$\overline{a} = m_{per}\, \widetilde{a}, \quad \overline{b} = -m_{per}\, \widetilde{b}.$$

To this end, we need to adjust the exponent q so that $\overline{a} \in L^{(q^*)'}$ and $\overline{b} \in L^{(q^*)'}$. Since m_{per} is bounded, a natural choice is $(q^*)' = \max(r, s)$. Thus

$$1 - \frac{1}{q^*} = \min\left(\frac{1}{r}, \frac{1}{s}\right), \quad \text{c'est-à-dire} \quad 1 - \frac{1}{q} + \frac{1}{d} = \min\left(\frac{1}{r}, \frac{1}{s}\right),$$

which indeed yields (6.48). For this particular value q, this proves the existence of $\widetilde{m} = T^*\left(m_{per}\, \widetilde{a}, -m_{per}\widetilde{b}\right) \in \left(L^{q'} + L^{(q^*)'}\right)(\mathbb{R}^d)$, solution to (6.45). Therefore, $m = m_{per} + \widetilde{m}$ is solution to (6.41).

Elliptic regularity applied to the equations satisfied by $\widetilde{m}$ and m_{per} (see Theorem A.15) then implies that both m_{per} on the one hand and $\widetilde{m}$, on the other hand, are uniformly Hölder continuous. In particular, $\widetilde{m} \in L^{(q^*)'} \cap L^\infty\left(\mathbb{R}^d\right)$ and

$$\|\widetilde{m}\|_{L^\infty(B_R^c)} \xrightarrow{R \to +\infty} 0.$$

Recall also that $\inf m_{per} > 0$ because of the Harnack inequality (Theorem A.16). Thus, for R sufficiently large, we have

$$\forall x \in B_R^c, \quad m(x) \geq \frac{1}{2}\inf m_{per} > 0. \qquad (6.51)$$

An application of the maximum principle on the ball B_R implies $m \geq 0$ therein. Applying the Harnack inequality in B_R, we prove that m is bounded away from 0 in B_R. This together with (6.51), implies that m is bounded away from 0 in the whole space $\mathbb{R}^d$.

Concluding our study would amount to now *proving* Proposition 6.1. The proof originally appeared in [BLL19]. As this proof is essentially a repetition of that of Proposition 4.1 in Sect. 4.1.2, we will skip it. Suffice it to say that it proceeds by continuation and uses the concentration-compactness method.

We eventually emphasize that the strategy of proof we have developed, on the one hand in the periodic setting in Sect. 6.1.2 and, on the other hand, in the case of a localized defect in the present Sect. 6.1.3, consists in using the invariant masure (the existence of which is preliminarily established) to put the original equation in divergence form. This strategy is classical and it carries over to various other settings, in particular the random stationary setting. We will not proceed in this direction and refer, e.g., to [ZKO94, Chapter 10] and [Yur82].

6.2 Energetic Interpretation and Related Techniques

6.2.1 Rewording of the Periodic Setting in Terms of Energy Functionals

Our first task here is to revisit the periodic setting of Chap. 3 and consider the associated energy functionals when the coefficient a_{per} is assumed symmetric. We start from the following observation. Given (3.1), Eqs. (1)–(2) is the *Euler-Lagrange equation* (a.k.a. *optimality equation*) of the minimization problem

$$\inf_{v \in H_0^1(\mathcal{D})} \frac{1}{2} \int_{\mathcal{D}} (a_\varepsilon(x) \nabla v(x)) \cdot \nabla v(x) \, dx - \int_{\mathcal{D}} v(x) \, f(x) \, dx. \qquad (6.52)$$

The unique minimizer of (6.52) is $v = u^\varepsilon$. Similarly, the homogenized equation (3.8) agrees with the Euler-Lagrange equation of the analogous problem to (6.52) where a^* is substituted for a_ε. This observation clearly leads to the intuition that it should be possible to homogenize the problem only in the language of energy functionals. We describe in the present Sect. 6.2.1 how this is achieved. The technical ingredients for the proofs will be mentioned in Sect. 6.2.2 below.

We first revisit the definition (3.54) of the homogenized matrix a^* using the *energetic* viewpoint. Since it is symmetric, the matrix a^* also reads as

$$\forall z \in \mathbb{R}^N, \quad z \, a^* z = \inf_{\substack{\nabla u \text{ periodic} \\ \int_Q \nabla u = z}} \int_Q (a(y) \nabla u(y)) \cdot \nabla u(y) \, dy. \qquad (6.53)$$

If, indeed, e_i denotes the i-th vector in the canonical basis, it is equivalent to consider all functions u that have a periodic gradient and that satisfy $\int_Q \nabla u = e_i$ or all the functions u that read as $u = x_i + w$ with a periodic function w. We leave to the reader the proof of this equivalence, which is easy in dimension $d = 1$ and somewhat trickier in higher dimensions. In any event, the minimization problem (6.53) then reads as

$$e_i a^* e_i = \inf_{w \text{ periodic}} \int_Q (a(y)(e_i + \nabla w(y))) \cdot (e_i + \nabla w(y)) \, dy. \qquad (6.54)$$

It is immediately seen that the minimizer of (6.54) is the periodic corrector $w_{e_i, per}$ solution to (3.67) for $p = e_i$, thus that

$$e_i a^* e_i = \inf_{w \text{ periodic}} \int_Q (a(y)(e_i + \nabla w(y))) \cdot (e_i + \nabla w(y)) \, dy$$

$$= \int_Q \left(a(y)(e_i + \nabla w_{e_i, per}(y)) \right) \cdot \left(e_i + \nabla w_{e_i, per}(y) \right) \, dy. \qquad (6.55)$$

This coincides with (3.54) for $i = j$. Using the polarization identity $2e_i a^* e_j = (e_i + e_j) a^* (e_i + e_j) - e_i a^* e_i - e_j a^* e_j$, we may recover all the coefficients i, j and again find (3.54).

The intuitive idea underlying the expression (6.53) of the homogenized matrix is illuminating, since u^ε solution to (1) is the minimizer of

$$\inf_{v \in H_0^1(\mathcal{D})} \frac{1}{2} \int_{\mathcal{D}} \left(a \left(\frac{x}{\varepsilon} \right) \nabla v(x) \right) \cdot \nabla v(x) \, dx - \int_{\mathcal{D}} f(x) v(x) \, dx. \qquad (6.56)$$

The homogenized solution u^*, approximation to u^ε minimizer of (6.56), is the minimizer of

$$\inf_{u \in H_0^1(\mathcal{D})} \frac{1}{2} \int_{\mathcal{D}} \left(\inf_{\substack{\nabla v \text{ periodic} \\ \int_Q \nabla v = \nabla u(x)}} \int_Q (a(y)\nabla v(y)) \cdot \nabla v(y) \, dy \right) dx - \int_{\mathcal{D}} f u.$$

$$(6.57)$$

The above is the *variational, or energetic interpretation* of the results we have previously obtained in the langage of partial differential equations.

This new interpretation of (6.57) is in fact *more general* than the langage of partial differential equations... *provided* of course the partial differential equation is associated to an underlying energy functional, which for a quadratic energy functional means a self-adjoint operator in the linear partial differential equation. The variational interpretation allows to address more difficult cases, with non-quadratic functionals, thus nonlinear equations.

A more general and even more illuminating, although somewhat formal rewriting of (6.57) is as follows:

$$\inf_{u \in H_0^1(\mathcal{D})} \frac{1}{2} \int_{\mathcal{D}} \left(\inf_{\langle \nabla v \rangle = \nabla u(x)} \langle \text{ Energy of } \nabla v \rangle \right) dx - \int_{\mathcal{D}} f u, \qquad (6.58)$$

where $\langle \cdot \rangle$ denotes the average over a representative element of the fine structure. In the periodic setting, this is the integral $\langle \cdot \rangle = \int_Q \cdot (y)\, dy$ over the periodicity cell.

Put under the form (6.58), the homogenization result and the approach are completely general and carry over to a huge variety of settings that admit an energetic interpretation. The *"microscopic"* energy ⟨ Energy of ∇v ⟩ need not be coming from a reinterpretation of a partial differential equation. It may refer to a discrete setting, for instance. We refer our readers who can read French to [Le 05, Chapitres 1 & 4] where many of the various modelling approaches in materials science presented actually, fit this type of general formalism.

The next section now briefly presents and summarizes the mathematical tools to proceed with homogenization results only in the language of energy functionals and not involving partial differential equations. The approach presented is Γ-*convergence* theory. It allows one to define the limit of a sequence of minimization problems as itself a minimization problem.

6.2.2 Γ-Convergence Theory

The interpretation of our earlier results in the langage of Sect. 6.2.1 has shown that, at least in the periodic setting, the minimizer of the minimization problem (6.56) converges to that of the minimization problem (6.57). It is thus a natural hope that an independent, autonomous mathematical strategy allows to pass from one minimization problem to the other. Consider a sequence of minimization problems

$$\inf \{ F_\varepsilon(u), \quad u \in X_\varepsilon \} \tag{6.59}$$

such that any corresponding minimizer u^ε converges to a minimizer u^* of the minimization problem

$$\inf \left\{ F^*(u), \quad u \in X^* \right\}, \tag{6.60}$$

in the sense that the following property holds

$$\lim_{\varepsilon \to 0} F_\varepsilon(u^\varepsilon) = F^*(u^*). \tag{6.61}$$

This notion of *variational* convergence, called Γ-convergence, has been introduced by Ennio De Giorgi. We briefly review it here. Our presentation, which is not to be mistaken for a genuine course on the topic, is inspired by the excellent monograph [Bra02] authored by Andrea Braides, an internationally recognized expert of the approach and a former student of De Giorgi himself. We could equally well refer to the earlier monograph [BD98].

To start with, we quote the very definition.

Definition 6.1 (Definition of Γ-Convergence, [Bra02, Definition 1.5, p 22])
Consider a metric space X and a sequence of functions $(F_n)_{n\in\mathbb{N}}$ defined on X and valued in $\overline{\mathbb{R}} = \mathbb{R} \cup \{-\infty, +\infty\}$. The sequence is said to Γ-*converge* to the function F_∞ from X to $\overline{\mathbb{R}}$ (and we denote this convergence by $F_\infty = \Gamma - \lim_{n\to+\infty} F_n$), when the following two properties are satisfied for all $x \in X$:

(i) (*liminf inequality*) *for all* sequence $(x_n)_{n\in\mathbb{N}}$ in X converging to x,

$$F_\infty(x) \leq \liminf_{n\to+\infty} F_n(x_n), \tag{6.62}$$

(ii) (*limsup inequality* **or** *existence of a recovery sequence*) *there exists a sequence* $(x_n)_{n\in\mathbb{N}}$ in X, converging to x, such that

$$F_\infty(x) \geq \limsup_{n\to+\infty} F_n(x_n). \tag{6.63}$$

The two properties (6.62)–(6.63) have to be compared with the expected convergence (6.61). We are clearly heading in the right direction. It is all the more convincing in view of the last assertion in of the following result:

Proposition 6.3 (Convergence of Minimizers, [Bra02, Theorem 1.21, p 29])
Consider a metric space X and a sequence of functions $(F_n)_{n\in\mathbb{N}}$ defined on X and valued in $\overline{\mathbb{R}}$, such that there exists a compact subset $K \subset X$ satisfying, for all $n \in \mathbb{N}$, $\inf_X F_n = \inf_K F_n$ in $\overline{\mathbb{R}}$.
Assume $F_\infty = \Gamma - \lim_{n\to+\infty} F_n$. Then

$$\min_X F_\infty = \lim_{n\to+\infty} \inf_X F_n \tag{6.64}$$

and this infimum is achieved. In addition, if $(x_n)_{n\in\mathbb{N}}$ is a precompact sequence in X (that is, the union $\bigcup_{n\in\mathbb{N}}\{x_n\}$ is included, for all $\varepsilon > 0$, in the union of finitely many balls of radius ε), such that $\lim_{n\to+\infty} F_n(x_n) = \lim_{n\to+\infty} \inf_X F_n$, then any adherent value x_∞ to the sequence $(x_n)_{n\in\mathbb{N}}$ is a minimizer of F_∞, that is $F_\infty(x_\infty) = \min_X F_\infty$.

We are now in position to state the following classical result of periodic homogenization using Γ-convergence theory.

Proposition 6.4 (Periodic Homogenization Result Using Γ-Convergence, [Bra02, Theorem 3.1, p 64]) *Take $1 < q < +\infty$ and a measurable function f from $\mathbb{R} \times \mathbb{R}^n$ to $\mathbb{R}_+$ satisfying, for some c_1, c_2, c_3 all positive, the* growth *condition*

$$c_1 |p|^q - c_2 \leq f(t, p) \leq c_3 (1 + |p|^q), \tag{6.65}$$

for all $(t, p) \in \mathbb{R} \times \mathbb{R}^n$. *Assume additionally that f is periodic with respect to its first argument. Denote by*

$$F_\varepsilon(u) = \int_0^1 f\left(\frac{t}{\varepsilon}, u'\right) dt \qquad (6.66)$$

for all functions $u \in W_0^{1,q} \left(]0, 1[\right)^n$. *Then there exists a function*

$$F^*(u) = \int_0^1 f^*(u'(t)) \, dt, \qquad (6.67)$$

that is the Γ-limit of the sequence F_ε, with, for all $p \in \mathbb{R}^n$

$$f^*(p) = \lim_{T \to +\infty} \inf \left\{ \frac{1}{T} \int_0^T f\left(y, u'(y) + p\right) dy; \quad u \in W_0^{1,q} \left(]0, T[\right)^n \right\}. \tag{6.68}$$

Remark 6.2 The function f^* defined in (6.68) may also be characterized as the *lower semi-continuous convex envelope* (that is, the largest lower semi-continuous and convex function everywhere bounded from above by the given function) of the function

$$g(p) = \liminf_{T \to +\infty} \inf_{x \in \mathbb{R}} \inf \left\{ \frac{1}{T} \int_x^{x+T} f\left(y, u'(y) + p\right) dy; \right.$$

$$\left. u \in W^{1,q} \left(]x, x + T[\right)^n, \ u \text{ is } T\text{-periodic} \right\}. \qquad (6.69)$$

$\square$

The power of Proposition 6.4 relies in the fact that it is not restricted to quadratic functions f. It is completely general. To better understand the nature of the result, it is nevertheless useful to consider the particular case $q = 2$, $n = 1$, $f(t, p) = a_{per}(t) \, p^2$. In that case, Proposition 6.4 claims that the Γ-limit of the sequence

$$F_\varepsilon(u) = \int_0^1 a_{per}\left(\frac{t}{\varepsilon}\right) \left(u'(t)\right)^2 dt$$

is the function (6.67) with

$$f^*(p) = \lim_{T \to +\infty} \inf \left\{ \frac{1}{T} \int_0^T a_{per}(y) \, (w'(y) + p)^2, dy; \quad w \in H_0^1 \left(]0, T[\right) \right\}.$$

It is readily seen that the latter function is actually

$$f^*(p) = \lim_{T \to +\infty} \left(\frac{1}{T} \int_0^T a_{per}^{-1}(y) \, dy \right)^{-1} p^2 = \langle a_{per}^{-1} \rangle^{-1} p^2.$$

It follows that

$$F^*(u) = \langle a_{per}^{-1}\rangle^{-1} \int_0^1 (u'(t))^2\, dt.$$

We therefore recover the result we know well from the previous chapters and which we saw again using the setting of energy functionals in the previous section.

Γ-convergence theory is well known for its versatility. It is useful and particularly well adapted for the *discrete* settings, such as the atomistic systems we have introduced in Sect. 1.3. Consider indeed a system of interacting particles, say on the real line to keep things simple. We assume there are N such particles, two contiguous particles being at a distance of order ε, with $N\varepsilon \approx 1$. In a simplified description, that is a truncated periodic system, the interatomic distance would be frozen at the exact value ε, but this is not what we consider here. Let us denote by u_n the position of the n-th particle when the system is such that its energy is minimal. We may describe the limit of this system, as $\varepsilon \to 0$ (or, equivalently, $N \to +\infty$)using Γ-convergence theory.

This is the purpose, for example, of the following result.

Proposition 6.5 (Limit of Discrete Systems, [Bra02, Theorem 4.3, p 79])
Take $1 < q < +\infty$ and a bounded function V satisfying, for c_1 and c_2 positive, $c_1 |p|^q - c_2 \le V(p)$ for all $p \in \mathbb{R}$. Then the energy functional

$$E_N(u) = \frac{1}{N} \sum_{n=1}^{N} V\left(\frac{u_n - u_{n-1}}{\frac{1}{N}}\right) \tag{6.70}$$

Γ-converges in L^q (]0, 1[) to

$$E_\infty(u) = \begin{cases} \int_0^1 V^{**}(u') \ \text{if } u \in W^{1,q} \ (]0, 1[), \\[2mm] +\infty \qquad \text{otherwise,} \end{cases} \tag{6.71}$$

where V^{**} denotes the lower semi-continuous convex envelope of V.

This result in essence expresses that, in the so-called *discrete-to-continuum* limit, the interaction potential is somewhat "convexified". A naive derivation passing to the limit in (6.70) while keeping u fixed would lead to (6.71) where V is substituted for V^{**}. The Γ-limit is more subtle. The non convexity of the interaction potential V creates, in the limit, degeneracies in the interaction potential V^{**} that do not originally exist in the potential V itself (see the flat region of the graph V^{**} of on Fig. 6.1).

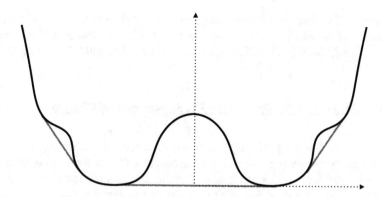

Fig. 6.1   An interaction potential V (black) and its envelope V^{**} obtained in the Γ-convergence limit. The red zones are those were $V^{**} < V$

6.3   Stochastic Interpretation of Deterministic Homogenization Theory

An alternative title for the present Sect. 6.3 could have been "*Stochastic homogenization*". If we had adopted this terminology (which we have not), we would have thereby perpetuated a certain tradition of ambiguity and confusion. The term "*Stochastic homogenization*" indeed indistinctly covers the following two very different topics:

(i) homogenization theory for equations that have random coefficients and/or random source terms,
(ii) homogenization theory using probabilistic type methods for deterministic, or random equations.

Sections 2.4 and 4.3 have presented topic **(i)**. We devote the present Sect. 6.3 to topic **(ii)**.

Homogenization theory for the elliptic equation (1) which has been our primary focus of interest throughout this textbook is amenable to probabilistic type techniques. In our opinion though, the elliptic setting is not the best setting to present a pedagogic introduction to those techniques. We are instead going to expose them in the parabolic setting. We will indicate on a regular basis how our arguments may be adapted to the original, elliptic setting. For this program to make sense, we need to first "transport" our results of the elliptic setting to the parabolic setting. It is the purpose of Sect. 6.3.1 to briefly expose homogenization theory for parabolic equations "as we have exposed" homogenization theory for elliptic equations previously in this textbook. Next, we need to take another preliminary step before we are in position to get to the heart of the matter. The reader may indeed be unfamiliar with the intimate links that relate *stochastic processes* and *stochastic differential equations* (two mathematical terms that will soon make sense...) on the one hand with *partial differential equations* on the other hand. As this link is instrumental for what we are discussing here, we need to recall a few elementary facts in this direction. This is the matter of Sect. 6.3.2. The final part of the present section, namely Sect. 6.3.3, will then, as announced, discuss the above topic **(ii)**.

6.3.1   Deterministic Parabolic Homogenization Theory

This section is devoted to the adaptation of our results on elliptic equations, and more precisely on the prototypical elliptic equation (1), to the case of parabolic equations (as well as, briefly, to the case of some other linear elliptic equations). To start with, we consider the following time-dependent equation, which is the exact analogous equation to (1),

$$\partial_t u^\varepsilon(t, x) - \operatorname{div}(a_\varepsilon(x) \nabla u^\varepsilon(t, x)) = f(t, x). \tag{6.72}$$

Like (1), this equation is supposed to be posed on a bounded domain $\mathcal{D} \subset \mathbb{R}^d$, for a right-hand side $f(t, x)$ possibly depending on time but not on the small parameter ε. Equation (6.72) is, like (1), supplied with the homogeneous Dirichlet boundary condition (2), which here reads as

$$u^\varepsilon(t, x) = 0, \quad x \in \partial \mathcal{D}, \quad \text{for all } t > 0. \tag{6.73}$$

In sharp contrast to (2), the equation, since it is an evolution equation, also requires an *initial condition*, that is

$$u^\varepsilon(t = 0, x) = u_0(x), \quad x \in \mathcal{D}, \tag{6.74}$$

for a given function u_0 defined on $\mathcal{D}$, and that may have to satisfy suitable regularity conditions on this domain or its boundary.

We will not say here why Eq. (6.72)–(6.73)–(6.74) is then well-posed, for all $\varepsilon > 0$. The mathematical techniques useful for proving this fact are relatively similar to those presented at the beginning of Sect. 3.1 for Eqs. (1) and (2). Th time-dependence of course introduces some additional technicalities but nothing unsurmountable and, anyway, this is definitely not our main focus here. We refer to the literature and for instance in particular to [Eva10, Section 7.1] for the case where a_ε is assumed symmetric and to [Eva10, Section 7.4] for the general case. The classical result is the following. Fix an arbitrary final time $T > 0$. Assume $f \in L^2\left([0, T], H^{-1}(\mathcal{D})\right)$. Assume also the usual boundedness and coerciveness conditions (3.1) on the coefficient a_ε. Then there exists a unique solution u^ε to (6.72)–(6.73)–(6.74), such that $u^\varepsilon \in L^2\left([0, T], H_0^1(\mathcal{D})\right)$ and $u^\varepsilon \in C^0\left([0, T], L^2(\mathcal{D})\right)$.

The following estimate is also satisfied by u^ε :

$$\|u^\varepsilon\|_{L^2([0,T],H_0^1(\mathcal{D}))} \leq C \left(\|f\|_{L^2([0,T],H^{-1}(\mathcal{D}))} + \|u_0\|_{L^2(\mathcal{D})} \right). \tag{6.75}$$

The proof may be found in [BLP11, pages 131–132]. Its proof proceeds by multiplication of Eq. (6.72) by u^ε and integration in both time and space.

The question of interest for us is the limit as $\varepsilon \to 0$ of the solution u^ε. In order to simplify as much as possible our brief summary of the topic, we assume that the coefficient a_ε in (6.72) is not only *time-independent* (a fact that had perhaps escaped the reader's attention so far), but also that it is of the form of the rescaled function

$$a_\varepsilon(x) = a_{per}\left(\frac{x}{\varepsilon}\right).$$

The homogenization of Eq. (6.72) could be made under much more general assumptions, but this restricted setting is far sufficient for what we intend to present here.

In this particular setting indeed, following [BLP11, Remark 1.8, p. 133–134], the homogenized limit of the parabolic equation (6.72) is easily identified as

$$\partial_t u^*(t, x) - \text{div}(a^* \nabla u^*(t, x)) = f(t, x), \tag{6.76}$$

where the homogenized coefficient a^* is identical to that for the elliptic case (3.8). To this end, we simply "recycle" the result established on (1) for the same coefficient a_ε.

It is indeed enough, for any function $\Phi \in C_0^\infty([0, T[)$, to multiply (6.72) by ϕ, and integrate in time from $t = 0$ to $t = T$. We find

$$- \text{div}(a_\varepsilon(x) \nabla(U^\varepsilon(\Phi))(x)) = (F(\Phi))(x) + (U^\varepsilon(\Phi'))(x) + u_0(x)\Phi(0), \tag{6.77}$$

where, for all $x \in \mathcal{D}$, we have denoted by

$$(U^\varepsilon(\Phi))(x) = \int_0^T u^\varepsilon(t, x)\,\Phi(t)\,dt, \quad F(\Phi)(x) = \int_0^T f(t, x)\,\Phi(t)\,dt,$$

and using an integration by parts in time

$$(U^\varepsilon(\Phi'))(x) + u_0(x)\Phi(0) = - \int_0^T \partial_t u^\varepsilon(t, x)\,\Phi(t)\,dt.$$

The point is, since the coefficient a_ε is time-independent, it commutes with the integral in time $\int_0^T dt$. The boundary condition (6.73) guarantees that $(U^\varepsilon(\Phi))$ vanishes on the boundary. All the above equations are actually to be understood in the weak sense, that is in the sense of their respective variational formulation as in Chap. 3. Our formal manipulations are in fact applied to the variational formulation of (6.72). We skip this technicality with which the reader is now familiar. By virtue of the bound (6.75), we may extract a subsequence, which we still denote by u^ε with a slight abuse of notation, which weakly converges in $L^2([0, T], H_0^1(\mathcal{D}))$ to some function u^*. By definition of weak convergence, this implies

$$U^\varepsilon(\Phi) \longrightarrow U^*(\Phi) \text{ in } H_0^1(\mathcal{D}),$$

where we have evidently denoted by $U^*(\phi) = \displaystyle\int_0^T u^*(t, x)\Phi(t)dt$. The right-hand side of (6.77) thus converges, in $L^2(\mathcal{D})$, to $F(\Phi) + U^*(\Phi')$. This allows us to pass to the limit in (6.77) precisely using homogenization theory as explained in Chap. 3. We obtain

$$- \operatorname{div}(a^*(x) \nabla(U^*(\Phi))(x)) = (F(\Phi))(x) + (U^*(\Phi'))(x) + u_0(x)\Phi(0). \tag{6.78}$$

In addition, the functional space $H_0^1(\mathcal{D})$ being closed for the weak H^1 topology, $U^*(\Phi) \in H_0^1(\mathcal{D})$. All this holds for any function $\Phi \in C_0^\infty([0, T[)$. This shows that u^* solves (6.74)–(6.76).

The reader may have noticed that we have applied the elliptic homogenization result on Eq. (6.77), despite the fact that the right-hand side of this equation varies and depends on ε, because of the term $U^\varepsilon(\Phi')$. It turns out that, in view of the bounds uniform in ε that we may establish on u^ε in $L^2([0, T], H_0^1(\mathcal{D}))$, the right-hand side strongly converges in $L^2(\mathcal{D})$ (up to an extraction, but since the final outcome of the argument is unique, we may discard this extraction eventually). It is therefore legitimate to pass to the limit in (6.77). This is a consequence of Lemma 3.1. The latter lemma, although stated for a fixed right-hand side f carries over to the case of a right-hand side that is a strongly converging sequence f_n in $H_{loc}^{-1}(\mathcal{D})$. The proof of the lemma actually only makes use of this property (in order to apply Lemma 3.2). Such a generalization argument has already been mentioned in Sect. 5.2.2.2.

The time-independent character of the coefficient a_ε allows us to bypass a complete study of the homogenization limit for a parabolic problem. We simply put it under the form of an elliptic problem. What we just did above with $- \operatorname{div}(a_\varepsilon(x) \nabla.)$ could be completed with other types of elliptic operator, such as the advection-diffusion operator studied in Sect. 6.1, or even more elaborate linear operators. Similarly, what we did for a *periodic* coefficient could be performed with the large variety of coefficients considered in Chaps. 3 and 4: stationary, periodic with localized defects, etc. We skip all these variants.

When the coefficient depends on time, the study proceeds by mimicking in the parabolic case our developments of Chap. 3. A two-scale expansion is postulated and it is next rigorously proven. We do not proceed in this direction either. We refer to [BLP11] (at least for the periodic setting) for homogenization theory of general parabolic problems.

An alternative option is to proceed using *probabilistic* arguments. This is precisely what we intend to do shortly.

One last remark is in order before we get to this. For a specific variant of Eq. (6.72), we may proceed even more economically than above, although this variant explicitly depends on time. This equation will be particularly useful as an element of comparison in Sect. 6.3.3 below. We consider

$$\partial_t u^\varepsilon(t, x) - a_{per}\left(\frac{x}{\varepsilon}\right) \Delta u^\varepsilon(t, x) = f(t, x), \qquad (6.79)$$

where the coefficient a_ε is assumed *scalar*, *time-independent* and of the form $a_\varepsilon(x) = a_{per}\left(\frac{x}{\varepsilon}\right)$. Equation (6.79) is supplied with the same boundary and initial conditions as (6.79), that is (6.73) and (6.74) respectively.

Applying to Eq. (6.79) the same string of manipulations as above, we transform it into the equation

$$- a_\varepsilon(x)\, \Delta(U^\varepsilon(\Phi))(x)) = (F(\Phi))(x) + (U^\varepsilon(\Phi'))(x) + u_0(x)\Phi(0), \qquad (6.80)$$

Finding the homogenized limit of this equation is straightforward. Upon "dividing" both sides of Eq. $- a_\varepsilon \Delta u^\varepsilon = f$, we immediately realize that the homogenized limit of the operator $- a_\varepsilon(x)\, \Delta$ is the operator $- a^* \Delta$, where the scalar constant reads

$$a^* = \left(\int_Q a_{per}^{-1}(y)\, dy\right)^{-1}.$$

This is indeed a direct application of the results of Chap. 1. It is then easy to prove (we leave this exercise to the reader but recommend that one should be cautious because of some technicalities that vary from the elliptic case to (6.72)) that the homogenized limit of Eq. (6.79) is thus

$$\partial_t u^*(t, x) - a^* \Delta u^*(t, x) = f(t, x), \qquad (6.81)$$

where

$$a^* = \left(\int_Q a_{per}^{-1}(y)\, dy\right)^{-1}. \qquad (6.82)$$

We will bear this in mind.

6.3.2 A Short Reminder of the Links Between PDEs and SDEs

To start with, we wish to complement the probabilistic setting that we have introduced in Sect. 1.6.1 of Chap. 1 in order to be able to address *time-dependent* random variables. We will now introduce the notion of stochastic differential equation, Itô integral, Itô calculus, etc. Needless to say, as we repeatedly mentioned throughout this textbook, the present section does not pretend to be a course on *stochastic calculus*. The topic deserves a full exposition, for which we *e.g.* refer to [Le 13] (in French) or to [Øks03, KS91, RW00a, RW00b]. We only scratch the surface. We provide as much as possible rigorous details, but we also allow ourselves to drastically simplify notions and bypass arguments.

A *stochastic process* (here continuous in time and valued in $\mathbb{R}^d$) is a family $(\mathbf{X}_t)_{t \geq 0}$ of random variables indexed by time, each of which defined on a probability space $(\Omega, \mathcal{T}, \mathbb{P})$, that is a triple we have defined in Sect. 1.6.1 of Chap. 1. A *filtration* $(\mathcal{T}_t, t \geq 0)$ is an increasing sequence, also indexed by time, of sub-tribes of the tribe $\mathcal{T}$. A stochastic process is said $(\mathcal{T}_t)_{t \geq 0}$-*adapted* if, for each time $t \geq 0$, the random variable $\mathbf{X}_t$ is measurable with respect to $\mathcal{T}_t$. Conversely, a stochastic process $\mathbf{X}_t$ being fixed, the *natural filtration* associated to $\mathbf{X}_t$ is the filtration $\mathcal{T}_t$ consisting, for each time $t \geq 0$, of the smallest tribe making the applications $\omega \to \mathbf{X}_s(\omega)$ measurable for all $0 \leq s \leq t$.

Given this formalism (which we may briefly and boldly summarize as the existence of a right setting so that all the manipulations we are going to make below make rigorous sense), we are in position to introduce the definition of *brownian motion*. Let us at once note that there exists many equivalent definitions. We only pick one of them. A process $\mathbf{X}_t(\omega)$ valued in the real line $\mathbb{R}$ is a (one-dimensional) *brownian motion* when this process

(i) *almost surely has continuous trajectories*, that is, almost surely, the applications $t \to \mathbf{X}_t(\omega)$ are continuous in the time variable $t \geq 0$;

(ii) *has independent increments*, that is, for all $0 \leq s \leq t$, the random variable $\mathbf{X}_t(\omega) - \mathbf{X}_s(\omega)$ is independent of the natural tribe $\mathcal{T}_s$, which reads as: for all $A \in \mathcal{T}_s$ and all given measurable function f,

$$\mathbb{E}\left(\mathbf{1}_A(\omega) \, f\left(\mathbf{X}_t(\omega) - \mathbf{X}_s(\omega)\right)\right) = \mathbb{E}\left(f\left(\mathbf{X}_t(\omega) - \mathbf{X}_s(\omega)\right)\right) \mathbb{P}(A);$$

(iii) *has stationary increments*, that is, for all $0 \leq s \leq t$, the law of $\mathbf{X}_t(\omega) - \mathbf{X}_s(\omega)$ is identical to that of $\mathbf{X}_{t-s}(\omega) - \mathbf{X}_0(\omega)$.

The three properties **(i)–(ii)–(iii)** jointly characterize a brownian motion and imply that for all $t \geq s \geq 0$, $\mathbf{X}_t(\omega) - \mathbf{X}_0(\omega)$ has gaussian law, with mean $r(t - s)$, for a certain r, and variance $\sigma^2(t - s)$ for a certain σ. It is neither evident that a brownian motion exists, nor that **(i)–(ii)–(iii)** imply the above properties. All these facts require a proof. In what follows, we will always assume that the brownian motion considered satisfies $r = 0$ and $\sigma = 1$, that is, the law of $\mathbf{X}_{t+1}(\omega) - \mathbf{X}_t(\omega)$ is a standard gaussian, for all $t \geq 0$.

Given the above one-dimensional brownian motion, it is evidently possible to construct a *multidimensional* brownian motion, that is a process $\mathbf{X}_t(\omega) = \left(\mathbf{X}_t^1(\omega), \ldots, \mathbf{X}_t^d(\omega)\right)$ valued in $\mathbb{R}^d$, such that the components $\mathbf{X}_t^k(\omega)$, $k = 1, \ldots, d$, are independent one-dimensional brownian motions.

The two classical notations for a brownian motion are $\mathbf{B}_t(\omega)$ (after the botanist Robert Brown who observed it experimentally) and $\mathbf{W}_t(\omega)$ (after the mathematician Norbert Wiener who formalized the notion under the name of a *Wiener process*). We favor the latter notation here.

We next introduce, using the multidimensional brownian motion $\mathbf{W}_t$, the notion of *stochastic differential equation*, often abridged in *SDE*, a notion which we will also connect to the partial differential equations we have been studying throughout this textbook.

A stochastic differential equation, for now posed on the real line $\mathbb{R}$ and assumed simple, is an equation of the form

$$d\,\mathbf{X}_t(\omega) = \mathbf{b}(\mathbf{X}_t(\omega))\,dt + \sigma\,d\,\mathbf{W}_t(\omega)\,, \qquad (6.83)$$

where the function $\mathbf{b}$, sufficiently regular so that the equation makes sense, and the constant σ are given. In our example (6.83), $\mathbf{b}$ does not explicitly depend upon time, but it could. The brownian motion $\mathbf{W}_t$ is assumed valued in $\mathbb{R}$ and *standard*, which means that we assume that, almost surely, $\mathbf{W}_0 = 0$. The equation is supplied with the following initial condition

$$\mathbf{X}_{t=0}(\omega) = \mathbf{X}^0(\omega), \qquad (6.84)$$

for a fixed real-valued random variable $\mathbf{X}^0(\omega)$. The writing (6.83) and (6.84) is purely *symbolic*. It must be understood in the following integral form

$$\mathbf{X}_t(\omega) = \mathbf{X}^0(\omega) + \int_0^t \mathbf{b}(\mathbf{X}_s(\omega))\,ds + \sigma\,\mathbf{W}_t(\omega). \qquad (6.85)$$

which has a rigorous meaning. For instance when $\mathbf{b}$ is continuous, we may assume (and this is indeed the case) that the trajectories $t \to \mathbf{X}_t(\omega)$ inherit from the almost sure continuity of the brownian trajectories $t \to \mathbf{W}_t(\omega)$, so that the integral in (6.85) is the integral of a continuous in time function. It is perfectly well defined. All this is of course to be proven, but it is intuitive this makes sense.

The above formal definition may be extended to the higher-dimensional setting, for instance (again to keep things simple) with

$$d\,\mathbf{X}_t(\omega) = \mathbf{b}(\mathbf{X}_t(\omega))\,dt + \sigma\,d\,\mathbf{W}_t(\omega)\,, \qquad (6.86)$$

where $\mathbf{b}$ is defined on $\mathbb{R}^d$ and valued in $\mathbb{R}^d$, where σ is a $d \times n$ matrix and $\mathbf{W}_t$ is an n-dimensional brownian motion. The initial condition analogous to (6.84) reads as

$$\mathbf{X}_{t=0}(\omega) = \mathbf{X}^0(\omega), \tag{6.87}$$

for a given random variable $\mathbf{X}^0(\omega)$ valued in $\mathbb{R}^d$. Multiple generalizations are possible regarding the coefficients $\mathbf{b}$ and σ. An important such generalization is the case when the coefficient σ is not constant any longer, but is a (bounded and) sufficiently regular function defined on $\mathbb{R}^d$ and even possibly time-dependent. In that case, and this is already seen in dimension $d = 1$, the equation reads as

$$d\,\mathbf{X}_t(\omega) = \mathbf{b}(\mathbf{X}_t(\omega))\,dt + \sigma(\mathbf{X}_t(\omega))\,d\,\mathbf{W}_t(\omega), \tag{6.88}$$

and corresponds to

$$\mathbf{X}_t(\omega) = \mathbf{X}^0(\omega) + \int_0^t \mathbf{b}(\mathbf{X}_s(\omega))\,ds + \int_0^t \sigma(\mathbf{X}_s)\,d\,\mathbf{W}_s(\omega). \tag{6.89}$$

where the rightmost integral is not a Lebesgue integral. It is an *Itô integral*. We refer to the bibliography for the mathematical construction of this integral. Suffice it to mention here that, under suitable assumptions, this integral defines a random variable the expectation of which vanishes. The multidimensional version of (6.88) and (6.89) is easy to obtain.

We now connect the stochastic differential equations with the partial differential equations. We again proceed formally. Our arguments may be made rigorous using the correct mathematical ingredients and assuming the necessary regularity of the coefficients at play. When we define

$$a_{ij} = \sum_{k=1}^n \sigma_{ik}\,\sigma_{jk}, \tag{6.90}$$

which we may summarize in the more concise form $\mathbf{a} = \sigma\,\sigma^T$, and u_0 denotes the law of the random variable $\mathbf{X}^0$, then the stochastic differential equation (6.87)–(6.88) is related to both the parabolic equation

$$\partial_t u - b_i \cdot \partial_i u - \frac{1}{2}\,a_{ij}\,\partial_{ij}^2\,u = 0, \tag{6.91}$$

and the adjoint equation to (6.91), that is

$$\partial_t p + \partial_i\,(b_i\,p) - \frac{1}{2}\,\partial_{ij}^2\,(a_{ij}\,p) = 0. \tag{6.92}$$

These two equations are posed on the entire space. We may alternatively pose them on a bounded domain and supply them with the suitable boundary conditions. The arguments and results below would be modified correspondingly, but nothing essential would change.

The relation between (6.88) and (6.91) is provided by the famous *Feynman-Kac formula* that allows one to express the solution $u(t, x)$ to (6.91) starting from the initial condition

$$u_{t=0} = u_0,$$

as

$$u(t, x) = \mathbb{E}_x \left(u_0 \left(\mathbf{X}_t(x, \omega) \right) \right), \tag{6.93}$$

where, to avoid any confusion, we have denoted by $\mathbf{X}_t(x, \omega)$ the solution to (6.88) starting from the initial condition

$$\mathbf{X}_{t=0}(\omega) = x,$$

corresponding to an initial law u_0 which would be a Dirac mass sitting at x. We have also denoted by $\mathbb{E}_x$ the expectation value *for x fixed*, that is the expectation value taken on all brownian trajectories only (with no randomness on the initial condition, which is frozen at x).

On the other hand, (6.88) is also related to Eq. (6.92), this time because if we denote by $p(t, x)$ the law of the random variable $X_t(\omega)$ solution to (6.88) for the initial condition (6.87) with $\mathbf{X}_{t=0}$ of law p_0, then $p(t, x)$ is the solution to (6.92) starting from the initial condition p_0. A calculation similar to (6.96) and consisting in evaluating, for an arbitrary, say bounded continuous on $\mathbb{R}^d$, function φ,

$$\int \varphi(x) \, \partial_t \, p(t, x) \, dx = \frac{d}{dt} \int \varphi(x) \, p(t, x) \, dx$$

$$= \frac{d}{dt} \mathbb{E} \left(\varphi(X_t) \right) = \mathbb{E} \left(\frac{d}{dt} \varphi(X_t) \right),$$

and in expressing the rightmost term using Itô calculus, yields the desired equation on $p(t, x)$.

In this setting, (6.92) is called the *Fokker-Planck equation*. Equations (6.91) and (6.92) are the *Kolmogorov equations* associated to the process $\mathbf{X}_t$.

Let us briefly sketch the formal proof of the Feynman-Kac formula (6.93). The key argument is an adaptation to the random setting of an argument the reader is presumably familiar with in the deterministic setting, namely the *method of lines*. Assume indeed that we wish to represent the solution to the linear transport equation

$$\partial_t u - b_i \cdot \partial_i \, u = 0, \tag{6.94}$$

starting from

$$u(t = 0, x) = u_0(x),$$

in terms of the solution $\mathbf{X}(s, x)$ of the (*ordinary*, as opposed to *stochastic*) differential equation

$$\begin{cases} \dfrac{d\mathbf{X}}{dt}(t, x) = \mathbf{b}(\mathbf{X}(t, x)), \\ \mathbf{X}(t = 0, x) = x. \end{cases}$$

We differentiate

$$\partial_s u(t - s, \mathbf{X}(s, x)) = -\partial_t u(t - s, \mathbf{X}(s, x)) + \dfrac{d\mathbf{X}}{dt}(s, x) \cdot \nabla u(t - s, \mathbf{X}(s, x))$$

$$= -\partial_t u(t - s, \mathbf{X}(s, x)) + \mathbf{b}(\mathbf{X}(s, x)) \cdot \nabla u(t - s, \mathbf{X}(s, x))$$

$$= 0,$$

since u is solution to (6.94), and we obtain, integrating from $s = 0$ to $s = t$,

$$u(t, x) = u_0(\mathbf{X}(t, x)). \tag{6.95}$$

In the random setting, we proceed similarly. The argument this time uses Itô calculus, which we admit here and provides us, when $\mathbf{X}_t(\omega)$ is the solution to (6.88), with

$$d\left(u(t - s, \mathbf{X}_s)\right) = -\partial_t u(t - s, \mathbf{X}_s)\, ds + \mathbf{b}(\mathbf{X}_s) . \nabla u(t - s, \mathbf{X}_s)\, ds$$

$$+ \frac{1}{2} \mathbf{a}(\mathbf{X}_s) : D^2 u(t - s, \mathbf{X}_s)\, ds + \left(\sigma^T(\mathbf{X}_s)\nabla u(t - s, \mathbf{X}_s)\right) . d\mathbf{W}_s, \tag{6.96}$$

where $A : B$ denotes the contracted product of two matrices A and B. Again integrating from $s = 0$ to $s = t$, and next taking the expectation value of the two sides and using both that u solves (6.91) and that the Itô integral has zero expectation, we obtain the Feynman-Kac formula (6.93), that is a representation formula analogous to (6.95).

What we have just done to relate stochastic differential equations and parabolic equations could be equally well done to relate stochastic differential equations and *elliptic* equations. The right-hand side f of the equation and possibly its boundary conditions then play the role that the initial condition u_0 has played above. Since the reader is familiar with a typical argument to pass from parabolic equations to elliptic equation, namely the Laplace transform seen in Sect. 2.5.1 of Chap. 2, this should be no surprise.

For instance, a Feynman-Kac formula connects the solution to

$$\begin{cases} -\frac{1}{2}\operatorname{div}(\mathbf{a}(x)\,\nabla u(x)) = f(x) \text{ in } \mathcal{D}, \\ \qquad\quad u(x) \qquad\quad = 0 \qquad \text{on } \partial\mathcal{D}, \end{cases} \tag{6.97}$$

to the solution of the stochastic differential equation (6.88). If we indeed take $\mathbf{a} = \sigma\sigma^T$, $\mathbf{b} = \operatorname{div}\mathbf{a}$ (that is, componentwise, $b_j = \partial_i a_{ij}$), and if we consider the solution to Eq. (6.88), the solution to (6.97) reads as

$$u(x) = \mathbb{E}\left(\int_0^{\tau_x(\omega)} f\,(\mathbf{X}_t(x,\omega))\,dt\right), \tag{6.98}$$

where the random time $\tau_x(\omega)$, is defined, for each x, as the first time when the trajectory $\mathbf{X}_t(x,\omega)$ reaches the boundary $\partial\mathcal{D}$ of the domain. Such a time is almost surely finite and is a stopping time. In order to intuitively figure out why such a formula holds true, it suffices to consider the specific case when σ and $\mathbf{a}$ are both the Identity matrix, thus $\mathbf{b} = 0$. The process $\mathbf{X}_t(x,\omega)$ then reads

$$\mathbf{X}_t(x,\omega) = x + \mathbf{W}_t(\omega).$$

Define $\mathbf{Y}_t(x,\omega) = u\,(x + \mathbf{W}_t(\omega))$ and differentiate it in time (which of course amounts to mimicking the calculus (6.96) of the parabolic setting to the specific case at hand):

$$d\,\mathbf{Y}_t(x,\omega) = \nabla u\,(x + \mathbf{W}_t(\omega))\,.d\,\mathbf{W}_t + \frac{1}{2}\Delta u\,(x + \mathbf{W}_t(\omega))\,dt$$
$$= \nabla u\,(x + \mathbf{W}_t(\omega))\,.d\,\mathbf{W}_t - f\,(x + \mathbf{W}_t(\omega))\,dt.$$

The integrated in time form reads as

$$\mathbf{Y}_T(x,\omega) - \mathbf{Y}_0(x,\omega) = \int_0^T \nabla u\,(x + \mathbf{W}_t(\omega))\,.d\,\mathbf{W}_t$$
$$- \int_0^T f\,(x + \mathbf{W}_t(\omega))\,dt.$$

Evaluate now this expression at time $T = \tau_x(\omega)$ (we admit that all these manipulations make sense) and substitute for $\mathbf{Y}$ its explicit value. We find

$$u\,\big(x + \mathbf{W}_{\tau_x(\omega)}(\omega)\big) - u(x) = \int_0^{\tau_x(\omega)} \nabla u\,(x + \mathbf{W}_t(\omega))\,.d\,\mathbf{W}_t$$
$$- \int_0^{\tau_x(\omega)} f\,(x + \mathbf{W}_t(\omega))\,dt. \tag{6.99}$$

As $\tau_x(\omega)$ is precisely a time when $x + \mathbf{W}_{\tau_x(\omega)}(\omega) \in \partial \mathcal{D}$, we have

$$u\left(x + \mathbf{W}_{\tau_x(\omega)}(\omega)\right) = 0,$$

by virtue of the second line of (6.97). The first term on the left-hand side therefore vanishes. To make things even simpler, assume f and a_{per} are sufficiently regular so that $\nabla u \in L^\infty(\mathcal{D})$. The argument may be adapted otherwise. This implies that, for fixed x, the function $\phi(t, \omega) = \nabla u(x + \mathbf{W}_t) \mathbf{1}_{t < \tau_x(\omega)}$ is square integrable, that is

$$\mathbb{E} \int_0^{+\infty} |\phi(t, \omega)|^2 dt = \mathbb{E} \int_0^{\tau_x(\omega)} |\nabla u(x + \mathbf{W}_t(\omega))|^2 dt$$

$$\leq \|\nabla u\|_{L^\infty(\mathcal{D})}^2 \mathbb{E}(\tau_x) < +\infty, \qquad (6.100)$$

where we have used that $\mathbb{E}(\tau_x) < +\infty$. This may be established using that $|\mathbf{W}_t|^2 - d \times t$ is a martingale (see [RY99, Proposition (1.2), page 52]), and that τ_x is a stopping time (see [RY99, Definition (4.4), page 42]). We deduce, applying [RY99, Proposition (1.4), page 52], that $\mathbb{E}(|\mathbf{W}_{\tau_x}|^2 - d\tau_x) = \mathbb{E}(|\mathbf{W}_0|^2)$. Therefore, $d\,\mathbb{E}(\tau_x) \leq E(|\mathbf{W}_{\tau_x}|^2) \leq \mathrm{diam}(\mathcal{D})^2 < +\infty$. The integrability (6.100) is known to imply that $\mathbb{E} \int_0^{+\infty} \phi(t, \omega) d\mathbf{W}_t(\omega) = 0$. See for instance [Øks03, Theorem 3.2.1, page 30]. This integral is exactly the expectation value of the first term on the right-hand side of (6.99). We obtain (6.98).

In short, both in the parabolic and in the elliptic setting, there exists an expression, respectively (6.93) et (6.98), of the solution of the partial differential equation in terms of an expectation value involving that of the stochastic differential equation.

Remark 6.3 The reader may be surprised by the coefficient $\dfrac{1}{2}$ that appears in the Eqs. (6.91), (6.92), and now (6.97). It originates from the Itô calculus. It is indeed the second order coefficient in the expansion within (6.96) and has, evidently, no particular relevance. It is sometimes the reason why, ironically, experts in analysis call the operator $\dfrac{1}{2}\Delta$ the *"probabilists' Laplacian"*.   □

This sort of viaticum describing the relationships between partial differential equations and stochastic differential equations suggests that, if we have been able to consider partial differential equations with oscillatory coefficients such as (1) in Chaps. 2 through 4 of this textbook or the parabolic version (6.72) of this equation in Sect. 6.3.1, and identify their homogenized limit as $\varepsilon \to 0$, then we *should* be able to find similar arguments for stochastic differential equations of the form

$$d\,\mathbf{X}_t(x, \omega) = \mathbf{b}\left(\frac{\mathbf{X}_t(x, \omega)}{\varepsilon}\right) dt + \sigma\left(\frac{\mathbf{X}_t(x, \omega)}{\varepsilon}\right) d\,\mathbf{W}_t(\omega). \qquad (6.101)$$

We *should* likewise be able to connect all these results together, or even to make the proofs of the former ones using the latter ones and conversely. This program is completed in the bibliography and we again refer to [BLP11], with, for instance, [BLP11, equation (4.4.3) & Theorem 5.2] for a general result of elliptic homogenization using probabilistic techniques.

In the next section, we present, in a relatively simple representative setting, a "*prototypical*" proof in this direction.

6.3.3 Deterministic Parabolic Homogenization Using Stochastic Processes

Throughout this section, we will be considering the specific case of Eq. (6.79) introduced in Sect. 6.3.1, that is

$$\partial_t u^\varepsilon(t, x) - a_{per}\left(\frac{x}{\varepsilon}\right) \Delta u^\varepsilon(t, x) = f(t, x),$$

for the *periodic, scalar, time-independent* coefficient a_{per}, posed on $[0, T] \times \mathcal{D}$ and supplied with conditions (6.73) and (6.74). This is the one and only example we will give in this textbook of a homogenization proof using the probabilistic interpretation of a deterministic partial differential equation.

We denote by

$$\sigma_{per} = \sqrt{2 a_{per}} \tag{6.102}$$

so that $\frac{1}{2}\sigma_{per}^2 = a_{per}$, which is the scalar version of the *matrix* equality (6.90), up to the usual prefactor $\frac{1}{2}$ which we have discussed in Remark 6.3. The stochastic process $\mathbf{X}_t$ solution to

$$d\mathbf{X}_t^\varepsilon(x, \omega) = \sigma_{per}\left(\frac{\mathbf{X}_t^\varepsilon(x, \omega)}{\varepsilon}\right) d\mathbf{W}_t(\omega), \tag{6.103}$$

is that we should consider, in view of (6.101), in our proof of the homogenization of Eq. (6.79) into Eq. (6.81), that is

$$\partial_t u^*(t, x) - a^* \Delta u^*(t, x) = f(t, x),$$

with the homogenized coefficient $a^* = \left(\int_Q a_{per}^{-1}(y)\, dy\right)^{-1}$ given in (6.82).

It should be borne in mind that what we are going to do could carry over to Eq. (6.79) for a *non-scalar* coefficient, as well as to any arbitrary elliptic or parabolic equation, up to technicalities we do not want to bring up. For instance, for Eq. (1) or its parabolic variant (6.72), the process to consider instead of (6.103) would be the particular case

$$d\,\mathbf{X}_t^\varepsilon\,(x\,,\,\omega) = \mathbf{b}_{per}\left(\frac{\mathbf{X}_t^\varepsilon\,(x\,,\,\omega)}{\varepsilon}\right)\,dt + \sigma_{per}\left(\frac{\mathbf{X}_t^\varepsilon\,(x\,,\,\omega)}{\varepsilon}\right)\,d\,\mathbf{W}_t(\omega), \qquad (6.104)$$

of the general setting (6.101), for the specific choice $\left(\mathbf{b}_{per}\right)_j = \nabla a_{per}$, provided the regularity is sufficient. The homogenized coefficient obtained in the limit would not then be the simple coefficient (6.82). Its explicit calculation would involve the invariant measure m_{per} associated to the operator $-\operatorname{div}(a_{per}(x)\,\nabla\,.)$ and which we have introduced in (6.27) of Sect. 6.1.2. This invariant measure, which in the particular case (6.79) reads as

$$m_{per} = \left(\int_Q a_{per}^{-1}(y)\,dy\right)^{-1} a_{per}^{-1},$$

and which will not *explicitly* show up in the argument below, would then play a role in the limit $\varepsilon \to 0$ of the process (6.104).

In any event, the essential ideas are already present in the setting (6.79)–(6.102)–(6.103). We will only describe the headlines of the study, so that the reader grasps these essential ideas, and we will spare the reader all the unnecessary technicalities. We refer to [BLP11, Sect. 4.1 & 4.2, p. 209-215] for the complete proof with all its necessary details.

Corrrespondence for Fixed ε

For each fixed $\varepsilon > 0$, the Feynman-Kac (6.93) formula expresses as

$$u^\varepsilon\,(t, x) = \mathbb{E}_x\left(u_0\left(\mathbf{X}_t^\varepsilon(x, \omega)\right)\right), \qquad (6.105)$$

the solution to (6.79)–(6.73)–(6.74). Understanding the limit $\varepsilon \to 0$ of this solution thus amounts to understanding the *limit in law* of the process $\mathbf{X}_t^\varepsilon$ solution to (6.103).

We now briefly formalize this.

The trajectories of the process $\mathbf{X}_t^\varepsilon$ solution to (6.103) are all continous, provided the coefficient a_{per}, and thus σ_{per}, have some minimal regularity. The classical theory, called the *Itô Theory* and analogous to the Cauchy-Lipschitz Theory for the deterministic setting, guarantees the existence and uniqueness of such a process that has continuous trajectories as soon as, in our case, σ_{per} is Lipschitz continuous. Some more general settings also exist for less regular coefficients.

Each trajectory is an element of the space $C^0\left([0,\,T],\,\mathbb{R}^d\right)$, called in this specific context the *Wiener space*. This space may then be *probabilized* using $\mathbf{X}_t^\varepsilon$. This defines a probability measure, which we denote by $\mathbb{P}^\varepsilon$. The Wiener space is for instance rigorously constructed in [KS91, Section 2.4]. The issue is to determine, in a suitable sense that is, in law (more precisely, the notion is that of *weak convergence of measures*), the limit $\varepsilon \to 0$ of this sequence of measures $\mathbb{P}^{\varepsilon}$. It is important to note that this notion is not simple, since the space $C^0\left([0, T], \mathbb{R}^d\right)$ is *infinite-dimensional*. We might thus expect a few substantial difficulties... which we will elude and for which we will refer to the bibliography.

The formalization of the argument uses *Prokhorov Theorem*. The idea is that we need a suitably adapted version of the classical Ascoli-Arzela Theorem: formally, the sequence converges if it converges at "at all points" and is "equicontinuous". If we indeed have these properties, then the sequence of continuous processes $\mathbf{X}_t^{\varepsilon}$ converges in law to some process $\mathbf{X}_t^*$, which will imply that the expectation values of the type (6.105), that is $\mathbb{E}_x\left(u_0\left(\mathbf{X}_t^{\varepsilon}\right)\right)$, converge to the analogous expectation values with $\mathbf{X}_t^*$, at least for all bounded continuous functions u_0.

The rigorous way to expresses the first condition, that is convergence "at all points", is the convergence of *finite-dimensional marginals*, that is of the laws of the "photographs at a finite number of times". By this, we more precisely mean that, for all $N \in \mathbb{N}$, and all *finite* set of times $t_1, \ldots, t_N$, the law of the N-tuple $\left(\mathbf{X}_{t_1}^{\varepsilon}, \ldots, \mathbf{X}_{t_N}^{\varepsilon}\right)$ converges (in the usual sense of random variables defined on $\mathbb{R}^{N \times d}$), to the law of the N-tuple $\left(\mathbf{X}_{t_1}^*, \ldots, \mathbf{X}_{t_N}^*\right)$.

The second condition, that of "equicontinuity", is related to the fact that the sequence of processes should be *tight*, a notion which we do not wish to introduce further here. For our specific case, it suffices to realize that this second condition is fulfilled since the processes $\mathbf{X}_t^{\varepsilon}$ in particular satisfy the so-called *Kolmogorov's criterion*, because, for $0 \le s \le t \le T$, in view of the properties of the stochastic integral and of the boundedness of a_{per} thus of σ_{per}, we have $\mathbb{E}\left\|\mathbf{X}_t^{\varepsilon} - \mathbf{X}_s^{\varepsilon}\right\|^4 \le C(t-s)^2$, where the constant C is independent of ε.

Proving the expected convergence therefore amounts to proving that of the finite-dimensional marginals.

Passage to the Limit $\varepsilon \to 0$ in the Finite-Dimensional Marginals

We are going to show, in terms of finite-dimensional marginals, that the process $\mathbf{X}_t^{\varepsilon}$ solution to (6.103) converges to the process $\mathbf{X}_t^*$ solution to

$$d\mathbf{X}_t^*(x, \omega) = \sqrt{2a^*} \; d\mathbf{W}_t(\omega), \tag{6.106}$$

that is

$$\mathbf{X}_t^*(x, \omega) = x + \sqrt{2a^*} \; \mathbf{W}_t(\omega), \tag{6.107}$$

where $a^* = \left(\int_Q a_{per}^{-1}(y)\, dy\right)^{-1}$ is indeed both the expected coefficient (6.82) for the homogenized limit (6.81) of (6.79) and the coefficient of the stochastic differential equation corresponding to that partial differential equation since we recover $\sigma^* = \sqrt{2a^*}$ starting from (6.102), that is $\sigma_{per} = \sqrt{2a_{per}}$.

In view of the properties in the previous paragraph, this will allow us to infer the convergence in law of the process $\mathbf{X}_t^{\varepsilon}$ to the process $\mathbf{X}_t^*$. Our sketch of proof using probabilistic arguments of the homogenization of (6.79) into (6.81) will then be completed.

In order to show convergence in law of the finite-dimensional marginals, we are going to prove, and we know this is equivalent, the convergence of the associated characteristic functions. We recall from our brief introduction to the theory of random variables seen in Sect. 1.6.4 of Chap. 1, that the *characteristic function* of a random variable X valued in $\mathbb{R}^d$ is the Fourier transform of its law, that is the function $\Phi_X(z) = \mathbb{E}\left(e^{i\,z\cdot X}\right)$.

It is for instance useful to recall that the characteristic functions may be used to prove the Central Limit Theorem. We have seen such an example in Sect. 1.6.4. It is therefore not unexpected that they are useful to establish the convergence in law we are aiming at here.

It is well known that the characteristic function of a centered gaussian law with variance σ^2 is $\Phi(z) = e^{-\frac{1}{2}\sigma^2 z^2}$. It is then easy to deduce that the characteristic function of the standard real brownian motion $\mathbf{W}_t$ is

$$\Phi_{\mathbf{W}_t}(z) = e^{-\frac{1}{2}t z^2},$$

and that, in addition, the increments of the standard brownian motion $\mathbf{W}_t - \mathbf{W}_s$ being gaussian, independent and of variance $t - s$, the characteristic function of the N-dimensional marginal at times $0 \le t_1 < \dots < t_N$ of the process $\mathbf{X}_t^*$, valued in $\mathbb{R}^d$, is the function defined on $\mathbb{R}^{d \times N}$ by

$$\Phi_{\mathbf{X}^*}(z_1, \dots, z_N) = \mathbb{E}\left(e^{i\left(z_1 \cdot \mathbf{X}_{t_1}^* + \dots + z_N \cdot \mathbf{X}_{t_N}^*\right)}\right)$$

$$= e^{i((z_1 + \dots + z_N)\cdot x)} \exp\left(-\sum_{1 \le i,j \le N} a^* z_i \cdot z_j \min(t_i, t_j)\right).$$

For $N = 1$, we have already proven this fact, since this amounts to writing

$$\Phi_{\mathbf{X}^*}(z_1) = e^{i z_1 \cdot x} \exp\left(-a^* |z_1|^2 t_1\right),$$

which is the adaptation to the d-dimensional case (6.106) of the elementary result which we have just recalled for the standard scalar brownian motion. For $N = 2$ (and we will leave all the other cases to the reader...), we write

$$e^{i\left(z_1 \cdot \mathbf{X}_{t_1}^* + z_2 \cdot \mathbf{X}_{t_2}^*\right)} = e^{i\,(z_1 + z_2)\cdot x + i\,(z_1 + z_2)\cdot(\mathbf{X}_{t_1}^* - x) + i z_2\,(\mathbf{X}_{t_2}^* - \mathbf{X}_{t_1}^*)}, \qquad (6.108)$$

and use independence of the increments to calculate the expectation value, which yields

$$\Phi_{\mathbf{X}^*}(z_1, z_2) = e^{i(z_1 + z_2)x}\,\mathbb{E}\left(e^{i(z_1 + z_2)(\mathbf{X}_{t_1}^* - x)}\right)\mathbb{E}\left(e^{i z_2(\mathbf{X}_{t_2}^* - \mathbf{X}_{t_1}^*)}\right).$$

Therefore

$$\Phi_{\mathbf{X}^*}(z_1, z_2) = e^{i(z_1+z_2)x} \, \exp\left(-a^*|z_1 + z_2|^2 t_1\right) \exp\left(-a^*|z_2|^2(t_2 - t_1)\right)$$

$$= e^{i(z_1+z_2)x} \, \exp\left[-a^*\left(|z_1|^2 t_1 + 2z_1 \cdot z_2 t_1 \right.\right.$$

$$\left.\left. +|z_2|^2 t_1 + |z_2|^2 t_2 - |z_2|^2 t_1\right)\right],$$

thus the expected result.

In order to conclude, we need to establish that

$$\lim_{\varepsilon \to 0} \mathbb{E}\left(e^{i\, z_1 \cdot \mathbf{X}_{t_1}^\varepsilon + \dots + z_N \cdot \mathbf{X}_{t_N}^\varepsilon}\right) = \Phi_{\mathbf{X}^*}(z_1, \dots, z_N) \qquad (6.109)$$

It is relatively intuitive that the convergence (6.109) may be established using an induction argument on N. We will not give the proof explicitly here, but just mention the two key ingredients: the proof for $N = 1$ and the outline of the passage from N to $N + 1$.

The first ingredient consists in proving that, for all $t \geq 0$ et $z \in \mathbb{R}^d$,

$$\lim_{\varepsilon \to 0} \mathbb{E}\left(e^{i\, z \cdot \mathbf{X}_t^\varepsilon}\right) = e^{i\, z \cdot x} \, \exp\left(-a^* \, |z|^2 t\right),$$

that is,

$$\lim_{\varepsilon \to 0} \mathbb{E}\left[\exp\left(i \int_0^t \sigma_{per}\left(\frac{\mathbf{X}_s^\varepsilon(x, \omega)}{\varepsilon}\right) z \cdot d\,\mathbf{W}_s(\omega)\right)\right] = \exp\left(-a^* \, |z|^2 t\right) \qquad (6.110)$$

The mathematical substance of the proof of (6.110) is not immediate to guess. We will briefly expose it below.

Given what we did in (6.108), the second ingredient requires to "start again from scratch" at each time, and takes the form

$$\lim_{\varepsilon \to 0} \mathbb{E}\left[\left(\int_{t_1}^{t_2} \Phi_{per}\left(\frac{\mathbf{X}_t^\varepsilon(x, \omega)}{\varepsilon}\right) dt\right)^2 \middle| \mathcal{T}_{t_1}\right] = 0, \qquad (6.111)$$

for all $0 \leq t_1 \leq t_2$ and for all *periodic* function Φ_{per} such that

$$\int_Q \Phi_{per}(y) \, a_{per}^{-1}(y) \, dy = 0. \qquad (6.112)$$

Condition (6.111) involves the *conditional* expectation for the filtration $\mathcal{T}_{t_1}$ of the ambient probability space (see Sect. 6.3.2). Put differently, the expectation is taken *"assuming the state at time t_1 is known"*. This formalizes the above mentioned somewhat vague notion of "starting again from scratch".

It is not immediate to realize why the restriction (6.112) should be imposed to the test functions for (6.111). Ils is in fact a *necessary* condition, which will be somewhat clearer once we have sketched the proof of (6.111). Consider indeed a periodic function φ_{per}, solution to

$$- a_{per} \, \Delta\varphi_{per} \; = \; \Phi_{per},$$

on the periodic cell Q. This equation only admits a solution when the condition (6.112) is satisfied, since

$$0 = \int_Q \Delta\varphi_{per}(y) \, dy = \int_Q a_{per}^{-1}(y) \, \Phi_{per}(y) \, dy,$$

by periodicity. We then calculate, from (6.103),

$$\varphi_{per} \left(\frac{\mathbf{X}_{t_2}^\varepsilon (x \, , \, \omega)}{\varepsilon} \right) - \varphi_{per} \left(\frac{\mathbf{X}_{t_1}^\varepsilon (x \, , \, \omega)}{\varepsilon} \right)$$

$$= \frac{1}{\varepsilon} \int_{t_1}^{t_2} \nabla\varphi_{per} \cdot \sigma_{per} \left(\frac{\mathbf{X}_t^\varepsilon (x \, , \, \omega)}{\varepsilon} \right) d\,\mathbf{W}_t(\omega)$$

$$+ \frac{1}{\varepsilon^2} \int_{t_1}^{t_2} a_{per} \, \Delta\varphi_{per} \left(\frac{\mathbf{X}_t^\varepsilon (x \, , \, \omega)}{\varepsilon} \right) dt,$$

from which we infer

$$\int_{t_1}^{t_2} \Phi_{per} \left(\frac{\mathbf{X}_t^\varepsilon (x \, , \, \omega)}{\varepsilon} \right) dt = - \varepsilon^2 \, \varphi_{per} \left(\frac{\mathbf{X}_{t_2}^\varepsilon (x \, , \, \omega)}{\varepsilon} \right) + \varepsilon^2 \, \varphi_{per} \left(\frac{\mathbf{X}_{t_1}^\varepsilon (x \, , \, \omega)}{\varepsilon} \right)$$

$$+ \varepsilon \int_{t_1}^{t_2} \nabla\varphi_{per} \cdot \sigma_{per} \left(\frac{\mathbf{X}_t^\varepsilon (x \, , \, \omega)}{\varepsilon} \right) d\,\mathbf{W}_t(\omega).$$

The functions on the right-hand side being bounded, we may then deduce that

$$\mathbb{E} \left[\left(\int_{t_1}^{t_2} \Phi_{per} \left(\frac{\mathbf{X}_s^\varepsilon (x \, , \, \omega)}{\varepsilon} \right) ds \right)^2 \middle| \mathcal{T}_{t_1} \right] = O(\varepsilon^2) + O(\varepsilon),$$

thus (6.111).

As the reader may have now understood, (6.111) is an analogous notion, in the setting of random processes and stochastic integrals, of the notion of average of an oscillatory periodic function, which we have developed for the classical notion of

Lebesgue integral in Chap. 1 and Proposition 1.1. To realize this, pick ($t_1 = 0$, $t_2 = t$), in which case (6.111) reads as

$$\lim_{\varepsilon \to 0} \mathbb{E} \left[\left(\int_0^t \Phi_{per} \left(\frac{\mathbf{X}_s^\varepsilon (x, \omega)}{\varepsilon} \right) ds \right)^2 \right] = 0, \tag{6.113}$$

Formally, the trajectories of the process $\dfrac{\mathbf{X}_s^\varepsilon (x, \omega)}{\varepsilon}$ progressively fill in the periodic cell Q as $\varepsilon \to 0$. This is an ergodicity phenomenon, which we will comment upon in the next paragraph. The average

$$\frac{1}{t} \int_0^t \Phi_{per} \left(\frac{\mathbf{X}_s^\varepsilon (x, \omega)}{\varepsilon} \right) ds$$

thus approaches the average of the function Φ_{per} over the cell. The subtlety is that different regions of the cell Q are weighted differently, and proportionally to the number of times they are visited by the process. This is exactly what we have briefly mentioned when constructing the probability measure on the Wiener space. The "number of visits" is precisely quantified by the value of the invariant measure, that is the solution to $\Delta(m_{per} \, a_{per}) = 0$, which agrees, up to an irrelevant normalization factor, with a_{per}^{-1}. The limit obtained is therefore the weighted integral

$$\int_Q \Phi_{per} (y) \, a_{per}^{-1} (y) \, dy,$$

which vanishes because of the constraint (6.112), thus (6.111). In Chap. 1, when we have addressed the integral $\int_0^t \Phi_{per} \left(\dfrac{s}{\varepsilon} \right) ds$, all the points within the periodic cell were equally weighted, thus the limit was the mere average of Φ_{per}. The similarity and the differences are evident.

We finally need to say some words on the proof of (6.110). We first observe that

$$\mathbb{E} \left[\exp \left(i \int_0^t \sigma_{per} \left(\frac{\mathbf{X}_s^\varepsilon (x, \omega)}{\varepsilon} \right) z \, . \, d \, \mathbf{W}_s(\omega) \right) \right.$$

$$\left. \cdot \exp \left(\frac{1}{2} \int_0^t \sigma_{per}^2 \left(\frac{\mathbf{X}_s^\varepsilon (x, \omega)}{\varepsilon} \right) |z|^2 \, ds \right) \right] = 1 \tag{6.114}$$

Indeed, if we denote by Z_t^ε the random variable within the expectation value, Itô calculus leads to

$$d \, Z_t^\varepsilon = i \, Z_t^\varepsilon \, \sigma_{per} \left(\frac{\mathbf{X}_t^\varepsilon}{\varepsilon} \right) z \, . \, d \, \mathbf{W}_t,$$

that is, since by construction, $Z_{t=0}^\varepsilon = 1$,

$$Z_t^\varepsilon = 1 + i \int_0^t Z_s^\varepsilon \sigma_{per} \left(\frac{\mathbf{X}_s^\varepsilon}{\varepsilon} \right) z \, . \, d \mathbf{W}_s. \tag{6.115}$$

Given the bounds we have on the functions, we obtain

$$\mathbb{E} \int_0^t \left| Z_s^\varepsilon \sigma_{per} \left(\frac{\mathbf{X}_s^\varepsilon}{\varepsilon} \right) \right|^2 ds < +\infty.$$

This again implies, by virtue of[Øks03, Theorem 3.2.1, page 30], that the rightmost term of (6.115) vanishes in expectation. We thus have $\mathbb{E}(Z_t^\varepsilon) = 1$, for all time $t \geq 0$, whence (6.114).

On the other hand, we know the behaviour of the second factor in (6.114) in the limit $\varepsilon \to 0$. Indeed, since

$$\int_Q \left(\sigma_{per}^2 (y) - 2 a^* \right) a_{per}^{-1} (y) \, dy = 2 \int_Q \left(a_{per}(y) - a^* \right) a_{per}^{-1} (y) \, dy = 0,$$

this periodic function satisfies condition (6.112). By (6.111) for $(t_1 = 0, t_2 = t)$, we thus have

$$\lim_{\varepsilon \to 0} \mathbb{E} \left[\left(\frac{1}{2} \int_0^t \sigma_{per}^2 \left(\frac{\mathbf{X}_s^\varepsilon (x, \omega)}{\varepsilon} \right) dt - a^* t \right)^2 \right] = 0.$$

If we formally pretend that this identity holds for all $\varepsilon > 0$ and not only in the limit $\varepsilon \to 0$, we may thus substitute $\exp\left(a^* |z|^2 t \right)$ for the second factor of (6.114). We obtain (6.110). Of course a rigorous proof is in order. Is it in fact not difficult once the essential ideas seen above have been collected. We refer to [BLP11] for this proof.

On Ergodicity
The study of the limit $\varepsilon \to 0$ of the sequence of processes $\mathbf{X}_t^\varepsilon$ may be conducted somewhat differently. The advantage of this different viewpoint is to emphasize a central phenomenon, which we have hardly approached so far: the relation between homogenization theory for (at least elliptic and parabolic) partial differential equations and *ergodicity theory*, that is the property of a dynamical system to explore, along its trajectory from time $t = 0$ to time $t = +\infty$, the entirety of the accessible space.

With a view to more specifically explaining the phenomenon in our context, we again consider the stochastic differential equation (6.103), that is

$$d \mathbf{X}_t^\varepsilon (x, \omega) = \sigma_{per} \left(\frac{\mathbf{X}_t^\varepsilon (x, \omega)}{\varepsilon} \right) d \mathbf{W}_t(\omega)$$

but we substitute for the process $\mathbf{X}_t^\varepsilon$ the process

$$\widetilde{\mathbf{X}}_t^\varepsilon = \frac{1}{\varepsilon} \mathbf{X}_{\varepsilon^2 t}^\varepsilon \qquad (6.116)$$

An elementary exercise of change of time in Eq. (6.103) shows that $\widetilde{\mathbf{X}}_t^\varepsilon$ solves

$$d\widetilde{\mathbf{X}}_t^\varepsilon (x, \omega) = \sigma_{per} \left(\widetilde{\mathbf{X}}_t^\varepsilon (x, \omega) \right) d\widetilde{\mathbf{W}}_t(\omega), \qquad (6.117)$$

where $\widetilde{\mathbf{W}}_t$ is another standard brownian motion (actually equal to $\frac{1}{\varepsilon} \mathbf{W}_{\varepsilon^2 t}$ in law, but its explicit dependency upon ε is irrelevant, only the law of its trajectories matter).
    Since (6.116) also reads as

$$\mathbf{X}_t^\varepsilon = \varepsilon \widetilde{\mathbf{X}}_{\frac{t}{\varepsilon^2}}^\varepsilon, \qquad (6.118)$$

the identification of the limit in law of the process $\mathbf{X}_t^\varepsilon$ as $\varepsilon \to 0$ is related to that of the behaviour in law *in the long time limit* of the process $\widetilde{\mathbf{X}}_t^\varepsilon$. The latter process being the solution to (6.117), we know its law is the solution to the Fokker-Planck equation associated to this stochastic differential equation, that is (6.92) which here reads as

$$\partial_t p - \frac{1}{2} \Delta (a_{per} \, p) = 0. \qquad (6.119)$$

At least formally, and all this may of course be made rigorous, we read on (6.119) that its solution $p(t, x)$ converges, as $t \to +\infty$, to the solution $p_\infty(x)$ to the associated stationary equation

$$- \frac{1}{2} \Delta (a_{per} \, p_\infty) = 0. \qquad (6.120)$$

obtained by eliminating the time derivative.   Equation (6.120) is satisfied the invariant measure m_{per} (we recall (6.27) from Sect. 6.1). We thus obtain $p_\infty = m_{per} = a^* a_{per}^{-1}$. This allows one to establish the homogenization result in a slightly different manner.
    We note in passing that this strategy of proof makes use of *averaging* results in the vein of (6.113). It is for instance possible to show that

$$\lim_{\varepsilon \to 0} \frac{1}{t} \int_0^t \Phi_{per} \left(\frac{\mathbf{X}_s^\varepsilon (x, \omega)}{\varepsilon} \right) ds = a^* \int_Q \Phi_{per} (y) \, a_{per}^{-1} (y) \, dy. \qquad (6.121)$$

The proof is as follows. We first rewrite the left-hand side under the form

$$\frac{1}{t} \int_0^t \Phi_{per}\left(\frac{\mathbf{X}_s^\varepsilon (x , \omega)}{\varepsilon}\right) ds = \frac{1}{t/\varepsilon^2} \int_0^{t/\varepsilon^2} \Phi_{per}\left(\widetilde{\mathbf{X}}_s^\varepsilon (x , \omega)\right) ds.$$

Proving (6.121) hence amounts to studying in the long time limit the average of a time integral of the form

$$\lim_{T \to +\infty} \frac{1}{T} \int_0^T \Psi (s) \, ds,$$

where, if everything goes according to plan, only the values of Ψ for large times should matter. In our setting, it all depends on whether the dynamics $\widetilde{\mathbf{X}}^\varepsilon$ fills in the periodicity cell Q of Φ_{per} for large times. We then naturally obtain

$$\int_Q \Phi_{per} (y) \, m_{per} (y) \, dy,$$

that is the right-hand side of (6.121). We at last understand the reason why the measure m_{per} is called *invariant*. It is the asymptotic limit in large times of the law associated to the considered process, thus it is necessarily left invariant.

The method of proof presented in this paragraph and throughout Sect. 6.3 taught us something fundamental. It suggests that, *for linear partial differential equations, the existence of a homogenized limit is equivalent to that of an invariant measure*, since the latter measure provides and characterizes the long time limit of the underlying random processes. It identifies the limit as $\varepsilon \to 0$ of the solution of the partial differential equation, after a change of time and application of the Feynman-Kac formula. The rest is technicalities. Such a key role of the invariant measure was already suspected in Sect. 6.1 when we insisted upon the *"algebraic"* role of that measure that allowed us, using the right manipulations, to proceed in our proofs as for the case (1). We now know this was not only a technical *ad hoc* recipe, but a general fundamental phenomenon.

6.4 To Infinity... and Beyond

This final section of the textbook collects some examples of situations radically different from those we have seen so far and for which we will even less dwell on the details than for the first sections of this Chap. 6 (provided this is even possible...). Their complete study requires technicals tools that are far too remote from our present discussion and which we definitely leave to the experts. In any event, we are not the best authors to introduce those tools to the reader.

We only wish to show, using a selection of examples, that theory and numerics constantly advance to contexts and settings that involve always more sophisticated equations. This constitutes an active research field of contemporary mathematics in the broad sense.

6.4.1 The Fully Nonlinear Setting

We approach in this Sect. 6.4.1 a whole different *universe*, that of the so-called *"fully nonlinear"* equations. The terminology originates from the fact that, in those equations, the differential operator of highest order is nonlinear. This is in contrast to, for instance, an equation such as $-\Delta u + u^2 = f$, which is called a *semilinear* equation since the nonlinearity only affects the term of order zero and not the Laplacian.

As a prototype of fully nonlinear equation, we choose to present the following oscillatory *first order Hamilton-Jacobi equation*

$$\partial_t u^\varepsilon (t, x) + H \left(\frac{x}{\varepsilon}, \nabla u^\varepsilon (t, x) \right) = 0. \qquad (6.122)$$

This equation is a particular case of the more general family

$$\partial_t u^\varepsilon (t, x) + H \left(\frac{x}{\varepsilon}, u^\varepsilon, \nabla u^\varepsilon (t, x), D^2 u^\varepsilon (t, x) \right) = 0.$$

We have considered its time-dependent form, but the equation could equally well be considered under one of its many stationary variants, namely

$$u^\varepsilon (x) + H \left(\frac{x}{\varepsilon}, \nabla u^\varepsilon (x) \right) = 0.$$

We have actually already briefly considered such an equation in Sect. 2.5.3.

Equation (6.122) is supplied with the initial condition

$$u^\varepsilon (0, x) = u^0 (x). \qquad (6.123)$$

The function H appearing in (6.122) is called the *Hamiltonian* and is classically denoted by $H(x, p)$ for $x \in \mathbb{R}^d$, $p \in \mathbb{R}^d$. We will see below in (6.127) the prototypical example of the quadratic Hamaltonian

$$H (x, p) = |p|^2 - V (x),$$

where the function V is called the *potential*. This particular Hamiltonian originates from Classical Mechanics. The term $|p|^2$ models, up to a missing prefactor $\frac{1}{2}$, the kinetic energy of the physical system considered, while the term V models the potential of the forces the system is subjected to.

Well beyond the quadratic Hamiltonian and the context of Classical Mechanics, (6.122) and its many variants are relevant to a large variety of contexts. It has in particular an interpretation in terms of Control Theory. It also models many practically relevant problems such as wave propagation phenomena, etc.

The purpose of homogenization theory in this context is to show that the solution u^ε to (6.122) and (6.123) converges, as $\varepsilon \to 0$, to the solution u^* to the homogenized equation

$$\partial_t u^* (t, x) + H^* \left(x, \nabla u^*(t, x)\right) = 0, \tag{6.124}$$

starting from the same initial condition (6.123). The *homogenized Hamiltonian H^** plays the same role in (6.124) as the homogenized coefficient a^* for (1).

The key questions are of course on the one hand the identification of H^*, which should be as explicit as possible, and, on the other hand, the proof of the convergence of u^ε to u^*, in the right topologies and with, possibly, rates of convergence. The properties classically assumed on the Hamiltonian to achieve this program are growth conditions at infinity (and perhaps convexity) in the p variable, structure assumptions such as periodicity or stationarity in the x variable, and some minimal regularity (that is often continuity) in both variables.

As far as we are concerned, we are going to assume here that the Hamiltonian $H = H_{per}$ is *periodic* in x. We will briefly mention some variants of this setting toward the end of the section. The study of the homogenization of Eq. (6.122) for such a periodic Hamiltonian began with the pioneering and seminal work [LPV96]. It is interesting to note that this work is one of the most famous works, that remained a preprint and was never published in a regular journal.

We now summarize the headlines of this particular work, in order to give a flavour of homogenization theory for fully nonlinear equations. A complete justification of all our arguments below (in particular when the ambient dimension is higher than 1 or the Hamiltonian is not quadratic) would require the notion of *viscosity solutions*. We only proceed formally here and refer to the bibliography for a comprehensive analysis.

As in Chap. 3, we start from the formal two-scale expansion of the solution u^ε in function of ε. We postulate that

$$u^\varepsilon (t, x) = u^* (t, x) + \varepsilon u^1 \left(t, \frac{x}{\varepsilon}\right) + \varepsilon^2 u^2 \left(t, \frac{x}{\varepsilon}\right) + \dots,$$

where the successive functions u^k are assumed periodic in the x argument, as the Hamiltonian H. Inserting this expansion into (6.122), we obtain at the leading order

$$\partial_t u^* (t, x) + H_{per} \left(y, \nabla_x u^*(t, x) + \nabla_y u^1(t, y)\right) = 0,$$

where y denotes the fast variable $\dfrac{x}{\varepsilon}$, decoupled from x. This equation may be put in the form (6.124), if, for each $p \in \mathbb{R}^d$ (symbolizing $\nabla_x u^*(t, x)$) we are able to solve

$$H_{per}^* (p) = H_{per} \left(y, p + \nabla_y w_p (y)\right) \quad \text{in} \quad \mathbb{R}^d, \quad w_p \text{ periodic.} \tag{6.125}$$

This equation is, not unexpectedly, called the *corrector equation* associated to the Eq. (6.122). There are, in this equation, *two* unknown quantities for each fixed $p \in \mathbb{R}^d$, namely the constant $H_{per}^* (p)$ and the periodic function w_p. Proving that there exists a solution to the equation, that the constant $H_{per}^* (p)$ is *unique*, which is necessary to obtain a unique homogenized equation (6.124), and study uniqueness of w_p (which often does not hold, even up to the addition of a constant) are all tasks that are far from evident. We cannot complete this program here so we are going to restrict our attention to a particular setting.

The One-Dimensional Periodic Quadratic Setting
Equation (6.122) reads as

$$\partial_t u^\varepsilon + H_{per} \left(\frac{x}{\varepsilon}, \partial_x u^\varepsilon\right) = 0, \tag{6.126}$$

in the one-dimensional setting. In spite of the apparent simplicity of this setting, this equation is still delicate to study with the only tools at our disposal. We are going to simplify it further. We consider the *quadratic Hamiltonian with periodic potential*

$$H_{per} (x, p) = |p|^2 - V_{per} (x), \tag{6.127}$$

so that we contemplate

$$\partial_t u^\varepsilon + \left(\partial_x u^\varepsilon\right)^2 - V_{per} \left(\frac{x}{\varepsilon}\right) = 0, \tag{6.128}$$

posed on the real line $\mathbb{R}$ and supplied with the initial condition (6.123). In (6.127), we assume the periodic potential V_{per} of period 1, *regular* and such that

$$\inf_{\mathbb{R}} V_{per} = 0. \tag{6.129}$$

Assumption (6.129) is a simple normalization and does not restrict generality. It is a simple adaptation of our arguments below to eliminate this normalization. Given the above specific setting, the general corrector equation (6.125) now reads as

$$H_{per}^* (p) = \left(p + (w_{p,per})' (y)\right)^2 - V_{per}(y) \quad \text{in} \quad \mathbb{R}, \quad w_{p,per} \text{ periodic,} \tag{6.130}$$

for all $p \in \mathbb{R}$, It is actually possible to explicitly compute a solution to (6.130). Denote by $x_0 \in \mathbb{R}$ a point where V_{per} attains its minimum, that is $V_{per}(x_0) = 0$. Pick next $p \in \mathbb{R}$ so that

$$|p| \leq \langle \sqrt{V_{per}} \rangle.$$

Since the function

$$y \longmapsto \int_{x_0}^{y} (\sqrt{V_{per}} - p) - \int_{y}^{1+x_0} (\sqrt{V_{per}} + p),$$

is continuous, takes the values $-\int_{x_0}^{1+x_0} (\sqrt{V_{per}} + p) \leq 0$ at x_0 $\int_{x_0}^{1+x_0} (\sqrt{V_{per}} - p) \geq 0$ at $1 + x_0$, we know there exists a point $\overline{x}$ where it vanishes, that is

$$\int_{x_0}^{\overline{x}} (\sqrt{V_{per}} - p) = \int_{\overline{x}}^{1+x_0} (\sqrt{V_{per}} + p).$$

We then define

$$w_{p,per}(y) = \begin{cases} \int_{x_0}^{y} (\sqrt{V_{per}} - p) & \text{si } x_0 \leq y \leq \overline{x} \\ \\ \int_{y}^{1+x_0} (\sqrt{V_{per}} + p) & \text{si } \overline{x} \leq y \leq 1 + x_0 \end{cases} \tag{6.131}$$

which we extend by periodicity. The function $w_{p,per}$ is such that

$$p + (w_{p,per})'(y) = \pm \sqrt{V_{per}(y)}, \tag{6.132}$$

the explicit sign depending on the various zones. The function $w_{p,per}$ constructed is indeed a viscosity solution of Eq. (6.130) (see our related comment on page 411 of the current section). Inserting (6.132) into (6.130), we obtain $H^*_{per}(p) = 0$ when $|p| \leq \langle \sqrt{V_{per}} \rangle$. A similar construction is performed for $|p| \geq \langle \sqrt{V_{per}} \rangle$. We obtain (see Fig. 6.2)

$$H^*_{per}(p) = \begin{cases} 0 \text{ if } |p| \leq \langle \sqrt{V_{per}} \rangle \text{ and otherwise,} \\ \\ \lambda \text{ such that } \lambda \geq 0 \quad \text{solves} \quad |p| = \langle \sqrt{\lambda + V_{per}} \rangle, \end{cases} \tag{6.133}$$

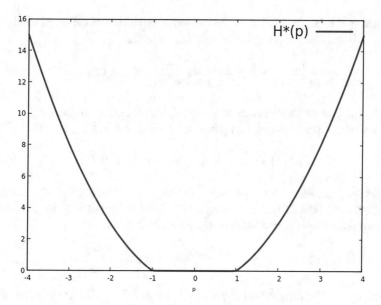

Fig. 6.2 Example of H^*_{per} when H_{per} is defined by (6.127) avec $V_{per} \geq 0$ and $\langle \sqrt{V_{per}} \rangle = 1$. We do have $H^*_{per}(p) = 0$ if $|p| \leq 1$, and, at infinity, $H^*_{per}(p)$ behaves as p^2

Put differently, when $|p| \geq \langle \sqrt{V_{per}} \rangle$, $H^*_{per}(p)$ is the non-negative solution to $|p| = \langle \sqrt{H^*_{per}(p) + V_{per}} \rangle$.

We have thus constructed, for all $p \in \mathbb{R}$, a couple $(H^*_{per}(p), w_{p,per})$ solution to (6.130). Uniqueness of the function $w_{p,per}$ is an issue. Often it does not hold, even in the simple one-dimensional setting. More important for us is uniqueness of the homogenized Hamiltonian H^*_{per}, since the latter entirely determines the homogenized equation. It is possible to show it is indeed unique in a large variety of circumstances. In our simple case, we are indeed going to show in the next paragraph that the Hamiltonian H^*_{per} given by (6.133) defines a homogenized equation (6.124) and thus a solution u^* *such that* the sequence u^ε actually converges to u^*. This will also prove that H^*_{per} is unique.

Proof of the Convergence
Starting from the corrector $w_{p,per}$ solution to (6.130), we construct the following expansion

$$\widetilde{u^\varepsilon}(t, x) = u^*(t, x) + \varepsilon\, w_{\partial_x u^*(x), per}\left(\frac{x}{\varepsilon}\right) \tag{6.134}$$

mimicking the first two terms of the two-scale expansion of the exact solution u^ε to (6.128)–(6.123). We are going to show the estimate

$$\sup_{[0,\infty[\,\times\,\mathbb{R}} |u^\varepsilon - u^*| \le 2\,\varepsilon \sup_{x\in\mathbb{R},\,y\in\mathbb{R}} |w_{\partial_x u^*(x),per}(y)|, \tag{6.135}$$

which thus proves convergence of u^ε to the homogenized solution u^*. For this convergence, we only consider an affine initial condition (6.123), that is

$$u^0(x) = \alpha + \beta x, \quad \text{for fixed } \alpha,\ \beta \in \mathbb{R}. \tag{6.136}$$

It is indeed possible to show, but we will not do so here, that such a particular initial condition is sufficient to recover generality. The advantage of such an affine initial condition (6.136) is that the function $\widetilde{u}^\varepsilon$, which here reads as

$$\widetilde{u}^\varepsilon(t,\,x) = \alpha + \beta x - t\,H^*_{per}(\beta) + \varepsilon\,w_{\beta,per}\left(\frac{x}{\varepsilon}\right), \tag{6.137}$$

(where $w_{\beta,per}$ of course solves (6.130) for $p = \beta$) is indeed *also* a solution to

$$\partial_t \widetilde{u}^\varepsilon + \left(\partial_x \widetilde{u}^\varepsilon\right)^2 - V_{per}\left(\frac{x}{\varepsilon}\right) = 0. \tag{6.138}$$

This is the same equation as (6.128) solved by u^ε, but this time for the initial condition

$$u^0(x) = \alpha + \beta x + \varepsilon\,w_{\beta,per}\left(\frac{x}{\varepsilon}\right). \tag{6.139}$$

This will allow us to easily estimate the difference

$$v^\varepsilon = u^\varepsilon - \widetilde{u}^\varepsilon,$$

which solves

$$\partial_t v^\varepsilon + \left(\partial_x u^\varepsilon\right)^2 - \left(\partial_x \widetilde{u}^\varepsilon\right)^2 = 0,$$

that is, explicitly exploiting the one-dimensional setting and the fact that the Hamiltonian is quadratic,

$$\partial_t v^\varepsilon + \left(\partial_x u^\varepsilon + \partial_x \widetilde{u}^\varepsilon\right)\partial_x v^\varepsilon = 0, \tag{6.140}$$

starting from the initial condition

$$v^\varepsilon(t = 0,\,x) = -\varepsilon\,w_{\beta,per}\left(\frac{x}{\varepsilon}\right). \tag{6.141}$$

For fixed ε, v^ε therefore solves the linear transport equation

$$\partial_t v^\varepsilon + b(t, x)\, \partial_x v^\varepsilon = 0, \qquad (6.142)$$

for $b = \partial_x u^\varepsilon + \partial_x \widetilde{u^\varepsilon}$, considered here as a datum, and the initial condition (6.141).

Such a transport equation satisfies, at least formally, the maximum principle. For instance, if the field b were sufficiently regular, this would be an immediate consequence of the representation of the solution using the method of lines (seen in another context in Sect. 6.3.2). A possible rigorous argument would require the notion of viscosity solution. In any event, given (6.141), we obtain that

$$\left| v^\varepsilon\,(t\,,\,x) \right| \leq \varepsilon \sup_{\mathbb{R}} \left| w_{\beta,per} \right|. \qquad (6.143)$$

We conclude that

$$\sup_{[0,\infty[\,\times\,\mathbb{R}} \left| u^\varepsilon - \widetilde{u^\varepsilon} \right| \leq \varepsilon \sup_{\mathbb{R}} \left| w_{\beta,per} \right|, \qquad (6.144)$$

which, in view of the expression (6.134) of $\widetilde{u^\varepsilon}$, implies estimate (6.135).

The homogenized equation (6.124) obtained by the formal two-scale expansion is thus correct mathematically at the dominant order, at least in this simple setting. We observe that, even though this equation looks like the original equation (6.122), the Hamiltonian therein is substantially different. The Hamiltonian (6.127) is indeed strictly convex in the p variable (a parabola is the prototype of a strictly convex function) while the homogenized Hamiltonian (6.133) is not strictly convex, except in the trivial case when $V_{per} \equiv 0$, since it is flat on a large region of values of p surrounding $p = 0$, as shown on Fig. 6.2.

The homogenization process therefore intimately modifies a nonlinear equation such as (6.122).

What About the Other Cases?

We keep our one-dimensional quadratic example (6.126) but now consider it with a nonperiodic potential V. The above explicit calculations are still valid in the case of a *stationary* potential V. The only minor modification (and we leave to the reader this easy exercise) is that the homogenized Hamiltonian has the same expression as in (6.133) where expectation values $\mathbb{E} \int_0^1 \sqrt{\lambda + V(.,\omega)}$ are substituted for the averages $\left\langle \sqrt{\lambda + V_{per}} \right\rangle$.

When the potential V is, on the other hand, a periodic potential perturbed by a localized defect, say for instance

$$V(x) = V_{per} + \widetilde{V}, \quad \text{for } \widetilde{V} \in C_0^\infty(\mathbb{R}),$$

solving the corrector problem (6.130) which now reads as

$$H^*(p) = \left(p + (w_p)'(y)\right)^2 - V_{per}(y) - \widetilde{V}(y), \quad \text{in} \quad \mathbb{R}, \tag{6.145}$$

becomes delicate. Things might indeed go wrong, as we will now see...

To start with, we easily notice that, since the perturbation potential $\widetilde{V}$ vanishes at infinity, we may translate Eq. (6.145) to infinity and discover that, *if there exists a solution*, then its derivative is bounded, and its translate at infinity $w_{p,\infty} = \lim_{|x| \to +\infty} w_p(. + x)$ should satisfy

$$H^*(p) = \left(p + (w_{p,\infty})'(y)\right)^2 - V_{per}(y).$$

It follows that $H^*(p) = H^*_{per}(p)$ and that $w_{p,\infty}$ is a periodic corrector. We then look for w_p solution to (6.145) under the form $w_p = w_{p,per} + \widetilde{w}_p$ with the latter function solution to

$$H^*_{per}(p) = \left(p + (w_{p,per})'(y) + (\widetilde{w}_p)'(y)\right)^2 - V_{per}(y) - \widetilde{V}(y), \quad \text{in} \quad \mathbb{R},$$

which also reads as

$$0 = \left((\widetilde{w}_p)'(y)\right)^2 + 2\left(p + (w_{p,per})'(y)\right)(\widetilde{w}_p)'(y) - \widetilde{V}(y), \tag{6.146}$$

This equation is, for fixed y, a second order polynomial equation with respect to the unknown $(\widetilde{w}_p)'(y)$. It is solvable under suitable conditions, and is not solvable otherwise.

To realize the difficulty, it is enough to consider the case $(p = 0, V_{per} \equiv 0)$, for which $(H^*_{per}(p) = 0, w_{p,per} \equiv 0)$. In that case, Eq. (6.146) trivially reads as

$$0 = \left((\widetilde{w}_p)'(y)\right)^2 - \widetilde{V}(y).$$

For a compactly supported potential $\widetilde{V}$ such that

$$\widetilde{V} \geq 0 \quad \text{sur} \quad \mathbb{R}, \tag{6.147}$$

it is solvable and the derivative of its solution $\widetilde{w}_p$ does vanish at infinity. We may then proceed backwards in the calculation performed in the previous paragraph and indeed verify that there is homogenization, with the same Hamiltonian as in the purely periodic setting. The argument carries over to any $p \in \mathbb{R}$ and all periodic potential V_{per}, provided at least condition (6.147) holds true.

On the other hand, for a potential that takes negative value somewhere, there is no solution to (6.146). There even exist cases when homogenization does not hold. For instance, the defect may induce, in the limit $\varepsilon \to 0$, an additional Dirichlet condition. The reader may interestingly recall our study of Sect. 2.5.3!

Beyond the above cases and for more general settings, homogenization theory is much more involved and many (many) questions remain unsolved.

Using a different approach based on ideas from G−convergence and the subadditive ergodic theorem, Gianni Dal Maso and Luciano Modica studied in [DMM86] the homogenization of variational problems and, hence, of *quasilinear* divergence form elliptic equations.

The next step in the development of the theory concerns nonlinear first- and second-order problems, that is, Hamilton-Jacobi equations like

$$H(Du^\varepsilon, u^\varepsilon, x, x/\varepsilon) = 0 \text{ in } \mathbb{R}^d, \tag{6.148}$$

and Hamilton-Jacobi-Isaacs-Bellman equations like

$$F(D^2 u^\varepsilon, Du^\varepsilon, u^\varepsilon, x, x/\varepsilon) = 0 \text{ in } \mathbb{R}^d. \tag{6.149}$$

In the periodic setting, (6.148) was studied in an unpublished paper [LPV96] of Pierre-Louis Lions, George Papanicolaou and Srinivasa Varadhan while (6.149) was considered by Lawrence C. Evans in [Eva92]. In both cases, it was possible to find bounded correctors and solve the associated cell problems.

The qualitative theory of stochastic homogenization for nonlinear first- and second-order PDEs began with the works of Fraydoun Rezakhanlou and James Tarver [RT00] for convex/concave HJ-equations, and, later, Pierre-Louis Lions and Panagiotis Souganidis [LS05] and Elena Kosygina, Fraydoun Rezakhanlou and Srinivasa Varadhan [KRV06], who considered viscous HJ-equations convex/concave Hamiltonians. Luis Caffarelli, Panagiotis Souganidis and Lihe Wang [CSW05] addressed the case of nonlinear second-order equations.

For either problems, in general, there are no bounded correctors. The asymptotic expansion suggests that, for (6.148), it is necessary to have strictly sublinear and, for (6.149), strictly subquadratic at infinity correctors, which again, in general, do not exist. This is in sharp contrast with the linear divergence form setting (1) where the linearity allows for the existence of strictly sublinear at infinity correctors.

This fundamental difficulty is circumvented by identifying a quantity that "controls" the solution and has an almost sure limit, by the (sub-additive) ergodic theorem. For (6.148), such a quantity may be found using the convexity of the Hamiltonian. For the second-order problem, the quantity is the measure of the contact set of an appropriate obstacle problem.

The first quantitative result —logarithmic error estimate for the convergence of u^ε to u^*— for the homogenization of nonlinear second-order equations in strongly mixing environments was obtained in [CS10] by Luis Caffarelli and Panagiotis Souganidis. Scott Armstrong, Pierre Cardaliaguet and Panagiotis Souganidis obtained in [ACS14] rates for stochastic homogenization of HJ-equations in essentially i.i.d. (independent and identically distributed) environments and, later, Scott

Armstrong and Pierre Cardaliaguet adapted in [AC15] the methods of [ACS14] and introduced some new arguments to prove rates for viscous HJ-equations, again in i.i.d. environments. Scott Armstrong and Charles Smart later found in [AS14] a better quantity to control solutions of the fully nonlinear second-order problem. Using the method of [CS10], they improved the rate of convergence from logarithmic to algebraic. There have also been many 'almost' sharp quantitative results for linear divergence and non-divergence form equations established on the one hand by Felix Otto and his collaborators and on the other hand by Scott Armstrong and Charles Smart and their collaborators; listing all of them is beyond the scope of this introductory section.

Two important issues for the random homogenization of HJ-equations are the existence of correctors and whether HJ-equations homogenize without convexity. The first example of non-existence of corrector was given by Pierre-Louis Lions and Panagiotis Souganidis in [LS03]. Pierre Cardaliaguet and Panagiotis Souganidis then showed in [CS17] that there is a corrector for convex viscous HJ-equations in the directions of the extreme points of the sub-level sets of the ergodic constant. The first example of non-homogenization for very particular nonconvex Hamiltonians was given by Bruno Ziliotto [Zil17]. Motivated by this, William Feldman and Panagiotis Souganidis identified in [FS17] a general class of Hamiltonians, namely H's with non-degenerate saddle points, for which there is a medium and a direction along which the equation does not homogenize. Then, William Feldman, Jean-Baptiste Fermanian and Bruno Ziliotto obtained in [FFZ21] a result similar to [FS17] for viscous HJ-equations.

6.4.2 In Lieu of Conclusion

We have opened this textbook mentioning that homogenization theory is now half-a-century old. This is a respectable age. We now close it by offering an assessment of its present status and a perspective on what we imagine its foreseeable future can be.

The engineering sciences, physics, the life sciences are an endless source of multiscale problems. It is highly unlikely that the source will run dry soon. For a large variety of problems that, some years ago, used to be addressed at one single scale, or for which at most a crude information was passed from a finer scale using an ad hoc parameter, it is now essential to make use at the larger scale of an accurate and explicit information at the finer scale. It is even likely that the spectrum of problems for which several scales matter at once will constantly increase. Materials science, climate science, fundamental physics, epidemiology, fluid mechanics, traffic, are some examples of possible application fields.

If homogenization theory has a future, this future will come from modelling.

Homogenization theory as presented in this textbook, and the computational equipment that accompanies it and which we have briefly exposed in Chap. 5, are now part of the corpus of mathematical sciences and the toolbox of computational sciences, respectively. Homogenization is not a passing fad. It will be around forever. But modelling complex multiscale problems will not only require homogenization. This theory will not have the monopole of all theoretical and practical answers. Other approaches will be created. This is science.

Why do we think so?

Homogenization is, as we have seen it repeatedly throughout this textbook, a remarkably powerful approach. It has allowed for considerable advances, transforming problems that were *a priori* too complex to be addressed only with brute force computational approaches into problems amenable to mathematical arguments and efficient numerical methods. Complex physical phenomena are now better understood. Practical problems of importance have been solved.

The computer power has been constantly improving, but the complexity of the problems that one wishes to address has grown concurrently. The level of accuracy expected and the level of certainty guaranteed has increased likewise. A certain equilibration follows. It is the never ending game of cat and mouse between the complexity of problems and the performance of the tools used to solve them. Even if some problems that used to be treatable only using the homogenization approach are now accessible using a brute force approach, it is a dangerous gamble to only capitalize on computer power for solving all practically relevant problems for the future. There will always be the need of sophisticated mathematical theories to solve the most challenging problems out of reach of the brute force approach.

Homogenization theory has demonstrated that some beliefs based on mere intuition or naive practical hunches are simply incorrect. The simplest and perhaps most illustrative example is that of the harmonic mean replacing the wrong, yet intuitively "natural" arithmetic mean. Homogenization theory has also, on the other hand, rigorously justified some other intuitive claims. One example is the proof, under good circumstances, of the *a priori* only formal two-scale expansion. It has also provided the basis for a rigorous error analysis, for the certification of numerical results, for the creation of ambitious and efficient computational techniques.

However powerful it is, homogenization theory has also its own limitations. One limitation originates from the necessary assumptions one need to make. Indeed, the periodicity assumption, which we have described and exploited at length in some parts of this textbook, and which we have also, in other parts of this textbook, literally fought, is a very restrictive and idealistic assumption. Another limitation is also practical in nature. Stochastic homogenization, although very general and theoretically appealing, is often prohibitively expensive computationally and delicate to put in action because in too many practical situations the law it requires as a starting point remain essentially unknown. Our own feeling is that the future of homogenization presumably does not reside in the quest for a never ending sophistication of the equations considered and the methods employed to address them, but rather in their *simplification*, not to say their deliberate and undertaken degradation.

As suggested by some of the developments we have mentioned in the previous chapters, considering perturbations of simple settings, such as random or "weakly" random perturbations, and looking for "approximate" answers is a nice and attractive option and an appealing compromise that may constitute, particularly in the mid-term, a solid course for the future.

Away from our practical developments above, homogenization theory is also, on a purely abstract and theoretical level, a remarkable mean to test mathematical models and more intimately understand the hidden structure behind equations. A first example in this vein was provided by the selection of equations we presented in Sect. 2.5 where, in the homogenized limit, each equation may change in form and feature new terms. A second example, which on the other hand we have not at all explored, is to supply a time-dependent equation with an oscillatory initial condition so as to investigate the evolution of that condition, following an idea originally due to Luc Tartar. A third, much more recent example is the extension of mean-field games theory with the works of Pierre-Louis Lions and Panagiotis Souganidis showing that considering the homogenized limit enlarges the class of systems under study, creating systems no less interesting mathematically than the systems originally examined.

There remains a lot to do, both in a practical perspective and in a theoretical perspective. It is thus amply justified that we conclude this textbook passing on the baton to our readers. May they now write their own part of the story.

Appendix A
Some Basic Elements of PDE Analysis

We collect in this Chapter some classical results in functional analysis and in the analysis of partial differential equations (PDE in short). For the proofs of these results, the reader is referred to the textbooks we cite below.

A.1 A Few Results in Functional Analysis

We start with the Morrey inequality.

Theorem A.1 (Morrey) *Assume* $q > d$. *Then* $W^{1,q}(\mathbb{R}^d) \subset L^\infty(\mathbb{R}^d)$, *with continuous injection. Moreover, there exists a constant C depending only on d and q, such that, for all* $u \in W^{1,q}(\mathbb{R}^d)$,

$$|u(x) - u(y)| \leq C|x - y|^{1-\frac{d}{p}} \|\nabla u\|_{L^q(\mathbb{R}^d)},$$

almost everywhere in x and y.

Theorem A.1 implies in particular that any function in $W^{1,q}(\mathbb{R}^d)$ with $q > d$ is continuous. Its proof may be read, for instance, in [Bre11, Theorem 9.12], [Eva10, Theorem 4, p 282], or [GT01, Theorem 7.17, p 163]. It is well-known that if $q \leq d$, functions in $W^{1,q}$ are not continuous in general. We nevertheless have the following result (see [Bre11, Theorem 9.9] or [Eva10, Theorem 1, p 279]):

Theorem A.2 (Gagliardo-Nirenberg-Sobolev) *Assume that q satisfies* $1 \leq q < d$. *Then*

$$W^{1,q}(\mathbb{R}^d) \subset L^{q^*}(\mathbb{R}^d),$$

© The Author(s), under exclusive license to Springer Nature Switzerland AG 2023
X. Blanc, C. Le Bris, *Homogenization Theory for Multiscale Problems*, MS&A 21,
https://doi.org/10.1007/978-3-031-21833-0

with continuous injection, where q^* is defined by

$$\frac{1}{q^*} = \frac{1}{q} - \frac{1}{d}. \tag{A.1}$$

Moreover, there exists a constant C depending only on d and q, such that

$$\forall u \in W^{1,q}(\mathbb{R}^d), \quad \|u\|_{L^{q^*}(\mathbb{R}^d)} \leq C\|\nabla u\|_{L^q(\mathbb{R}^d)}. \tag{A.2}$$

A consequence of Theorem A.2 is the following:

Corollary A.1 *Assume that $u \in L^1_{loc}(\mathbb{R}^d)$ satisfies $\nabla u \in L^q(\mathbb{R}^d)$, with $1 \leq q < d$. Then there exists a constant $M \in \mathbb{R}$ and a constant C such that*

$$\|u - M\|_{L^{q^*}(\mathbb{R}^d)} \leq C\|\nabla u\|_{L^q(\mathbb{R}^d)}, \tag{A.3}$$

where q^* is defined by (A.1) and the constant C depends only on q and d.

Proof For all $R > 0$, we define

$$M_R = \fint_{B_{2R} \setminus B_R} u = \frac{1}{|B_{2R} \setminus B_R|} \int_{B_{2R} \setminus B_R} u,$$

where B_R denotes the ball of $\mathbb{R}^d$ of radius R centered at the origin, and $|B_{2R} \setminus B_R|$ the Lebesgue measure of the annulus $B_{2R} \setminus B_R$. Let χ be a smooth cut-off function such that:

$$0 \leq \chi \leq 1, \quad |\nabla \chi| \leq 2, \quad \chi = 1 \text{ in } B_1, \quad \chi = 0 \text{ in } B_2^c.$$

Consider

$$u_R(x) = (u(x) - M_R) \chi \left(\frac{x}{R}\right).$$

The support of this function u_R is included in B_{2R}. Hence, according to the assumptions we have made on u, $u_R \in W^{1,q}(\mathbb{R}^d)$. We may thus apply inequality (A.2) to u_R:

$$\|u_R\|^q_{L^{q^*}(\mathbb{R}^d)} \leq C^q \|\nabla u_R\|^q_{L^q(\mathbb{R}^d)}$$

$$= C^q \int_{B_{2R}} |\nabla u|^q(x)\chi^q\left(\frac{x}{R}\right) dx + \frac{C^q}{R^q} \int_{B_{2R} \setminus B_R} |u(x) - M_R|^q |\nabla \chi|^q\left(\frac{x}{R}\right) dx$$

$$\leq C^q \int_{\mathbb{R}^d} |\nabla u|^q + \frac{2^q C^q}{R^q} \int_{B_{2R} \setminus B_R} |u - M_R|^q, \tag{A.4}$$

where we have used the bounds $0 \leq \chi \leq 1$ and $|\nabla \chi| \leq 2$. Next, we apply the Poincaré-Wirtinger inequality (see for instance [Eva10, Theorem 1, page 292]), which reads as

$$\int_{B_{2R} \setminus B_R} |u - M_R|^q \leq C(R) \int_{B_{2R} \setminus B_R} |\nabla u|^q, \qquad (A.5)$$

where the constant $C(R)$ does not depend on u. A scaling argument allows to prove that $C(R) = C_1 R^q$, where C_1 is the constant associated with the case $R = 1$, hence depends only on d and q. This argument has already been used to establish estimate (3.140), and to show (4.14). We then insert (A.5) into (A.4), and obtain

$$\|u_R\|_{L^{q^*}(\mathbb{R}^d)}^q \leq C^q \int_{\mathbb{R}^d} |\nabla u|^q + 2^q C^q C_1 \int_{B_{2R} \setminus B_R} |\nabla u|^q \leq C' \int_{\mathbb{R}^d} |\nabla u|^q, \qquad (A.6)$$

with a constant C' that depends only on q and d. Hence, u_R is bounded in $L^{q^*}(\mathbb{R}^d)$. Up to an extraction that we do not make explicit, we thus have weak convergence to some $v \in L^{q^*}(\mathbb{R}^d)$:

$$u_R \xrightarrow[R \to +\infty]{} v \text{ in } L^{q^*}(\mathbb{R}^d). \qquad (A.7)$$

On the other hand, $\nabla u_R = \nabla u \chi_R + R^{-1}(u - M_R)\nabla\chi(\cdot/R)$ converges to ∇u in the sense of distributions. Since all differential operators are continuous in this topology, (A.7) implies that $\nabla v = \nabla u$. Thus, $v = u + M$, for some constant M. The weak convergence (A.7) allows to pass to the liminf in estimate (A.6). This implies (A.3). $\qquad\square$

Note that the assumptions of Corollary A.1 rule out the case $d = 1$. Actually, the result is not valid in this case. Indeed, consider the function

$$u(x) = \begin{cases} 0 & \text{if } x < 0, \\ x & \text{if } 0 < x < 1, \\ 1 & \text{if } 1 < x. \end{cases}$$

Clearly, u satisfies $\nabla u \in L^q(\mathbb{R})$, for all $q \geq 1$, but whatever the value of M, the function $u - M$ cannot belong to any L^p, $p < +\infty$, since its limits in $+\infty$ and $-\infty$ do not agree.

We now give a characterization of Hölder regularity due to Campanato [Cam63]. To this end, we first recall the definition of the $C^{0,\alpha}$ semi-norm on $\mathcal{D}$, for $\alpha \in]0, 1[$:

$$[u]_{C^{0,\alpha}(\mathcal{D})} = \sup_{x \neq y \in \mathcal{D}} \frac{|u(x) - u(y)|}{|x - y|^\alpha}.$$

The following result is a consequence of [Gia83, p 70] (see also [GM12, Theorem 5.5]):

Theorem A.3 (Campanato) *Assume that $\mathcal{D}$ is an open set with Lipschitz boundary. The semi-norm $[\cdot]_{C^{0,\alpha}(\mathcal{D})}$ is equivalent to the so-called* Campanato *semi-norm:*

$$[u]_{\mathcal{L}^{q,\lambda}(\mathcal{D})} = \sup_{x_0 \in \mathcal{D}} \sup_{r>0} \left(\frac{1}{r^\lambda} \int_{\mathcal{D} \cap B(x_0,r)} \left| u - \fint_{\mathcal{D} \cap B(x_0,r)} u \right|^q \right)^{\frac{1}{q}}$$

where $\lambda = q\,\alpha + d$.

Recall that $\fint_{\mathcal{D} \cap B(x_0,r)} u$ denotes the average of u in the set $\mathcal{D} \cap B(x_0, r)$. The assumption that the boundary of $\mathcal{D}$ is Lipschitz continuous implies that

$$|\mathcal{D} \cap B(x_0, r)| \geq Ar^d,$$

for some constant A independent of r and of x_0. Actually, the proofs in the textbooks we cited above are given under this weaker assumption. Our point here is that, in view of the link between λ and α, one easily proves that the Campanato semi-norm is equivalent to the following semi-norm:

$$|u|_{\mathcal{L}^{q,\lambda}(\mathcal{D})} = \sup_{x_0 \in \mathcal{D}} \sup_{r>0} \frac{1}{r^\alpha} \left(\fint_{\mathcal{D} \cap B(x_0,r)} \left| u - \fint_{\mathcal{D} \cap B(x_0,r)} u \right|^q \right)^{\frac{1}{q}}.$$

It is a simple exercise to prove that the latter expression is equal, in the case $q = 2$, to the semi-norm used in (3.143).

We next give a trace theorem. It corresponds to [LM72, Théorème 9.4, page 41], in the particular case $j = 0$, or [Eva10, Theorem 1, p 274].

Theorem A.4 (Trace Theorem) *Assume that $\mathcal{D}$ is a bounded Lipschitz open set. Then there exists a bounded linear operator*

$$\gamma : H^1(\mathcal{D}) \longrightarrow H^{1/2}(\partial\mathcal{D})$$

such that, for all $u \in H^1(\mathcal{D}) \cap C^0(\overline{\mathcal{D}})$, $\gamma(u) = u_{|\partial\mathcal{D}}$. Moreover, the map γ is onto, and there exists a continuous linear operator (called a lift *operator)*

$$R : H^{1/2}(\partial\mathcal{D}) \longrightarrow H^1(\mathcal{D})$$

such that

$$\forall g \in H^{1/2}(\partial\mathcal{D}), \quad \gamma(R(g)) = g.$$

Note that in [LM72, Théorème 9.4, page 41] the boundary $\mathcal{D}$ is in fact assumed to be of class C^∞. This allows to prove a stronger result than Theorem A.4. The reader may read the proof in the case of a domain $\mathcal{D}$ of class C^1 in [Eva10, Theorem 1, p 274]. For the case of a Lipschitz domain, we refer to [Din96].

The following result is the so-called *Aubin-Lions Lemma* (in a slightly simplified version). For its exact statement and proof, we refer to [Lio69, Théorème 5.1, page 58] (in French) or [Sim87, Corollary 4].

Lemma A.1 (Aubin-Lions Lemma) *Consider three reflexive Banach spaces $V_0 \subset V \subset V_1$, where all inclusions are continuous. Assume that the inclusion $V_0 \subset V$ is compact. Let $q \in]1, +\infty[$, and define, for a fixed $T > 0$,*

$$W = \left\{ u \in L^q(]0, T[, V_0), \quad \partial_t u \in L^q(]0, T[, V_1) \right\},$$

equipped with the natural norm

$$\|u\|_W = \|u\|_{L^q(]0,T[,V_0)} + \|\partial_t u\|_{L^q(]0,T[,V_1)},$$

which makes W a Banach space. Then $W \subset L^q(]0, T[, V)$, with compact injection.

A.2   Elliptic Partial Differential Equations: Existence and Uniqueness of Solutions

We recall in this section the basics of elliptic PDE theory in the Hilbert setting. Our aim here is of course not to give a course on elliptic PDEs, but to collect all the results we need in the body of the text. The reader interested in this subject is referred to reference textbooks such as [Eva10] or [GT01].

We consider the particular case of a divergence-form elliptic PDE:

$$-\operatorname{div}(a\nabla u) = f, \tag{A.8}$$

posed in a bounded open set $\mathcal{D} \subset \mathbb{R}^d$, with the homogeneous Dirichlet boundary condition

$$u = 0 \text{ on } \partial\mathcal{D}. \tag{A.9}$$

We assume that the matrix-valued coefficient $a : \mathcal{D} \longrightarrow \mathbb{R}^{d \times d}$ is bounded. We also assume that a is coercive, that is, there exists $\mu > 0$ such that

$$\forall \xi \in \mathbb{R}^d, \quad (a(x)\xi) \cdot \xi \geq \mu |\xi|^2, \tag{A.10}$$

almost everywhere in $x \in \mathcal{D}$. The above assumptions do not allow to write Eq. (A.8) in the classical sense, since a is not differentiable a priori. We thus need a weak formulation of this equation. In order to define it, we temporarily assume that the equation is defined in a classical way, multiply it by a test function $\varphi \in C^\infty(\mathcal{D})$, with compact support, and integrate by parts. We obtain

$$\int_{\mathcal{D}} (a\nabla u) \cdot \nabla \varphi = \int_{\mathcal{D}} f\varphi.$$

For now, this formula is only valid in the regular case. But it actually makes sense provided that u and φ belong to $H^1(\mathcal{D})$. If the domain $\mathcal{D}$ is Lipschitz continuous, Theorem A.4 defines a trace operator γ. The boundary condition then reads as $\gamma(u) = 0$. This is equivalent to the fact that $u \in H_0^1(\mathcal{D})$, which is the closure in $H^1(\mathcal{D})$ of the space of smooth compactly supported functions in $\mathcal{D}$ (see [Eva10, Theorem 2, page 275]). This leads us to the

Definition A.1 (Weak Solutions) For all $f \in H^{-1}(\mathcal{D})$, a weak solution of (A.8)–(A.9) is a function $u \in H_0^1(\mathcal{D})$ such that

$$\forall v \in H_0^1(\mathcal{D}), \quad \int_{\mathcal{D}} (a\nabla u) \cdot \nabla v = \langle f, v \rangle_{H^{-1}(\mathcal{D}), H_0^1(\mathcal{D})}. \tag{A.11}$$

This formulation is also called the variational formulation of (A.8) and (A.9), as we will explain below. Of course, when $f \in L^2(\mathcal{D})$, the right-hand side of (A.11) reads as $\int_{\mathcal{D}} fv$.

We now state the Lax-Milgram Lemma. It is a fundamental tool to study problems of the form (A.11).

Lemma A.2 (Lax-Milgram) *Let H be a Hilbert space, and let $B : H \times H \longrightarrow \mathbb{R}$ be a continuous bilinear form. Assume that B is coercive, that is, it satisfies*

$$\exists \mu > 0, \quad \forall u \in H, \quad B(u, u) \geq \mu \|u\|_H^2. \tag{A.12}$$

Consider $L : H \longrightarrow \mathbb{R}$ a linear continuous map. Then there exists a unique $u \in H$ satisfying

$$\forall v \in H, \quad B(u, v) = L(v). \tag{A.13}$$

Moreover, this solution satisfies

$$\|u\|_H \leq \frac{1}{\mu}\|L\|_{H'}. \tag{A.14}$$

A simple application of this Lemma allows to prove that the weak formulation (A.11) has a unique solution $u \in H_0^1(\mathcal{D})$, for all $f \in H^{-1}(\mathcal{D})$ (recall that $H^{-1}(\mathcal{D})$ is the dual space of $H_0^1(\mathcal{D})$). In order to do so, we define $H = H_0^1(\mathcal{D})$, equipped with the scalar product $(u, v) \longmapsto \int_{\mathcal{D}} \nabla u . \nabla v$. Owing to the Poincaré inequality, H is a Hilbert space. We define

$$B(u, v) = \int_{\mathcal{D}} (a \nabla u) . \nabla v,$$

which is a bilinear form. It is continuous because a is bounded, and coercive (in the sense of (A.12)) because a is coercive (in the sense of (A.10)). We also define

$$L(v) = \langle f, v \rangle_{H^{-1}(\mathcal{D}), H_0^1(\mathcal{D})},$$

which, by definition, is linear and continuous over the Hilbert space H. This allows to apply Lemma A.2, and obtain

Corollary A.2 *Assume that a is bounded and coercive. Then, for all $f \in H^{-1}(\mathcal{D})$, there exists a unique solution $u \in H_0^1(\mathcal{D})$ to the weak formulation (A.11). Moreover, this solution satisfies*

$$\|u\|_{H_0^1(\mathcal{D})} \le \frac{1}{\mu} \|f\|_{H^{-1}(\mathcal{D})}. \tag{A.15}$$

Of course, the map $f \mapsto u$ defined in Corollary A.2 is linear. Inequality (A.15) implies that it is continuous from $H^{-1}(\mathcal{D})$ to $H_0^1(\mathcal{D})$.

We now make the link with the original equation, namely (A.8)–(A.9). First, the variational formulation (A.13) implies that (A.8) holds in the sense of distributions. If we assume higher regularity, as, say, if $u \in H^2(\mathcal{D})$, $a \in W^{1,\infty}(\mathcal{D})$, and $f \in L^2(\mathcal{D})$, then Eq. (A.8) holds almost everywhere in $\mathcal{D}$. We may indeed use the weak formulation (A.11) and perform the integration by parts in reverse order. This gives

$$\int_{\mathcal{D}} (\text{div}(a \nabla u) + f) \, v = 0, \tag{A.16}$$

for all $v \in C^1(\mathcal{D}) \cap C^0(\overline{\mathcal{D}})$ that vanishes on ∂D. Hence, (A.8) holds almost everywhere in $\mathcal{D}$. In such a case, the boundary condition holds only in the sense $u \in H_0^1(\mathcal{D})$. In order to give a *classical* sense to (A.8) and (A.9), we need to assume higher regularity. For instance, if the domain $\mathcal{D}$ is Lipschitz, if a is differentiable, and if u is twice differentiable, we are able to proceed. The latter regularity is in fact implied by the former, provided f is regular enough (say, Hölder continuous), as we will see below. Under these regularity assumptions, we may repeat the above integration by parts in reverse order, and obtain (A.16) again, for all $v \in C^1(\mathcal{D}) \cap C^0(\overline{\mathcal{D}})$ vanishing at the boundary. All the functions involved here are continuous, hence we infer that (A.8) holds *everywhere*. Finally, we recover (A.9) by using that u is in the kernel of the trace operator γ, and that for continuous functions, $\gamma(u)$ is the restriction of u to the boundary $\partial \mathcal{D}$.

In the case where a is a symmetric matrix, one easily shows that solving the weak formulation is equivalent to minimizing the functional $J : H_0^1(\mathcal{D}) \longrightarrow \mathbb{R}$ defined by:

$$J(u) = \frac{1}{2} \int_{\mathcal{D}} (a\nabla u) \cdot \nabla u - \langle f, u \rangle_{H^{-1}(\mathcal{D}), H_0^1(\mathcal{D})}.$$

We therefore consider the following minimization problem

$$\inf \left\{ J(u), \quad u \in H_0^1(\mathcal{D}) \right\}. \tag{A.17}$$

The energy functional J is strictly convex, and satisfies

$$\lim_{\|u\|_{H_0^1(\mathcal{D})} \to +\infty} J(u) = +\infty.$$

This allows to prove that any minimizing sequence of problem (A.17) is bounded, hence converges, up to an extraction, weakly in $H_0^1(\mathcal{D})$, to some limit u. Convexity allows to pass to the liminf, hence u is a solution to (A.17). This solution is unique since J is strictly convex. The differential of J at u reads as

$$\forall v \in H_0^1(\mathcal{D}), \quad dJ_u(v) = \int_{\mathcal{D}} (a\nabla u) \cdot \nabla v - \langle f, v \rangle_{H^{-1}(\mathcal{D}), H_0^1(\mathcal{D})}. \tag{A.18}$$

The Euler-Lagrange equation associated with the minimization problem (A.17) writes $dJ_u = 0$, that is, (A.11). Conversely, if u is solution to (A.11), then it is a critical point of J, hence a minimizer since J is strictly convex.

This argument makes clear why (A.11) is also called the variational formulation of (A.8) and (A.9). Note that such an interpretation as a minimization problem requires that a is symmetric, contrary to the approach using the Lax-Milgram Lemma. When a is not symmetric, only the symmetric part of a appears in (A.18). On the other hand, such a minimization approach also allows to conveniently address some nonlinear equations (that is, energy functionals J that are not quadratic, but still strictly convex).

We conclude this Section with *the Caccioppoli inequality*. This estimate allows to control the L^2 norm of ∇u with the L^2 norm of u for a solution to (A.8) with $f = 0$. The result and its proof may be found in [GM12, Proposition 2.1, page 76], and [Gia83, Theorem 4.4, page 63].

Theorem A.5 (Caccioppoli Inequality)   *Assume that a is bounded and coercive. Let $u \in H^1(B_{2R}(x))$ be a solution to $-\operatorname{div}(a\nabla u) = 0$ in the ball $B_{2R}(x)$, for $x \in \mathbb{R}^d$. Then there exists a constant C that depends only on the coercivity constant of a and on $\|a\|_{L^\infty}$ such that*

$$\int_{B_R(x)} |\nabla u|^2 \leq \frac{C}{R^2} \int_{B_{2R}(x)} u^2. \tag{A.19}$$

A.3 Maximum Principle

For an equation of the form (A.8), the maximum principle reads as (see for instance [Eva10, §6.4, Theorem 1, page 346]):

Theorem A.6 (Weak Maximum Principle: Strong Solutions) *Let $\mathcal{D}$ be a bounded open set, and assume that the matrix-valued coefficient $a \in C^1(\mathcal{D})$ is bounded and coercive. Let $u \in C^2(\mathcal{D}) \cap C^0(\overline{\mathcal{D}})$.*

(i) If $- \operatorname{div}(a\nabla u) \leq 0$ in $\mathcal{D}$, then $\max_{\overline{\mathcal{D}}} u = \max_{\partial \mathcal{D}} u$.

(ii) If $- \operatorname{div}(a\nabla u) \geq 0$ in $\mathcal{D}$, then $\min_{\overline{\mathcal{D}}} u = \min_{\partial \mathcal{D}} u$.

In order to state more general results, we rephrase $- \operatorname{div}(a\nabla u) \leq 0$ in the sense of distributions, that is:

$$\forall \varphi \in C_c^\infty(\mathcal{D}), \quad \varphi \geq 0, \quad \int_{\mathcal{D}} (a\nabla u) . \nabla \varphi \leq 0.$$

Similarly, we understand $- \operatorname{div}(a\nabla u) \geq 0$ as

$$\forall \varphi \in C_c^\infty(\mathcal{D}), \quad \varphi \geq 0, \quad \int_{\mathcal{D}} (a\nabla u) . \nabla \varphi \geq 0.$$

We then have

Theorem A.7 (Weak Maximum Principle: Weak Solutions) *Let $\mathcal{D}$ be a bounded open set, and assume that a is a bounded coercive matrix over $\mathcal{D}$. Let $u \in H^1(\mathcal{D})$.*

(i) If $- \operatorname{div}(a\nabla u) \leq 0$ in $\mathcal{D}$, then $\operatorname*{ess\,sup}_{\overline{\mathcal{D}}} u \leq \max \left(\operatorname*{ess\,sup}_{\partial \mathcal{D}} u, 0 \right).$

(ii) If $- \operatorname{div}(a\nabla u) \geq 0$ in $\mathcal{D}$, then $\operatorname*{ess\,inf}_{\overline{\mathcal{D}}} u \geq \min \left(\operatorname*{ess\,inf}_{\partial \mathcal{D}} u, 0 \right).$

For the proof of Theorem A.7, we refer to the original article [Sta65] or to [GT01, Theorem 8.1, p 179].

A.4 Harnack Inequality

The *Harnack inequality* is a fundamental tool in the analysis of elliptic PDEs. It allows in particular to prove that any solution to (A.8) with $f = 0$ is Hölder continuous without any regularity assumption on the coefficients (the matrix a is only assumed to be bounded and coercive. Some of the intermediate results in the proof of the Harnack inequality are interesting *per se*, so we cite them below.

In particular, Theorem 8.16 (page 191) of [GT01] implies the following result:

Theorem A.8 *Let $\mathcal{D}$ be a bounded open Lipbshitz subset of $\mathbb{R}^d$, and let $a \in (L^\infty(\mathcal{D}))^{d \times d}$ be a coercive matrix. Assume that $f \in L^{q/2}(\mathcal{D})$ and $F \in (L^q(\mathcal{D}))^d$, for some $q > d$. Let $u \in H^1(\mathcal{D})$ be a solution in the sense of distributions to*

$$-\operatorname{div}(a\nabla u) = f + \operatorname{div}(F), \quad in \quad \mathcal{D}. \tag{A.20}$$

Then

$$\|u\|_{L^\infty(\mathcal{D})} \le \|u\|_{L^\infty(\partial\mathcal{D})} + \frac{C}{\mu}\left(\|f\|_{L^{q/2}(\mathcal{D})} + \|F\|_{(L^q(\mathcal{D}))^d}\right), \tag{A.21}$$

where the constant C depends only on the volume $|\mathcal{D}|$ of the open set $\mathcal{D}$, on the dimension d and on q.

Another preliminary result useful in the proof of Harnack inequality is the following result, which is a consequence de [GT01, Theorem 8.17, p 194]:

Theorem A.9 *Let $\mathcal{D}$ be a bounded open subset of $\mathbb{R}^d$, and let $a \in (L^\infty(\mathcal{D}))^{d \times d}$ be a coercive matrix. Assume that $f \in L^{q/2}(\mathcal{D})$ and $F \in (L^q(\mathcal{D}))^d$, for some $q > d$. If $u \in H^1(\mathcal{D})$ satisfies $-\operatorname{div}(a\nabla u) \le f + \operatorname{div}(F)$ in $\mathcal{D}$, then, for all ball $B_{2R}(y) \subset \mathcal{D}$,*

$$\operatorname*{ess\,sup}_{B_R(y)} u \le C\left(\frac{1}{R^{d/2}}\|u^+\|_{L^2(B_{2R}(y))} + R^{1-d/q}\|F\|_{(L^q(\mathcal{D}))^d}\right.$$
$$\left. + R^{2-2d/q}\|f\|_{L^{q/2}(\mathcal{D})}\right), \tag{A.22}$$

where C depends only on the dimension d, on $\|a\|_{L^\infty}$, on the coercivity constant of a and on q.

Similarly, if u satisfies $-\operatorname{div}(a\nabla u) \ge f + \operatorname{div}(F)$ in $\mathcal{D}$, then, for all ball $B_{2R}(y) \subset \mathcal{D}$,

$$\operatorname*{ess\,sup}_{B_R(y)}(-u) \le C\left(\frac{1}{R^{d/2}}\|u^-\|_{L^2(B_{2R}(y))} + R^{1-d/q}\|F\|_{(L^q(\mathcal{D}))^d}\right.$$
$$\left. + R^{2-2d/q}\|f\|_{L^{q/2}(\mathcal{D})}\right), \tag{A.23}$$

where C depends only on the dimension d, on $\|a\|_{L^\infty}$, on the coercivity constant of a and on q.

In this result, the inequality $-\operatorname{div}(a\nabla u) \leq f + \operatorname{div}(F)$ is to be understood in the sense of distributions, that is,

$$\forall \varphi \in C_c^\infty(\mathcal{D}), \quad \varphi \geq 0, \quad \int_{\mathcal{D}} (a\nabla u) \cdot \nabla\varphi \leq \int_{\mathcal{D}} f\varphi - \int_{\mathcal{D}} F \cdot \nabla\varphi.$$

Similarly, the inequality $-\operatorname{div}(a\nabla u) \geq f + \operatorname{div}(F)$ means

$$\forall \varphi \in C_c^\infty(\mathcal{D}), \quad \varphi \geq 0, \quad \int_{\mathcal{D}} (a\nabla u) \cdot \nabla\varphi \geq \int_{\mathcal{D}} f\varphi - \int_{\mathcal{D}} F \cdot \nabla\varphi.$$

When u is solution to $-\operatorname{div}(a\nabla u) = 0$, we may apply both inequalities (A.22) and (A.23), and obtain

Corollary A.3 *Let $\mathcal{D}$ be a bounded open subset of $\mathbb{R}^d$, and let $a \in L^\infty(\mathcal{D})^{d\times d}$ be an coercive matrix. If $u \in H^1(\mathcal{D})$ satisfies $-\operatorname{div}(a\nabla u) = 0$ in the sense of distributions in $\mathcal{D}$, then, for all ball $B_{2R}(y) \subset \mathcal{D}$,*

$$\sup_{B_R(y)} |u| \leq \frac{C}{R^{d/2}} \|u\|_{L^2(B_{2R}(y))}, \tag{A.24}$$

where C depends only on the dimension d, on $\|a\|_{L^\infty}$, and on the coercivity constant of a.

The above results allow to prove the Harnack inequality. It reads as follows (see for instance [GT01, Theorem 8.20, p. 199] or [Eva10, Theorem 5, p. 353]):

Theorem A.10 (Harnack Inequality) *Let $\mathcal{D}$ be a bounded open subset of $\mathbb{R}^d$, and let $a \in L^\infty(\mathcal{D})^{d\times d}$ be a coercive matrix. Assume that $u \in H^1(\mathcal{D})$ satisfies $u \geq 0$ and $-\operatorname{div}(a\nabla u) = 0$ in the sense of distributions in $\mathcal{D}$. Then there exists a constant C depending only on d, on the coercivity constant of a and on $\|a\|_{L^\infty(\mathcal{D})}$, but not on u, such that, if $y \in \mathbb{R}^d$ and $R > 0$ are such that $B_{4R}(y) \subset \mathcal{D}$, then*

$$\sup_{B_R(y)} u \leq C \inf_{B_R(y)} u. \tag{A.25}$$

Scrutinizing [GT01, Theorem 8.20, p. 199], the reader will see that in fact, in a more general case, the constant in (A.25) also depends on R. The reason is, the equation considered in [GT01] is of the more general form

$$-\operatorname{div}(a\nabla u + bu) + c \cdot \nabla u + gu = 0,$$

where $b, c : \mathcal{D} \longrightarrow \mathbb{R}^d$ and $g : \mathcal{D} \longrightarrow \mathbb{R}$ satisfy $|b|^2 + |c|^2 + \mu|g| \leq \nu^2\mu^2$, for some $\nu \geq 0$, where μ is the coercivity constant of the matrix a. The constant C in (A.25) depends on R only through νR. Since in our case, $\nu = 0$, the constant in fact does not depend on R.

In the original article by Jürgen Moser [Mos61], the following Liouville type theorem is proved, as a consequence of the Harnack inequality:

Corollary A.4 *Let a be a bounded coercive matrix defined on $\mathbb{R}^d$. Assume that u is a solution to $-\operatorname{div}(a\nabla u) = 0$ in the sense of distributions in $\mathbb{R}^d$. If u is not a constant, then there exists $\gamma > 0$ such that*

$$\forall r > 1, \quad \sup_{|x|=r} u(x) - \inf_{|x|=r} u(x) \geq r^{\gamma}.$$

In the case of the Laplace operator, the following stronger result is classical:

Proposition A.1 *Let $u \in L^1_{loc}(\mathbb{R}^d)$ be such that $-\Delta u = 0$ in the sense of distributions in $\mathbb{R}^d$. Then u is a harmonic polynomial. In particular, if it is strictly sublinear at infinity, u is constant.*

One may prove this result using the Fourier transform: equation $-\Delta u = 0$ reads as $|\xi|^2\widehat{u}(\xi) = 0$, hence $\widehat{u}$ is a distribution the support of which is reduced to $\{0\}$. Such a distribution is a sum of derivatives of the Dirac mass at the origin, hence u is a polynomial.

Note that this proof of Proposition A.1 is also valid for any divergence form elliptic operator with constant coefficients.

A.5   Elliptic Regularity

We give in this Section some elliptic regularity results that are useful in the main body of the text.

We start with *Nash-Moser regularity* theory, also called *De Giorgi-Nash* or *De Giorgi-Nash-Moser* theory. This theory is a consequence of the Harnack inequality (see Theorem A.10). The main result concerns a homogeneous equation and may be stated as follows

Theorem A.11 (Nash-Moser) *Let a be a coercive and bounded matrix on the ball B_{2R}. Assume that $u \in H^1(B_{2R})$ is a solution to*

$$-\operatorname{div}(a\nabla u) = 0$$

in the sense of distributions in B_{2R}. Then there exists $\alpha \in]0, 1[$ depending only on the coercivity constant of a and on $\|a\|_{L^\infty(B_{2R})}$ such that $u \in C^{0,\alpha}(B_R)$.

This result has been originally proved by Ennio De Giorgi [DG57], Jürgen Moser [Mos60, Mos61] and John Nash [Nas58]. It is also stated and proved in [Mor08, Theorem 5.3.3, p 139] and [BE97, Theorem 3.4, p 236], or, in a more general form, in [GT01, Theorem 8.24, p 202].

We next give elliptic regularity results for an inhomogeneous equation with a right-hand side in the form of a divergence, first, and next a general right-hand side. The first result is borrowed from [GM12, Theorem 5.19, p 87] (see also [GT01, Theorem 8.32, p 210]):

Theorem A.12 (Schauder Interior Elliptic Regularity) *Let $\mathcal{D}$ be a bounded open subset of $\mathbb{R}^d$, let $a : \mathcal{D} \longrightarrow \mathbb{R}^{d \times d}$ be a bounded coercive matrix, and $f : \mathcal{D} \to \mathbb{R}^d$ a vector field. Assume that $f \in C^{0,\sigma}(\mathcal{D})$ and $a \in C^{0,\sigma}(\mathcal{D})$ for some $\sigma \in]0, 1[$. Then, for any compact set $K \subset\subset \mathcal{D}$, there exists a constant C such that, if $u \in H^1(\mathcal{D})$ is solution to*

$$- \operatorname{div}(a\nabla u) = \operatorname{div}(f)$$

in the sense of distributions in $\mathcal{D}$, then,

$$\|\nabla u\|_{C^{0,\sigma}(K)} \leq C \left(\|\nabla u\|_{L^2(\mathcal{D})} + \|f\|_{C^{0,\sigma}(\mathcal{D})} \right),$$

and C depends only on K, $\mathcal{D}$, on the coercivity constant of a, and on $\|a\|_{C^{0,\sigma}(\mathcal{D})}$.

In Theorem A.12, we have assumed not only that the matrix a is coercive and bounded, but also that it is Hölder continuous, contrary to Theorem A.11. On the other hand, Theorem A.12 is valid for systems, and not only for scalar equations, as it is stated in [GM12, Theorem 5.19, p 87]. It is not the case of the De Giorgi-Nash-Moser regularity theory, which holds only if the unknown u is scalar-valued.

We next state elliptic regularity results in Sobolev norms. The following results are proved for instance in [GM12, Theorem 7.2, p 140].

Theorem A.13 (Sobolev Interior Elliptic Regularity) *Let $\mathcal{D}$ be an open bounded subset of $\mathbb{R}^d$, let $a : \mathcal{D} \longrightarrow \mathbb{R}^{d \times d}$ be a coercive and bounded matrix, and $f : \mathcal{D} \to \mathbb{R}^d$ a vector field. Assume that a is uniformly continuous in $\overline{\mathcal{D}}$, and that $f \in L^q(\mathcal{D})$ for some $q \geq 2$. Then, for all compact set $K \subset\subset \mathcal{D}$, there exists a constant C such that, if $u \in H^1(\mathcal{D})$ is solution to*

$$- \operatorname{div}(a\nabla u) = \operatorname{div}(f)$$

in the sense of distributions in $\mathcal{D}$, then

$$\|\nabla u\|_{L^q(K)} \leq C \left(\|\nabla u\|_{L^2(\mathcal{D})} + \|f\|_{L^q(\mathcal{D})} \right),$$

and C depends only on K, $\mathcal{D}$, q on the coercivity constant of a, on $\|a\|_{L^\infty(\mathcal{D})}$, and on the continuity modulus of a.

This theorem is, like Theorems A.11 and A.12, an *interior* regularity result. We now state an elliptic regularity result "up to the boundary", with a right-hand side that is no more in divergence form, contrary to Theorems A.11 and A.12. It is a consequence of [GT01, Lemma 9.17] and of [GT01, Theorem 9.19]:

Theorem A.14 (Elliptic Regularity up to the Boundary) *Let $\mathcal{D} \subset \mathbb{R}^d$ be a bounded open subset of $\mathbb{R}^d$, with C^2 boundary. Let $a : \mathcal{D} \longrightarrow \mathbb{R}^{d \times d}$ be a coercive bounded matrix, such that $a \in W^{1,\infty}(\mathcal{D})$. Assume that $f \in L^q(\mathcal{D})$, for some $q \in]1, +\infty[$, and consider $u \in H_0^1(\mathcal{D})$ a solution to*

$$- \operatorname{div}(a \nabla u) = f,$$

in the sense of distributions in $\mathcal{D}$. Then, $u \in W^{2,q}(\mathcal{D})$, and there exists a constant C depending only on $\mathcal{D}$, q and a such that

$$\|u\|_{W^{2,q}(\mathcal{D})} \le C \|f\|_{L^q(\mathcal{D})}.$$

All the above results are stated for a divergence-form equation. However, in Chap. 6, we also need regularity results for the following equation:

$$- \partial_{ij}^2 \left(a_{ij} u\right) + \partial_i \left(b_i u\right) = 0, \tag{A.26}$$

where we use the summation convention on repeated indices. This equation is sometimes called the stationary Fokker-Planck-Kolmogorov (FPK) equation. As we have pointed out in Chap. 6, the fact that the matrix $\partial_{ij}^2 u$ is symmetric implies that we may always assume that a is a symmetric matrix. In addition, as for equations of the form (A.8), the solution to (A.26) is to be understood in a weak sense, that is,

$$\forall \varphi \in C_c^\infty(\mathcal{D}), \quad \int_{\mathcal{D}} \left(-a_{ij} \partial_{ij}^2 \varphi - b_i \partial_i \varphi\right) u = 0. \tag{A.27}$$

For this formulation to make sense, if a and b are continuous, it is sufficient to assume that u is a measure.

The following result is borrowed from [BS17, Theorem 3.1].

Theorem A.15 *Assume that the matrix a is coercive and bounded, that $a \in C^{0,\alpha}(\mathcal{D})$ for some $\alpha \in]0, 1[$, and that $b \in (L^q(\mathcal{D}))^d$ for some $q > d$. Then any solution $u \in L^1(\mathcal{D})$ to (A.27) satisfies $u \in C^{0,\alpha}(\mathcal{D})$.*

Actually, this result holds true even if we only assume that u is a measure. In such a case, we have that, as a measure, u is absolutely continuous with respect to the Lebesgue measure, and its density is in $C^{0,\alpha}(\mathcal{D})$.

We conclude with the Harnack inequality for equations of the form (A.26).

Theorem A.16 *Let a be a bounded coercive matrix, and $b \in (L^\infty(\mathcal{D}))^d$. Let $u \in C^0(\mathcal{D})$ be a nonnegative solution to (A.27) in $\mathcal{D}$. Assume that $x_0 \in \mathcal{D}$ and $R > 0$ are such that $B(x_0, 2R) \subset \mathcal{D}$. Then there exists a constant C depending only on R,*

$\|b\|_{L^\infty(\mathcal{D})}$, $\|a\|_{L^\infty(\mathcal{D})}$, *and on the coercivity constant of a, such that*

$$\sup_{B_R(x_0)} u \leq C \inf_{B_R(x_0)} u.$$

This result is stated and proved in [BS17, Theorem 3.1], and in [BKRS15, Theorem 3.4.2, p 100].

References

[AAP19] Assyr Abdulle, Doghonay Arjmand, and Edoardo Paganoni. Exponential decay of the resonance error in numerical homogenization via parabolic and elliptic cell problems. *C. R. Math. Acad. Sci. Paris*, 357(6):545–551, 2019.

[AAP21a] Assyr Abdulle, Doghonay Arjmand, and Edoardo Paganoni. An elliptic local problem with exponential decay of the resonance error for numerical homogenization. arXiv:2001.06315, 2021.

[AAP21b] Assyr Abdulle, Doghonay Arjmand, and Edoardo Paganoni. A parabolic local problem with exponential decay of the resonance error for numerical homogenization. *Math. Models Methods Appl. Sci.*, 31(13):2733–2772, 2021.

[AB05] Grégoire Allaire and Robert Brizzi. A multiscale finite element method for numerical homogenization. *Multiscale Model. Simul. Journal*, 4(3):790–812, 2005.

[AC15] Scott N. Armstrong and Pierre Cardaliaguet. Quantitative stochastic homogenization of viscous Hamilton-Jacobi equations. *Comm. Partial Differential Equations*, 40(3):540–600, 2015.

[ACS14] Scott N. Armstrong, Pierre Cardaliaguet, and Panagiotis E. Souganidis. Error estimates and convergence rates for the stochastic homogenization of Hamilton-Jacobi equations. *J. Amer. Math. Soc.*, 27(2):479–540, 2014.

[AD16] Scott Armstrong and Jean-Paul Daniel. Calderón-Zygmund estimates for stochastic homogenization. *J. Funct. Anal.*, 270(1):312–329, 2016.

[AEEVE12] Assyr Abdulle, Weinan E, Björn Engquist, and Eric Vanden-Eijnden. The heterogeneous multiscale method. *Acta Numerica*, 21:1–87, 2012.

[AF09] Luigi Ambrosio and Hermano Frid. Multiscale Young measures in almost periodic homogenization and applications. *Arch. Ration. Mech. Anal.*, 192(1):37–85, 2009.

[AGK16] Scott Armstrong, Antoine Gloria, and Tuomo Kuusi. Bounded correctors in almost periodic homogenization. *Arch. Ration. Mech. Anal.*, 222(1):393–426, 2016.

[AHP21] Robert Altmann, Patrick Henning, and Daniel Peterseim. Numerical homogenization beyond scale separation. *Acta Numer.*, 30:1–86, 2021.

[AKM19] Scott Armstrong, Tuomo Kuusi, and Jean-Christophe Mourrat. *Quantitative stochastic homogenization and large-scale regularity.*, volume 352. Cham: Springer, 2019.

[AL87] Marco Avellaneda and Fang-Hua Lin. Compactness methods in the theory of homogenization. *Commun. Pure Appl. Math.*, 40(6):803–847, 1987.

[AL89] Marco Avellaneda and Fang-Hua Lin. Un théorème de Liouville pour des équations elliptiques à coefficients périodiques. *C. R. Acad. Sci. Paris Sér. I Math.*, 309(5):245–250, 1989.

[AL91] Marco Avellaneda and Fang Hua Lin. L^p bounds on singular integrals in homogenization. *Commun. Pure Appl. Math.*, 44(8–9):897–910, 1991.

[ALB11] Arnaud Anantharaman and Claude Le Bris. A numerical approach related to defect-type theories for some weakly random problems in homogenization. *Multiscale Model. Simul.*, 9(2):513–544, 2011.

[All92] Grégoire Allaire. Homogenization and two-scale convergence. *SIAM J. Math. Anal.*, 23(6):1482–1518, 1992.

[All02] Grégoire Allaire. *Shape optimization by the homogenization method.*, volume 146. New York, NY: Springer, 2002.

[AR16] Doghonay Arjmand and Olof Runborg. A time dependent approach for removing the cell boundary error in elliptic homogenization problems. *J. Comput. Phys.*, 314:206–227, 2016.

[AS14] Scott N. Armstrong and Charles K. Smart. Quantitative stochastic homogenization of elliptic equations in nondivergence form. *Arch. Ration. Mech. Anal.*, 214(3):867–911, 2014.

[AT15] Yves Achdou and Nicoletta Tchou. Hamilton-Jacobi equations on networks as limits of singularly perturbed problems in optimal control: dimension reduction. *Comm. Partial Differential Equations*, 40(4):652–693, 2015.

[AT19] Yves Achdou and Nicoletta Tchou. Homogenization of a transmission problem with Hamilton-Jacobi equations and a two-scale interface. Effective transmission conditions. *J. Math. Pures Appl. (9)*, 122:164–197, 2019.

[Bar94] Guy Barles. *Solutions de viscosité des équations de Hamilton-Jacobi.*, volume 17. Paris: Springer-Verlag, 1994.

[BBMM05] Andriy Bondarenko, Guy Bouchitté, Luísa Mascarenhas, and Rajesh Mahadevan. Rate of convergence for correctors in almost periodic homogenization. *Discrete Contin. Dyn. Syst.*, 13(2):503–514, 2005.

[BCE97] Martino Bardi, Michael G. Crandall, Lawrence C. Evans, Halil Mete Soner, and Panagiotis E. Souganidis. *Viscosity solutions and applications*, volume 1660 of *Lecture Notes in Mathematics.* Springer-Verlag, Berlin; Centro Internazionale Matematico Estivo (C.I.M.E.), Florence, 1997.

[BD98] Andrea Braides and Anneliese Defranceschi. *Homogenization of multiple integrals*, volume 12 of *Oxford Lecture Series in Mathematics and its Applications.* The Clarendon Press, Oxford University Press, New York, 1998.

[BE97] Piero Bassanini and Alan R. Elcrat. *Theory and applications of partial differential equations*, volume 46 of *Mathematical Concepts and Methods in Science and Engineering.* Plenum Press, New York, 1997.

[Bec81] Maria E. Becker. Multiparameter groups of measure-preserving transformations: a simple proof of Wiener's ergodic theorem. *Ann. Probab.*, 9:504–509, 1981.

[Ber01] Michel Bernadou. *Le calcul scientifique*, volume 1357 of *Collection Que sais-je ?* Presses universitaires de France, 2001.

[Bes32] Abram S. Besicovitch. Almost periodic functions. Cambridge: Univ. Press. XIII, 180 p, 1932.

[BGL14] Dominique Bakry, Ivan Gentil, and Michel Ledoux. *Analysis and geometry of Markov diffusion operators*, volume 348 of *Grundlehren der Mathematischen Wissenschaften [Fundamental Principles of Mathematical Sciences].* Springer, Cham, 2014.

[BGMP08] Guillaume Bal, Josselin Garnier, Sébastien Motsch, and Vincent Perrier. Random integrals and correctors in homogenization. *Asymptotic Anal.*, 59(1–2):1–26, 2008.

[Bil95] Patrick Billingsley. *Probability and measure.* Wiley Series in Probability and Mathematical Statistics. John Wiley & Sons, Inc., New York, third edition, 1995.

[BJL20] Xavier Blanc, Marc Josien, and Claude Le Bris. Precised approximations in elliptic homogenization beyond the periodic setting. *Asymptotic Analysis*, 116(2):93–137, 2020.

[BKRS15] Vladimir I. Bogachev, Nicolai V. Krylov, Michael Röckner, and Stanislav V. Shaposhnikov. *Fokker-Planck-Kolmogorov equations*, volume 207 of *Mathematical Surveys and Monographs*. American Mathematical Society, Providence, RI, 2015.

[BL76] Jöran Bergh and Jörgen Löfström. *Interpolation spaces. An introduction.*, volume 223. Springer, Berlin, 1976.

[BL11] Ivo Babuska and Robert Lipton. Optimal local approximation spaces for generalized finite element methods with application to multiscale problems. *Multiscale Model. Simul.*, 9(1):373–406, 2011.

[BL22] Xavier Blanc and Claude Le Bris. *Homogénéisation en milieu périodique . . . ou non : une introduction.* Cham: Springer, 2023. ISBN:978-3-031-12800-4.

[BLA13] Xavier Blanc, Frédéric Legoll, and Arnaud Anantharaman. Asymptotic behavior of Green functions of divergence form operators with periodic coefficients. *AMRX, Appl. Math. Res. Express*, 2013(1):79–101, 2013.

[BLB10] Xavier Blanc and Claude Le Bris. Improving on computation of homogenized coefficients in the periodic and quasi-periodic settings. *Netw. Heterog. Media*, 5(1):1–29, 2010.

[BLBL16] Xavier Blanc, Claude Le Bris, and Frédéric Legoll. Some variance reduction methods for numerical stochastic homogenization. *Philos. Trans. Roy. Soc. A*, 374(2066):20150168, 15, 2016.

[BLBL20] Xavier Blanc, Claude Le Bris, and Pierre-Louis Lions. Erratum to the article "On correctors for linear elliptic homogenization in the presence of local defects: the case of advection-diffusion". https://www.ljll.math.upmc.fr/~blanc/erratum_jmpa.pdf, 2020.

[BLL03] Xavier Blanc, Claude Le Bris, and Pierre-Louis Lions. A definition of the ground state energy for systems composed of infinitely many particles. *Commun. Partial Differ. Equations*, 28(1–2):439–475, 2003.

[BLL07a] Xavier Blanc, Claude Le Bris, and Pierre-Louis Lions. Stochastic homogenization and random lattices. *J. Math. Pures Appl. (9)*, 88(1):34–63, 2007.

[BLL07b] Xavier Blanc, Claude Le Bris, and Pierre-Louis Lions. The energy of some microscopic stochastic lattices. *Arch. Ration. Mech. Anal.*, 184(2):303–339, 2007.

[BLL12] Xavier Blanc, Claude Le Bris, and Pierre-Louis Lions. A possible homogenization approach for the numerical simulation of periodic microstructures with defects. *Milan J. Math.*, 80(2):351–367, 2012.

[BLL15] Xavier Blanc, Claude Le Bris, and Pierre-Louis Lions. Local profiles for elliptic problems at different scales: defects in, and interfaces between periodic structures. *Commun. Partial Differ. Equations*, 40(12):2173–2236, 2015.

[BLL18] Xavier Blanc, Claude Le Bris, and Pierre-Louis Lions. On correctors for linear elliptic homogenization in the presence of local defects. *Commun. Partial Differ. Equations*, 43(6):965–997, 2018.

[BLL19] Xavier Blanc, Claude Le Bris, and Pierre-Louis Lions. On correctors for linear elliptic homogenization in the presence of local defects: the case of advection-diffusion. *J. Math. Pures Appl. (9)*, 124:106–122, 2019.

[BLLL22] Rutger Biezemans, Claude Le Bris, Frédéric Legoll, and Alexei Lozinski. Nonintrusive implementation of multiscale finite element methods: an illustrative example. *J. Comput. Phys.*, 477:10, 2023.

[BLP11] Alain Bensoussan, Jacques-Louis Lions, and George Papanicolaou. *Asymptotic analysis for periodic structures. Reprint of the 1978 original with corrections and bibliographical additions.* Providence, RI: AMS Chelsea Publishing, 2011.

[BM85] Ivo M. Babuška and Richard C. Morgan. Composites with a periodic structure: mathematical analysis and numerical treatment. *Comput. Math. Appl.*, 11(10):995–1005, 1985.

[BMW94] Alain Bourgeat, Andro Mikelić, and Steve Wright. Stochastic two-scale convergence in the mean and applications. *J. Reine Angew. Math.*, 456:19–51, 1994.

[BO00] Ivo Babuška and John E. Osborn. Can a finite element method perform arbitrarily badly? *Math. Comp.*, 69(230):443–462, 2000.

[Boh18] Harald Bohr. *Almost periodic functions. Reprint of the 1947 English edition published by Chelsea Publishing Company.* Mineola, NY: Dover Publications, 2018.

[BP99] Alain Bourgeat and Andrey Piatnitski. Estimates in probability of the residual between the random and the homogenized solutions of one-dimensional second-order operator. *Asymptotic Anal.*, 21(3–4):303–315, 1999.

[BP04] Alain Bourgeat and Andrey Piatnitski. Approximations of effective coefficients in stochastic homogenization. *Ann. Inst. H. Poincaré Probab. Statist.*, 40(2):153–165, 2004.

[BR18] Leonid Berlyand and Volodymyr Rybalko. *Getting acquainted with homogenization and multiscale.* Compact Textbooks in Mathematics. Birkhäuser/Springer, Cham, 2018.

[Bra02] Andrea Braides. *Γ-convergence for beginners.*, volume 22. Oxford: Oxford University Press, 2002.

[Bre11] Haim Brezis. *Functional analysis, Sobolev spaces and partial differential equations.* New York, NY: Springer, 2011.

[Bri87] William L. Briggs. *A multigrid tutorial.* Society for Industrial and Applied Mathematics (SIAM), Philadelphia, PA, 1987.

[BS88] Colin Bennett and Robert Sharpley. *Interpolation of operators*, volume 129 of *Pure and Applied Mathematics.* Academic Press, Inc., Boston, MA, 1988.

[BS08] Susanne C. Brenner and L. Ridgway Scott. *The mathematical theory of finite element methods. 3rd ed.*, volume 15. New York, NY: Springer, 2008.

[BS17] Vladimir I. Bogachev and Stanislav V. Shaposhnikov. Integrability and continuity of solutions to double divergence form equations. *Ann. Mat. Pura Appl. (4)*, 196(5):1609–1635, 2017.

[BT04] Edward B. Burger and Robert Tubbs. *Making transcendence transparent. An intuitive approach to classical transcendental number theory.* Springer-Verlag, New York, 2004.

[Cam63] Sergio Campanato. Proprietà di hölderianità di alcune classi di funzioni. *Ann. Scuola Norm. Sup. Pisa Cl. Sci. (3)*, 17:175–188, 1963.

[CCC05] Eric Cancès, François Castella, Philippe Chartier, Erwan Faou, Claude Le Bris, Frédéric Legoll, and Gabriel Turinici. Long-time averaging for integrable Hamiltonian dynamics. *Numer. Math.*, 100(2):211–232, 2005.

[CD99] Doina Cioranescu and Patrizia Donato. *An introduction to homogenization.*, volume 17. Oxford: Oxford University Press, 1999.

[CDG18] Doina Cioranescu, Alain Damlamian, and Georges Griso. *The periodic unfolding method. Theory and applications to partial differential problems.*, volume 3. Singapore: Springer, 2018.

[CEL19] Matteo Cicuttin, Alexandre Ern, and Simon Lemaire. A hybrid high-order method for highly oscillatory elliptic problems. *Comput. Methods Appl. Math.*, 19(4):723–748, 2019.

[CEL20a] Eric Cancès, Virginie Ehrlacher, Frédéric Legoll, Benjamin Stamm, and Shuyang Xiang. An embedded corrector problem for homogenization. I: Theory. *Multiscale Model. Simul.*, 18(3):1179–1209, 2020.

[CEL20b] Eric Cancès, Virginie Ehrlacher, Frédéric Legoll, Benjamin Stamm, and Shuyang Xiang. An embedded corrector problem for homogenization. Part II: Algorithms and discretization. *J. Comput. Phys.*, 407:109254, 26, 2020.

[CELS15] Éric Cancès, Virginie Ehrlacher, Frédéric Legoll, and Benjamin Stamm. Un problème d'inclusion pour approcher les coefficients homogénéisés d'une équation elliptique [An embedded corrector problem to approximate the homogenized coefficients of an elliptic equation] . *C. R., Math., Acad. Sci. Paris*, 353(9):801–806, 2015.

[Cia78] Philippe G. Ciarlet. *The finite element method for elliptic problems.* Studies in Mathematics and its Applications, Vol. 4. North-Holland Publishing Co., Amsterdam-New York-Oxford, 1978.

[CLBL10a] Ronan Costaouec, Claude Le Bris, and Frédéric Legoll. Approximation numérique d'une classe de problèmes en homogénéisation stochastique[Numerical approximation of a class of problems in stochastic homogenization] . *C. R. Math. Acad. Sci. Paris*, 348(1–2):99–103, 2010.

[CLBL10b] Ronan Costaouec, Claude Le Bris, and Frédéric Legoll. Variance reduction in stochastic homogenization: proof of concept, using antithetic variables. *Bol. Soc. Esp. Mat. Apl. SeMA*, 50:9–26, 2010.

[CLL98] Isabelle Catto, Claude Le Bris, and Pierre-Louis Lions. *The mathematical theory of thermodynamic limits: Thomas-Fermi type models.* Oxford: Clarendon Press, 1998.

[CM82] Doina Cioranescu and François Murat. Un terme étrange venu d'ailleurs. Nonlinear partial differential equations and their applications, Coll. de France Semin., Vol. II, Res. Notes Math. 60, 98–138, 1982.

[CS10] Luis A. Caffarelli and Panagiotis E. Souganidis. Rates of convergence for the homogenization of fully nonlinear uniformly elliptic pde in random media. *Invent. Math.*, 180(2):301–360, 2010.

[CS17] Pierre Cardaliaguet and Panagiotis E. Souganidis. On the existence of correctors for the stochastic homogenization of viscous Hamilton-Jacobi equations. *C. R. Math. Acad. Sci. Paris*, 355(7):786–794, 2017.

[CSW05] Luis A. Caffarelli, Panagiotis E. Souganidis, and L. Wang. Homogenization of fully nonlinear, uniformly elliptic and parabolic partial differential equations in stationary ergodic media. *Comm. Pure Appl. Math.*, 58(3):319–361, 2005.

[DG57] Ennio De Giorgi. Sulla differenziabilità e l'analiticità delle estremali degli integrali multipli regolari. *Mem. Accad. Sci. Torino. Cl. Sci. Fis. Mat. Nat. (3)*, 3:25–43, 1957.

[DG16] Mitia Duerinckx and Antoine Gloria. Analyticity of homogenized coefficients under Bernoulli perturbations and the Clausius-Mossotti formulas. *Arch. Ration. Mech. Anal.*, 220(1):297–361, 2016.

[Din96] Zhonghai Ding. A proof of the trace theorem of Sobolev spaces on Lipschitz domains. *Proc. Amer. Math. Soc.*, 124(2):591–600, 1996.

[DJ84] Guy David and Jean-Lin Journé. A boundedness criterion for generalized Calderón-Zygmund operators. *Ann. of Math. (2)*, 120(2):371–397, 1984.

[DM95] Georg Dolzmann and Stefan Müller. Estimates for Green's matrices of elliptic systems by L^p theory. *Manuscr. Math.*, 88(2):261–273, 1995.

[DMM86] Gianni Dal Maso and Luciano Modica. Nonlinear stochastic homogenization and ergodic theory. *J. Reine Angew. Math.*, 368:28–42, 1986.

[DS88a] Nelson Dunford and Jacob T. Schwartz. *Linear operators. Part I.* Wiley Classics Library. John Wiley & Sons, Inc., New York, 1988.

[DS88b] Nelson Dunford and Jacob T. Schwartz. *Linear operators. Part II.* Wiley Classics Library. John Wiley & Sons, Inc., New York, 1988.

[Dud02] Richard M. Dudley. *Real analysis and probability*, volume 74 of *Cambridge Studies in Advanced Mathematics*. Cambridge University Press, Cambridge, 2002.

[Dur19] Rick Durrett. *Probability—theory and examples*, volume 49 of *Cambridge Series in Statistical and Probabilistic Mathematics*. Cambridge University Press, Cambridge, 2019.

[EE03] Weinan E and Björn Engquist. The heterogeneous multiscale methods. *Commun. Math. Sci.*, 1(1):87–132, 2003.

[EG04] Alexandre Ern and Jean-Luc Guermond. *Theory and practice of finite elements.*, volume 159. New York, NY: Springer, 2004.

[EH09] Yalchin Efendiev and Thomas Y. Hou. *Multiscale finite element methods. Theory and applications.* New York, NY: Springer, 2009.

[ES08] Björn Engquist and Panagiotis E. Souganidis. Asymptotic and numerical homogenization. *Acta Numerica*, 17:147–190, 2008.

[Eva92] Lawrence C. Evans. Periodic homogenisation of certain fully nonlinear partial differential equations. *Proc. Roy. Soc. Edinburgh Sect. A*, 120(3–4):245–265, 1992.

[Eva10] Lawrence C. Evans. *Partial differential equations. 2nd ed.*, volume 19. Providence, RI: American Mathematical Society (AMS), 2010.

[FAO22] Qingqing Feng, Gregoire Allaire, and Pascal Omnes. Enriched nonconforming multiscale finite element method for Stokes flows in heterogeneous media based on high-order weighting functions. *Multiscale Model. Simul.*, 20(1):462–492, 2022.

[FC00] Frédéric Feyel and Jean-Louis Chaboche. FE^2 multiscale approach for modelling the elastoviscoplastic behaviour of long fibre SiC/Ti composite materials. *Comput. Methods Appl. Mech. Eng.*, 183(3–4):309–330, 2000.

[FFZ21] William M. Feldman, Jean-Baptiste Fermanian, and Bruno Ziliotto. An example of failure of stochastic homogenization for viscous Hamilton-Jacobi equations without convexity. *J. Differential Equations*, 280:464–476, 2021.

[Fis19] Julian Fischer. The choice of representative volumes in the approximation of effective properties of random materials. *Arch. Ration. Mech. Anal.*, 234(2):635–726, 2019.

[FM87] Gilles A. Francfort and François Murat. Optimal bounds for conduction in two-dimensional, two-phase, anisotropic media. Non-classical continuum mechanics, Proc. Symp., Durham/Engl. 1985, Lond. Math. Soc. Lect. Note Ser. 122, 197–212, 1987.

[FM94] Gilles A. Francfort and Graeme W. Milton. Sets of conductivity and elasticity tensors stable under lamination. *Commun. Pure Appl. Math.*, 47(3):257–279, 1994.

[Fol95] Gerald B. Folland. *A course in abstract harmonic analysis.* Boca Raton, FL: CRC Press, 1995.

[Fra63] Joel N. Franklin. Deterministic simulation of random processes. *Math. Comp.*, 17:28–59, 1963.

[FS17] William M. Feldman and Panagiotis E. Souganidis. Homogenization and non-homogenization of certain non-convex Hamilton-Jacobi equations. *J. Math. Pures Appl. (9)*, 108(5):751–782, 2017.

[GH16] Antoine Gloria and Zakaria Habibi. Reduction in the resonance error in numerical homogenization II: Correctors and extrapolation. *Found. Comput. Math.*, 16(1):217–296, 2016.

[Gia83] Mariano Giaquinta. *Multiple integrals in the calculus of variations and nonlinear elliptic systems.*, volume 105. Princeton University Press, Princeton, NJ, 1983.

[Glo11] Antoine Gloria. Reduction of the resonance error—Part 1: Approximation of homogenized coefficients. *Math. Models Methods Appl. Sci.*, 21(8):1601–1630, 2011.

[GM12] Mariano Giaquinta and Luca Martinazzi. *An introduction to the regularity theory for elliptic systems, harmonic maps and minimal graphs. 2nd ed.*, volume 11. Pisa: Edizioni della Normale, 2012.

[GMS00] Yury Grabovsky, Graeme W. Milton, and Daniel S. Sage. Exact relations for effective tensors of composites: necessary conditions and sufficient conditions. *Commun. Pure Appl. Math.*, 53(3):300–353, 2000.

[GNO20] Antoine Gloria, Stefan Neukamm, and Felix Otto. A regularity theory for random elliptic operators. *Milan J. Math.*, 88(1):99–170, 2020.

[GO17] Antoine Gloria and Felix Otto. Quantitative results on the corrector equation in stochastic homogenization. *J. Eur. Math. Soc. (JEMS)*, 19(11):3489–3548, 2017.

[Gol91] François Golse. Particle transport in nonhomogeneous media. In *Mathematical aspects of fluid and plasma dynamics (Salice Terme, 1988)*, volume 1460 of *Lecture Notes in Math.*, pages 152–170. Springer, Berlin, 1991.

[Gou22] Rémi Goudey. A periodic homogenization problem with defects rare at infinity. *Netw. Heterog. Media*, 17(4):547–592, 2022.

[Gra93] Yury Grabovsky. The G-closure of two well-ordered, anisotropic conductors. *Proc. R. Soc. Edinb., Sect. A, Math.*, 123(3):423–432, 1993.

[Gra14] Loukas Grafakos. *Classical Fourier analysis*, volume 249 of *Graduate Texts in Mathematics*. Springer, New York, third edition, 2014.

[Gri11] Pierre Grisvard. *Elliptic problems in nonsmooth domains*, volume 69 of *Classics in Applied Mathematics*. Society for Industrial and Applied Mathematics (SIAM), Philadelphia, PA, 2011.

[GT01] David Gilbarg and Neil S. Trudinger. *Elliptic partial differential equations of second order. Reprint of the 1998 ed.* Berlin: Springer, 2001.

[GW82] Michael Grueter and Kjell-Ove Widman. The Green function for uniformly elliptic equations. *Manuscr. Math.*, 37:303–342, 1982.

[HÖ3] Lars Hörmander. *The analysis of linear partial differential operators. I.* Classics in Mathematics. Springer-Verlag, Berlin, 2003.

[HFMQ98] Thomas J. R. Hughes, Gonzalo R. Feijóo, Luca Mazzei, and Jean-Baptiste Quincy. The variational multiscale method—a paradigm for computational mechanics. *Comput. Methods Appl. Mech. Engrg.*, 166(1–2):3–24, 1998.

[HKM20] Fredrik Hellman, Tim Keil, and Axel Målqvist. Numerical upscaling of perturbed diffusion problems. *SIAM J. Sci. Comput.*, 42(4):a2014–a2036, 2020.

[HM19] Fredrik Hellman and Axel Målqvist. Numerical homogenization of elliptic PDEs with similar coefficients. *Multiscale Model. Simul.*, 17(2):650–674, 2019.

[HP13] Patrick Henning and Daniel Peterseim. Oversampling for the multiscale finite element method. *Multiscale Model. Simul.*, 11(4):1149–1175, 2013.

[HW97] Thomas Y. Hou and Xiao-Hui Wu. A multiscale finite element method for elliptic problems in composite materials and porous media. *J. Comput. Phys.*, 134(1):169–189, 1997.

[Iwa83] Tadeusz Iwaniec. Projections onto gradient fields and L^p-estimates for degenerated elliptic operators. *Stud. Math.*, 75:293–312, 1983.

[JL18] Gaspard Jankowiak and Alexei Lozinski. Non-conforming multiscale finite element method for stokes flows in heterogeneous media. part II: error estimates for periodic microstructure. arXiv:1802.04389[v1], 2018.

[Jol90] Pascal Joly. *Mise en œuvre de la méthode des éléments finis*, volume 2. Paris: Ellipses, 1990.

[KLS14] Carlos Kenig, Fanghua Lin, and Zhongwei Shen. Periodic homogenization of Green and Neumann functions. *Commun. Pure Appl. Math.*, 67(8):1219–1262, 2014.

[Koz79] Sergei M. Kozlov. Averaging differential operators with almost periodic, rapidly oscillating coefficients. *Math. USSR, Sb.*, 35:481–498, 1979.

[Koz80] Sergei M. Kozlov. Averaging of random operators. *Mathematics of the USSR-Sbornik*, 37(2):167–180, feb 1980.

[KPY17] Ralf Kornhuber, Joscha Podlesny, and Harry Yserentant. Direct and iterative methods for numerical homogenization. In *Domain decomposition methods in science and engineering XXIII*, volume 116 of *Lect. Notes Comput. Sci. Eng.*, pages 217–225. Springer, Cham, 2017.

[KPY18] Ralf Kornhuber, Daniel Peterseim, and Harry Yserentant. An analysis of a class of variational multiscale methods based on subspace decomposition. *Math. Comp.*, 87(314):2765–2774, 2018.

[Kre85] Ulrich Krengel. *Ergodic theorems.*, volume 6. Walter de Gruyter, Berlin, 1985.

[KRV06] Elena Kosygina, Fraydoun Rezakhanlou, and S. R. S. Varadhan. Stochastic homogenization of Hamilton-Jacobi-Bellman equations. *Comm. Pure Appl. Math.*, 59(10):1489–1521, 2006.

[KS91] Ioannis Karatzas and Steven E. Shreve. *Brownian motion and stochastic calculus. 2nd ed.*, volume 113. New York etc.: Springer-Verlag, 1991.

[LBLL13] Claude Le Bris, Frédéric Legoll, and Alexei Lozinski. MsFEM à la Crouzeix-Raviart for highly oscillatory elliptic problems. *Chin. Ann. Math. Ser. B*, 34(1):113–138, 2013.

[LBLL14] Claude Le Bris, Frédéric Legoll, and Alexei Lozinski. An MsFEM type approach for perforated domains. *Multiscale Model. Simul.*, 12(3):1046–1077, 2014.

[LBLM16] Claude Le Bris, Frédéric Legoll, and William Minvielle. Special quasirandom structures: a selection approach for stochastic homogenization. *Monte Carlo Methods Appl.*, 22(1):25–54, 2016.

[LBT12] Claude Le Bris and Florian Thomines. A reduced basis approach for some weakly stochastic multiscale problems. *Chin. Ann. Math. Ser. B*, 33(5):657–672, 2012.

[LC84a] Konstantin A. Lurie and Andrej V. Cherkaev. G-closure of a set of anisotropically conducting media in the two-dimensional case. *J. Optim. Theory Appl.*, 42:283–304, 1984.

[LC84b] Konstantin A. Lurie and Andrej V. Cherkaev. G-closure of some particular sets of admissible material characteristics for the problem of bending of thin elastic plates. *J. Optim. Theory Appl.*, 42:305–316, 1984.

[LC87] Konstantin A. Lurie and Andrej V. Cherkaev. On G-closure (Erratum). *J. Optim. Theory Appl.*, 53:319–339, 1987.

[Le 05] Claude Le Bris. *Systèmes multi-échelles. Modélisation et simulation.*, volume 47. Berlin: Springer, 2005.

[Le 13] Jean-François Le Gall. *Mouvement brownien, martingales et calcul stochastique.*, volume 71. Paris: Springer, 2013.

[Lio69] Jacques-Louis Lions. Quelques méthodes de résolution des problèmes aux limites non linéaires. Etudes mathématiques. Paris: Dunod; Paris: Gauthier-Villars. XX, 554 p, 1969.

[Lio78] Jacques-Louis Lions. Some aspects of modelling problems in distributed parameter systems. Distrib. Param. Syst.: Model. Identif., Proc. IFIP Conf., Rome 1976, Lect. Notes Control Inf. Sci. 1, 11–41, 1978.

[Lio84] Pierre-Louis Lions. The concentration-compactness principle in the calculus of variations. The locally compact case. I & II. *Ann. Inst. Henri Poincaré, Anal. Non Linéaire*, 1:109–145 & 223–283, 1984.

[Lio85] Pierre-Louis Lions. The concentration-compactness principle in the calculus of variations. The limit case. I & II. *Rev. Mat. Iberoam.*, 1(1–2):45–121 & 145–201, 1985.

[LK10] Xuefeng Liu and Fumio Kikuchi. Analysis and estimation of error constants for P_0 and P_1 interpolations over triangular finite elements. *J. Math. Sci. Univ. Tokyo*, 17(1):27–78, 2010.

[LM72] J.-L. Lions and E. Magenes. *Non-homogeneous boundary value problems and applications. Vol. I.* Die Grundlehren der mathematischen Wissenschaften, Band 181. Springer-Verlag, New York-Heidelberg, 1972. Translated from the French by P. Kenneth.

[LM15] Frédéric Legoll and William Minvielle. A control variate approach based on a defect-type theory for variance reduction in stochastic homogenization. *Multiscale Model. Simul.*, 13(2):519–550, 2015.

[LN03] Yanyan Li and Louis Nirenberg. Estimates for elliptic systems from composite material. *Comm. Pure Appl. Math.*, 56(7):892–925, 2003.

[LNNW09] Dag Lukkassen, Gabriel Nguetseng, Hubert Nnang, and Peter Wall. Reiterated homogenization of nonlinear monotone operators in a general deterministic setting. *J. Funct. Spaces Appl.*, 7(2):121–152, 2009.

[LPV96] Pierre-Louis Lions, George Papanicolaou, and S. R. Srinivasa Varadhan. Homogenization of Hamilton-Jacobi equations. Unpublished, 1996.

[LS03] Pierre-Louis Lions and Panagiotis E. Souganidis. Correctors for the homogenization of Hamilton-Jacobi equations in the stationary ergodic setting. *Comm. Pure Appl. Math.*, 56(10):1501–1524, 2003.

[LS05] Pierre-Louis Lions and Panagiotis E. Souganidis. Homogenization of "viscous" Hamilton-Jacobi equations in stationary ergodic media. *Comm. Partial Differential Equations*, 30(1–3):335–375, 2005.

[LV00] Yan Yan Li and Michael Vogelius. Gradient estimates for solutions to divergence form elliptic equations with discontinuous coefficients. *Arch. Ration. Mech. Anal.*, 153(2):91–151, 2000.

[Mey63] Norman G. Meyers. An L^p-estimate for the gradient of solutions of second order elliptic divergence equations. *Ann. Sc. Norm. Super. Pisa, Sci. Fis. Mat., III. Ser.*, 17:189–206, 1963.

[Mey90] Yves Meyer. *Ondelettes et opérateurs. II: Opérateurs de Calderón-Zygmund.* Paris: Hermann, Éditeurs des Sciences et des Arts, 1990.

[Mey92] Yves Meyer. *Wavelets and operators*, volume 37 of *Cambridge Studies in Advanced Mathematics.* Cambridge University Press, Cambridge, 1992. Translated from the 1990 French original by D. H. Salinger.

[Mil90] Graeme W. Milton. On characterizing the set of possible effective tensors of composites: The variational method and the translation method. *Commun. Pure Appl. Math.*, 43(1):63–125, 1990.

[MNLD15] Bagus Putra Muljadi, Jacek Narski, Alexei Lozinski, and Pierre Degond. Non-conforming multiscale finite element method for Stokes flows in heterogeneous media. Part I: Methodologies and numerical experiments. *Multiscale Model. Simul.*, 13(4):1146–1172, 2015.

[Mor08] Charles B. Jr. Morrey. *Multiple integrals in the calculus of variations.* Classics in Mathematics. Springer-Verlag, Berlin, 2008.

[Mos60] Jürgen Moser. A new proof of De Giorgi's theorem concerning the regularity problem for elliptic differential equations. *Comm. Pure Appl. Math.*, 13:457–468, 1960.

[Mos61] Jürgen Moser. On Harnack's theorem for elliptic differential equations. *Comm. Pure Appl. Math.*, 14:577–591, 1961.

[MP14] Axel Målqvist and Daniel Peterseim. Localization of elliptic multiscale problems. *Math. Comp.*, 83(290):2583–2603, 2014.

[MP21] Axel Målqvist and Daniel Peterseim. *Numerical homogenization by localized orthogonal decomposition*, volume 5 of *SIAM Spotlights.* Society for Industrial and Applied Mathematics (SIAM), Philadelphia, PA, 2021.

[MT97] François Murat and Luc Tartar. H-convergence. In *Topics in the mathematical modelling of composite materials*, volume 31 of *Progr. Nonlinear Differential Equations Appl.*, pages 21–43. Birkhäuser Boston, Boston, MA, 1997.

[Mur78] François Murat. Compacité par compensation. *Ann. Sc. Norm. Super. Pisa, Cl. Sci., IV. Ser.*, 5:489–507, 1978.

[MV22] Axel Målqvist and Barbara Verfürth. An offline-online strategy for multiscale problems with random defects. *ESAIM Math. Model. Numer. Anal.*, 56(1):237–260, 2022.

[Nas58] John Nash. Continuity of solutions of parabolic and elliptic equations. *Amer. J. Math.*, 80:931–954, 1958.

[Nat96] Melvyn B. Nathanson. *Additive number theory*, volume 164 of *Graduate Texts in Mathematics.* Springer-Verlag, New York, 1996. The classical bases.

[Ngu89] Gabriel Nguetseng. A general convergence result for a functional related to the theory of homogenization. *SIAM J. Math. Anal.*, 20(3):608–623, 1989.

[Ngu03a] Gabriel Nguetseng. Homogenization structures and applications. I. *Z. Anal. Anwend.*, 22(1):73–107, 2003.

[Ngu03b] Gabriel Nguetseng. Mean value on locally compact abelian groups. *Acta Sci. Math.*, 69(1–2):203–221, 2003.

[Ngu04] Gabriel Nguetseng. Homogenization in perforated domains beyond the periodic setting. *J. Math. Anal. Appl.*, 289(2):608–628, 2004.

[Ngu06] Gabriel Nguetseng. Deterministic homogenization. In *Multi-scale problems and asymptotic analysis. Proceedings of the midnight sun Narvik conference (satellite conference of the fourth European congress of mathematics), Narvik, Norway, June 22–26, 2004*, pages 233–248. Tokyo: Gakkōtosho, 2006.

[NS11] Gabriel Nguetseng and Nils Svanstedt. Σ-convergence. *Banach J. Math. Anal.*, 5(1):101–135, 2011.

[Øks03] Bernt Øksendal. *Stochastic differential equations.* Universitext. Springer-Verlag, Berlin, sixth edition, 2003.

[Owh17] Houman Owhadi. Multigrid with rough coefficients and multiresolution operator decomposition from hierarchical information games. *SIAM Rev.*, 59(1):99–149, 2017.

[Pan96] Alexander Pankov. Almost periodic functions, Bohr compactification, and differential equations. *Rend. Sem. Mat. Fis. Milano*, 66:149–158 (1998), 1996.

[Pra16] Christophe Prange. Weak and strong convergence methods for Partial Differential Equations, graduate course, Lecture 6: Regularity theory by compactness methods. http://prange.perso.math.cnrs.fr/documents/coursEDMI2016_lecture6.pdf, 2016.

[PS08] Grigorios A. Pavliotis and Andrew M. Stuart. *Multiscale methods. Averaging and homogenization.*, volume 53. New York, NY: Springer, 2008.

[PV81] George C. Papanicolaou and S. R. Srinivasa Varadhan. Boundary value problems with rapidly oscillating random coefficients. Random fields. Rigorous results in statistical mechanics and quantum field theory, Esztergom 1979, Colloq. Math. Soc. Janos Bolyai 27, 835–873, 1981.

[Qua18] Alfio Quarteroni. *Numerical models for differential problems. 3rd edition.*, volume 16. Cham: Springer, 2018.

[QV99] Alfio Quarteroni and Alberto Valli. *Domain decomposition methods for partial differential equations.* Numerical Mathematics and Scientific Computation. The Clarendon Press, Oxford University Press, New York, 1999.

[QV08] Alfio Quarteroni and Alberto Valli. *Numerical approximation of partial differential equations. 1st softcover printing.*, volume 23. Berlin: Springer, 2008.

[RBD98] Michel Rappaz, Michel Bellet, and Michel Deville. *Modélisation numérique en science et génie des matériaux*, volume 10 of *Traité des Matériaux [The Science of Materials]*. Presses Polytechniques et Universitaires Romandes, Lausanne, 1998.

[RT00] Fraydoun Rezakhanlou and James E. Tarver. Homogenization for stochastic Hamilton-Jacobi equations. *Arch. Ration. Mech. Anal.*, 151(4):277–309, 2000.

[RW00a] Leonard C. G. Rogers and David Williams. *Diffusions, Markov processes and martingales. Vol. 1: Foundations. 2nd ed.* Cambridge: Cambridge University Press, 2000.

[RW00b] Leonard C. G. Rogers and David Williams. *Diffusions, Markov processes, and martingales. Vol. 2: Itô calculus. 2nd ed.* Cambridge: Cambridge University Press, 2000.

[RY99] Daniel Revuz and Marc Yor. *Continuous martingales and Brownian motion*, volume 293 of *Grundlehren der Mathematischen Wissenschaften*. Springer-Verlag, Berlin, third edition, 1999.

[Saa03] Yousef Saad. *Iterative methods for sparse linear systems*. Society for Industrial and Applied Mathematics, Philadelphia, PA, second edition, 2003.

[Sen95] Marjorie Senechal. *Quasicrystals and geometry*. Cambridge: Cambridge Univ. Press, 1995.

[She18] Zhongwei Shen. *Periodic homogenization of elliptic systems*, volume 269 of *Operator Theory: Advances and Applications*. Birkhäuser/Springer, Cham, 2018.

[Shi95] Albert N. Shiryaev. *Probability. 2nd ed.*, volume 95. New York, NY: Springer-Verlag, 1995.

[Sim87] Jacques Simon. Compact sets in the space $L^p(0, T; B)$. *Ann. Mat. Pura Appl. (4)*, 146:65–96, 1987.

[SPSH92] Evariste Sanchez-Palencia and Jacqueline Sanchez-Hubert. *Introduction aux méthodes asymptotiques et à l'homogénéisation: Application à la Mécanique des milieux continus*. Masson, Paris, 1992.

[Sta65] Guido Stampacchia. Le problème de Dirichlet pour les équations elliptiques du second ordre à coefficients discontinus. *Ann. Inst. Fourier (Grenoble)*, 15(fasc. 1):189–258, 1965.

[Ste93] Elias M. Stein. *Harmonic analysis: Real-variable methods, orthogonality, and oscillatory integrals*. Princeton, NJ: Princeton University Press, 1993.

[Tar79] Luc Tartar. Compensated compactness and applications to partial differential equations. Nonlinear analysis and mechanics: Heriot-Watt Symp., Vol. 4, Edinburgh 1979, Res. Notes Math. 39, 136–212, 1979.

[Tar89] Luc Tartar. Nonlocal effects induced by homogenization. Partial differential equations and the calculus of variations. Essays in Honor of Ennio de Giorgi, 925–938, 1989.

[Tar09] Luc Tartar. *The general theory of homogenization. A personalized introduction.*, volume 7. Berlin: Springer, 2009.

[Tem79] Roger Temam. Navier-Stokes equations. Theory and numerical analysis. Studies in Mathematics and its Applications. Vol. 2. Amsterdam - New York - Oxford: North-Holland Publ. Co., 1979.

[Tho08] Matthieu Thomas. *Propriétés thermiques de matériaux composites : caractérisation expérimentale et approche microstructurale.* PhD thesis, Université de Nantes, Laboratoire de Thermocinétique, CNRS - UMR 6607, 2008.

[Tho12] Florian Thomines. *Méthodes mathématiques et techniques numériques de changement d'échelle : application aux matériaux aléatoires.* PhD thesis, Université Paris Est, 2012.

[Yur82] Vadim V. Yurinskii. On the averaging of non-divergent equations of second order with random coefficients. *Sib. Mat. Zh.*, 23(2):176–188, 1982.

[Zil17] Bruno Ziliotto. Stochastic homogenization of nonconvex Hamilton-Jacobi equations: a counterexample. *Comm. Pure Appl. Math.*, 70(9):1798–1809, 2017.

[ZKO94] Vasilii V. Zhikov, Sergei M. Kozlov, and Olga A. Olejnik. *Homogenization of differential operators and integral functionals.* Berlin: Springer-Verlag, 1994.

[Zyg02] Antoni Zygmund. *Trigonometric series. Volumes I and II combined. 3rd ed.* Cambridge: Cambridge University Press, 2002.

Index

© The Author(s), under exclusive license to Springer Nature Switzerland AG 2023   451
X. Blanc, C. Le Bris, *Homogenization Theory for Multiscale Problems*, MS&A 21,
https://doi.org/10.1007/978-3-031-21833-0

Printed in the United States
by Baker & Taylor Publisher Services